W0005989

The Practice of Algebraic Curves

GRADUATE STUDIES
IN MATHEMATICS **250**

The Practice of Algebraic Curves

A Second Course in Algebraic Geometry

David Eisenbud
Joe Harris

AMERICAN
MATHEMATICAL
SOCIETY
Providence, Rhode Island

EDITORIAL COMMITTEE

Matthew Baker
Marco Gualtieri (Chair)
Sean T. Paul
Natasa Pavlovic
Rachel Ward

2020 *Mathematics Subject Classification.* Primary 14–01, 14–03, 14H10, 14H40, 14H50, 14H55, 13Dxx.

For additional information and updates on this book, visit
www.ams.org/bookpages/gsm-250

Library of Congress Cataloging-in-Publication Data

Cataloging-in-Publication Data has been applied for by the AMS.
See http://www.loc.gov/publish/cip/.
Graduate Studies in Mathematics ISSN: 1065-7339 (print)
DOI: https://doi.org/10.1090/gsm/250

Copying and reprinting. Individual readers of this publication, and nonprofit libraries acting for them, are permitted to make fair use of the material, such as to copy select pages for use in teaching or research. Permission is granted to quote brief passages from this publication in reviews, provided the customary acknowledgment of the source is given.

Republication, systematic copying, or multiple reproduction of any material in this publication is permitted only under license from the American Mathematical Society. Requests for permission to reuse portions of AMS publication content are handled by the Copyright Clearance Center. For more information, please visit www.ams.org/publications/pubpermissions.

Send requests for translation rights and licensed reprints to reprint-permission@ams.org.

© 2024 by the American Mathematical Society. All rights reserved.
The American Mathematical Society retains all rights
except those granted to the United States Government.
Printed in the United States of America.

∞ The paper used in this book is acid-free and falls within the guidelines
established to ensure permanence and durability.
Visit the AMS home page at https://www.ams.org/

10 9 8 7 6 5 4 3 2 1   29 28 27 26 25 24

For our children, biological and mathematical

Contents

Preface

In writing this book we have been keenly aware of the influence of our great teachers in this subject, David Buchsbaum, Phillip Griffiths, David Mumford, and Antonius van de Ven. Great stylists as well as great mathematicians, we can only hope that, consciously or not, the reader will occasionally hear their voices in our writing.

We have taught large parts of this material at Harvard and Berkeley, and our students' contributions to whatever virtues our text may have are considerable. İzzet Coşkun, in particular, read an earlier version of this manuscript in detail, and made myriad useful comments.

We have had the good fortune to have, in the role of copy-editor, a mathematician and friend, Silvio Levy, whose sense of style and publishing expertise informs the layout of this book. Silvio also converted our sketches, done (to quote James Thurber) with a dull pencil, into professional graphics that are mathematically accurate when possible.

Writing this book was an enterprise of years that involved the patience of more than just the authors, and we have been fortunate to have had the support of Monika and Vicky, who understood and valued our work during the hours "away" that were necessary.

To all these, our heartfelt thanks!

David and Joe, August, 2024

Introduction

Why you want to read this book

Algebraic geometry is an old subject. Descartes' introduction in 1637 of co-ordinates in the plane and in space made it possible to relate the algebra of polynomials to the geometry of their zero loci. Gauss' proof of the fundamental theorem of algebra showed that the degree of a polynomial, an algebraic invariant, was equal to the number of roots counted with multiplicity, a geometric invariant. Mathematicians have been doing what is recognizably algebraic geometry for more than two centuries.

Within algebraic geometry, the study of algebraic curves is the oldest topic. Newton already classified all the possible types of real affine cubics. By the middle of the nineteenth century a rich theory of curves in the complex projective plane was a central topic, overturned by Riemann's work in mid-century — what became the theory of Riemann surfaces, which introduced new complex-analytic techniques to the field. This was taken up and made algebraic as a theory of plane curves by Alexander Brill, Max Noether, F. S. Macaulay and many others. By the end of the century Halphen and others were interested in the classification of space curves as well. The interested reader will find more about the early work on algebraic curves in the historical appendix to this book written by historian Jeremy Gray (page 353).

The development continues unabated today: in the second half of the twentieth century Grothendieck's foundations led to the solution of many classical problems and, in particular, to a firm foundation for the theory of moduli, allowing mathematicians to exploit the fact that algebraic varieties generally come in families parametrized by other algebraic varieties — something that we will return to often in this book. Algebraic geometry has merged more and

1

more with number theory, and the moduli space of curves plays an important role in string theory, responding to the needs of physicists.

For these reasons, the subject of algebraic curves is one of the richest in algebraic geometry, if not in all of mathematics. If you want to know whether a conjecture is plausible, you can generally find well-understood special cases on which to test it. Some of the fundamental constructions of algebraic geometry, like the construction of moduli spaces and their description, can be carried out in the setting of algebraic curves with a degree of precision and detail far beyond what has been possible in higher dimensions.

Already the geometry of plane curves of degrees 3 and 4 shows some of the promise of the subject:

Every smooth plane cubic has exactly 9 flexes — points where the tangent line has contact of order 3 with the curves — and these points form a remarkable configuration; they're the only known example of a finite, noncollinear set of points in the plane such that the line joining any two contains a third. Generalizing the notion of the flexes of a plane cubic leads us to the subject of inflectionary behavior of linear series on curves in general, which arose separately in the theory of Weierstrass points on Riemann surfaces, and has become a powerful tool.

Every smooth quartic curve has exactly 28 bitangents, also forming a beautiful and mysterious configuration; see Figures 6.4 and A.5. Salmon [1852, p. 197] computed the number (315) of 4-tuples of bitangents whose eight points of tangency lie on a conic. The extension of these ideas to curves of higher genus leads us to the rich theory of theta characteristics, bringing together algebra (in the form of a bilinear pairing on the points of order 2 in the Jacobian of a curve), analysis (in the form of theta-functions) and of course projective geometry.

The richest subject in what is arguably the richest branch of mathematics[1] — of course you want to read this book!

Why we wrote this book

The wealth of beauty, both in theory and in examples, certainly makes the study of algebraic curves an attractive prospect. But it comes at a price: to absorb in detail all the things we've learned over the centuries about algebraic curves would take years, if not decades. This is, in essence, the conundrum facing anyone who undertakes to write a book on the subject: how to convey the wealth of information (and the many many ways in which our knowledge is incomplete) without writing an encyclopedia. We have chosen to try to be useful and broad but not necessarily complete.

---

[1]Number theorists may quibble…

When we introduce a technique or a construction without full proofs, we do so as a "cheerful fact:"[2]

I'm very well acquainted, too, with matters mathematical,
I understand equations, both the simple and quadratical,
About binomial theorem I am teeming with a lot o' news,
With many cheerful facts about the square of the hypotenuse.

— Gilbert and Sullivan, *Pirates of Penzance*, Major General's Song

Our intended audience is a graduate student considering working in the field of algebraic curves, or a researcher in a related field whose work has led them to questions about algebraic curves. Our goal is to equip the reader with the understanding of both the techniques and the state of our knowledge necessary to read the current literature and work on open problems.

What's with the title?

Be simple by being concrete. Listeners are prepared to accept unstated (but hinted) generalizations much more than they are able, on the spur of the moment, to decode a precisely stated abstraction and to re-invent the special cases that motivated it in the first place.

–Paul Halmos, *How to Talk Mathematics*

This book aims to present those ideas and methods from the theory of algebraic curves that are used *in practice* by mathematicians working in a variety of fields of mathematical research.

Although mathematicians aspire to understand their subjects deeply, we feel that we learn in stages: in early stages we accept large and difficult results as black boxes and explore the rich examples that they yield. That is how we have tried to organize this book: We begin with two chapters that we hope will bridge the gap between first courses in algebraic geometry/commutative algebra at the level of Fulton's [1969] or Reid's [1988] well-known books, and the professional language of invertible sheaves, cohomology and linear series. An ideal background would be Hartshorne's textbook [1977] or Vakil's about to be published book [2023], but very much less will suffice if the reader is willing to accept some advanced ideas or look them up at leisure; we have tried to give precise references where this might be required. Subsequent chapters roughly alternate between expositions of basic techniques (partly without proofs) and families of examples, treated in detail.

---

[2]Hats off to the "Many Cheerful Facts" seminar run by the U. C. Berkeley graduate students, which gave us this idea!

What's in this book

In organizing this book we faced a common problem of mathematical exposition:

> We are dealing here with a fundamental and almost paradoxical difficulty. Stated briefly, it is that learning is sequential but knowledge is not. A branch of mathematics... consists of an intricate network of interrelated facts, each of which contributes to the understanding of those around it. When confronted with this network for the first time, we are forced to follow a particular path, which involves a somewhat arbitrary ordering of the facts.
> –Robert Osserman [2011]

In Chapters 1 and 2 we lay out the central objects of algebraic curve theory: invertible sheaves, linear series, canonical sheaves and the Riemann–Roch theorem, as well as Hurwitz's theorem, the adjunction formula and some elementary facts about the geometry of surfaces. We prove or sketch the proofs of many of these basic results. These chapters may serve as review for someone who has been exposed already to material roughly equivalent to Chapter IV of [Hartshorne 1977].

Thereafter, we alternate between chapters focused on special cases and chapters developing more of the theory. Chapter 3 describes the geometry of curves of genus 0 in projective space. We emphasize some of the things that make rational normal curves so special, and take the opportunity to introduce the conditions implying that a curve is *arithmetically Cohen–Macaulay*.

In Chapter 4 We begin by explaining the Riemann–Roch theorem and its consequences for smooth plane curves, in the style of the late nineteenth century, computing the canonical series and showing algorithmically how to compute the complete linear series of effective divisors linearly equivalent to a given (not necessarily effective) divisor, and we use this to describe the low-degree linear series on curves of genus 1, and demonstrate how counts of parameters suggest the presence of a moduli space.

Because most curves cannot be represented as smooth plane curves, a general treatment requires dealing with singular curves and their normalizations. This requires somewhat more algebraic technique, so we postpone the general case to Chapter 15.

In genus ≥ 2 there is a fundamental shift: not all invertible sheaves of a given degree on a curve C of genus $g \geq 2$ are congruent modulo the automorphism group of the curve. It is a salient feature of algebraic geometry that families of similar algebro-geometric objects are often naturally parametrized by algebraic varieties. This idea goes back to the beginning of algebraic geometry and the family of plane curves of degree d, but was only fully clarified in the work of Grothendieck, Deligne and Mumford.

Chapter 5 introduces our first examples of this phenomenon: the *fine moduli spaces* of divisors on a curve, as well as the Jacobian, for which we give the classic analytic construction, and the Picard varieties $\text{Pic}_d(C)$ parametrizing isomorphism classes of invertible sheaves. We also introduce the subvarieties $W_d^r(C)$ parametrizing invertible sheaves with many sections.

Since the Jacobian is irreducible, we can speak meaningfully of a *general* invertible sheaf, and of the dimension of various families of special invertible sheaves. With this in hand we prove that every curve of genus $g > 1$ can be embedded in $\mathbb{P}^3$ as a curve of degree $g + 3$.

Equipped with information about the Jacobian, we proceed in Chapter 6 to study curves of genus 2 and 3, describing in particular how the geometry of a map of the curve to projective space, given by sections of an invertible sheaf, depends on the sheaf. We show how a hyperelliptic curve of genus g can be described by a set of $2g + 2$ points in $\mathbb{P}^1$, and we describe the canonical maps. Again, this suggests the existence of moduli spaces.

We spend Chapters 7 and 8 on other moduli spaces as they appear in the world of curves, though we do not give complete proofs of all the assertions made. We start in Chapter 7 with some central examples of fine moduli spaces, primarily the Hilbert scheme. Using properties of the Hilbert scheme we show that curves of genus ≥ 2 can have only finitely many automorphisms, and more generally that there can only be finitely many morphisms from one such curve to another.

The most interesting example of a moduli space, the moduli of smooth (or stable) curves of genus g, is not a fine moduli space, and we spend Chapter 8 describing what it is and isn't. In that chapter we also describe the Hurwitz space of coverings and the Severi variety of nodal plane curves.

After this, it's back to examples: in Chapter 9 we analyze aspects of the geometry of curves of genus 4 and 5.

In the following two chapters we take up the properties of the points of a general hyperplane sections of a curve in projective space. In Chapter 10 we give Rathmann's proof that the points of a general hyperplane section of a reduced irreducible curve are in linearly general position (independent of the characteristic). Consequences include Castelnuovo's bound on the maximal genus of curves of a given degree. We show that the curves that achieve the maximum genus, called Castelnuovo curves, are arithmetically Cohen–Macaulay. These include all plane curves, as well as canonical curves and linearly normal curves of high degree compared to their genera.

We also use the linear general position result to prove the strong forms of Clifford's and Martens' theorems on special linear series; to show that every curve has a nodal plane model; and to show that a general invertible sheaf of degree $g+2$ on a curve of genus g maps the curve to a nodal curve in the plane — unless the curve is hyperelliptic, in which case the image has a single multiple point of multiplicity g.

In characteristic 0 an even stronger statement than linearly general position is true: the monodromy of the hyperplane divisors is the full symmetric group. We take up this and some other monodromy questions in Chapter 11 and prove consequences for secant planes and for sums of linear series.

In Chapter 12 we return to the general question: "What linear series exist on curves of genus g?" We give three possible interpretations of this question, and the answers to each: Clifford's theorem, the Castelnuovo bound, and the Brill–Noether theorem, postponing the proof of the latter. We apply the Brill–Noether result to explore various special classes of curves of genus 6.

Chapter 13 prepares for the proof of the Brill–Noether theorem with a discussion of inflection points of linear series, generalizing the flexes of plane curves. The Plücker formula counts the number of inflection points, with appropriate weights. In this chapter we explain the remarkable connection between the study of inflections on a rational curve and the Schubert calculus of cycles in the Grassmannian. In Chapter 14 we use this material, together with a degeneration to cuspidal rational curves, to give a relatively short proof of the Brill–Noether theorem.

In Chapter 15 we take the classical point of view of Brill and Noether, and describe an explicit algorithm for finding the complete linear series on a smooth curve, using a singular plane model. This involves the result classically known as the completeness of the adjoint linear series. For the general case we use a simple case of the theory of dualizing sheaves, introduced in more generality in the next chapter.

Returning to curves in $\mathbb{P}^3$ in Chapter 16 we explain the theory of Hartshorne and Rao classifying curves up to linkage. This theory applies to all purely 1-dimensional subschemes of $\mathbb{P}^3$, and to explain this we detour to discuss dualizing sheaves and Grothendieck's $f^!$ operation.

Curves that lie on a quadric in $\mathbb{P}^3$ are easy to understand. In Chapter 17 we systematically describe a natural generalization: rational normal scrolls and the curves that lie on them. This includes all the Castelnuovo curves of high degree.

Chapter 18 presents some aspects of the theory of syzygies of the homogeneous ideals of curves and the famous — and, as of this writing, open — conjecture of Mark Green that connects syzygies of canonical curves to the Clifford index and the theory of rational normal scrolls. A table reproduced from work of Frank-Olaf Schreyer shows the sensitivity of the numerical information in the free resolution of a canonical curve to questions of the existence of special linear series on the curve.

Chapter 19 is in some ways the culmination of the book. It is concerned with *Hilbert schemes*, schemes $\mathcal{H}_{g,3,d}$ that parametrize smooth curves of genus g and degree d in $\mathbb{P}^3$. We work out examples up to degree 7, define the "principal component" — the only one dominating the moduli space of curves — and derive dimension estimates from deformation theory and from the Brill–Noether theorems using most of the ideas we have introduced.

The historical appendix written by Jeremy Gray (page 353) is a survey of some of the work on algebraic curves up to that of Brill and Noether, and we are grateful to him for allowing us to include it.

Exercises and hints. There are many exercises, clumped at the end of the chapters. Hints, and occasionally solutions, are given at the end of the book for exercises (or parts thereof) marked with ♦.

Relation of this book to other texts. Chapter IV of [Hartshorne 1977] has a similar flavor to that of this book, and contains details of most of the results in our Chapters 1 and 2. A beautiful (and brief) account of a number of topics in a style we particularly admire is found in [Mumford 1975].

A far more extensive treatment, partially overlapping that of the present book and containing many other topics, can be found in the important and encyclopaedic [Arbarello et al. 1985] and [Arbarello et al. 2011], and yet even these works do not cover all the major topics in the field. One of the topics we do not cover is the construction of the moduli space of curves, which can be found in [Arbarello et al. 2011]. Though not a complete account, [Harris and Morrison 1998] deals directly with this topic.

There are more elementary accounts of some of our material, in [Fulton 1969] and [Walker 1950] (who goes farther than we do into local resolution of singularities) as well as [Griffiths 1989]. The book [Kunz 2005] treats plane curves and their normalizations via the theory of valuations, and contains a detailed account of the ideas we treat in Chapter 15 from this more algebraic point of view. There is a comprehensive treatment of the local topological theory of plane curves and their singularities in [Brieskorn and Knörrer 1986]. The topological questions there are developed in different directions in [Milnor 1968] and [Eisenbud and Neumann 1985]. An idiosyncratic collection of interesting topics is presented in [Clemens 2003].

The Riemann surface point of view is well represented in the books [Forster 1981], [Gunning 1966], [Hulek 1995], [Kirwan 1992], and [Miranda 1995].

Prerequisites, notation and conventions

The reader should be familiar with the (Krull) dimension of rings and varieties, and their primary decomposition at the level of [Atiyah and Macdonald 1969]. Ideally the reader will already have some familiarity with the geometry of curves and surfaces at the level of Chapter IV and the beginning of Chapter V of [Hartshorne 1977], though our summary of the necessary material in the first two chapters of this book may suffice for the intrepid.

Unless otherwise mentioned, we assume that the ground field is the field of complex numbers $\mathbb{C}$, though much of what we do could be done over any algebraically closed field.

Commutative algebra. All the rings we consider are commutative with unit and Noetherian.

Since we are working over a field of characteristic 0, we use the terms smooth and nonsingular interchangeably.

Some results that we use:

Theorem 0.1 (Lasker's theorem). *If $f_1, \ldots, f_c \subset \mathbb{C}[x_0, \ldots, x_n]$ generates an ideal of codimension c, then the ideal $(f_1, \ldots, f_c)$ is unmixed (all its primary components have codimension c).*

Theorem 0.2. *If R is a domain that is a finitely generated algebra over a field or a localization of such an algebra, then the normalization (= integral closure) of R is a finitely generated R-module. If R is 1-dimensional, then its normalization is nonsingular.*

Projective geometry. Schemes are assumed quasiprojective, and *varieties* (including curves) are reduced and irreducible schemes unless otherwise stated. "Points" will always be closed points unless we say otherwise. **Note to the patient reader: We usually use the term curve to refer to a smooth irreducible projective purely 1-dimensional scheme, but in some of the later chapters we explicitly allow more general 1-dimensional schemes.**

Though we occasionally use the classical topology, the term "open set" refers to the Zariski topology unless otherwise stated.

We adopt the Grothendieck convention that points of projective space $\mathbb{P}(V)$ are 1-dimensional quotients of V, or hyperplanes in V, so that $\mathrm{Sym}(V)$ is the homogeneous coordinate ring of $\mathbb{P}(V)$ and lines in V correspond to points of $\mathbb{P}(V^*)$.

However we write $G(k, V)$ for the variety of k-dimensional linear subspaces of V, so that in particular $G(1, V) = \mathbb{P}(V^*)$. We also write $\mathbb{G}(k, r)$ for k-dimensional projective subspaces of $\mathbb{P}^r$.

The results we assume are well represented by the following classical theorems:

Theorem 0.3 (Bézout's theorem). *If $X, Y \subset \mathbb{P}^r$ are subvarieties satisfying the condition $\operatorname{codim}(X \cap Y) = \operatorname{codim} X + \operatorname{codim} Y$, then $\deg(X \cap Y) = \deg X \deg Y$.*

Theorem 0.4 (Bertini's theorem). *If $X \subset \mathbb{P}^r$ is a nonsingular quasiprojective variety, and $\{H_\lambda \mid \lambda \in \Lambda\}$ is a linear family of hyperplanes of $\mathbb{P}^r$, then for an open subset of $\lambda \in \Lambda$ the scheme $H_\lambda \cap X$ is nonsingular away from the union of the base locus $\bigcap_{\lambda \in \Lambda} H_\lambda$ and the singular locus of X.*

Theorem 0.5 (main theorem of elimination theory). *Any morphism $\phi : X \to Y$ of projective varieties (or schemes) is closed: if $X' \subset X$ is a Zariski closed subset, then $\phi(X') \subset Y$ is also closed.*

Corollary 0.6. *If $\phi : C \to D$ is a nonconstant morphism of (projective) curves, then ϕ is finite and surjective.*

If D is a smooth curve then the local ring of D at any point is a discrete valuation ring, so any torsion free module is flat. Thus:

Proposition 0.7. *If $\phi : C \to D$ is a nonconstant morphism of smooth curves, then ϕ is finite, surjective, and flat.*

If $X \subset \mathbb{P}^r$ is any scheme, we define the *homogeneous coordinate ring* of $X \subset \mathbb{P}^r$ to be $R_X = R_{X/\mathbb{P}^r} := S/I(X)$, where $S = \mathbb{C}[x_0, \ldots, x_r]$ is the homogeneous coordinate ring of $\mathbb{P}^r$. We emphasize that, unlike the coordinate ring of an affine variety, this is not an intrinsic invariant of X, but depends on the embedding in $\mathbb{P}^r$.

When X is reduced and irreducible we write $\kappa(X)$ for the field of rational functions on X. In particular, if p is a closed point we write $\kappa(p) \cong \mathbb{C}$ for the residue class field at p.

Sheaves and cohomology. Some familiarity with coherent sheaves is recommended; a possible source is the first chapter of [Eisenbud and Harris 2000]. As for cohomology, it is probably enough if the reader can write down H^i and exact sequences without blushing. In any case we review some of the theory of coherent sheaves and their cohomology theory, and that of divisors on projective varieties, in the first two Chapters.

We occasionally use the bijection between algebraic and analytic sheaves on smooth projective curves, which preserves cohomology and exact sequences. This is a special case of the results in [Serre 1955/56].

If $\mathcal{F}$ is a sheaf on X and $X \subset Y$ then we will identify $\mathcal{F}$ with the sheaf usually written $\iota_* \mathcal{F}$, where $\iota : X \to Y$ is the inclusion map, and thus regard $\mathcal{F}$ as a sheaf on Y as well. The cohomology $H^i(X, \mathcal{F}) = H^i(Y, \iota_* \mathcal{F})$ canonically, so we will simply and unambiguously write $H^i(\mathcal{F})$ for either of these. We write $h^i(\mathcal{F})$ or (if D is a divisor) $h^i(D)$ for $\dim_{\mathbb{C}} H^i(\mathcal{F})$ or $\dim_{\mathbb{C}} H^i(\mathcal{O}_X(D))$.

If $\mathcal{F}$ is a sheaf on projective space (perhaps supported on a subvariety) we write $H^i_*(\mathcal{F})$ for $\bigoplus_{m \in \mathbb{Z}} H^i(\mathcal{F}(m))$.

Chapter 1

Linear series and morphisms to projective space

We start by laying out four major definitions: divisors, linear series, invertible sheaves, and maps to projective space. These ideas are used throughout this book. In the last section we explore some special cases, culminating with the condition for an invertible sheaf to provide an embedding of a curve in projective space.

We prove only some of our assertions; a reader who wants to see all the proofs should keep handy a copy of [Hartshorne 1977] or the equivalent. A more experienced reader, could instead skip ahead to Chapter 3.

As an analytic space, a complex projective smooth curve is a compact Riemann surface, a compact 1-dimensional complex manifold. In this sense its local structure is trivial, but its global structure can be hard to visualize. Any two Riemann surfaces with the same genus[1] are C^∞ isomorphic, but except for genus 0, where the sphere has a unique complex structure, there are continuous families of nonisomorphic global structures. The differences among Riemann surfaces or algebraic curves are revealed in the geometry of their

---

[1]The notion of genus was introduced by Riemann in his great paper [1857], though he never used the term. Instead, he defined the *connectivity* (Zusammenhang) of a Riemann surface X to be the number of disjoint closed curves (one-dimensional real submanifolds) that can be drawn on X such that the complement of their union is connected. The term genus (Geschlecht) was first used by Clebsch. Both uses were subtly different from their modern counterparts: whereas we speak of genus as a property of a given surface, Riemann and Clebsch used this idea as a way of partitioning the family of all Riemann surfaces into groups. For a discussion, see [Lê 2020].

maps to projective spaces. Throughout this book we will study curves in this way. It turns out that in characteristic 0 the algebraic and complex analytic theories are equivalent.

If U is a Zariski open set of a projective variety, then a *regular* function on U is a rational function whose denominator does not vanish on U. Because the only global regular functions on a projective variety are the constant functions, interesting maps must be described and studied using a different approach, and a large part of this chapter is devoted to the necessary machinery for doing this. The idea is simple: a point in projective space is the intersection of the hyperplanes containing it, so a map $\phi : C \to \mathbb{P}^r$ from a curve to projective space can be described set-theoretically by the set of preimages of these hyperplanes. In other words, a point $p \in C$ is sent to the point that is the intersection of those hyperplanes whose preimages contain p, as in Figure 1.1.

Some hyperplanes in $\mathbb{P}^r$ will be tangent to the image $\phi(C)$ at a point $\phi(p)$, and in that case the point p should be counted with higher multiplicity in the preimage of that hyperplane; in this way we arrive at a notion of divisor as a sum of points with multiplicities.

1.1. Divisors

On a smooth projective curve C, we define a *divisor* to be a finite formal sum of points of C with integer coefficients, compactly written as $\sum_{p \in C} m_p \cdot p$ with all but finitely many $m_p = 0$. The coefficient m_p is called the *multiplicity* of the point p in the divisor D; if all coefficients m_p are nonnegative we say that D is *effective*. Thus the group of divisors $\operatorname{Div} C$ is the free abelian group whose generators are the points of C.

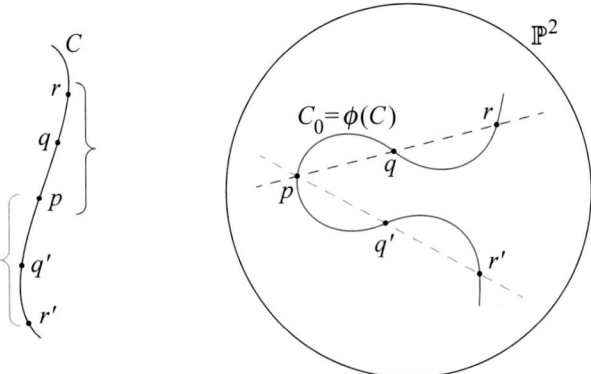

Figure 1.1. Divisors of the linear series corresponding to the map ϕ from C to $\mathbb{P}^2$ are preimages of the hyperplanes — in this case the dashed lines — in the plane. The image of p is determined by the intersection of the two lines.

The *degree* of $D = \sum_{p \in C} m_p \cdot p$ is by definition the sum $\sum m_p$ of its coefficients. Because the curve is smooth each local ring is a discrete valuation ring, so two elements of $\mathcal{O}_{C,p}$ with the same order of vanishing at p generate the same ideal in the local ring. Thus an effective divisor can be identified with a finite subscheme of C, and vice versa. If $D = \sum m_p \cdot p$ and $E = \sum n_p \cdot p$ are any two divisors, we write $E \geq D$ to mean $n_p \geq m_p$ for all $p \in C$.

1.2. Divisors and rational functions

Let C be a smooth projective curve, $U \subset C$ an open subset and $f \in \mathcal{O}_C(U)$ a regular function, not identically zero. At every point $p \in U$, we define the *order* of f at p, denoted $\mathrm{ord}_p(f)$, to be the highest power of the maximal ideal $m_p \subset \mathcal{O}_p$ containing f, and we define the divisor (f) associated to the function f by

$$(f) := \sum_{p \in U} \mathrm{ord}_p(f) \cdot p.$$

More generally, if h is any rational function on C, given locally as a quotient f/g of regular functions f and g, we define the divisor of h to be

$$(h) := (f) - (g);$$

this is independent of the choice of f and g. The divisor of a rational function h is called a *principal divisor*; principal divisors form a subgroup $\mathrm{Div}_0(C)$ of the group $\mathrm{Div}(C)$ of all divisors. We say that two divisors D and E on C are *linearly equivalent* if their difference $D - E$ is a principal divisor. The group $\mathrm{Div}(C)/\mathrm{Div}_0(C)$ of linear equivalence classes of divisors is called the *Picard group* of C, denoted $\mathrm{Pic}\, C$.

Linear equivalence is important in the description of a map $C \to \mathbb{P}^r$ by divisors because in $\mathbb{P}^r$ any two hyperplanes, defined by the vanishing of linear forms $\ell_1 = 0$ and $\ell_2 = 0$, differ by the globally defined rational function ℓ_1/ℓ_2. The preimages of these hyperplanes differ by the pullback of this rational function, and are thus linearly equivalent divisors.

Generalizations. We can generalize the notion of a divisor on a smooth curve in two respects.

First, we can extend the notion to higher-dimensional smooth varieties X, defining a divisor to be a finite formal linear combination of irreducible subvarieties of codimension 1 in X. In this setting, most of the notions introduced above have natural analogues; for example, a divisor is called *effective* if all the irreducible varieties in its expression as a sum appear with nonnegative coefficients. We can similarly define the divisor of a rational function on X, and the notion of linear equivalence of divisors. (One notion that does not have an obvious analogue is the degree of a divisor.)

We can also extend the notion to possibly singular varieties X, though this requires more care. We can mimic the definition above, defining a divisor to be a finite formal linear combination of irreducible subvarieties of codimension 1 in X; this is called a *Weil divisor* on X. But for most purposes we have to introduce a slightly different notion of divisor, called a *Cartier divisor*. We first define an *effective Cartier divisor* to be a subscheme locally defined on some open covering U_i by the vanishing of one nonzerodivisor in $\mathcal{O}_X(U_i)$; and a *Cartier divisor* to be a difference of effective Cartier divisors.

A simple example will serve to illustrate this. Suppose $C \subset \mathbb{A}^2$ is a plane curve with a node at a point $p = (0,0) \in C$. The point p itself (that is, the reduced subscheme) is *not* a Cartier divisor, since its ideal (x, y) is not generated by a single element. Likewise, the *fat point* — that is, the subscheme $\Gamma = V(x^2, xy, y^2) \subset C$ — is not Cartier. But a subscheme of degree 2 supported at p,

$$\Gamma_{\alpha,\beta} := V(\alpha x + \beta y, x^2, xy, y^2) \subset C,$$

— is in general Cartier, since it is defined locally by the equation $\alpha x + \beta y = 0$. For distinct lines L the schemes Γ_L are also distinct, since $\Gamma_L \subset L$. The two exceptions, when Γ_L is not Cartier, occur when the line L is tangent to either branch of C at the node. In these cases the scheme $\Gamma_{\alpha,\beta}$ is not a Cartier divisor on C: the intersection $L \cap C$ has multiplicity 3 at p, and is not equal to $\Gamma_{\alpha,\beta}$. For all these assertions see Exercise 1.7 at the end of the chapter; the reader not already familiar with the notion of Cartier divisor is encouraged to carry out the verifications.

Here is another way to characterize Cartier divisors on a scheme X: for each open affine set $U \subset X$ we define $K_X(U)$ to be the field of fractions of $\mathcal{O}_X(U)$, and write K_X for the associated sheaf. We write $\mathcal{O}_X^*$ and K_X^* for the sheaves of groups of units in $\mathcal{O}_X(U)$ and $K_X(U)$. A Cartier divisor on X is then by definition a global section of $K_X^*/\mathcal{O}_X^*$. Moreover, if X is a *normal* scheme (so that its local rings at the generic points of codimension-1 subvarieties are discrete valuation rings) then Cartier divisors are the same as locally principal Weil divisors. See [Hartshorne 1977, Section II.6] for more information.

In this book, we deal for the most part with smooth curves; even when a singular curve C_0 arises, it will be viewed as the image of its normalization C and its geometry analyzed in terms of that of C. Thus for the most part we will be dealing with divisors in the simplest setting of smooth curves. But there will be occasions when we want to extend our analysis to singular curves or to higher-dimensional varieties, and in those cases the notion of Cartier divisor is called for.

Divisors of functions. Returning to the case of smooth curves, we have:

Theorem 1.1. *Let C be a smooth projective curve. If $f \in K(C)$, the degree of the divisor (f) is 0. Thus any two linearly equivalent divisors on C have the same degree.*

Proof. The result is evident on $\mathbb{P}^1$, where a rational function is the ratio of two forms of the same degree. In general, a rational function ϕ on a smooth curve C defines a map $\pi : C \to \mathbb{P}^1$, such that

$$(\phi) = \pi^{-1}(0) - \pi^{-1}(\infty).$$

If $C \to D$ is any map of smooth projective curves, then restricting to an affine open subset D' of D and its preimage $C' \subset C$ the map is represented by a homomorphism of rings $\mathcal{O}_D(D') \to \mathcal{O}_C(C')$ in such a way that $\mathcal{O}_C(C')$ becomes a finitely generated module over $\mathcal{O}_D(D')$, which is torsion-free because C is reduced and irreducible. Since D is smooth, $\mathcal{O}_D(D')$ is a Dedekind domain, so $\mathcal{O}_C(C')$ is free of constant rank equal to the degree of the extension of the corresponding fields of rational functions $\kappa(D) \subset \kappa(C)$. This implies that the degree of the fibers is constant.

Returning to the case of the map $C \to \mathbb{P}^1$ defined by a rational function, we see that $\pi^{-1}(0)$ and $\pi^{-1}(\infty)$ have the same degree, so (ϕ) has degree 0 as required. □

In particular, we see that if $C \to \mathbb{P}^r$ is a morphism then the pullbacks of different hyperplanes have the same degree, and we define this to be the *degree of the morphism*.

It follows from Theorem 1.1 that we can write Pic C as a disjoint union

$$\operatorname{Pic} C = \bigsqcup_{d \in \mathbb{Z}} \operatorname{Pic}_d C,$$

where $\operatorname{Pic}_d C$ is the set of linear equivalence classes of divisors of degree d.

Invertible sheaves. To deal with families of linearly equivalent divisors we will use the language of invertible sheaves.

Recall first that a *coherent sheaf* $\mathcal{L}$ on X may be defined by specifying

- an open affine cover $\{U_i\}$ of X;
- for each i, a finitely generated $\mathcal{O}_X(U_i)$-module L_i;
- for each i, j, an isomorphism $\sigma_{i,j} : L_i|_{U_i \cap U_j} \to L_j|_{U_i \cap U_j}$ satisfying the compatibility conditions $\sigma_{j,k}\sigma_{i,j} = \sigma_{i,k}$.

A *global section* of $\mathcal{L}$ is a family of elements $t_i \in L_i$ such that $\sigma_{i,j}t_i = t_j$. Such a section may be realized as the image of the constant function 1 under a homomorphism of sheaves $\mathcal{O}_X \to \mathcal{L}$.

A coherent sheaf $\mathcal{L}$ is said to be *locally free* if the modules L_i are all free; when X is irreducible, the ranks of the free modules L_i are all the same, and this is called the *rank* of $\mathcal{L}$. An *invertible sheaf* is a locally free coherent sheaf of rank 1; that is, $L|_U \cong \mathcal{O}_U$, for every open set in some covering of X.

Invertible sheaves form a group: the tensor product $\mathcal{L}_1 \otimes_{\mathcal{O}_X} \mathcal{L}_2$ of two invertible sheaves is again an invertible sheaf. The inverse of an invertible sheaf $\mathcal{L}$ is the dual invertible sheaf $\mathcal{H}om_{\mathcal{O}_X}(\mathcal{L}, \mathcal{O}_X)$ (Proof: Evaluation of functions defines a natural map $\mathcal{F} \otimes \mathcal{H}om_{\mathcal{O}_X}(\mathcal{F}, \mathcal{O}_X)$, and when $\mathcal{F} = \mathcal{L}$ is locally free of rank 1 this is locally an isomorphism.) Invertible sheaves can be defined from divisors on a smooth scheme (or more generally from Cartier divisors on any scheme). In case C is a smooth projective curve, this is concrete: if $D = \sum m_p \cdot p$, then we define the sections of $\mathcal{O}_C(D)$ on any open set U by

$$\mathcal{O}_C(D)(U) := \{\text{rational functions } f \mid \mathrm{ord}_p(f) + m_p \geq 0 \text{ for all } p \in U\}.$$

Thus, at a point p with $m_p > 0$, we are allowing f to have a pole of order at most m_p; if $m_p \leq 0$ then f must be regular at p, with a zero of order at least $-m_p$.

More generally, we can associate an invertible sheaf to any Cartier divisor on a scheme X. First, if $D \subset X$ is an effective Cartier divisor (and thus locally defined by the vanishing of a nonzerodivisor) then D corresponds to a subscheme of X whose ideal sheaf $\mathcal{J}_{D/X}$ is invertible. We define $\mathcal{O}_X(-D)$ to be the ideal sheaf $\mathcal{J}_{D/X}$, and define $\mathcal{O}_X(D)$ to be its inverse. The dual of the inclusion of $\mathcal{O}_X(-D) \subset \mathcal{O}_X$ is a homomorphism $\sigma := \mathcal{O}_X \to \mathcal{O}_X(D)$, which we regard as a global section $\sigma \in H^0(\mathcal{O}_X(D))$, the image of $1 \in H^0(\mathcal{O}_X)$. It vanishes precisely along D.

For a general (Cartier) divisor $D = E - F$, we define

$$\mathcal{O}_X(D) := \mathcal{O}_X(E) \otimes \mathcal{O}_X(F)^{-1}.$$

If D and E are linearly equivalent divisors on C — that is, there is a rational function f on C with $(f) = D - E$ — then multiplication by f defines an isomorphism $\mathcal{O}_C(E) \to \mathcal{O}_C(D)$. Thus the isomorphism class of the invertible sheaf $\mathcal{O}_C(D)$ corresponds to the linear equivalence class of D. We will see in Corollary 2.2 that every invertible sheaf on a projective scheme has this form.

If $\sigma \in H^0(\mathcal{L})$ is a global section of an invertible sheaf on X and $p \in X$ is a point, then the *value* $\sigma(p)$ *of* σ *at* p is the image of σ under the natural map $H^0(\mathcal{L})$ to the fiber $\kappa(p) \otimes \mathcal{L}_p \cong \mathbb{C}$. Since the isomorphism is not canonical, σ does not define a function on X at p; but since any two isomorphisms differ by a unit in $\mathcal{O}_{X,p}$, the vanishing locus of σ, denoted $(\sigma)_0$, is a well-defined subscheme of X.[2] Moreover, if X is reduced and irreducible, the ratio σ/τ of two global sections of the same invertible sheaf is a well-defined rational function $\sigma(p)/\tau(p)$ at all the points where the denominator $\tau(p)$ is not 0, so the divisor class of $(\sigma)_0$ is independent of the choice of $\sigma \in H^0(\mathcal{L})$.

---

[2]When we say that σ vanishes at p, or $\sigma(p) = 0$, we mean that the image of σ in the stalk $\mathcal{L}_p = \mathcal{L} \otimes_{\mathcal{O}_X} \mathcal{O}_{X,p}$ is in the maximal ideal $\mathfrak{m}_{X,p}$ of $\mathcal{O}_{X,p}$ times $\mathcal{L}_p$, and *not* that σ is zero as an element of the stalk $\mathcal{L}_p$ itself. Similarly, we say that σ vanishes to order m at p if $\sigma(p)$ lies in $\mathfrak{m}_{X,p}^m \mathcal{L}_p$.

The tensor product of (rational) sections σ of $\mathcal{L}$ and σ' of $\mathcal{L}'$ is a rational section of $\mathcal{L} \otimes \mathcal{L}'$ whose divisor is the sum of the divisors of σ and σ'. Thus the group of divisor classes $\operatorname{Pic} X$ is naturally isomorphic to the group of invertible sheaves under $\otimes$, and we will identify the two.

Invertible sheaves and line bundles. If $\mathcal{L}$ is an invertible sheaf on a variety X and $p \in X$ is a point, the *stalk* $\mathcal{L}_p$ of $\mathcal{L}$ is isomorphic to the local ring $\mathcal{O}_{X,p}$. We write $\mathfrak{m}_{X,p}$ for the maximal ideal of $\mathcal{O}_{X,p}$. By definition the *fiber* of $\mathcal{L}$ is

$$\kappa(p) \otimes \mathcal{L} = \mathcal{L}_p / \mathfrak{m}_{X,p} \mathcal{L}_p \cong \mathbb{C},$$

where $\kappa(p) := \mathcal{O}_{X,p} / \mathfrak{m}_{X,p}$ is the residue field of X at p. It is not hard to prove that the collection of these fibers forms a line bundle on X; that is, a morphism of schemes $L \to X$ whose fibers have the structure of 1-dimensional vector spaces such that $L|_{U_i} \cong U_i \times \mathbb{C}^1$ for some open covering $\{U_i\}$ of X.

Given a line bundle L on X, we can recover an invertible sheaf $\mathcal{L}$ associated to L by defining $\mathcal{L}(U)$ to be the set of sections of L defined over U. These two processes are inverse to one another, and allow us to think of invertible sheaves and line bundles interchangeably.

Though we generally use the invertible sheaf terminology, there are at least two points in which the line bundle approach is more natural. First, the vanishing of a section of an invertible sheaf at a point p is genuinely the vanishing of the section of the line bundle as a function. Second, and more serious, given a morphism $f : Y \to X$ of schemes, the pullback $f^*(\mathcal{L})$ of an invertible sheaf $\mathcal{L}$ on X is defined as the tensor product of $\mathcal{O}_Y$ with a sort of naive pullback, whereas the pullback of a line bundle is a straightforward set-theoretic operation.

Example 1.2 (Invertible sheaves on $\mathbb{P}^r$). Since $\mathbb{C}[x_0, \ldots, x_r]$ is a unique factorization domain, the ideal of any codimension 1 subvariety of $\mathbb{P}^r$ is generated by one nonzero element, which is thus a nonzerodivisor — it is a hypersurface. As we explained above, any two hypersurfaces of degree d differ by the divisor of a rational function, so the group of divisor classes on $\mathbb{P}^r$ is $\mathbb{Z}$. In other words, the class of a divisor is defined by its degree. Thus if $D = V(F) \subset \mathbb{P}^r$ is a hypersurface defined by the vanishing of a form F of degree d, it is natural to use the name $\mathcal{O}_{\mathbb{P}^r}(d)$ for $\mathcal{O}_{\mathbb{P}^r}(D)$.

If D is an effective divisor other than 0, then $H^0(\mathcal{O}_{\mathbb{P}^r}(-D)) = 0$, since there are no globally defined functions vanishing on D except 0.

To compute $H^0(\mathcal{O}_{\mathbb{P}^1}(d))$ directly, let $D = z_1 + z_2 + \cdots + z_d$ be a divisor of degree d and suppose that the coordinates are chosen so that none of the z_i are at infinity. The sections of $\mathcal{O}_{\mathbb{P}^1}(D)$ are the rational functions with poles in $\mathbb{P}^1$

only at the z_i. Identifying $\mathbb{P}^1 \setminus \{\infty\} = \mathbb{A}^1$ with $\mathbb{C}$ these can each be written as

$$\frac{g(z)}{(z - z_1)(z - z_2)\cdots(z - z_d)},$$

where g is a polynomial. The condition that the point at infinity is not a pole is the condition $\deg g \leq d$. With this condition, these rational functions form a vector space of dimension $d + 1$.

More generally, because every rational function on $\mathbb{P}^r$ has degree 0, and any two global sections differ by a rational function, every global section of $\mathcal{O}_{\mathbb{P}^r}(d)$ vanishes on a divisor of degree d. Thus we may identify $H^0(\mathcal{O}_{\mathbb{P}^r}(d))$ with the $\binom{r+d}{r}$-dimensional vector space of forms of degree d on $\mathbb{P}^r$.

Putting this together for future reference we have:

Proposition 1.3. *Every invertible sheaf $\mathcal{L}$ on $\mathbb{P}^r$ has the form $\mathcal{L} \cong \mathcal{O}_{\mathbb{P}^r}(m)$ for a unique $m = \deg \mathcal{L} \in \mathbb{Z}$; and we have*

$$H^0(\mathcal{O}_{\mathbb{P}^r}(m)) = \mathbb{C}[x_0,\dots,x_r]_m,$$

the space of forms of degree m in r + 1 variables.

1.3. Linear series and maps to projective space

We will use invertible sheaves to describe maps of a given variety X to projective space. For this we add the notion of linear series (sometimes called linear system).

Definition 1.4. A *linear series* on a scheme X is a pair $\mathcal{V} = (\mathcal{L}, V)$ where $\mathcal{L}$ is an invertible sheaf on X and V is a nonzero vector space of global sections of $\mathcal{L}$. We defa *dimension* of the linear series to be

$$\dim \mathcal{V} := \dim_{\mathbb{C}} V - 1.$$

To every global section σ of an invertible sheaf $\mathcal{L}$ on a variety X we can associate an effective divisor $(\sigma) = (\sigma)_0$ defined by the vanishing of σ. If τ is a scalar multiple of σ, it has the same divisor; and if $H^0(\mathcal{O}_X) = \mathbb{C}$ (for example if X is reduced, connected, and projective) then the converse is true: two sections of $\mathcal{L}$ with the same divisor differ by multiplication by a scalar.

We sometimes write $|\mathcal{L}|$ for the complete linear series $(\mathcal{L}, H^0(\mathcal{L}))$. If $\mathcal{L} = \mathcal{O}(D)$ we often write this as $|D|$.

Thus a linear series $\mathcal{V} = (\mathcal{L}, V)$ gives rise to a family of effective divisors on X, all in the same linear equivalence class, parametrized by the projective space $\mathbb{P}V^*$ of nonzero $\sigma \in V$ mod scalars. This is indeed the way we think of a linear series: as a family of divisors parametrized by a projective space. The definition of the dimension of a linear series reflects this: it's not the dimension of V as a vector space, but the dimension of the corresponding projective space.

Similarly, we speak of "the divisors of a linear series V" or as divisors "moving" in a linear series.

The intersection of the vanishing loci of all the sections in V is called the *base locus* of V. It is in general a subscheme of C. The points in its support are called *basepoints* of V. If the vector space V equals $H^0(\mathcal{L})$, the linear series is said to be *complete*; from the viewpoint of the *family of divisors*, this is the same as saying the linear series includes every effective divisor in the linear equivalence class.

Suppose that C is a smooth projective curve. Since the divisors in a linear series are linearly equivalent, they all have the same degree, called the degree of the linear series. If V is a linear series of degree d and dimension r we say that V is a g_d^r. This classical language means that V represents a group of d points moving within a linear equivalence class with r degrees of freedom.[3]

The next result is fundamental:

Theorem 1.5. *For any scheme X there is a natural bijection between the set of nondegenerate morphisms $\phi : X \to \mathbb{P}^r$ modulo PGL_{r+1} and base-point free linear series of dimension r on X up to isomorphism.*

Here *nondegenerate* means the image of the morphism ϕ is not contained in any hyperplane. The phrase "modulo PGL_{r+1}" is needed because the notation $\mathbb{P}^r$ supposes a choice of projective coordinates, and PGL_{r+1} is the group of linear coordinate transformations (actually all automorphisms, by Exercise 1.4). To get a correspondence without the dependence on a basis we could think of morphisms to $\mathbb{P}V$, the set of 1-quotients of V.

Suppose that $(\mathcal{L}, V)$ is a linear series on a smooth projective curve C that does have a base locus D_0. We can then subtract D_0 from all the divisors of the linear series, replacing $\mathcal{L}$ by $\mathcal{L}(-D_0)$ and dividing each section in V by the section σ of $\mathcal{O}_C(D_0)$ vanishing on D_0. This yields a new, base-point free linear series of the same dimension but lower degree.

We prove Theorem 1.5 by describing the correspondence in both directions:

From morphisms to linear series. Let $f : X \to \mathbb{P}^r$ be any nondegenerate morphism. The associated linear series $V = (\mathcal{L}, V)$ on X has $\mathcal{L} = f^*\mathcal{O}_{\mathbb{P}^r}(1)$, the pullback of the invertible sheaf $\mathcal{O}_{\mathbb{P}^r}(1)$, and

$$V = f^*H^0(\mathcal{O}_{\mathbb{P}^r}(1)) \subset H^0(\mathcal{L}).$$

In geometric terms, if we think of a linear series as a family of effective divisors, this is the linear series on X consisting of preimages of hyperplanes in $\mathbb{P}^r$. Nondegeneracy assures us that the preimage of a hyperplane in $\mathbb{P}^r$ is indeed a divisor on X.

---

[3]This "g" is unrelated to the genus; it's short for *gruppi di punti*, Italian for "divisors."

From linear series to morphisms. Suppose that X is any scheme, and $\mathcal{V} = (\mathcal{L}, V)$ is a base-point free linear series of dimension r on X; we want to describe a corresponding morphism $f : X \to \mathbb{P}^r$. Choose a basis $\sigma_0, \ldots, \sigma_r$ for V. If we let $D_i = (\sigma_i) \subset X$ be the divisor of zeroes of σ_i and set $U_i := X \setminus D_i$, the ratio σ_j / σ_i is a regular function on U_i, and we can define a map $f_i : U_i \to \mathbb{P}^r$ by

$$f_i : p \mapsto \left(\frac{\sigma_0}{\sigma_i}(p), \ldots, \frac{\sigma_r}{\sigma_i}(p) \right),$$

where the component $\sigma_i / \sigma_i = 1$ ensures that not all the components are 0, so that the image point is well-defined. The maps f_i and f_j agree on the overlap $U_i \cap U_j$, and by the hypothesis that $\mathcal{V}$ is base-point free the U_i cover X; so together they define a regular map f from X to $\mathbb{P}^r$.

We can describe this map set-theoretically without having to choose a basis: since $\mathcal{V}$ is assumed base-point free, for any point $p \in X$ the subspace $H_p := \{\sigma \in V \mid \sigma(p) = 0\}$ is a hyperplane in V; thus we get a map $f : X \to \mathbb{P}V$.

Example 1.6. The morphism from $\mathbb{P}^r$ defined by the complete linear series $(\mathcal{O}_{\mathbb{P}^r}(d), H^0(\mathcal{O}_{\mathbb{P}^r}(d)))$ has target $\mathbb{P}^{\binom{r+d}{r}-1}$, and takes a point $(a_0, \ldots a_r)$ to the point whose coordinates are all the monomials of degree d in $x_0, \ldots x_r$. It is called the *d-th Veronese morphism* from $\mathbb{P}^r$. For example, on $\mathbb{P}^1$ this has the form

$$(x_0, x_1) \mapsto (x_0^d, \ x_0^{d-1} x_1, \ \ldots, x_1^d).$$

The image of $\mathbb{P}^1$ under this morphism is called the *rational normal curve* of degree d; in the case $d = 2$ is the *plane conic*, and in the case $d = 3$ it is called the *twisted cubic*. Veronese himself studied the image of $\mathbb{P}^2$ by the Veronese morphism of degree 2 now simply called the *Veronese surface*.

1.4. The geometry of linear series

An upper bound on $h^0(\mathcal{L})$. We will develop sophisticated ways of estimating the dimensions of linear series. We begin with an elementary bound:

Theorem 1.7. *Let C be a smooth projective curve. If $\mathcal{L}$ is an invertible sheaf of degree $d \geq 0$ on C, then $h^0(\mathcal{L}) \leq d + 1$; and equality holds if and only if $C \cong \mathbb{P}^1$.*

Proof. Let $p_1, \ldots, p_{d+1}$ be points of C. If $h^0(\mathcal{L})$ exceeded $d + 1$, there would be a nonzero section $\sigma \in H^0(\mathcal{L})$ vanishing at $p_1, \ldots, p_{d+1}$; the divisor (σ) would then have degree $\geq d + 1$, contradicting the hypothesis that $\deg \mathcal{L} = d$.

For the second part, suppose that $h^0(\mathcal{L}) = d + 1$. If $p_1, \ldots, p_d$ are points of C, there is a nonzero section $\sigma \in H^0(\mathcal{L})$ vanishing at $p_1, \ldots, p_d$; by degree considerations, it cannot vanish anywhere else. It follows that any two divisors of degree d on C are linearly equivalent, hence that any two points $p, q \in C$ are linearly equivalent. Thus there is a rational function f on C with exactly one zero and one pole, giving an isomorphism $C \cong \mathbb{P}^1$. $\qquad \square$

From the correspondence between invertible sheaves and maps to projective space, we now get:

Corollary 1.8. *If $C \subset \mathbb{P}^d$ is a nondegenerate curve, then the degree of C is at least d, with equality only in the case that C is a rational normal curve.* □

Rational normal curves arise often in the literature because they have many extremal properties, such as those of Corollaries 1.8 and 1.9. For a related result see Corollary 13.3. Since there is only one invertible sheaf of degree d on $\mathbb{P}^1$, any two rational normal curves of degree d differ by a transformation in PGL_{d+1}, and we will therefore often speak of *the* rational normal curve.

Corollary 1.9. *Let $C \subset \mathbb{P}^d$ be the rational normal curve of degree d. If E is an effective divisor on C of degree $e \leq d + 1$, then the* span *of E (that is, the smallest linear space containing the subscheme E) has dimension $e - 1$.*

Less formally: any finite set or subscheme of C is as linearly independent as possible.

Proof. Since E imposes at most e conditions on hyperplanes, the span of E has dimension at most $e - 1$.

On the other hand, the hyperplanes containing E meet C in a divisor of the form $E + E'$, where $\deg E' = d - e$. Thus the projection of C from E is a nondegenerate curve of degree $d - e$ in $\mathbb{P}^{d-(\dim \operatorname{span} E)-1}$, so from Corollary 1.8 we get $d - e \geq d - \dim \operatorname{span} E - 1$, as required. □

In the case of distinct points on a rational normal curve it is easy to make a direct argument why they are as independent as possible: Choose coordinates so that none of the points are at infinity. We can identify the points $\lambda_1, \ldots, \lambda_{d+1} \in C \cong \mathbb{P}^1$ with distinct complex numbers, and the independence (for $\ell = d + 1$) is equivalent to the the nonvanishing of the Vandermonde determinant

$$\begin{vmatrix} 1 & \lambda_1 & \lambda_1^2 & \cdots & \lambda_1^d \\ 1 & \lambda_2 & \lambda_2^2 & \cdots & \lambda_2^d \\ \vdots & & & & \vdots \\ 1 & \lambda_{d+1} & \lambda_{d+1}^2 & \cdots & \lambda_{d+1}^d \end{vmatrix} = \prod_{1 \leq i < j \leq d+1} (\lambda_j - \lambda_i).$$

Incomplete linear series. In classical algebraic geometry a linear series of dimension 1 is called a *pencil*,[4] a linear series of dimension 2 is called a *net* and, less commonly, a three-dimensional linear series is called a *web*. We will use only the first of these terms.

---

[4]This usage harks back to the early meaning of "pencil" in English, an artist's fine brush, borrowed before 1400 from (old) French "pincel" and (late) Latin "pincellus". In the geometric sense the term initially referred to a set of lines (light rays, in a 1665 example attested in the OED) meeting in a point.

The morphism associated to an incomplete linear series $V \subset H^0(\mathcal{L})$ is the composition of the morphism associated to the complete linear series $|\mathcal{L}|$ with a linear projection. In general, if $V \subset W \subset H^0(\mathcal{L})$ are nested linear series, then a 1-dimensional quotient of W restricts to a 1-dimensional quotient of V unless it vanishes on V. Thus we have a partially defined linear morphism $\pi : \mathbb{P}W \to \mathbb{P}V$. The *indeterminacy locus* of the map consists of the set of 1-quotients vanishing on V, that is, $\mathbb{P}(W/V) \subset \mathbb{P}W$; we will call it the *center of the projection* π. (It is sometimes useful to think of the dual picture: lines in W^* map to lines in V^* except when they lie in the subspace $(W/V)^* = \operatorname{Ann} V \subset W^*$.) Thus there is a commutative diagram

$$
\begin{array}{ccc}
 & & \mathbb{P}W^* \\
 & {\phi_W}\nearrow & \big| \\
 & & \big| \pi \\
 & & \downarrow \\
C & \xrightarrow{\phi_V} & \mathbb{P}V^*.
\end{array}
$$

If W is base-point free, then V is base-point free if and only if the center of the projection π is disjoint from $\phi_W(C)$. We then say that π is *regular* on C. An example is illustrated in Figure 1.2.

By way of language, we will say that a curve $C \subset \mathbb{P}^r$ embedded by a complete linear series $|\mathcal{L}|$ is *linearly normal*; this is equivalent to saying that the pullback map

$$
H^0(\mathcal{O}_{\mathbb{P}^r}(1)) \to H^0(\mathcal{L})
$$

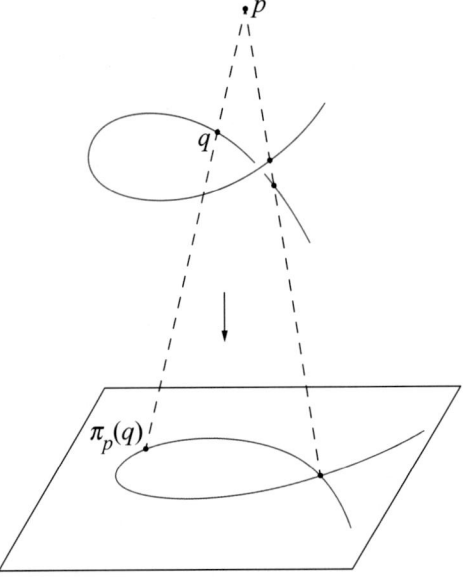

Figure 1.2. Projection of a space curve from a general point p to $\mathbb{P}^2$.

is surjective. Since regular projections of a curve correspond to subseries, this is equivalent to saying that C is *not* the regular projection of a nondegenerate curve $\tilde{C} \subset \mathbb{P}^{r+1}$.

Sums of linear series. If $\mathcal{D} = (\mathcal{L}, V)$ and $\mathcal{E} = (\mathcal{M}, W)$ be two linear series on a curve C. By the *sum* $\mathcal{D} + \mathcal{E}$ of $\mathcal{D}$ and $\mathcal{E}$ we will mean the pair

$$\mathcal{D} + \mathcal{E} = (\mathcal{L} \otimes \mathcal{M}, U)$$

where $U \subset H^0(\mathcal{L} \otimes \mathcal{M})$ is the subspace generated by the image of $V \otimes W$, under the multiplication/cup product map $H^0(\mathcal{L}) \otimes H^0(\mathcal{M}) \to H^0(\mathcal{L} \otimes \mathcal{M})$. In other words, it's the subspace of the complete linear series $|\mathcal{L} \otimes \mathcal{M}|$ spanned by divisors of the form $D + E$, with effective divisors $D \in \mathcal{D}$ and $E \in \mathcal{E}$.

The sum of two base-point free linear series is clearly base-point free; if the two series correspond to maps $\phi : C \to \mathbb{P}^r$ and $\psi : C \to \mathbb{P}^s$ then the sum corresponds to the composition of the map $(\phi, \psi) : C \to \mathbb{P}^r \times \mathbb{P}^s$ with the *Segre embedding* $\sigma : \mathbb{P}^r \times \mathbb{P}^s \to \mathbb{P}^{(r+1)(s+1)-1}$ given in coordinates by

$$((x_0, \ldots, x_r), (y_0, \ldots, y_s)) \mapsto (x_0 y_0, \ldots, x_i y_j, \ldots, x_r y_s).$$

Note that this says nothing about the dimension of $\mathcal{D} + \mathcal{E}$, since the composite map $\sigma \circ (\phi, \psi)$ will typically not be nondegenerate. In fact, the best we can do in general is the following proposition.

Proposition 1.10. *If $\mathcal{D}$ and $\mathcal{E}$ are two linear series that contain effective divisors on a curve C, then*

$$\dim(\mathcal{D} + \mathcal{E}) \geq \dim \mathcal{D} + \dim \mathcal{E}.$$

Proof. Saying that $\dim \mathcal{D} \geq m$ is equivalent to saying that we can find a divisor $D \in \mathcal{D}$ containing any given m points of C; since $\mathcal{D} + \mathcal{E}$ contains all pairwise sums $D + E$ with $D \in \mathcal{D}$ and $E \in \mathcal{E}$, we can certainly find a divisor $F \in \mathcal{D} + \mathcal{E}$ containing any given $\dim \mathcal{D} + \dim \mathcal{E}$ points of C. $\square$

Which linear series define embeddings? A linear series $\mathcal{V} = (\mathcal{L}, V)$ on a projective variety is called *very ample* if it is base-point free and defines an embedding (what Hartshorne calls a "closed immersion"). If D is a Cartier divisor on X, then we say that D is *very ample* if the complete linear series $|D|$ is very ample, and we say that D is *ample* if mD is very ample for some integer $m > 0$.

Similarly, $\mathcal{V}$ or D are called *birationally very ample* if V or D are base-point free and define a map that is generically one-to-one or, equivalently, an embedding when restricted to an open set. This arises often when speaking of a map from a smooth curve C onto a curve C_0 in the plane, since the latter frequently has singularities, as in Figure 1.2.

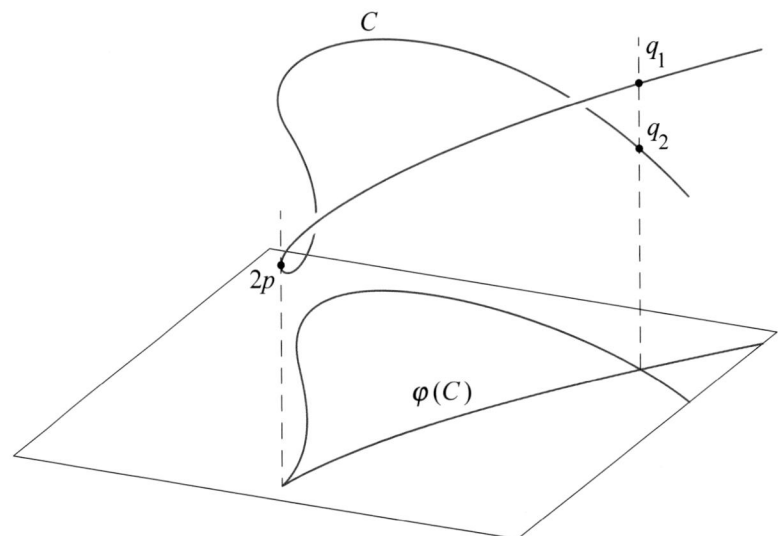

Figure 1.3. Two distinct points or one double point might impose just one condition on the linear series defined by projection from a point *a* (here illustrated by vertical projection).

Given a linear series $V = (\mathcal{L}, V)$ and an effective divisor D on C, we set $V(-D) = (\mathcal{L}(-D), V(-D))$ where

$$\mathcal{L}(-D) := \mathcal{L} \otimes \mathcal{O}(-D) \text{ and } V(-D) := \{\sigma \in V \mid \sigma(D) = 0\}.$$

The difference $\dim V - \dim V(-D)$ is called the *number of conditions imposed by D on the linear series V*; we say that *D imposes independent conditions* on V if $\dim V - \dim V(-D) = \deg D$.

Via the correspondence of Theorem 1.5, statements about the geometry of a morphism $\phi : C \to \mathbb{P}^r$ can be formulated as statements about the relevant linear series. In the case of complete series, these are statements about the vector space $H^0(\mathcal{L})$ of global sections of $\mathcal{L}$. We write $h^0(\mathcal{L})$ for the vector space dimension of $H^0(\mathcal{L})$ (and similarly for other cohomology groups). It is useful to have criteria in these terms for when a linear series defines an embedding, or even to be base-point free so that it defines a morphism:

Proposition 1.11. [Hartshorne 1977, Theorem IV.3.1] *Let $\mathcal{L}$ be an invertible sheaf on a smooth curve C. The complete linear series $|\mathcal{L}|$ is base-point free if and only if*

$$h^0(\mathcal{L}(-p)) = h^0(\mathcal{L}) - 1 \quad \forall p \in C;$$

and $\mathcal{L}$ is very ample if and only if

$$h^0(\mathcal{L}(-p-q)) = h^0(\mathcal{L}) - 2 \quad \forall p, q \in C.$$

Proof. First, if $\mathcal{L}$ is base-point free, then vanishing at a point imposes one linear condition on sections of $\mathcal{L}$, so $h^0(\mathcal{L}(-D)) \geq h^0(\mathcal{L}) - \deg D$ for any effective divisor D.

To say that $|\mathcal{L}|$ is base-point free means that for every point $p \in C$ there is a section of $\mathcal{L}$ that does not vanish at p; thus vanishing at p is a nontrivial linear condition on $H^0(\mathcal{L})$. Conversely, if $h^0(\mathcal{L}(-p)) = h^0(\mathcal{L}) - 1$ then p imposes a nontrivial condition, so some section of $\mathcal{L}$ does not vanish at p.

Since a divisor of degree d cannot impose more than d conditions on a linear series, the statement $h^0(\mathcal{L}(-p-q)) = h^0(\mathcal{L}) - 2$ for all p, q implies the condition for base-point freeness; and saying that $\phi_{\mathcal{L}}(p) \neq \phi_{\mathcal{L}}(q)$ implies that the linear series defines a set-theoretic injection.

Let $\phi : C \to \mathbb{P}^r$ be the map defined by $\mathcal{L}$. To say that ϕ is an embedding locally at a point $p \in C$, we need to know in addition that the map of local rings

$$\phi^* : \mathcal{O}_{\mathbb{P}^r, \phi(p)} \to \mathcal{O}_{C,p}$$

is surjective.

Since C is projective, the map $C \to \phi(C)$ is finite, so ϕ^* makes $\mathcal{O}_{C,p}$ into a finitely generated $\mathcal{O}_{\mathbb{P}^r, \phi(p)}$-module. By Nakayama's lemma it suffices to show that $\mathcal{O}_{C,p}/\phi^*(\mathfrak{m}_{\mathbb{P}^r, \phi(p)})$ is generated by the image of $\mathcal{O}_{\mathbb{P}^r, \phi(p)}/\mathfrak{m}_{\mathbb{P}^r, \phi(p)} = \mathbb{C}$.

Since the constants $\mathbb{C} = \mathcal{O}_{\mathbb{P}^r, \phi(p)}/\mathfrak{m}_{\mathbb{P}^r, \phi(p)}$ pull back to the constants in $\mathcal{O}_{C,p}/\mathfrak{m}_{C,p}$, surjectivity will follow if

$$\frac{\mathcal{O}_{C,p}}{\phi^*(\mathfrak{m}_{\mathbb{P}^r, \phi(p)})\mathcal{O}_{C,p}}$$

is 1-dimensional; that is, if the linear series contains a section that vanishes to order exactly 1 at p, which is equivalent to the condition that $h^0(\mathcal{L}(-2p)) \neq h^0(\mathcal{L}(-p))$. $\qquad\square$

A more geometric version of the last part of the proof would be to say that the condition of the existence of a section vanishing to order exactly 1 implies that ϕ is an injection on the tangent space to C at p. This implies that the map is analytically an isomorphism onto its image, locally on the source; and together with the finiteness and set-theoretic injectivity of the map, this suffices. Figure 1.3 illustrates the two ways in which a linear series on a smooth curve can fail to be very ample.

If $\phi : X \to \mathbb{P}^r$ is a generically finite morphism, then the *degree* of ϕ is the number of points in the preimage of a general point of $\phi(X)$. It follows that if $D := \sum_{p \in C} n_p p$ is a divisor on a smooth curve, and the linear series $|D|$ is base-point free, then the degree of the morphism associated to $|D|$ is $\deg D := \sum_{p \in C} n_p$.

Exercises

Exercise 1.1. Show that there is no nonconstant morphism $\mathbb{P}^r \to \mathbb{P}^s$ when $s < r$ by showing that any nontrivial linear series of dimension $< r$ on $\mathbb{P}^r$ has a nonempty base locus. ◆

Exercise 1.2. Let

$$V = \langle s^{a_0} t^{d-a_0}, \ldots, s^{a_r} t^{d-a_r} \rangle \subset \mathbb{C}[s,t]_d,$$

and let $\mathcal{V} = (\mathcal{O}_{\mathbb{P}^1}(d), V)$. Determine the conditions for each of the following properties to hold:

(1) $\mathcal{V}$ is base-point free.

(2) $\mathcal{V}$ is ample.

(3) $\mathcal{V}$ is very ample.

(4) $\mathcal{V}$ is complete.

Exercise 1.3. Extend the statement of Proposition 1.11 to incomplete linear series; that is, prove that the morphism associated to a linear series $(\mathcal{L}, V)$ on a smooth curve is an embedding if and only if

$$\dim \left(V \cap H^0(\mathcal{L}(-p-q)) \right) = \dim V - 2 \quad \forall p, q \in C.$$

Exercise 1.4. Show that an automorphism of $\mathbb{P}^r$ takes hyperplanes to hyperplanes. Deduce that it is given by the linear series $\mathcal{V} = (\mathcal{O}_{\mathbb{P}^r}(1), H^0(\mathcal{O}_{\mathbb{P}^r}(1)))$, and use this to show that $\mathrm{Aut}\,\mathbb{P}^r = \mathrm{PGL}(r+1)$. ◆

Exercise 1.5. Let $C \subset \mathbb{P}^r$ be any linearly normal curve and $\phi : C \to C$ an automorphism. Show that ϕ is induced by an automorphism of $\mathbb{P}^r$ if and only if ϕ carries the invertible sheaf $\mathcal{O}_C(1)$ to itself; that is, $\phi^*(\mathcal{O}_C(1)) \cong \mathcal{O}_C(1)$. In this case we say that the automorphism is *projective*. Show that every automorphism of a rational normal curve $C \subset \mathbb{P}^d$ extends to $\mathbb{P}^d$. Since the automorphism group PGL_2 of $\mathbb{P}^1$ acts transitively on $\mathbb{P}^1$, we say that C is *projectively homogeneous*.

Exercise 1.6. Show that the ring $\mathbb{C}[s^d, s^{d-1}t, \ldots, t^d]$ is normal (= integrally closed) by noting that its integral closure must be contained in $\mathbb{C}[s,t]$ and then showing that if f is any polynomial in the integral closure then the homogeneous components of f are also in the integral closure.

Exercise 1.7. Suppose $C = V(y^2 - x^2 - x^3) \subset \mathbb{A}^2$ (so that C is a plane curve with a node at the point $p = (0,0) \in C$).

(1) Show that the point p itself (that is, the reduced subscheme) is not a Cartier divisor, since its ideal (x, y) is not generated by a single element.

(2) Show that the fat point, that is, the subscheme $\Gamma = V(x^2, xy, y^2)$, is contained in C but is not a Cartier divisor on C.

(3) Let $\Gamma_{\alpha,\beta} := V(\alpha x + \beta y, x^2, xy, y^2) \subset C$. Show that if $\beta \neq \pm\alpha$ then $\Gamma_{\alpha,\beta}$ is a Cartier divisor on C.

(4) With $\Gamma_{\alpha,\beta}$ as above, show that if $\beta = \pm\alpha$ then $\Gamma_{\alpha,\beta}$ is not a Cartier divisor on C. ◆

Chapter 2

The Riemann–Roch theorem

2.1. How many sections?

To study curves via their maps to projective spaces, we want to estimate the dimension of the space of global sections of an invertible sheaf $\mathcal{L}$. The beginning of the story is the Riemann–Roch theorem.

Though we would like to be able to compute $h^0(\mathcal{L})$, it is much easier to compute the Euler characteristic

$$\chi(\mathcal{L}) := \sum_{i \geq 0}(-1)^i h^i(\mathcal{L}).$$

This computes $h^0(\mathcal{L})$ itself in many cases, by virtue of the following result:

Theorem 2.1 (Serre–Grothendieck vanishing theorem). *If $\mathcal{F}$ is a coherent sheaf on a projective scheme X of dimension n, then for any i, the vector space $H^i(\mathcal{F})$ is finite-dimensional, and is 0 if $i > n$. Moreover, if $X \subset \mathbb{P}^m$ then for $d \gg 0$, $\mathcal{F}(d)$ is generated by its global sections and $H^i(\mathcal{F}(d)) = 0$ for all $i > 0$.* □

Proof. This is a combination of theorems due to Grothendieck and Serre. See [Hartshorne 1977, Theorems III.2.7 and III.5.2], and also [Serre 1955] for a reasonably concrete proof. □

A shortcoming of this vanishing theorem is the lack of a bound on the number d needed to achieve the second assertion. For smooth curves and invertible sheaves this is corrected by Theorem 2.16, which gives a bound in terms of the genus and the degree.

One easy consequence of Theorem 2.1 is that on a smooth variety the groups of invertible sheaves and divisor classes are the same (see also [Hartshorne 1977, Proposition II.6.13]).

Corollary 2.2. *If X is a projective variety that is nonsingular in codimension 1, every invertible sheaf $\mathcal{L}$ on X is of the form $\mathcal{L} = \mathcal{O}_C(D)$ for some Cartier divisor D on X. Thus if X is a smooth projective variety the map $\div$ is an isomorphism from the group of invertible sheaves to the group of divisor classes.*

Proof. Let $H \subset \mathbb{P}^r$ be a general hyperplane, and E the divisor of intersection of C with H. We know that for $n \gg 0$, $\mathcal{L}(n)$ has sections; and if F is the divisor of zeroes of one such section, we have

$$\mathcal{L} = \mathcal{O}_C(F - nE).$$

If X is smooth, then, since a regular local ring is a unique factorization domain, every codimension 1 subvariety is defined locally by a single nonzerodivisor, and thus corresponds to a Cartier divisor. This implies that $\div$ is surjective. Furthermore any isomorphism between invertible sheaves is defined by multiplication with a global rational function, so that invertible sheaves defining linearly equivalent divisors are isomorphic. Thus $\div$ is injective as well. $\square$

Riemann–Roch without duality. It follows from Theorem 2.1 that on any scheme $X \subset \mathbb{P}^r$ we have $\chi(\mathcal{L}(d)) = h^0(\mathcal{L}(d))$ for large d, and that $\chi(\mathcal{L}) = h^0(\mathcal{L}) - h^1(\mathcal{L})$ in the case of a curve.

Theorem 2.3 (easy Riemann–Roch). *If C is a smooth projective curve, and $\mathcal{L}$ is an invertible sheaf on C, then $\chi(\mathcal{L}) = \deg \mathcal{L} + \chi(\mathcal{O}_C)$.*

Proof. The result is tautological if $\mathcal{L} = \mathcal{O}_C$. Every invertible sheaf on C has the form $\mathcal{L} = \mathcal{O}_C(D)$ for some divisor D. If $p \in C$, then writing $\kappa(p)$ for the structure sheaf of the subscheme $p \in C$, the long exact sequence in cohomology associated to the short exact sequence

$$0 \to \mathcal{L}(-p) \to \mathcal{L} \to \mathcal{L} \otimes \kappa(p) \to 0$$

together with the isomorphism $\mathcal{L} \otimes \kappa(p) \cong \kappa(p)$ and the vanishing of higher cohomology of a sheaf with zero-dimensional support allows us to compute

$$\chi(\mathcal{L}) = \chi(\mathcal{L}(-p)) + \chi(\kappa(p)) = \chi(\mathcal{L}(-p)) + 1.$$

Since every divisor on C can be reached by adding and subtracting points, this suffices. $\square$

Since the Euler characteristic of a sheaf is well-behaved, we can extend the result of Theorem 2.3 to invertible sheaves on any one-dimensional scheme C, by defining $\deg \mathcal{L} := \chi(\mathcal{L}) - \chi(\mathcal{O}_C)$. We will use this definition to express the self-intersection of a divisor on a surface in Section 2.6.

We can make the Riemann–Roch theorem still more useful by understanding the error term $h^1(\mathcal{L})$. This requires the canonical divisor and Serre duality, to which we now turn.

2.2. The most interesting linear series

The most important vector bundles on a manifold are the tangent and cotangent bundles. For reasons that will become clear, the focus in algebraic geometry is on the cotangent bundle or, equivalently, the sheaf of differential 1-forms. On a smooth curve C the *canonical sheaf* is the sheaf of differentials, which is an invertible sheaf; on a smooth variety of dimension n we define the canonical sheaf to be the n-th exterior power of the sheaf of differentials. A section of ω_C is thus a differential form, and the class of the divisor of such a form is usually denoted K_C.

Cheerful Fact 2.4. Canonical sheaves are defined for any projective scheme; see Definition 16.6. They are usually called *dualizing sheaves* in that generality. A scheme is said to be *Gorenstein* when its dualizing sheaf is invertible, something that is true, for example, for any subscheme of $\mathbb{P}^r$ that is locally a complete intersection (see Section 16.6), and in particular for plane curves.

On projective space we can compute the canonical sheaf directly; other computations of the canonical sheaf will usually reduce to this central case.

Theorem 2.5. *The canonical sheaf of $\mathbb{P}^r$ is $\mathcal{O}_{\mathbb{P}^r}(-r-1)$.*

Proof. Let $x_0, \ldots, x_r$ be the projective coordinates on $\mathbb{P}^r$ and let $U = \mathbb{P}^r \setminus H$ be the affine open set where $x_0 \neq 0$. Thus $U \cong \mathbb{A}^r$ with coordinates $z_1 := x_1/x_0$, $\ldots, z_r := x_r/x_0$. The space of r-dimensional differential forms on U is spanned by $d(x_1/x_0) \wedge \cdots \wedge d(x_r/x_0)$, which is regular everywhere in U. In view of the formula

$$d\frac{x_i}{x_0} = \frac{x_0\, dx_i - x_i\, dx_0}{x_0^2}$$

we get

$$d\frac{x_1}{x_0} \wedge \cdots \wedge d\frac{x_r}{x_0} = \frac{dx_1 \wedge \cdots \wedge dx_r}{x_0^r} - \sum_{i=1}^{r} x_i \frac{dx_1 \wedge \cdots \wedge \widehat{dx_i} \wedge \cdots \wedge dx_r}{x_0^{r+1}}$$

which has a pole of order $r+1$ along the locus H defined by x_0. Thus the divisor of this differential form is $-(r+1)H$, and this is the canonical class. $\qquad\square$

Cheerful Fact 2.6. A different derivation: there is a short exact sequence of sheaves, called the Euler sequence [Hartshorne 1977, Chapter II, §8]:

$$0 \to \Omega_{\mathbb{P}^r} \to \mathcal{O}_{\mathbb{P}^r}^{r+1}(-1) \to \mathcal{O}_{\mathbb{P}^r} \to 0.$$

Summing over all twists, and taking global sections, that is, applying H_*^0, we see that $H_*^0(\Omega_{\mathbb{P}^r})$ fits into an exact sequence:

$$0 \to H_*^0(\Omega_{\mathbb{P}^r}) \to S^{r+1}(-1) \xrightarrow{\delta_1} S \to \mathbb{C} \to 0,$$

where S is the homogeneous coordinate ring of $\mathbb{P}^r$ and δ_1 sends the i-th basis vector of $S^{r+1}(-1)$ to the i-th variable of S; that is, $H_*^0(\Omega_{\mathbb{P}^r})$ is the second syzygy of the residue field $\mathbb{C}$ of S. We can extend this sequence to the Koszul complex that is the free resolution of $\mathbb{C}$, [Eisenbud 1995, §17.5]:

$$0 \to S(-r-1) \xrightarrow{\delta_{r+1}} \bigwedge^r S^{r+1}(-r) \xrightarrow{\delta_r} \cdots \to S^{r+1}(-1) \to S \to \mathbb{C} \to 0.$$

For each i, the i-th exterior power of the map $H_*^0(\Omega_{\mathbb{P}^r}) \to S^{r+1}(-1)$ is an inclusion, and represents $\bigwedge^i(\Omega_{\mathbb{P}^r})$ as the sheaf associated to the graded module that is the $(i+1)$-st syzygy of $\mathbb{C}$. In particular, the canonical module $\omega_{\mathbb{P}^r} = \bigwedge^r(\Omega_{\mathbb{P}^r})$ is the sheaf associated to the $(r+1)$-st syzygy, $S(-r-1)$.

For more on syzygies, see Chapter 18.

The most important invariant of a smooth curve can be defined in terms of the canonical sheaf:

Definition 2.7. If C is an irreducible smooth curve we define the genus $g(C)$ to be the dimension of $H^0(\omega_C)$.

Computations of the canonical sheaf on a variety usually involve comparing the variety to a variety whose canonical sheaf is already known. The most useful results of this type are the *adjunction formula* and *Hurwitz's theorem*.

The adjunction formula. In the simplest case, the adjunction formula says that the canonical divisors of a smooth plane curve C of degree d are the intersections of C with curves of degree $d - 3$ (see Figure 2.1). More generally, for a divisor X on a smooth variety Y, it says that the canonical sheaf on X is $\omega_Y(X)|_X$. This is an immediate consequence of the still more general formula below because the normal bundle of X is $\mathcal{O}_Y(X)$.

In general, the adjunction formula describes the difference between the canonical divisor of a subscheme and the restriction of the canonical divisor from the ambient variety. If $X \subset Y$ we define the *conormal sheaf* of X in Y to be $\mathcal{I}_{X/Y}/\mathcal{I}_{X/Y}^2$, and the *normal sheaf* of X in Y to be its dual,

$$\mathcal{N}_{X/Y} = \mathcal{H}om(\mathcal{I}_{X/Y}/\mathcal{I}_{X/Y}^2, \mathcal{O}_Y).$$

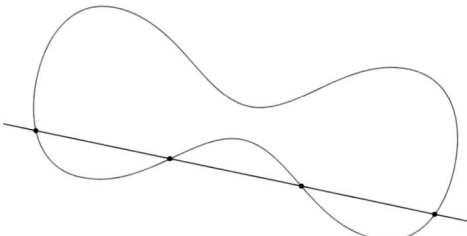

Figure 2.1. On a smooth plane quartic, the canonical divisors are its intersections with lines.

If X and Y are smooth, X is locally a complete intersection in Y, so $\mathcal{I}_{X/Y}/\mathcal{I}_{X/Y}^2$ is
a vector bundle on X of rank equal to the codimension, $\dim Y - \dim X$. When,
in addition, the codimension is 1, so that X is a divisor and $\mathcal{I}_X = \mathcal{O}_Y(-X)$, we
get

$$\mathcal{N}_{X/Y} = \mathcal{O}_X(X).$$

Proposition 2.8 (adjunction formula). *Let $X \subset Y$ a smooth subscheme of codimension c in a smooth variety Y, and let K_Y be the canonical class of Y. The
canonical class K_X of X is*

$$\omega_X = \bigwedge^c \mathcal{N}_{X/Y} \otimes \omega_Y.$$

*In particular, when X is a divisor, K_X is the restriction to X of the divisor $K_Y + X$
on Y.*

Proof. Because X is locally a complete intersection in Y there is an exact sequence of sheaves

$$0 \to \mathcal{I}_{X/Y}/\mathcal{I}_{X/Y}^2 \to \Omega_Y|_X \to \Omega_X \to 0,$$

where Ω_X is the sheaf of differential forms on X (see [Eisenbud 1995, Proposi-
tion 16.3]), and $\mathcal{I}_{X/Y}|_X = \mathcal{O}_Y(-X)|_X = \mathcal{O}_X(-X)$. The proposition follows by
taking top exterior powers, as in Lemma 2.9. $\qquad\square$

Lemma 2.9. *If*

$$0 \to \mathcal{E} \to \mathcal{F} \to \mathcal{G} \to 0$$

*is a short exact sequence of locally free sheaves of ranks e, f, g on a scheme X, then
there is a natural isomorphism*

$$\bigwedge^e \mathcal{E} \otimes \bigwedge^g \mathcal{G} \to \bigwedge^f \mathcal{F}.$$

Proof of Lemma 2.9. We may define a map $\bigwedge^e \mathcal{E} \otimes \bigwedge^g \mathcal{G} \to \bigwedge^f \mathcal{F}$ in terms of
local sections as

$$(\epsilon_1 \wedge \cdots \wedge \epsilon_e) \otimes (\gamma_1 \wedge \cdots \wedge \gamma_g) \mapsto \epsilon_1 \wedge \cdots \wedge \epsilon_e \wedge \gamma_1 \wedge \cdots \wedge \gamma_g.$$

This is globally well-defined because changing one of the γ_i by a local section of $\mathcal{E}$ would not change the exterior product. To check that the map is an isomorphism, it is enough to show that this is true locally.

Because $\mathcal{G}$ is locally free, there is a covering of X by open sets U so that the sequence

$$0 \to \mathcal{E}|_U \to \mathcal{F}|_U \to \mathcal{G}|_U \to 0$$

is a split exact sequence of free modules, $\mathcal{F}|_U = \mathcal{E}|_U \oplus \mathcal{G}|_U$. It follows that

$$\bigwedge^f \mathcal{F}|_U = \bigoplus_{i+j=f} \bigwedge^i \mathcal{E}|_U \otimes \bigwedge^j \mathcal{G}|_U.$$

In our case all the exterior powers of $\mathcal{E}$ vanish above the e-th, and all the exterior powers of $\mathcal{G}$ vanish above the g-th, so

$$\bigwedge^f \mathcal{F}|_U = \bigwedge^e \mathcal{E}|_U \otimes \bigwedge^g \mathcal{G}|_U,$$

with isomorphism given as above.   □

Corollary 2.10. *If $C \subset \mathbb{P}^2$ is a smooth plane curve of degree d, then $\omega_C = \mathcal{O}_C(d-3)$; more generally, if $X \subset \mathbb{P}^r$ is a smooth complete intersection of hypersurfaces of degrees $d_1, \ldots, d_c$ in $\mathbb{P}^r$ then $\omega_X = \mathcal{O}_X(\sum d_i - r - 1)$.*

Proof. Since $\mathcal{N}_{X/Y} = \bigoplus_{i=1}^{c} \mathcal{O}_X(d_i)$, the result follows from Theorem 2.8.   □

Hurwitz's theorem. Given a (nonconstant) morphism $f : C \to X$ of smooth projective curves, the Riemann–Hurwitz formula computes the canonical sheaf C in terms of that of X and the local geometry of f. To do this we define the *ramification index* of f at $p \in C$, denoted $\mathrm{ram}(f, p)$, by the formula

$$f^{-1}(q) = \sum_{\substack{p \in C \\ f(p)=q}} (\mathrm{ram}(f, p) + 1) \cdot p$$

for any point $q \in X$.

Proposition 2.11. *If $f : C \to X$ is a (nonconstant) morphism of smooth projective curves, there are only finitely many points $p \in C$ such that $\mathrm{ram}(f, p) > 0$.*

In light of this result we define the *ramification divisor* of f to be the divisor

$$R = \sum_{p \in C} \mathrm{ram}(f, p) \cdot p \in \mathrm{Div}(C).$$

and the *branch divisor* to be

$$B = \sum_{q \in X} \left(\sum_{p \in f^{-1}(q)} \mathrm{ram}(f, p) \right) \cdot q \in \mathrm{Div}(X).$$

Note that R and B have the same degree, which is $\sum_{p \in C} \mathrm{ram}(f, p)$.

Proof. The result follows from the separability of the map of fields of rational functions, $K(X) \to K(C)$, which holds because we are in characteristic 0 (in characteristic p the Frobenius map provides a counterexample). A proof using separability is given in [Hartshorne 1977, Section IV.2]. Here is an analytic version:

In terms of local parameters z on C around p and w on X around $f(p)$, we can write the morphism as $z \mapsto w = z^m$ for some integer $m > 0$; that is, if w is a local parameter on X and z is a local parameter in the source, then the map

$$\mathbb{C}\{\{w\}\} \cong \hat{\mathcal{O}}_{X,f(p)} \xrightarrow{\hat{f^*}} \hat{\mathcal{O}}_{C,p} \cong \mathbb{C}\{\{z\}\}$$

of convergent power series rings induced by f^* sends w to uz^m, where u is a power series with nonvanishing constant term. In this case $\mathrm{ram}(f, p) = m - 1$. These power series expansions are valid in a neighborhood of p, and the derivative of f vanishes at the ramification points in this neighborhood. Since the zeros of a nonconstant analytic function are isolated, the ramification points are isolated. Since C is compact in the classical topology, there are only finitely many. □

Hurwitz's theorem describes the difference between the canonical divisor of C and the pullback of the canonical divisor of X.

Theorem 2.12 (Hurwitz's theorem). [Hartshorne 1977, Proposition IV.2.3] *If $f : C \to X$ is a nonconstant morphism of smooth curves, with ramification divisor R, then*

$$K_C = f^*(K_X) + R,$$

or equivalently $\omega_C = (f^*\omega_X)(R)$.

Proof. Let B be the branch divisor of f. Choose a rational 1-form ω on X, and let $\eta = f^*(\omega)$ be its pullback to C. Since we have the freedom to multiply by any rational function on X, we can arrange for the zeroes and poles of ω to avoid B, so that ω is regular and nonzero at each branch point. (Actually the calculation goes through even without this assumption, albeit with more complicated notation.)

With this arrangement, for every zero of ω of multiplicity m we have exactly d zeroes of η, each with multiplicity m; and likewise for the poles of ω. At a point p where (locally) f has the form $z \mapsto w = z^e$ and $\omega = dw$, we have $\eta = z^{e-1} dz$; that is, η has a zero of multiplicity $\mathrm{ram}(f, p)$ at p. Thus the divisor K_C of η is $K_C = f^*(K_X) + R$. □

Example 2.13. Let $C \subset \mathbb{P}^2$ be a smooth plane curve and let p be a point of $\mathbb{P}^2$ not on C. Suppose that the coordinates on $\mathbb{P}^2$ are chosen so that the ideal sheaf of p is generated by the vector space of linear forms $W = \langle x_0, x_1 \rangle$. The linear

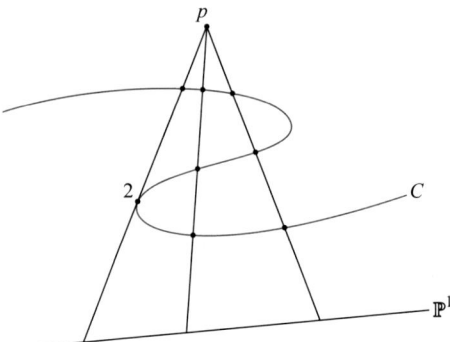

Figure 2.2. Projection of a plane cubic from a general point p to $\mathbb{P}^1$ is a three-to-one map.

series $(\mathcal{O}_C(1), W)$ defines the projection of C from p to $\mathbb{P}^1$, a map of degree $d = \deg C$ (see Figure 2.2).

The canonical sheaf of $\mathbb{P}^1$ has degree -2, so by Hurwitz's theorem K_C has degree $-2d + \deg R$, where R is the ramification divisor. We may choose coordinates so that none of the branch points lie on the line $x_0 = 0$. Taking this to be the line at infinity, we may compute R after passing to the affine open set $x_0 \neq 0$, where the projection map is given by the function $z = x_1/x_0$. Suppose that C is defined, in this open set, by the equation $f(x, y) = 0$. A point $q \in C$ is a ramification point if the tangent line to C at q passes through p, that is, if dx and

$$df = \frac{\partial f}{\partial x} dx + \frac{\partial f}{\partial y} dy$$

are linearly dependent. Since C is smooth, $\partial f/\partial x$ and $\partial f/\partial y$ cannot vanish simultaneously, so this happens if and only if $\partial f/\partial y$ vanishes at q. The intersection of C with the curve defined by $\partial f/\partial y = 0$ has degree $d(d-1)$ by Bézout's theorem, so the degree of the ramification divisor R is $d(d-1)$. Thus the degree of the canonical divisor on C is $\deg K_C = -2d + d(d-1) = d(d-3)$, which is in accord with Corollary 2.10.

Example 2.14. Let $V = H^0(\mathcal{O}_{\mathbb{P}^1}(d))$ be the vector space of homogeneous polynomials of degree d in two variables. In the projectivization $\mathbb{P}(V^*) \cong \mathbb{P}^d$, let Δ be the locus of polynomials with a repeated factor. Since Δ is defined by the vanishing of the discriminant, it is a hypersurface. What is its degree?

To answer this, we intersect Δ with a general line; the degree of Δ is the degree of the intersection. Let $W \subset V$ be a general 2-dimensional linear subspace, that is, a general pencil of forms of degree d on $\mathbb{P}^1$. The linear series $W = (\mathcal{O}_{\mathbb{P}^1}, W)$ defines a morphism $\phi_W : \mathbb{P}^1 \to \mathbb{P}(W) \cong \mathbb{P}^1$ and the fiber over the point of $\mathbb{P}(W)$ corresponding to a form f of degree d is the divisor $\{f = 0\} \subset \mathbb{P}^1$. Thus the intersection of Δ with the line is the locus of polynomials in W with a multiple root; that is, the branch locus of ϕ_W, where we would count an m-fold root $m-1$ times if there were multiple roots. By Hurwitz's formula, the degree of the branch locus B of a degree d morphism from $\mathbb{P}^1$ to $\mathbb{P}^1$ is

$$\deg B = \deg \omega_{\mathbb{P}^1} - d \deg \omega_{\mathbb{P}^1} = 2d - 2.$$

Thus $\deg \Delta = 2d - 2$.

2.3. Riemann–Roch with duality

We now return to the task of understanding $h^0(\mathcal{L})$ for an invertible sheaf $\mathcal{L}$ on a smooth curve. Since $\chi(\mathcal{L}) = h^0(\mathcal{L}) - h^1(\mathcal{L})$ is easier to compute, we would like to understand $h^1(\mathcal{L})$ in a more concrete way. The key is duality:

Theorem 2.15 (Serre duality). *If C is a smooth curve and D is a divisor on C, then*

$$H^1(D) = H^0(K_C - D)^* := \mathrm{Hom}_{\mathbb{C}}(H^0(K_C - D), \mathbb{C}),$$

and thus $h^1(D) = h^0(K_C - D)$.

For proofs see [Hartshorne 1977, Theorem III.5.2 and III.7.6].

For example we see that if C is a smooth connected curve then $h^1(\mathcal{O}_C) = h^0(K_C) = g(C)$ and thus $\chi(\mathcal{O}_C) = 1 - g(C)$. Using this we can recast Theorem 2.3 in the more useful form:

Theorem 2.16 (Riemann–Roch). *If D is any divisor on C, then*

$$h^0(D) - h^0(K_C - D) = \deg D - g(C) + 1.$$

In particular $\deg K_C = 2g(C) - 2$.

Proof. Combine Theorem 2.3 with Theorem 2.15. For the second statement, apply the formula with $D = K_C$. □

See Sections 16.5 and Theorem 16.22 for the corresponding results on singular curves.

We can now explain the relationship between the genus of a smooth curve, as we have defined it and the topological genus, the "number of holes" in the Riemann surface (Figure 2.3):

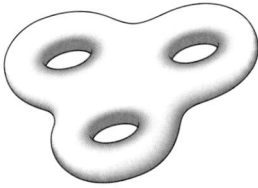

Figure 2.3. A Riemann surface of genus 3.

Cheerful Fact 2.17 (Hodge theory). The sole topological invariant of a smooth projective curve C, viewed as an analytic space, is its genus. As a manifold it is a compact, oriented surface, and its genus is half the rank of its first singular cohomology, $H^1(C; \mathbb{C})$, which is equal to its first de Rham cohomology. Breaking up the de Rham cohomology of any smooth projective complex variety X in terms of holomorphic and antiholomorphic differential forms we get the *Hodge decomposition*

$$H^i(X, \mathbb{C}) = H^i_{\text{de Rham}}(X) = \bigoplus_{j=0}^{i} H^j(\textstyle\bigwedge^{i-j} \Omega_X).$$

For a smooth curve C, this says in particular that

$$H^1(C; \mathbb{C}) = H^0(\omega_C) \oplus H^1(\mathcal{O}_C) = H^0(\omega_C) \oplus (H^0(\omega_C))^\vee,$$

so $h^0(\omega_C)$ is half the rank of the middle singular cohomology group, justifying the name "genus". For details, see [Griffiths and Harris 1978, p. 116].

A divisor E of negative degree satisfies the equation $H^0(E) = 0$, so we get the form of the Riemann–Roch theorem originally proved by Riemann:

Corollary 2.18. *For any divisor D of degree d we have*

$$h^0(D) \geq d - g + 1,$$

with equality if $d > 2g - 2$.

It was Gustav Roch, a student of Riemann's, who supplied the correction term $h^0(K_C - D)$ for divisors of lower degree [Roch 1865]. The dimension $h^0(K_C - D) = h^1(D)$ was called the *superabundance* of D: the "expected" number of sections was $d - g + 1$, and $h^1(\mathcal{L})$ reflected how much larger the actual number was.

Corollary 2.18 and Proposition 1.11 together show that all high degree divisors come from hyperplane sections in suitable embeddings; and unlike the general vanishing theorems, they give a bound on the degree necessary for vanishing of cohomology and for global generation:

Corollary 2.19. *Let D be a divisor of degree d on a smooth, connected projective curve of genus g.*

(1) *If $d > 2g - 2$ then $H^1(\mathcal{O}_C(D)) = 0$.*

(2) *If $d \geq 2g$ then $\mathcal{O}_C(D)$ is generated by global sections; that is, the complete linear series $|D|$ is basepoint free; and $\mathcal{O}_C(D)$ is very ample unless $D = K_C + E$ for some divisor E of degree 2.*

(3) *If $d \geq 2g + 1$ then $\mathcal{O}_C(D)$ is very ample; that is, the associated morphism $\phi_D : C \to \mathbb{P}^{d-g}$ is an embedding, and D is the preimage of the intersection of C with a hyperplane in $\mathbb{P}^{d-g}$.*

Proof. If $d > 2g - 2$ then $K - D$ has negative degree, and thus

$$h^1(D) = h^0(K - D) = 0.$$

The last two parts follow from the Riemann–Roch theorem and Proposition 1.11.

$\square$

Since the complement of a hyperplane in projective space is an affine space, we get an affine embedding result too:

Corollary 2.20. *If C is any smooth projective curve and $\Gamma \subset C$ a nonempty finite subset then $C \setminus \Gamma$ is affine (that is, isomorphic to a closed subscheme of an affine space).*

Proof. Let D be the divisor defined by Γ. By Corollary 2.19 a high multiple of D is very ample, and gives an embedding $\phi : C \to \mathbb{P}^n$ such that the preimage of the intersection of C with some hyperplane H is a multiple of D. It follows that $C \setminus \Gamma$ is embedded in $\mathbb{A}^n = \mathbb{P}^n \setminus H$.

$\square$

We can use Corollary 2.18 to determine the Hilbert polynomial of a projective curve. To do this, let $C \subset \mathbb{P}^r$ be a smooth curve of degree d and genus g, and consider the exact sequence of sheaves

$$0 \to \mathcal{I}_{C/\mathbb{P}^r}(m) \to \mathcal{O}_{\mathbb{P}^r}(m) \to \mathcal{O}_C(m) \to 0$$

and the corresponding exact sequence

$$H^0(\mathcal{O}_{\mathbb{P}^r}(m)) \xrightarrow{\rho_m} H^0(\mathcal{O}_C(m)) \to H^1(\mathcal{I}_{C/\mathbb{P}^r}(m)) \to 0.$$

The *Hilbert function* $h_C(m)$ of C is defined in terms of the homogeneous coordinate ring R_C of C by

$$h_C(m) = \dim_{\mathbb{C}}(R_C)_m = \operatorname{rank} \rho_m,$$

where $(R_C)_m$ is the degree m component of the homogeneous coordinate ring of C in $\mathbb{P}^n$.

By Theorem 2.1 we have $H^1(\mathcal{I}_{C/\mathbb{P}^r}(m)) = 0$ for large m so, for large m, $h_C(m) = h^0(\mathcal{O}_C(m))$. Using Theorem 2.1 again, we see that, for large m,

$$h^0(\mathcal{O}_C(m)) = \chi(\mathcal{O}_C(m)).$$

Finally, by the Riemann–Roch theorem,

$$\chi(\mathcal{O}_C(m)) = dm - g + 1,$$

so, for large m, the Hilbert function $h_C(m) = dm - g + 1$ is in agreement with the Hilbert polynomial $p_C(m) := \chi(\mathcal{O}_C(m))$.

More generally, we define the *arithmetic genus* and *geometric genus* as follows:

Definition 2.21. If $C \subset \mathbb{P}^n$ is a 1-dimensional projective scheme with Hilbert polynomial $p_C(m) = \chi(\mathcal{O}_C(m))$, the *arithmetic genus* $p_a(C)$ of C is $1 - \chi(\mathcal{O}_C) = 1 - p_C(0)$. If C is reduced and irreducible, then the *geometric genus* $g(C)$ is the genus of the normalization of C (see Proposition 2.24 for this term).

We see from the Riemann–Roch theorem that if C is smooth and connected, then $p_a(C) = g(C) = h^0(\omega_C)$, the genus of C. We will see that for reduced and irreducible curves $p_a(C) \geq g(C)$, with equality only when C is smooth. For some examples with curves that are not reduced and irreducible, see Exercise 2.8.

The Riemann–Roch theorem and Serre duality have extensions to arbitrary coherent sheaves in place of invertible sheaves and to singular curves, which we will explain in Chapter 16.

Divisors D for which $h^0(K_C - D) > 0$ are called *special divisors*. The existence or nonexistence of divisors D with given $h^0(D)$ and $h^1(D)$ often serves to distinguish one curve from another, and will be an important part of our study.

Residues. The Riemann–Roch theorem is so central to the study of curves that it is worth understanding from another point of view. We remarked at the beginning of Chapter 1 that a smooth projective curve over $\mathbb{C}$ is the same thing as a compact Riemann surface. We will briefly adopt the complex analytic viewpoint, and give an explanation of a special case of Theorem 2.16.

If $D = \sum a_i p_i$ is an effective divisor on a compact Riemann surface X then we write $L(D)$ for the vector space of meromorphic functions on X with poles of order at most a_i at p_i.

Theorem 2.22. *Let X be a compact Riemann surface of genus g, and let D be an effective divisor of degree d on X. Suppose that $K - D$ is also effective for some canonical divisor K. The dimension of $L(D)$ is $d - g + 1 + \dim_\mathbb{C} L(K - D)$.*

Because meromorphic functions on a Riemann surface are rational functions on the corresponding algebraic curve, and $L(D) = H^0(\mathcal{O}_C(D))$, the assertion is equivalent to Theorem 2.16.

Proof. Recall that the *residue* of a meromorphic 1-form ϕ at a point p on X is defined by an integral: choose a disc $\Delta \subset X$ containing p and in which ϕ is holomorphic except for its pole at p. The residue $\text{Res}_p(\phi)$ is $\frac{1}{2\pi i}$ times the integral of ϕ along the boundary of Δ. If z is a local coordinate on Δ zero at p and we write the differential ϕ as

$$\phi = \sum_{i=-n}^{\infty} a_i z^i \, dz$$

then by Cauchy's formula, the residue of ϕ at p is the coefficient a_{-1}.

Proposition 2.23. *If ϕ is a meromorphic differential on a compact Riemann surface X, then the sum of the residues of ϕ at all the poles of ϕ is 0.*

Proof. Apply Stokes' theorem to the complement of the union of small discs around each of the poles of ϕ. □

Let $D = \sum a_i p_i$ be an effective divisor on X, and set $d = \sum a_i = \deg D$. Locally, a function with a pole of order at most n at p may be written in terms of a local coordinate z at p as $\sum_{i=-n}^{\infty} a_i z^i$; the sum $\sum_{i=-n}^{-1} a_i z^i$ is called its *polar part*. By the maximum principle, a meromorphic function in $L(D)$ is determined, up to the addition of a constant, by its polar parts at the points p_i. Thus we have $\dim L(D) \leq d + 1$.

When is a given collection $c_1, \ldots, c_d \in \mathbb{C}[z^{-1}]$ the polar parts of a global meromorphic function f on X? A necessary condition is that if $\phi \in L(K)$ is a holomorphic differential on X, then

$$\sum \mathrm{Res}_{p_i}(f \cdot \phi) = 0.$$

This gives g linear relations on the c_i. However, if ϕ is a holomorphic differential vanishing at all the points p_i then the corresponding relation is trivial. Thus the number of linearly independent relations on the polar parts of f is actually $g - \dim L(K - D)$; and we arrive at an inequality

$$\dim L(D) \leq d + 1 - g + \dim L(K - D).$$

This is a priori only an inequality. But we can apply the same logic to an effective divisor $K - D$, and we see that

$$
\begin{aligned}
\dim L(K - D) &\leq \deg(K - D) + 1 - g + L(K - (K - D)) \\
&= 2g - 2 - d + 1 - g + L(D) \\
&= g - d - 1 + L(D).
\end{aligned}
$$

Adding the two inequalities we have

$$L(D) + L(K - D) \leq L(D) + L(K - D).$$

Since the sum of the inequalities is an equality, we conclude that each inequality is also an equality; this is the Riemann–Roch formula in our special case. □

Arithmetic genus and geometric genus. Throughout this book, we will be primarily concerned with the geometry of smooth curves. Of course singular curves will arise — for example, as images of smooth curves under morphisms that are not embeddings. At least when C is a 1-dimensional variety (that is, a reduced irreducible 1-dimensional scheme) we can regard C as the image of a smooth curve in an optimal way:

Proposition 2.24. *If C_0 is a projective variety of dimension 1 then the* normalization $\nu : C \to C_0$ *of C_0 is a birational morphism from a smooth curve C. The curve C is again projective, and the pair (C, ν) is unique up to isomorphism. In particular, every birational map of smooth curves is an isomorphism.*

Proof. We use the result that the normalization (= integral closure) of a domain finitely generated over a field is again finitely generated, and nonsingular in codimension 1 [Eisenbud 1995, Theorem 4.14 and 11.5]. Thus, starting with a projective embedding of C we normalize the homogeneous coordinate ring of C. The resulting ring may have generators in many degrees, but a suitable Veronese subring will be generated in a single degree, and thus is the homogeneous coordinate ring of a smooth projective curve.

Localization commutes with normalization, and any map from a normal ring to a domain factors uniquely through any normalization. Therefore the normalization constructed above is independent of the choices.   □

A different procedure for finding a smooth curve birational to a given curve is explained in Exercises 2.13 to 2.15.

Still assuming that C_0 is reduced and irreducible, we can relate the arithmetic and geometric genera of C_0 using the map of sheaves

$$\mathcal{O}_{C_0} \to \nu_* \mathcal{O}_C.$$

Since the normalization map of rings is injective and finite, this map is injective. The cokernel $\mathcal{F}$ is a coherent sheaf supported on the singular points of C_0, and is thus finite over $\mathbb{C}$. The definition implies that there is a short exact sequence

$$0 \to \mathcal{O}_{C_0} \to \nu_* \mathcal{O}_C \to \mathcal{F} \to 0.$$

Proposition 2.25. *Suppose that C_0 is a reduced, irreducible curve and let $\nu : C \to C_0$ be its normalization. Let $\mathcal{F} = \nu_* \mathcal{O}_C / \mathcal{O}_{C_0}$. If we set $\delta(C_0) := h^0(\mathcal{F})$, then*

$$p_a(C_0) - g(C) = \delta(C_0)$$

Proof. Since the normalization map $\nu : C \to C_0$ is finite, the direct images $R^i \nu_* \mathcal{O}_C$ vanish for $i > 0$, so the Leray spectral sequence (see 2.26 just below) computes $\chi(\nu_* \mathcal{O}_C) = \chi(\mathcal{O}_C)$. Thus

$$p_a(C_0) - g(C) = \chi(\mathcal{O}_C) - \chi(\mathcal{O}_{C_0}) = \chi(\mathcal{F}) = h^0(\mathcal{F})$$

since the support of $\mathcal{F}$ is finite.   □

The operation of normalization localizes, so we can understand $\mathcal{F}$ by looking at it locally. Writing R for the affine ring of some affine subset $U \subset C_0$ we see that $\mathcal{O}_C|_U$ is the integral closure $\overline{R}$ of R, so $\mathcal{F}|_U$ is $\overline{R}/R$. Thus the annihilator of $\mathcal{F}|_U$, called the *conductor* $\mathfrak{f}_{\overline{R}/R}$, is an ideal of both $\overline{R}$ and R, and correspondingly $\nu^{-1}(\mathfrak{f}_{C/C_0})$ is also an ideal sheaf in $\mathcal{O}_C$.

Informally, we may say that $\delta(C_0)$ is the number of linear conditions a locally defined function f on C has to satisfy to be the pullback of a function from C_0. The length of the stalk of $\mathcal{F}$ at a particular singular point $p \in C_0$ is called the *δ invariant δ_p* of the singularity. Thus

$$p_a(C_0) - g(C) = \sum_{p \in (C_0)_{\text{sing}}} \delta_p.$$

Cheerful Fact 2.26. In general, the Leray spectral sequence addresses the situation of a morphism $f : X \to Y$ of varieties or schemes, and a coherent sheaf $\mathcal{G}$ on the source X; it says in this circumstance that there is a spectral sequence

$$H^p(R^q f_*(\mathcal{G})) \Longrightarrow H^{p+q}(\mathcal{G}).$$

(This is a special case of the spectral sequence for the derived functors of a composite functor (H^0 composed with f_*); see [Godement 1973, II.4.17.1] or [Gelfand and Manin 2003, Section III.7] for proofs.)

In the simplest cases the δ invariant is easy to compute:

(1) A *node* of a curve C_0 is a point p such that an analytic neighborhood of p in C_0 consists of two smooth arcs intersecting transversely at p (See Figure 2.4, left.) Equivalently, the completion of the local ring $\mathcal{O}_{C_0,p}$ is isomorphic to $k[\![x,y]\!]/(xy)$. If $p \in C_0$ is a node, its preimage in the normalization C of C_0 consists of two points $r, s \in C$. The condition for a function f on C to descend to C_0 — that is, to be the pullback of a function on C_0 — is that $f(r) = f(s)$; this is one linear condition, so $\delta_p = 1$.

(2) A *cusp* of a curve C_0 is a point p such that an analytic neighborhood of p in C_0 is given by the equation $y^2 = x^3$. (See Figure 2.4, second from left.) If $p \in C_0$ is a cusp then its preimage in the normalization C of C_0 will consist of one point $r \in C$. The condition for a function f on C is that the derivative $f'(r) = 0$; again, this is one linear condition, so $\delta_p = 1$.

We will give more examples at the end of Chapter 15.

2.4. The canonical morphism

We now return to the world of smooth curves.

The canonical sheaf on $\mathbb{P}^1$ has negative degree, so $|K_{\mathbb{P}^1}| = \emptyset$. If C is a curve of genus 1 then the canonical sheaf has a nonzero global section, and since the sheaf has degree $2g - 2 = 0$ this is nowhere vanishing, whence $\omega_C = \mathcal{O}_C$, and $K_C = 0$. Thus in studying the canonical series we restrict our attention to curves C of genus $g \geq 2$.

Theorem 2.27. *Let C be a smooth curve of genus ≥ 2.*

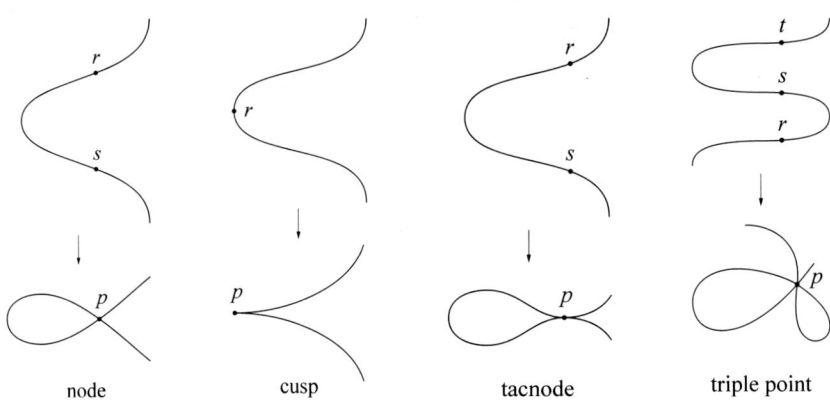

Figure 2.4. Simple planar curve singularities.

(1) $|K_C|$ *is basepoint free.*

(2) $|K_C|$ *is very ample if and only if C admits no map of degree 2 to $\mathbb{P}^1$.*

A curve of genus ≥ 2 is said to be *hyperelliptic* if there exists a morphism $f : C \to \mathbb{P}^1$ of degree 2. It is easy to describe this in terms of linear series:

Lemma 2.28. *Let C be a smooth, projective curve of genus $g \geq 2$. If C has an invertible sheaf $\mathcal{L}$ of degree ≤ 2 with two independent sections, then $\mathcal{L}$ has degree 2 and $|\mathcal{L}|$ defines a morphism of degree 2 to $\mathbb{P}^1$, so C is hyperelliptic. In particular, if $g(C) = 2$ then the canonical series $|K_C|$ defines a 2-to-1 morphism to $\mathbb{P}^1$, so C is hyperelliptic.*

Proof. Since $C \not\cong \mathbb{P}^1$, Theorem 1.7 shows that C cannot have a 1-dimensional linear series of degree < 2. Thus $\mathcal{L}$ has degree exactly 2, and no basepoints, so it defines a morphism of degree 2 to $\mathbb{P}^1$ as claimed. □

Proof of Theorem 2.27. A point p is a basepoint of $|K_C|$ if and only if

$$h^0(K_C - p) = h^0(K_C) = g.$$

By the Riemann–Roch theorem, this is equivalent to $h^0(p) = 2$, which would imply that $C \cong \mathbb{P}^1$ by Theorem 1.7. Thus K_C has no basepoints.

By Proposition 1.11 we must show that for any two points $p, q \in C$ we have

$$h^0(K_C(-p-q)) = h^0(K_C) - 2 = g - 2.$$

Applying the Riemann–Roch theorem we see that this fails if and only if $h^0(\mathcal{O}_C(p+q)) \geq 2$ for some $p, q \in C$. By Lemma 2.28, this implies that C is hyperelliptic. Conversely, if C is hyperelliptic then for some divisor $D = p + q$ we have $h^0(D) = 2$, whence $h^0(K - p - q) = h^0(K) - 1$ by the Riemann–Roch formula. □

The image of the canonical morphism of a nonhyperelliptic curve of genus $g > 2$ is called a *canonical curve*, and (for a nonhyperelliptic curve) we speak of the image as being *canonically embedded*.

Geometric Riemann–Roch. There is a useful way to express the Riemann–Roch formula in terms of the geometry of the canonical map. Suppose $C \subset \mathbb{P}^r$ is a smooth curve and D an effective divisor on C, thought of as a subscheme of C. We define the *span* $\overline{D}$ of D to be the intersection of the hyperplanes $H \subset \mathbb{P}^r$ containing D. More generally, if $\phi : C \to \mathbb{P}^r$ is a map and D an effective divisor on C, we define the span $\overline{\phi(D)} \subset \mathbb{P}^r$ to be the intersection

$$\overline{\phi(D)} := \bigcap_{\substack{H \subset \mathbb{P}^r \text{ a hyperplane} \\ D \subset \phi^{-1}(H)}} H.$$

Now suppose C is a smooth curve of genus $g \geq 2$ and $\phi_K : C \to \mathbb{P}^{g-1}$ is the canonical morphism. If D is an effective divisor on C of degree d, then since the hyperplanes in $\mathbb{P}^{g-1}$ containing $\phi_K(D)$ correspond (up to scalars) to sections of K_C vanishing on D, we see that

$$h^0(K_C - D) = \text{codim } \overline{\phi_K(D)} \subset \mathbb{P}^{g-1} = g - 1 - \dim \overline{\phi_K(D)}.$$

Applying the Riemann–Roch theorem we obtain the *geometric Riemann–Roch theorem*:

Corollary 2.29. *If D is a divisor on a smooth curve C of genus ≥ 2 then*

$$r(D) = d - 1 - \dim \overline{\phi_K(D)}.$$

In particular, if $D = \sum_{i=1}^{d} p_i$ is a sum of distinct points, then the dimension of the linear series $|D|$ is equal to the number of linear relations among the images of the p_i on the canonical image of C. Thus, for example, a divisor given as the sum $D = p + q + r$ of three points will move in a pencil if and only if the three points are collinear on the canonical model.

In general, the condition that a form of degree d vanishes at a given point p in $\mathbb{P}^r$ is expressed by one homogeneous linear equation on the coefficients of the form, obtained by evaluating the variables at the coordinates of p. Thus the condition of vanishing on a set Γ of γ points is given by γ linear equations and the same is true when Γ is a finite subscheme of length γ, as one sees from the exact sequence

$$0 \to H^0(\mathcal{I}_\Gamma(d)) \to H^0(\mathcal{O}_{\mathbb{P}^r}(d)) \xrightarrow{ev} H^0(\mathcal{O}_\Gamma(d)) \to 0,$$

bearing in mind that

$$H^0(\mathcal{O}_\Gamma(d)) \cong H^0(\mathcal{O}_\Gamma) \cong \mathbb{C}^\gamma.$$

However, the linear equations may be dependent; that is, the map marked *ev* above may have rank $< \gamma$. We will say that the points of Γ *fail to impose independent conditions on hypersurfaces of degree d* by the amount equal to $\gamma - \text{rank}(ev)$.

With this terminology, the geometric Riemann–Roch theorem says that the dimension of the complete linear series $|D|$ on a smooth curve C is the amount by which the points of D fail to impose independent conditions on canonical divisors, and the geometric version simply translates this into hyperplanes in the canonical embedding of C. The Cayley–Bacharach–Macaulay theorem 4.5 is a useful special case.

Remark 2.30. For a singular curve C_0 to have a canonical morphism it is necessary first of all that its canonical sheaf ω_{C_0} be invertible; when this is the case, we say that C_0 is (locally) Gorenstein, and then the theory above can be applied. See for example [Bayer and Eisenbud 1991] and [Bayer and Eisenbud 1995] for examples.

Linear series on a hyperelliptic curve. We can describe the canonical map of a hyperelliptic curve — and indeed all its special linear series — quite precisely:

Corollary 2.31. *Let C be a smooth hyperelliptic curve of genus* $g \geq 2$. *The curve C admits a unique degree 2 morphism* $\pi : C \to \mathbb{P}^1$, *and the canonical map* $\phi_K : C \to \mathbb{P}^{g-1}$ *is the composition of* π *with the Veronese map of degree* $g - 1$ *from* $\mathbb{P}^1$ *to the rational normal curve of degree* $g-1$; *in particular, every canonical divisor is a sum of* $g - 1$ *fibers of* π *and vice versa.*

Proof. If D is a general fiber of a degree 2 map $\pi : C \to \mathbb{P}^1$ so that $D = p + q$ and $r(D) = 1$, then by the Riemann–Roch theorem or its geometric version, D imposes only one condition on sections of K_C; that is, $\phi_K(p) = \phi_K(q)$. Consequently the degree of the canonical morphism ϕ_K is at least 2, and the image is thus a nondegenerate curve in $\mathbb{P}^{g-1}$ of degree $\leq (2g - 2)/2 = g - 1$. By Theorem 1.7, the image of ϕ_K is the rational normal curve of degree $g - 1$ and every fiber of ϕ_K is a divisor linearly equivalent to D. It follows that π is determined by K, so it is unique. Since every canonical divisor is the pullback of a hyperplane section of the rational normal curve, every canonical divisor is a sum of $g - 1$ fibers of π.   □

We can use these ideas to analyze all special divisors.

Corollary 2.32. *Let C be a smooth hyperelliptic curve with map* $\pi : C \to \mathbb{P}^1$ *of degree 2. Every special divisor D of degree d on C is a sum of s fibers of* π *and* $d - 2s$ *points* $p_1, \ldots, p_{d-2s}$ *with distinct images under the canonical map; and in this case we have* $r(D) = s$ *and the points* p_i *are basepoints of the linear series* $|D|$. *Thus no divisor on a hyperelliptic curve is both special and very ample.*

Proof. Suppose that D contains the sum E of s fibers of π and no more; and write $D = E + D'$, with $\deg D' = d - 2s$.

If $\phi_K : C \to \mathbb{P}^{g-1}$ is the canonical map, then $\phi_K(D')$ consists of $d - 2s$ distinct points, while $\phi_K(E)$ consists of s points. By Corollary 1.9 the span of $\phi_K(D)$ has dimension $\min\{g-1, d-s-1\}$, so $r(D) = d-1-\min\{g-1, d-s-1\} = \max\{s, d - g\}$. Since D is a special divisor, $r(D) > d - g$, so $r(D) = s$. Since $r(E) = s$, we see that E is basepoint free, so D' is the base locus of D. $\qquad\square$

2.5. Clifford's theorem

While the Riemann–Roch theorem gives a lower bound for the dimension of a linear series, $r(\mathcal{L}) := h^0(\mathcal{L}) - 1 \geq \deg \mathcal{L} - g$, Clifford's theorem gives an upper bound. If $\deg \mathcal{L} > 2g - 2$, then the Riemann–Roch inequality becomes an equality, so it is enough to treat the case $\deg \mathcal{L} \leq 2g - 2$. The bound is actually a corollary of the Riemann–Roch theorem.

Corollary 2.33. *Let C be a curve of genus g and $\mathcal{L}$ an invertible sheaf of degree $d \leq 2g - 2$. Then*

$$r(\mathcal{L}) \leq \frac{d}{2}.$$

Proof. If $\mathcal{L}$ is nonspecial then, since $g \geq d/2+1$, we have $r(\mathcal{L}) = d-g+1 \leq d/2$. Otherwise we have

$$r(K_C \otimes \mathcal{L}^{-1}) = r(\mathcal{L}) + g - d - 1$$

by the Riemann–Roch theorem, and so by Proposition 1.10

$$g = r(K_C) + 1 \geq r(\mathcal{L}) + r(K_C \otimes \mathcal{L}^{-1}) + 1 = 2r(\mathcal{L}) + g - d;$$

hence $r(\mathcal{L}) \leq d/2$. $\qquad\square$

The usual statement of Clifford's theorem includes a description of when equality can occur:

Theorem 2.34. *Let C be a curve of genus g and $\mathcal{L}$ an invertible sheaf of degree $d \leq 2g - 2$. If*

$$r(\mathcal{L}) = \frac{d}{2},$$

the largest possible value, then either

(1) $d = 0$ and $\mathcal{L} = \mathcal{O}_C$;

(2) $d = 2g - 2$ and $\mathcal{L} = K_C$; or

(3) C is hyperelliptic, and $|\mathcal{L}|$ is a multiple of the g_2^1 on C.

From Corollary 2.31 and Corollary 2.29 we see that the equality does indeed hold in each of the cases enumerated. See Corollary 10.14 for the converse and [Hartshorne 1977, IV.5.4] for a different proof.

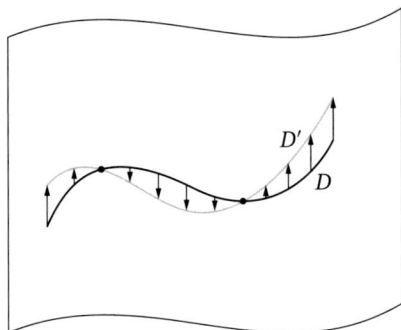

Figure 2.5. The result, D', of moving D infinitesimally along a normal vector field meets D twice.

2.6. Curves on surfaces

We will often analyze curves on a smooth surface. Compared to the theory of linear series on curves, there is a new element: the intersection pairing. We refer to [Hartshorne 1977, Chapter V] and [Beauville 1996, Chapter I] for proofs of the unproven statements in this section. We suppose for this section that S is a smooth projective surface, and write $\operatorname{Pic}(S)$ for the group of invertible sheaves on S.

The intersection pairing. When two codimension 1 subschemes D, E on S meet transversely we define $D{\cdot}E$ to be the number of points in which they meet. It is less obvious how one might define an intersection number $D \cdot E$ in more general cases, including the case $D = E$. However a codimension 1 subvariety of S has a fundamental class in $H^2(X, \mathbb{Z})$, and the product just defined agrees with the cup product pairing

$$H^2(S, \mathbb{Z}) \times H^2(S, \mathbb{Z}) \to H^4(S, \mathbb{Z}) \cong \mathbb{Z},$$

showing that it is possible. For example, the self-intersection $D \cdot D$ can be realized geometrically by choosing a normal vector field σ on D and using it to push a copy D' of D infinitesimally away from D; then one can count the points where σ vanishes (see Figure 2.5. Here one must take orientations into account, and the result is that $D \cdot D$ may be negative. In the previous case (and when σ is complex analytic) the multiplicities are all positive because a complex subvariety is canonically oriented using the orientation of $\mathbb{C}$ itself.

This pairing can be realized algebraically as follows:

First, if D and E have no common components, then the *intersection multiplicity* $\operatorname{mult}_p(D, E)$ at $p \in S$ is defined as the vector space dimension of the local ring $\mathcal{O}_{S,p}/(\mathcal{I}_{D,p} + \mathcal{I}_{E,p})$, and

$$D \cdot E = \sum_p \mathrm{mult}_p(D, E).$$

Quite generally, the intersection product can be computed using the (algebraic) Euler characteristic. This works even for Cartier divisors on a singular surface: Setting $\mathcal{L} := \mathcal{O}_S(D)$ and $\mathcal{M} := \mathcal{O}_S(E)$ to simplify the notation, we have

$$D \cdot E = \chi(\mathcal{O}_S) - \chi(\mathcal{L}^{-1}) - \chi(\mathcal{M}^{-1}) + \chi(\mathcal{L}^{-1} \otimes \mathcal{M}^{-1}).$$

Theorem 2.35. *The pairing $(D, E) \mapsto (D \cdot E)$ is the unique bilinear map* $\mathrm{Pic}(S) \times \mathrm{Pic}(S) \to \mathbb{Z}$ *extending the case of intersections of two curves meeting transversely at smooth points of S.*

An important special case is that of the self-intersection $D \cdot D = D^2$. The general formula reduces immediately to the case of an effective divisor, and in that case it resembles the C^∞ construction above:

Corollary 2.36. *If D is a codimension 1 subscheme of S, then the self-intersection $D \cdot D$ is the degree of the normal bundle $\mathcal{N}_{D/S} := \mathcal{O}_D(D)$.*

Proof. Substituting $\mathcal{O}_S(-D)$ for both $\mathcal{L}^{-1}$ and $\mathcal{M}^{-1}$ we have the exact sequences

$$0 \to \mathcal{O}_S(-D) \to \mathcal{O}_S \to \mathcal{O}_D \to 0,$$
$$0 \to \mathcal{O}_S(-2D) \to \mathcal{O}_S(-D) \to \mathcal{O}_D(-D) \to 0;$$

the Riemann–Roch theorem then gives

$$\chi(\mathcal{O}_S) - \chi(\mathcal{O}_S(-D)) = \chi(\mathcal{O}_D) = -p_a(D) + 1,$$
$$\chi(\mathcal{O}_S(-D)) - \chi(\mathcal{O}_S(-2D)) = \chi(\mathcal{O}_D(-D)) = -\deg \mathcal{O}_D(D) - p_a(D) + 1.$$

Subtracting the second of these equations from the first we see that $D \cdot D = \deg \mathcal{O}_D(D)$. $\square$

Using the intersection pairing we can turn the adjunction formula of Proposition 2.8 into a numerical formula:

Theorem 2.37 (adjunction formula). *If C is a Cartier divisor on a smooth surface S then*

$$p_a(C) = \frac{(K_X + C) \cdot C}{2} + 1.$$

The Riemann–Roch theorem for smooth surfaces. Often we wish to compute the dimension of the space of sections of an invertible sheaf, but as with the case of curves, the Euler characteristic is more accessible:

Theorem 2.38 (Riemann–Roch for surfaces). *Let $\mathcal{L}$ be an invertible sheaf on a smooth surface S. The Euler characteristic $\chi(D) := h^0(\mathcal{L}) - h^1(\mathcal{L}) + h^2(\mathcal{L})$, where $\mathcal{L} = \mathcal{O}_S(D)$, is given by*

$$\chi(D) = \chi(\mathcal{O}_S) + \frac{(D - K_S) \cdot D}{2} + 1.$$

Blowups of smooth surfaces. It is useful to know what happens under mappings of surfaces, particularly the case of the mapping corresponding to blowing up a point.

Theorem 2.39. *If $\pi : X \to Y$ is a birational map of smooth surfaces, then the pullback map on divisors preserves the intersection pairing. If X is the blowup of Y at a point p with exceptional divisor $E = \pi^{-1}(p)$, then:*

(1) *$\operatorname{Pic} X = \pi^*(\operatorname{Pic} Y) \oplus \mathbb{Z}E$.*

(2) *The canonical class on X is given by $K_X = \pi^*(K_Y) + E$.*

(3) *The intersection pairing on $\operatorname{Pic} X$ is given by:*
  - *$\pi^*(D) \cdot \pi^*(D') = D \cdot D'$ for all $D, D' \in \operatorname{Pic} Y$.*
  - *$\pi^*(D) \cdot E = 0$ for all $D \in \operatorname{Pic} Y$.*
  - *$E \cdot E = -1$.*
  - *$K_X \cdot E = -1$.*
  - *If C is a curve that has an m-fold point at p then $\pi^{-1}(C)$ contains E with multiplicity m, so that the proper transform $\tilde{C}$ — that is, the closure in X of the preimage of $C \setminus \{p\}$ — has class*
  $$\tilde{C} \sim \pi^*C - mE.$$

Using these facts, we can compare the adjunction formula applied to a curve C on a smooth surface S to the formula applied to the proper transform $\tilde{C} \subset \tilde{S}$: we have

$$\tilde{C} \cdot \tilde{C} = C \cdot C - m^2 \quad \text{and} \quad \tilde{C} \cdot K_{\tilde{S}} = C \cdot K_S + m.$$

Combining these, we arrive at the formula

$$p_a(\tilde{C}) = p_a(C) - \binom{m}{2},$$

and thus the δ invariant of $p \in C$ is $\binom{m}{2}$ plus the sum of the δ invariants of all the points of $\tilde{C}$ in the preimage of p. One can resolve the curve singularity by repeatedly blowing up in this way and thus compute the δ invariant of any singularity as a sum of binomial coefficients.

In the special case where $\tilde{C}$ is smooth — for example, if the point $p \in C$ is an (*ordinary*) *m-fold point*, consisting of m smooth branches with distinct tangent lines — we conclude that the δ invariant of the point p is $\binom{m}{2}$.

Blowups occur frequently in the theory of surfaces, and are characterized by *Castelnuovo's theorem* [Hartshorne 1977, Theorem V.5.7].

Theorem 2.40. *If $E \subset X$ is a curve on a smooth projective surface X and that $E^2 = E \cdot K_X = -1$ then E can be "blown down" in the sense that X is the blowup of a smooth surface Y at a point $p \in Y$, and E is the exceptional divisor.*

2.7. Quadrics in $\mathbb{P}^3$ and the curves they contain

We will frequently be interested in curves on a quadric surface in $\mathbb{P}^3$, and we can describe these curves and their intersections very concretely.

The classification of quadrics. The form defining a quadric $Q \subset \mathbb{P}^3$ can be written in suitable coordinates x_i as

$$q = \sum_{i=0}^{3} x_i^2,$$

where $r \leq 4$ is the *rank* of Q. (The corresponding statement is also true in $\mathbb{P}^r$, with $m \leq r+1$.) Equivalently, this is the rank of the symmetric matrix M representing the bilinear form $\frac{1}{2}(q(x+y) - q(x) - q(y))$.

(1) A quadric of rank 1 is a double plane.

(2) A quadric of rank 2 is the union of two distinct planes.

(3) A quadric Q of rank 3 is a cone over a smooth plane conic. The cone point is called the vertex. Every line on such a quadric passes through the vertex. The Picard group of Q is $\mathbb{Z}$ [Hartshorne 1977, Exercise II.6.5].

(4) A quadric Q of rank 4 is smooth and is isomorphic to $\mathbb{P}^1 \times \mathbb{P}^1 \subset \mathbb{P}^3$, embedded by the Segre embedding. Every line on such a quadric has the form $\mathbb{P}^1 \times \{p\}$ or $\{p\} \times \mathbb{P}^1$; the two families of such lines are called *rulings*. The Picard group of Q is $\mathbb{Z} \times \mathbb{Z}$, generated by the lines of the rulings [Hartshorne 1977, Example II.6.1].

(5) Quadrics in $\mathbb{P}^r$: Inside the space $\mathbb{P}(\mathcal{O}_{\mathbb{P}^r}(2)) = \mathbb{P}^{\binom{r+2}{2}-1}$ the space of singular quadrics is a hypersurface of degree $r+1$, called the *discriminant*, which is defined by the determinant of a generic symmetric $(r+1) \times (r+1)$ matrix M. More generally, the rank k locus is defined by the $(k+1) \times (k+1)$ minors of M.

It follows that an irreducible nondegenerate curve cannot be contained in a quadric of rank 1 or 2.

Some classes of curves on quadrics.

Example 2.41. Let C be a reduced curve on a quadric Q of rank 3 in $\mathbb{P}^3$. If C has even degree $2m$ then C is a complete intersection of the quadric with a hypersurface of degree m. Therefore if the curve is smooth it cannot contain the vertex of the quadric. The genus of C is $(m-1)^2$.

If C has odd degree $2m + 1$, then the union of C with a line on the quadric has even degree, so if L_1, L_2 are lines on the quadric, then

$$C \cup L_1 = Q \cap F_1 \quad \text{and} \quad C \cup L_2 = Q \cap F_2$$

are complete intersections of Q with forms of degree $m + 1$, and the homogeneous ideal of C is generated by the equations of Q, F_1 and F_2. The genus of C is $m(m-1)$. The curve C contains the vertex, and is a Weil divisor, not a Cartier divisor. These things follow from the description of the desingularization of Q in Corollary 17.24.

Example 2.42. If C is a reduced curve on a quadric Q of rank 4, then in terms of the isomorphism $Q \cong \mathbb{P}^1 \times \mathbb{P}^1$ we can express the divisor class of C as $(a, b) \in \text{Pic}(Q) = \mathbb{Z} \oplus \mathbb{Z}$. The intersection form on Q is determined by the fact that the lines of a ruling are all linearly equivalent, so that $(1, 0)^2 = (0, 1)^2 = 0$ and $(1, 0) \cdot (0, 1) = 1$; thus $(a, b) \cdot (d, e) = ae + bd$. The hyperplane section has class $(1, 1)$, and the canonical class is $(K_{\mathbb{P}^3} + Q)|_Q = (-2, -2)$ by the adjunction formula.

Thus if C has class (a, b) then the degree of C is $a + b$, while the genus of C is $(a - 1)(b - 1)$. Note that if $a = b$, the curve is a complete intersection, and if $b = a + 1$ then the curve is residual to a line in a complete intersection of the quadric with a surface of degree b. In these cases we get the same genus and degree as we would for a curve on a rank 3 quadric.

The cohomology of $\mathcal{O}_Q(a, b)$ is determined by the Künneth formula (or the Leray spectral sequence) as

$$H^i(\mathcal{O}_Q(a, b)) = \bigoplus_{i=s+t} H^s(\mathcal{O}_{\mathbb{P}^1}(a)) \otimes H^t(\mathcal{O}_{\mathbb{P}^1}(b)).$$

The degrees of the generators of the homogeneous ideal of C can be determined from the exact sequence

$$0 \to \mathcal{I}_{Q/\mathbb{P}^3} \to \mathcal{I}_{C/\mathbb{P}^3} \to \mathcal{I}_{C/Q} \to 0.$$

Since $\mathcal{I}_{C/Q}(m) = (m - a, m - b)$ we see that, supposing $0 < a \leq b$ and $1 < b$ there are no generators other than the equation q of Q having degree $< b$, and there are $b - a + 1$ generators of degree exactly b; with q these are the minimal generators of the homogeneous ideal (Exercise 2.12).

2.8. Exercises

Exercise 2.1. (1) The degree of a zero-dimensional subscheme $X \subset \mathbb{P}^r$ is by definition the value of the Hilbert polynomial of X, which is a constant. Show that this is the sum of the lengths of the components of X.

(2) The degree of a projective subscheme $X \subset \mathbb{P}^r$ of dimension n is defined to be the degree of the 0-dimensional scheme that is the intersection of X

with a general plane of degree $r-n$. Prove that this is the leading coefficient of the Hilbert polynomial, multiplied by $n!$.

(3) If $C \subset \mathbb{P}^r$ is a smooth curve of genus g and degree d, show that the Hilbert polynomial of C is $H_C(m) = dm - g + 1$. ◆

Exercise 2.2. If $C \subset \mathbb{P}^r$ is a 1-dimensional variety with normalization $\phi :$ $\widetilde{C} \to C$, and $X \subset \mathbb{P}^r$ is any subscheme that does not contain C we define the multiplicity of intersection $\mathrm{mult}_p(C,X)$ at the point $p \in X \cap C$ to be the sum of the lengths of the finite scheme $\phi^{-1}(X)$ at the points of $\phi^{-1}(p)$.

(1) Show that C is singular at p if and only if $\mathrm{mult}_p(C,X) \geq 2$ for all X containing p. Show further that if $H \subset \mathbb{P}^r$ is a hyperplane then $\deg C = \sum_{p \in C} \mathrm{mult}_p(C,H)$.

(2) Show that the degree of the image of C under projection from p is the degree of C minus $\mathrm{mult}_p(C,H)$ for a general hyperplane H. ◆

In the following series of exercises, we will work with the smooth projective curve birational to a possibly singular affine curve $C^\circ := V(f(x,y)) \subset \mathbb{A}^2$; this is the unique smooth projective curve containing the normalization of C° as a Zariski dense open subset.

Exercise 2.3. Let C be the smooth projective curve birational to the affine plane curve $C^\circ := V(y^3 + x^3 - 1)$, and let $\pi : C \to \mathbb{P}^1$ be the map given by the rational function x.

(1) Show that the closure $\overline{C^\circ}$ of C° in $\mathbb{P}^2$ is smooth (i.e., $C = \overline{C^\circ}$).

(2) Find the branch points and ramification points of π, and deduce that the genus of C is 1.

(3) For any two points $p, q \in C$ find the complete linear series $|p + q|$. ◆

(4) Find the unique map $\eta : C \to \mathbb{P}^1$ of degree 2 such that $\eta((1,0)) = \eta((0,1))$, and determine the ramification points of η.

(5) Show that C is isomorphic to the smooth projective curve associated to the affine plane curve $y^2 + x^3 = 1$.

For the next three exercises, let C° be the affine plane curve given as the zero locus of $y^2 - x^6 + 1$, and let C be the corresponding smooth projective curve. The map $C^\circ \to \mathbb{A}^1$ given by the projection $(x,y) \mapsto x$ extends to a map $\pi : C \to \mathbb{P}^1$, expressing C as a 2-sheeted cover of $\mathbb{P}^1$ branched over the points $1, \zeta, \ldots, \zeta^5$, where ζ is any primitive sixth root of unity. In addition, let p and $q \in C$ be the two points lying over the point $\infty \in \mathbb{P}^1$.

Exercise 2.4. With C° and C as above show that the map $C^\circ \to \mathbb{A}^1$ given by the projection $(x,y) \mapsto x$ extends to a map $\pi : C \to \mathbb{P}^1$, expressing C as a 2-sheeted cover of $\mathbb{P}^1$ branched over the points $1, \zeta, \ldots, \zeta^5$, where ζ is any primitive sixth root of unity. Show that there are two distinct points p and $q \in C$ lying over the point $\infty \in \mathbb{P}^1$, so that C is unramified over ∞.

What is the genus of C?

Exercise 2.5. With C as in Exercise 2.4:

(1) Let r_α be the ramification point over ζ^α. Show that

$$p + q \sim 2r_\alpha \quad \text{and} \quad \sum_{\alpha=0}^{5} r_\alpha \sim 3p + 3q.$$

(2) Find the vector space $H^0(\mathcal{O}_C(D))$ where $D = r_0 + r_2 + r_4$, and find the (unique) divisor E on C such that $E + r_1 \sim r_0 + r_2 + r_4$.

Exercise 2.6. With C as in Exercise 2.4: Let D be the divisor $D = p + q + r_0 + r_3$

(1) Find the vector space $H^0(\mathcal{O}_C(D))$. ◆

(2) Describe the map $\phi_{|D|} : C \to \mathbb{P}^2$.

(3) Find the equation of the image curve $\phi_{|D|}(C) \subset \mathbb{P}^2$, and describe its singularities.

Exercise 2.7. Let C be the smooth projective curve associated to the affine curve $y^3 = x^5 - 1$. The map $\pi : C \to \mathbb{P}^1$ given by the function x expresses C as a cyclic, 3-sheeted cover of $\mathbb{P}^1$, branched over the fifth roots of unity and the point at ∞. By way of notation, if we take $\eta = e^{2\pi i/5}$ a primitive fifth root of unity, we'll denote by r_α the point $(\eta^\alpha, 0) \in C$ lying over η^α, and by p the point lying over $\infty \in \mathbb{P}^1$.

(1) Verify that there is indeed a unique point $p \in C$ lying over $\infty \in \mathbb{P}^1$, and the map has ramification index 2 at p.

(2) Show that the genus of C is 4.

(3) Establish the linear equivalences

$$3p \sim 3r_\alpha \quad \text{and} \quad r_1 + \cdots + r_5 \sim 5p.$$

(4) Find a basis for the space $H^0(K_C)$ of regular differentials on C.

(5) Show that C is not hyperelliptic.

(6) Describe the canonical map $\phi_K : C \to \mathbb{P}^3$ and find the equations of the image.

(7) Let D be the divisor $D = r_1 + \cdots + r_5$. Show that $h^0(K_C(-D)) = 1$; deduce that $r(D) = 2$, and find a basis for $H^0(\mathcal{O}_C(D))$.

(8) If $E = 3p$, show that $r(E) = 1$; that $|E|$ is the unique g^1_3 on C and that $2E \sim K$.

Exercise 2.8. (1) Show that the arithmetic genus of the disjoint union of two lines is -1.

(2) Let $L \subset Q \subset \mathbb{P}^3$ be a line on a smooth quadric surface in $\mathbb{P}^3$. Show that the divisor $3L$, regarded as a 1-dimensional scheme, has $p_a(3L) = -2$.

(3) Compare these results with the result of simply applying the adjunction formula.

Exercise 2.9. Show that a curve of genus $g \geq 3$ cannot be simultaneously hyperelliptic and a three-sheeted cover of $\mathbb{P}^1$. See Proposition 9.5 for a generalization.

Exercise 2.10. Show that normalization of the affine curve

$$C = V(xy(x - y)) \subset \mathbb{A}^2$$

is the disjoint union of three affine lines (that is, $\mathrm{Spec}(\mathbb{C}[x] \times \mathbb{C}[y] \times \mathbb{C}[z])$). Compute the linear conditions on the values and derivatives of three polynomial functions f, g, h defined on these three lines that they "descend" to give a regular function on the planar triple point.

Exercise 2.11. Let $p \in C$ be a singular point of a reduced curve C. Show that if $\delta_p = 1$, then p must be either a node or a cusp.   ◆

Exercise 2.12. Suppose that $C \subset Q \subset \mathbb{P}^3$ is a curve on a smooth quadric Q, and that C lies in the class (a, b) with $0 \leq a \leq b$.

(1) Compute the generators of the homogeneous ideal of C in the case $a = 0$; and in the case $(a, b) = (1, 1)$ not treated in Example 2.42.

(2) Prove the assertion made at the end of Example 2.42 about the generation of the homogeneous ideal by forms of degrees 2 and b.

The following three exercises give a proof (in arbitrary characteristic) of resolution of singularities for curves; that is, every reduced curve is birational to a smooth curve.

Exercise 2.13. Let $C_0 \subset \mathbb{P}^2$ be an irreducible and reduced plane curve of degree d. Show that for large m, the m-th Veronese map $\mathbb{P}^2 \hookrightarrow \mathbb{P}^{\binom{m+2}{2}-1}$ embeds C_0 as a curve C of degree md spanning a linear space $\mathbb{P}^N$ of dimension $N = md - \binom{d-1}{2}$.   ◆

Exercise 2.14. We define the *multiplicity* $m_p(C)$ of a point $p \in C$ on a curve $C \subset \mathbb{P}^r$ to be the degree of the component of the intersection $H \cap C$, where $H \subset \mathbb{P}^r$ is a general hyperplane through p. (The Show that if the projection $\pi_p : C \to \mathbb{P}^{r-1}$ from p is birational onto its image, then the degree of the image curve $C_0 = \pi_p(C) \subset \mathbb{P}^{r-1}$ is $\deg C_0 = \deg C - m_p(C)$, and more generally if $\pi_p : C \to \mathbb{P}^{r-1}$ has degree k onto its image, then

$$\deg C_0 = \frac{\deg C - m_p(C)}{k}.$$

Note that the definition of multiplicity is trickier for varieties of dimension greater than 1; see [Eisenbud and Harris 2016].

Exercise 2.15. Returning to the situation of Exercise 2.13, suppose we take the curve $C \subset \mathbb{P}^N$ and project it from a singular point, then do the same to its image $C_1 \subset \mathbb{P}^{N-1}$ and continue in this fashion to produce a sequence of curves $C_n \subset \mathbb{P}^{N-n}$. Combining the last two exercises with the statement of Corollary 1.8, show that

(1) all the projection maps are birational onto their images; and

(2) for some n, the process terminates; that is, the image curve C_n is smooth.

Chapter 3

Curves of genus 0

We begin our project of describing curves in projective space with the simplest case, that of genus 0. Even here, there are interesting statements to make about the geometry of their embeddings in $\mathbb{P}^r$, and there are many open problems. Though we won't treat such questions, curves of genus 0 over other fields pose their own interesting problems: the study of such curves over $\mathbb{Q}$ was a major part of Gauss's work.

Using Theorem 1.7 and the Riemann–Roch theorem, we can show that (over an algebraically closed field) $\mathbb{P}^1$ is the only curve of arithmetic genus 0:

Corollary 3.1. *Every reduced irreducible projective curve C of arithmetic genus 0 over an algebraically closed field is isomorphic to $\mathbb{P}^1$.*

Proof. The curve C is smooth since otherwise its normalization would have negative genus. By the Riemann–Roch theorem, any linear series $\mathcal{L}$ of degree d on C has $h^0(\mathcal{L}) \geq d + 1$, so we may use Theorem 1.7 to conclude that $C \cong \mathbb{P}^1$. $\qquad\square$

Even images of genus 0 curves must have genus 0. The following proof works as well if we replace $\mathbb{C}$ by any algebraically closed field.

Theorem 3.2 [Lüroth 1875]. (1) *If $C \to D$ is a nonconstant map of reduced, irreducible projective curves, then the geometric genus of C must be at least that of D. In particular, if C has geometric genus 0, then so does D.*

(2) *If K is a field with $\mathbb{C} \subsetneq K \subset \mathbb{C}(x)$ then $K = \mathbb{C}(y)$ for some $y \in \mathbb{C}(x)$.*

Proof. (1) Normalizing, we get a map $\tilde{C} \to \tilde{D}$, and the first statement follows from Hurwitz's theorem.

(2) Since $\mathbb{C}$ is algebraically closed, K is a transcendental extension. Since $\mathbb{C}(x)$

has transcendence degree 1, if $z \in K \setminus \mathbb{C}$ then x is algebraic over $\mathbb{C}(z)$. Thus $\mathbb{C}(x)$ is finite over $\mathbb{C}(z)$, so K is finite over $\mathbb{C}(z)$ as well. In particular, K is finitely generated over $\mathbb{C}$. It follows that K is the field of rational functions on a curve D, and the inclusion $K \subset \mathbb{C}(x)$ corresponds to a finite map $\mathbb{P}^1 \to D$. Applying part (1), we see that $D \cong \mathbb{P}^1$ so $K \cong \mathbb{C}(y)$ for some y. $\square$

Cheerful Fact 3.3 (rational curves over other fields). *Lüroth's theorem* refers to statement (2) in Theorem 3.2. For an elementary proof of it, valid over any field, see [Jacobson 1989, Section 8.13].

Over a non-algebraically closed field, a curve C of genus 0 need not have any points, or any invertible sheaves of odd degree. Since the canonical sheaf K_C has degree -2, there necessarily exist invertible sheaves of every even degree; thus an arbitrary curve of genus 0 is isomorphic to a plane conic.

A projective curve of genus 0 over a field k is called a *form* of $\mathbb{P}^1$ if it becomes isomorphic to $\mathbb{P}^1$ after extension of scalars to the algebraic closure of k. The unique example with $k = \mathbb{R}$ that is not isomorphic over $\mathbb{R}$ to $\mathbb{P}^1$ is the conic with no $\mathbb{R}$-rational points, $x^2 + y^2 + z^2 = 0$. The classification of curves of genus 0 over $\mathbb{Q}$ is a subject that goes back to Gauss.

Noncommutative algebras enter the subject of forms of $\mathbb{P}^1$ (and $\mathbb{P}^n$ more generally) in a surprising way: The curve $\mathbb{P}^1_k$ itself may be described as the scheme of left ideals of k-vector-space dimension 2 in the ring of 2×2 matrices over k: such an ideal can be embedded in the matrix ring as a linear combination of the 2 columns in an appropriate sense. More generally, any scheme that is a form of $\mathbb{P}^1$ over k may be described as the scheme of 2-dimensional left ideals in a 4-dimensional central simple (= Azumaya) algebra over k. For example, the conic $x^2 + y^2 + z^2 = 0$ with no points over $\mathbb{R}$ is the scheme of left ideals in the algebra of quaternions. See [Serre 1979, Section X.6].

There is an analogue of the last statement of part (1) of Theorem 3.2 for rational surfaces, proved by Castelnuovo: every complex surface admitting a dominant rational map from $\mathbb{P}^2$ is rational (see [Beauville 1996, Corollary V.5], for example). However, the analogue in higher dimensions is false; for example, [Clemens and Griffiths 1972] shows that a smooth cubic threefold X admits a dominant rational map $\mathbb{P}^3 \to X$ but is not rational.

3.1. Rational normal curves

The homogeneous coordinate ring of a rational normal curve. If $\mathcal{V} = (V, \mathcal{L})$ is a linear series on a scheme X, then the inclusion $V \subset H^0(\mathcal{L})$ induces by multiplication a map

$$\rho_{\mathcal{V}} : \mathrm{Sym}(V) \to \bigoplus_{n \in \mathbb{Z}} H^0(\mathcal{L}^n)$$

from the symmetric algebra of V. When $\mathcal{V}$ embeds X in $\mathbb{P}^r = \mathbb{P}V$, the image R_X of this map is called the *homogeneous coordinate ring* of X.

The rational normal curve $C \subset \mathbb{P}^d$ of degree d is embedded by the complete linear series $\mathcal{V} = (\mathcal{O}_{\mathbb{P}^1}(d), \mathbb{C}[s,t]_d)$. Since any form of degree nd in $\mathbb{C}[s,t]$ is a sum of products of n forms of degree d, the corresponding homogeneous coordinate ring of the rational normal curve is

$$\mathbb{C}[s,t]_{(d)} := \bigoplus_n (\mathbb{C}[s,t]_{nd}),$$

and the map $\rho_{\mathcal{V}}$ is surjective. This is expressed by saying that C is *arithmetically Cohen–Macaulay*; more generally, if $C \subset \mathbb{P}^r$ is a one-dimensional scheme, we say that C is linearly (respectively, quadratically, ..., n-ically) normal if the natural maps

$$\rho_m : H^0(\mathcal{O}_{\mathbb{P}^r}(m)) \to H^0(\mathcal{O}_C(m))$$

are surjective for $m = 1$ (respectively $m = 2, \ldots, m = n$). We say that C is arithmetically Cohen–Macaulay (usually abbreviated ACM) when ρ_m is surjective for all m. We'll discuss the significance of this condition in Section 3.3, and we will prove that it is equivalent to the condition that R_C is a Cohen–Macaulay ring, justifying the name.

The equations defining a rational normal curve. Choosing a basis s, t for the linear forms on $\mathbb{P}^1$, we can write the d-th Veronese map $\mathbb{P}^1 \to \mathbb{P}^d$ as

$$\phi_d : (s,t) \mapsto (s^d, s^{d-1}t, \ldots, t^d),$$

from which we see that the image C of ϕ_d lies in the zero locus of the homogeneous quadratic polynomial $x_i x_j - x_{i+1} x_{j-1}$ for every i, j. We can realize these quadratic forms as the 2×2 minors of the matrix

$$M = \begin{pmatrix} x_0 & x_1 & \cdots & x_{d-1} \\ x_1 & x_2 & \cdots & x_d \end{pmatrix}.$$

Note that if we substitute $s^i t^{(d-i)}$ for x_i and identify $H^0(\mathcal{O}_{\mathbb{P}^1}(i))$ with $\mathbb{C}[s,t]_i$, this becomes the multiplication table

$$H^0(\mathcal{O}_{\mathbb{P}^1}(1)) \times H^0(\mathcal{O}_{\mathbb{P}^1}(d-1)) \to H^0(\mathcal{O}_{\mathbb{P}^1}(d)).$$

It is easy to see that C is set-theoretically defined by the 2×2 minors of M: the affine set $s = 1$ in $\mathbb{P}^1$ maps to the affine set $x_0 = 1$ in $\mathbb{P}^d$, and the affine form of the map is $t \mapsto (t, t^2, \ldots, t^d)$. But if $x_1 = t$ then from the equations $x_0 x_i = x_1 x_{i-1}$ we see successively that $x_i = t^i$; that is, the vanishing of the minors of M at a point p implies that p lies on C.

A much stronger statement is true:

Proposition 3.4. *The homogeneous ideal of the rational normal curve*

$$\mathbb{P}^1 \to C \subset \mathbb{P}^d, \quad (s,t) \mapsto (s^d, s^{d-1}t, \ldots, t^d),$$

is generated by the 2×2 minors of M.

Proof. Let $I \subset \mathbb{C}[x_0, \ldots, x_d]$ be the ideal generated by the 2×2 minors of M. As we have seen, the map

$$\phi : \mathbb{C}[x_0, \ldots, x_d]/I \to \mathbb{C}[s, t]_{(d)}, \quad x_i \mapsto s^{d-i}t^i,$$

is surjective, and we must show that this is a monomorphism, or equivalently that the source of ϕ in degree n has (at most) the same dimension $nd + 1$ as the target of ϕ in degree nd.

If $0 < i \le j < d$ then

$$x_i x_j \equiv x_{i-1} x_{j+1} \pmod{I}.$$

Thus every monomial in the x_i of degree t is equivalent, modulo I, to a monomial of the form

$$x_0^a x_1^{\epsilon_1} \cdots x_{d-1}^{\epsilon_{d-1}} x_d^b$$

where at most one ϵ_i is 1, and the rest are 0. There are $n + 1$ such elements of degree n with all the $\epsilon_i = 0$ and $n(d-1)$ elements with one of the $\epsilon_i = 1$. Thus there are $nd + 1$ such elements in all, proving that ϕ is an isomorphism. $\square$

Cheerful Fact 3.5. The description of the equations above can be considerably extended; see Proposition 17.6. For example, the equations of the Veronese surface, which is the image of $\mathbb{P}^2$ under the complete linear series of quadrics, is defined by the 2×2 minors of the generic symmetric matrix

$$\begin{pmatrix} x_0 & x_1 & x_2 \\ x_1 & x_3 & x_4 \\ x_2 & x_4 & x_5 \end{pmatrix},$$

which arises from the multiplication table of

$$H^0(\mathcal{O}_{\mathbb{P}^2}(1)) \otimes H^0(\mathcal{O}_{\mathbb{P}^2}(1)) \to H^0(\mathcal{O}_{\mathbb{P}^2}(2)).$$

Corollary 3.6. *The dimension of the degree n part of the homogeneous ideal of the rational normal curve of degree d is*

$$H^0(\mathcal{J}_{C/\mathbb{P}^d}(n)) = \binom{n+d}{n} - (nd + 1).$$

In particular the $\binom{d}{2}$ minors of the matrix M are linearly independent.

Proof. The homogeneous coordinate ring of the rational normal curve is

$$\mathbb{C}[s, t]_{(d)} \subset \mathbb{C}[s, t].$$

Comparing the dimension of $\mathbb{C}[x_0, \ldots, x_d]_n$ with the dimension of $\mathbb{C}[s, t]_{nd}$ gives the result. $\square$

The number of quadrics containing the rational normal curve is extremal:

Proposition 3.7. *If $C \subset \mathbb{P}^d$ is any irreducible, nondegenerate curve, then*

$$h^0(\mathcal{J}_{C/\mathbb{P}^d}(2)) \leq \binom{d}{2}.$$

Equality holds only if C is a rational normal curve.

In the proof we will use a general fact about hyperplane sections:

Proposition 3.8. *If $X \subset \mathbb{P}^r$ is a nondegenerate, reduced, irreducible variety, then $H^0(\mathcal{J}_X(1)) = H^1(\mathcal{J}_X) = 0$. Moreover if H is any hyperplane of $\mathbb{P}^r$, defined by a linear form h, then the natural sequence*

$$0 \to \mathcal{J}_{X/\mathbb{P}^r} \xrightarrow{h} \mathcal{J}_{X/\mathbb{P}^r}(1) \to \mathcal{J}_{(H \cap X)/H}(1) \to 0$$

arising from the restriction of the ideal sheaf $\mathcal{J}_{X/\mathbb{P}^r}$ to H is exact, and the subscheme $H \cap X$ spans H; that is, $H^0(\mathcal{J}_{(H \cap X)/H}(1)) = 0$.

See Exercise 3.5 for the necessity of the hypotheses and Exercise 3.6 for a generalization.

Corollary 3.9. *If $X \subset \mathbb{P}^r$ is a nondegenerate, reduced, irreducible variety of codimension c, then $\deg X \geq c + 1$.*

Proof. The intersection of X with a general plane Λ of dimension c is a finite scheme spanning Λ. □

Proof of Proposition 3.8. Since X is reduced, irreducible and nondegenerate, h is a nonzerodivisor modulo $\mathcal{J}_{X/\mathbb{P}^r}$, so $h\mathcal{J}_{X/\mathbb{P}^r} = (h) \cap \mathcal{J}_{X/\mathbb{P}^r}$ where $(h) = h\mathcal{O}_X$. Setting $\Gamma = H \cap X$ we have

$$
\begin{aligned}
\mathcal{J}_{X/\mathbb{P}^r}(1)/h\mathcal{J}_{X/\mathbb{P}^r} &= \mathcal{J}_{X/\mathbb{P}^r}(1)/((h) \cap \mathcal{J}_{X/\mathbb{P}^r}) \\
&= (\mathcal{J}_{X/\mathbb{P}^r}(1) + (h))/(h) \\
&= \mathcal{J}_{\Gamma/H}(1),
\end{aligned}
$$

from which the exactness assertion follows.

Again because X is reduced and irreducible, $H^0(\mathcal{O}_X)$ contains only the constant function, so the map $H^0(\mathcal{O}_{\mathbb{P}^r}) \to H^0(\mathcal{O}_X)$ is surjective, from which it follows that $H^1(\mathcal{J}_{X/\mathbb{P}^r}) = 0$. From the long exact sequence in cohomology it follows that the restriction map $H^0(\mathcal{J}_{X/\mathbb{P}^r}(1)) \to H^0(\mathcal{J}_{\Gamma/H}(1))$ is surjective. Since X is nondegenerate, $H^0(\mathcal{J}_{X/\mathbb{P}^r}(1)) = 0$, and the desired result follows. □

Proof of Proposition 3.7. Consider the restriction of the quadrics containing C to a general hyperplane $H \cong \mathbb{P}^{d-1} \subset \mathbb{P}^d$, and let $\Gamma = H \cap C$. There is an exact sequence:

$$0 \to \mathcal{J}_{C/\mathbb{P}^d}(1) \to \mathcal{J}_{C/\mathbb{P}^d}(2) \to \mathcal{J}_{\Gamma/\mathbb{P}^{d-1}}(2) \to 0.$$

Proposition 3.8 shows that $h^0(\mathcal{J}_{C/\mathbb{P}^d}(1)) = h^1(\mathcal{J}_{C/\mathbb{P}^d}(1)) = 0$, so Γ imposes the same number of conditions on quadrics as C does.

Proposition 3.8 also shows that H is the linear span of Γ. Therefore the hyperplane section Γ of C must contain at least d linearly independent points (this gives another proof that such a curve must have degree $\geq d$). Any set of linearly independent points imposes independent conditions (cf. Definition 10.5) on linear forms, and thus also on quadrics (see Proposition 10.6 for a stronger result). Thus

$$h^0(\mathcal{I}_{\Gamma/\mathbb{P}^{d-1}}(2)) \leq h^0(\mathcal{O}_{\mathbb{P}^{d-1}}(2)) - d = \binom{d+1}{2} - d = \binom{d}{2},$$

establishing the desired inequality.

To prove that equality can be achieved only with the rational normal curve, suppose that $C \subset \mathbb{P}^d$ is not a rational normal curve, so that $\deg C > d$. Let

$$\Gamma' \subset H \cong \mathbb{P}^{d-1}$$

be a subset of d linearly independent points of a general hyperplane section $H \cap C$ and let $\Gamma = \Gamma' \cup \{p\} \subset H$ be a subset containing one more point.

We will show that p imposes a nontrivial vanishing condition on quadrics containing Γ', since then C will lie on fewer quadrics than the rational normal curve. Since the points of Γ' are linearly independent, each subset of $d-1$ of them spans a $(d-2)$-plane inside H, and the intersection of these spans is empty. Thus one of the spans Λ does not contain p. The union of a general hyperplane H containing Λ and a general hyperplane containing the point $\Gamma' \setminus \Gamma' \cap \Lambda$ is a quadric containing Γ' but not p, as required.   □

For each m, rational normal curves lie on more hypersurfaces of degree m than any other irreducible, nondegenerate curve in $\mathbb{P}^d$; see Exercise 10.4.

Rational normal curves are projectively homogeneous. An important property of rational normal curves $C \subset \mathbb{P}^d$ is that they are *projectively homogeneous*, in the sense that the subgroup $G \subset \mathrm{PGL}_{d+1}$ of automorphisms of $\mathbb{P}^d$ that carry C to itself acts transitively on C. More generally:

Proposition 3.10. *If $X \subset \mathbb{P}^n$ is the image of $\mathbb{P}^r$ by the Veronese map associated to $|\mathcal{O}_{\mathbb{P}^r}(d)|$, then X is projectively homogeneous.*

Proof. $\mathbb{P}^r$ itself is a homogeneous variety in that $\mathrm{Aut}\,\mathbb{P}^r$ acts transitively. For any automorphism σ of $\mathbb{P}^r$, we have $\sigma^*\mathcal{O}_{\mathbb{P}^r}(d) = \mathcal{O}_{\mathbb{P}^r}(d)$ because $\mathcal{O}_{\mathbb{P}^r}(d)$ is the unique invertible sheaf of degree d on $\mathbb{P}^r$; thus σ induces an automorphism ϕ on $H^0(\mathcal{O}_{\mathbb{P}^r}(d))$ and an automorphism $\bar{\phi}$ on the ambient space $\mathbb{P}^N := \mathbb{P}H^0(\mathcal{O}_{\mathbb{P}^r}(d))$ of the target of the d-th Veronese map. If $\ell \in H^0(\mathcal{O}_{\mathbb{P}^r}(d))$ with divisor $D \subset X$, then $\phi(\ell) = \ell \circ \sigma$ has divisor $\sigma^{-1}(D)$, so $\bar{\phi}^{-1}$ induces σ on $\mathbb{P}^r$.   □

The rational normal curve $C \subset \mathbb{P}^d$ can be characterized among irreducible, nondegenerate curves as the unique projectively homogeneous curve in $\mathbb{P}^d$ (Corollary 13.3).

Interpolation for rational normal curves. Another striking property of rational normal curves is expressed in the following proposition. Recall that a collection of points (or a finite subscheme) of $\mathbb{P}^d$ is *in linearly general position* if no $k+1$ of them lie in a $(k-1)$-plane with $k \leq n$.

Proposition 3.11. *Any $d + 3$ points $p_1, \ldots, p_{d+3} \in \mathbb{P}^d$ in linearly general position lie on a unique rational normal curve $C \subset \mathbb{P}^d$.*

Proof. There is an automorphism $\Phi : \mathbb{P}^d \to \mathbb{P}^d$ carrying $p_1, \ldots, p_{d+1}$ to the coordinate points $(0, \ldots, 0, 1, 0, \ldots, 0) \in \mathbb{P}^d$ and the point p_{d+2} to the point $(1, 1, \ldots, 1)$. Denote the image of the remaining point p_{d+3} by $[\alpha_0, \ldots, \alpha_n]$. We consider maps $\mathbb{P}^1 \to \mathbb{P}^d$ given in terms of an inhomogeneous coordinate z on $\mathbb{P}^1$ by

$$ z \mapsto \left(\frac{1}{z - \nu_0}, \frac{1}{z - \nu_1}, \ldots, \frac{1}{z - \nu_d} \right) $$

with $\nu_0, \ldots, \nu_d$ any distinct scalars. Clearing denominators, we see that the image of such a map is a rational normal curve, and it passes through the $d + 1$ coordinate points of $\mathbb{P}^d$, which are the images of the points $z = \nu_0, \ldots, \nu_d \in \mathbb{P}^1$. Moreover, the image of the point at infinity is $(1, 1, \ldots, 1)$. Because the points are assumed linearly independent, the α_i must all be nonzero, and thus if we take $\nu_i = -1/\alpha_i$, the image of the point $z = 0$ is $(\alpha_0, \ldots, \alpha_d)$. This proves existence; we leave uniqueness to Exercise 3.11. $\qquad\square$

Cheerful Fact 3.12. A generalization of Proposition 3.11 is given, with a proof that is different even in our case, in [Harris 1982a, Proposition 3.19]. There is also a version replacing the set of $d + 3$ distinct points by a finite scheme of degree $d + 3$ satisfying necessary conditions, in [Eisenbud and Harris 1992].

Given some "natural" family of curves in projective space $\mathbb{P}^r$ (such as an open set of smooth curves in a component of the Hilbert scheme described in Chapters 7 and 19) and an integer m, we can ask whether there exists a curve in the family passing through m given general points of $\mathbb{P}^r$. Proposition 3.11 gives the only example we know of curves other than complete intersections for which there is a *unique* such curve.

3.2. Other rational curves

What about other rational curves in projective space?

Any linear series $\mathcal{D}$ of degree d on $\mathbb{P}^1$ is a subseries of the complete series $|\mathcal{O}_{\mathbb{P}^1}(d)|$, so any map $\phi : \mathbb{P}^1 \to \mathbb{P}^r$ of degree d may be given as the composition of the embedding $\phi_d : \mathbb{P}^1 \to \mathbb{P}^d$ of $\mathbb{P}^1$ as a rational normal curve with a linear projection $\pi : \mathbb{P}^d \to \mathbb{P}^r$. Since the natural degree d Veronese embedding of

$\mathbb{P}^1 = \mathbb{P}(V)$ as the rational normal curve is the map $\mathbb{P}V \to \mathbb{P}(\mathrm{Sym}^d(V))$, the projection is a map into $\mathbb{P}W$, where $W \subset \mathrm{Sym}^d(V)$.

We can make this more explicit in several ways: first, choosing a basis s, t for V and a basis of forms F_i of degree d on $\mathbb{P}^1$ for W, we may write the map as

$$(s, t) \mapsto (F_0(s, t), \ldots, F_r(s, t)).$$

If the F_i have a greatest common factor $G(s, t)$ then the set $G = 0$ will be the base locus of the linear series, and the map

$$(s, t) \mapsto \left(\frac{F_0(s, t)}{G(s, t)}, \ldots, \frac{F_r(s, t)}{G(s, t)} \right).$$

gives the same map, represented as a linear series of lower degree $d - \deg G$. Thus we will now assume that the forms in W have no common factor.

Perhaps more simply, we can pass to an open affine $\mathbb{A}^1 \subset \mathbb{P}^1$ given by $t = 1$, with coordinate function $z = s/t$, and dehomogenize the F_i to get a vector space of polynomials $f_i = F(s, 1)$ of degree $\leq d$. Then we may write the map as

$$z \mapsto (f_0(z), \ldots, f_r(z)).$$

Since the F_i have no common factor, and thus are not all divisible by t, at least one of the f_i will have degree exactly d. For example, the twisted cubic itself can be represented by the map $z \mapsto (1, z, z^2, z^3)$.

Here, we think of a polynomial $f(z) = f(s/t)$ of degree $\leq d$ as a rational function on $\mathbb{P}^1$ having a pole of order at most d at the point at infinity $(1, 0) \in \mathbb{P}^1$; but we could also take rational functions $F_i(s, t)/G_i(s, t)$ of total degree $\deg F_i - \deg G_i = d$ or, dehomogenizing, $\phi_i(z) = f_i(z)/g_i(z)$.

Smooth rational quartics. Given how easy it is to describe rational curves in projective space in this way, it is surprising how many open questions there are. We begin with one of the simplest cases: a smooth nondegenerate rational curve C of degree 4 in $\mathbb{P}^3$. As described above, such a curve is the image of $\mathbb{P}^1$ under a map given by 4 quartic forms, or, in the most coordinate-free formulation, a codimension 1 subspace of $\mathrm{Sym}^4(V)$, where $V \cong \mathbb{C}^2$.

Example 3.13. In Exercise 3.10 the reader is asked to prove that there is a 1-parameter family of PGL_4 orbits of smooth rational quartic curves in $\mathbb{P}^3$. Perhaps the simplest is given by the parametrization

$$\mathbb{P}^1 \ni (s, t) \mapsto (s^4, s^3 t, s t^3, t^4) \in \mathbb{P}^3,$$

or, more simply, $t \mapsto (t, t^3, t^4)$.

Proposition 3.14 shows that the homogeneous ideal of this curve requires 4 generators, and it is not hard to show that the ideal is generated scheme theoretically (that is, up to saturation) by 3 elements. Nevertheless it is one of the most famous open problems in the theory of curves to determine whether or not the ideal is generated up to radical by just 2 elements — that is, whether it can be written *set-theoretically* as the intersection of two surfaces. For a sample of the recent work on this, see [Hartshorne and Polini 2015].

Proposition 3.14. *If $C \subset \mathbb{P}^3$ is a smooth rational quartic curve, then C lies on a smooth quadric surface S in the divisor class $(1, 3)$, and the homogeneous ideal of C is minimally generated by the quadric defining S and three cubic forms.*

Proof. We first consider the restriction maps

$$\rho_e : H^0(\mathcal{O}_{\mathbb{P}^3}(e)) \to H^0(\mathcal{O}_C(e)).$$

Since $C \cong \mathbb{P}^1$, we may identify $H^0(\mathcal{O}_C(e))$ with $H^0(\mathcal{O}_{\mathbb{P}^1}(4e))$. Since we assume that C is nondegenerate, it does not lie on a hyperplane, so ρ_1 is a monomorphism.

The source $H^0(\mathcal{O}_{\mathbb{P}^3}(2))$ of ρ_2 is 10-dimensional, and the target $H^0(\mathcal{O}_{\mathbb{P}^1}(8))$ is 9-dimensional, so ρ_2 has a kernel of dimension at least 1; that is, C lies on a quadric surface, and since C is irreducible and nondegenerate, any quadric surface containing C is irreducible. This quadric is smooth; see Example 2.41.

Given that C lies on a smooth quadric surface Q, we consider its divisor class (a, b) in the Picard group $\operatorname{Pic}(Q) = \mathbb{Z} \oplus \mathbb{Z}$. As we saw in Example 2.42, a smooth curve in the class (a, b) has degree $a + b$ and genus $(a-1)(b-1)$; solving the equations $a + b = 4$ and $(a-1)(b-1) = 0$ we see that $C \sim (1, 3)$ or $C \sim (3, 1)$. Since the cases are symmetric we assume $C \sim (1, 3)$.

The source of ρ_3 is 20-dimensional and the target is 13-dimensional so there are at least 7 cubics in the ideal of $C \subset \mathbb{P}^3$. Four of these come from multiplying the quadric by the 4 variables on $\mathbb{P}^3$, so there are at least 3 more cubic generators in $I_{C/\mathbb{P}^3}$, the homogeneous ideal of C.

From the exact sequence

$$0 \to I_{Q/\mathbb{P}^3} \to I_{C/\mathbb{P}^3} \to I_{C/Q} \to 0$$

we see that $I_{C/\mathbb{P}^3}$ is generated by the form defining the quadric together with generators for $I_{C/Q}$. Moreover, the natural inclusion $I_{C/Q} \subset H^0_*(\mathcal{I}_{C/Q})$ is an equality; this follows from the exact sequence of sheaves

$$0 \to \mathcal{I}_{Q/\mathbb{P}^3} \to \mathcal{I}_{C/\mathbb{P}^3} \to \mathcal{I}_{C/Q} \to 0$$

because $H^1_*(\mathcal{I}_{Q/\mathbb{P}^3}) = H^1_*(\mathcal{O}_{\mathbb{P}^3}(-2)) = 0$. Thus we need only compute generators for $H^0_*(\mathcal{I}_{C/Q})$.

Since C lies in class $(1, 3)$ we see that

$$\mathcal{I}_{C/Q}(d) = \mathcal{O}_{\mathbb{P}^1 \times \mathbb{P}^1}(d-1, d-3).$$

By the Künneth formula,

$$h^0(\mathcal{I}_{C/Q}(3)) = h^0(\mathcal{O}_{\mathbb{P}^1 \times \mathbb{P}^1}(2, 0)) = h^0(\mathcal{O}_{\mathbb{P}^1}(2)) \cdot h^0(\mathcal{O}_{\mathbb{P}^1}(0)) = 3,$$

and we see that $I(C)$ contains just 3 cubic generators modulo the quadric defining Q.

We claim that the three generators of $h^0(\mathcal{O}_{\mathbb{P}^1\times\mathbb{P}^1}(2,0))$ generate all of

$$H^0_*(\mathcal{I}_{C/Q}(3)) = H^0_*(\mathcal{O}_{\mathbb{P}^1\times\mathbb{P}^1}(2,0)) = \bigoplus_{m\geq 0} H^0(\mathcal{O}_{\mathbb{P}^1\times\mathbb{P}^1}(m+2,m)).$$

For this we must show that the maps

$$\rho_m : H^0(\mathcal{O}_{\mathbb{P}^3}(1)) \otimes H^0(\mathcal{I}_{C/Q}(m+3)) \to H^0(\mathcal{I}_{C/Q}(m+1+3))$$

are surjective for $m \geq 0$.

Since the restriction of $H^0(\mathcal{O}_{\mathbb{P}^3}(1))$ to the quadric is

$$H^0(\mathcal{O}_{\mathbb{P}^1\times\mathbb{P}^1}(1,1)) = H^0(\mathcal{O}_{\mathbb{P}^1}(1)) \otimes H^0(\mathcal{O}_{\mathbb{P}^1}(1)),$$

the map ρ_m is the tensor product of the two maps

$$H^0(\mathcal{O}_{\mathbb{P}^1}(1)) \otimes H^0(\mathcal{O}_{\mathbb{P}^1}(m+2)) \to H^0(\mathcal{O}_{\mathbb{P}^1}(m+1+2))$$

and

$$H^0(\mathcal{O}_{\mathbb{P}^1}(1)) \otimes H^0(\mathcal{O}_{\mathbb{P}^1}(m)) \to H^0(\mathcal{O}_{\mathbb{P}^1}(m+1)),$$

both of which are surjective for $m \geq 0$. This completes the proof. $\square$

Some open problems about rational curves. We can say far less about rational curves of higher degree, even in $\mathbb{P}^3$. For example, when d is large, we don't know the possible Hilbert functions for curves of degree d in $\mathbb{P}^3$, and the situation in $\mathbb{P}^r$ for $r > 3$ is even worse. However, we do know the Hilbert function of a *general* rational curve $C \subset \mathbb{P}^r$ of degree d.

Since such statements will come up often, we pause to explain exactly what it means to say "A general X has property Y." This statement presupposes a choice of a parameter space for objects of type X, that is to say, an algebraic variety Z whose points each correspond to an object of type X. Thus, to be precise, the statement should be, "An object of type X that is general with respect to parameter space Z has property Y." The statement then means: inside Z there is a dense open set whose elements correspond to objects of type X that have property Y. Often Z is irreducible, and then it is enough to have a nonempty open set, since every such set is Zariski dense.

In the case of nondegenerate rational curves of degree d in $\mathbb{P}^r$ we could take for Z the space of $(r+1)$-tuples of independent elements of $H^0(\mathcal{O}_{\mathbb{P}^1}(d))$, or equivalently, taking the quotient by $\mathrm{PGL}(r+1)$, the Grassmannian of $(r+1)$-dimensional planes in $H^0(\mathcal{O}_{\mathbb{P}^1}(d))$. We will later make statements about general curves of genus g, referring either to a component of a Hilbert scheme (discussed in Chapter 7) or to the moduli space of curves that is discussed at length in Chapter 8.

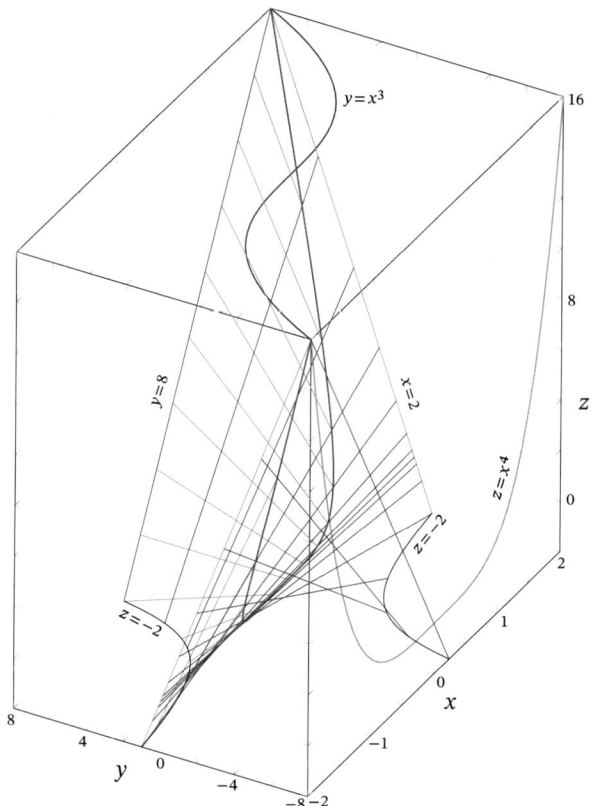

Figure 3.1. A rational quartic space curve of type $(1, 3)$ on a quadric surface.
The quartic, in red, is given by $t \mapsto (t, t^3, t^4)$, and so projects to the cubic and
quartic curves in the xy- and xz-planes (shown on the top and right surfaces
of the box). The quartic, $z = xy$, is illustrated by means of its rulings and its
intersections with the bounding planes. (Inspired by code of Enrique Acosta
Jaramillo for the twisted cubic.)

As in Proposition 3.14, knowing the Hilbert function of a curve $C \subset \mathbb{P}^r$ is
tantamount to knowing the ranks of the restriction maps

$$\rho_m : H^0(\mathcal{O}_{\mathbb{P}^r}(m)) \to H^0(\mathcal{O}_C(m)) = H^0(\mathcal{O}_{\mathbb{P}^1}(md)).$$

Equivalently we ask: if V is a general $(r+1)$-dimensional vector space of homogeneous polynomials of degree d, what is the dimension of the space of polynomials spanned by m-fold products of polynomials in V?

We might guess that the answer is, "as large as possible," meaning that the
rank of ρ_m is $\binom{m+r}{r}$ or $md + 1$, whichever is less — in other words, the map ρ_m
is either injective or surjective for each m. This was proved in [Ballico and Ellia
1983], and is a special case of Larson's maximal rank theorem [Larson 2017].

As we will see in subsequent chapters it is possible to speak of a general
curve of genus g and a general invertible sheaf of degree d on such a curve; and

the analogous statement about Hilbert functions was proven in [Larson 2017]; see Chapter 12.

Nevertheless, even the degrees of the generators of the homogeneous ideal of a general rational curve of degree d in $\mathbb{P}^3$ is unknown for larger d.

The secant plane conjecture. If $C \subset \mathbb{P}^r$, then an e-secant s-plane to C is an s-plane $\Lambda \cong \mathbb{P}^s \subset \mathbb{P}^r$ such that the intersection $\Lambda \cap C$ has degree $\geq e$.

Should you expect a curve $C \subset \mathbb{P}^r$ to have any e-secant s-planes? The set of s-planes in $\mathbb{P}^r$ is parametrized by the Grassmannian $\mathbb{G} = \mathbb{G}(s, r)$, which has dimension $(s + 1)(r - s)$. Inside $\mathbb{G}$, the locus of planes that meet C has codimension $r - s - 1$ (reason: the locus of planes containing a given point p is isomorphic by projection from p to $\mathbb{G}(s - 1, r - 1)$, and thus has codimension $r - s$). Thus one might conjecture that a curve $C \subset \mathbb{P}^r$ will have e-secant s-planes when

$$e \leq (s+1)\frac{r-s}{r-s-1},$$

perhaps with a few low-degree exceptions. Is this true of a general rational curve? For most e, r and s, we don't know.

3.3. The Cohen–Macaulay property

In Section 3.1 we defined a curve $C \subset \mathbb{P}^r$ to be arithmetically Cohen–Macaulay (ACM) if the natural maps

$$\rho_m : H^0(\mathcal{O}_{\mathbb{P}^r}(m)) \to H^0(\mathcal{O}_C(m))$$

are surjective for all m. When this is the case we can accurately predict the number of independent forms of each degree vanishing on the curve, and the condition has other important consequences that we will explore here and in Chapters 16 and 18. For general treatments of Cohen–Macaulay rings see [Eisenbud 1995, Chapter 18] or the book [Bruns and Herzog 1993]. One of the characterizations of the condition is in terms of regular sequences:

Definition 3.15. A sequence of elements $f_1, \ldots, f_c$ in a ring R is *regular* if f_{i+1} is a nonzerodivisor modulo $(f_1, \ldots, f_i)$ for $i = 0, \ldots, c - 1$ and the ideal $(f_1, \ldots, f_c)$ is not the unit ideal.

The *grade* of an ideal I in a ring R (sometimes called the depth of I on R) is the maximal length of a regular sequence contained in I, or ∞ if $I = R$.

A local ring $(R, \mathfrak{m})$ is *Cohen–Macaulay* if grade $\mathfrak{m} = \dim R$. A Noetherian ring R is called Cohen–Macaulay if every localization is Cohen–Macaulay. Likewise, a scheme is Cohen–Macaulay if all its local rings are Cohen–Macaulay.

Example 3.16.   • The sequence of elements $x_0, \ldots, x_r \subset S := \mathbb{C}[x_0 \ldots, x_r]$ is regular, proving from the definition that the localization of S at the homogeneous maximal ideal is Cohen–Macaulay.

- Regular local rings are Cohen–Macaulay. Thus every smooth scheme is Cohen–Macaulay, and in particular $S := \mathbb{C}[x_0 \dots, x_r]$ is Cohen–Macaulay.

- It follows from the definition that any complete intersection in a Cohen–Macaulay scheme is again Cohen–Macaulay. Thus every plane curve, and more generally every complete intersection curve, is ACM; and we shall show in Section 10.2 that every canonically embedded curve and every curve embedded by a complete linear series of sufficiently high degree is also ACM.

- Zero-dimensional rings are all Cohen–Macaulay. The set of zerodivisors in a Noetherian ring is the union of the associated primes of 0, so a 1-dimensional ring is Cohen–Macaulay if and only if it is pure-dimensional (sometimes called "unmixed") — that is, every associated prime ideal is 1-dimensional. Thus any purely 1-dimensional scheme is Cohen–Macaulay.

Cheerful Facts 3.17. • We defined the grade as the maximal length of a regular sequence in I; but in fact all maximal regular sequences have the same length, equal to the smallest integer k such that $\mathrm{Ext}^k(R/I, R) \neq 0$ [Eisenbud 1995, Theorem 17.4 and Proposition 18.4].

• For any proper ideal $I \subsetneq R$ we have grade $I \leq \mathrm{codim}\, I$. On the other hand, if R is Cohen–Macaulay, then grade $I = \mathrm{codim}\, I$ for every ideal I of R. In this sense the grade is an arithmetic approximation to the codimension.

• Every localization of a Cohen–Macaulay ring is Cohen–Macaulay. If $X \subset \mathbb{P}^r$ has homogeneous coordinate ring R_X, then we say that $X \subset \mathbb{P}^r$ is arithmetically Cohen–Macaulay if R_X is Cohen–Macaulay, a property that depends on the embedding, and implies the intrinsic property that X is Cohen–Macaulay (since the local rings of X are essentially localizations of R_X). Previously we defined a curve $C \subset \mathbb{P}^r$ to be arithmetically Cohen–Macaulay if the natural map $R_C \to H^0_*(\mathcal{O}_C)$ is surjective. We will prove that this is equivalent to R_C being Cohen–Macaulay in Proposition 3.18. The same definitions and theorem apply to a positively graded ring, and homogeneous ideals.

• By Serre's criterion [Eisenbud 1995, Section 11.2] the homogeneous coordinate ring R_C of a curve C is normal (that is, integrally closed) if and only if C is both nonsingular and ACM, and sometimes C is said to be *projectively normal* in this case. (This is also the excuse for the terminology "linearly normal" for a curve embedded by a complete linear series, quadratically normal if ρ_2 is surjective, and so on.)

Proposition 3.18. *Suppose that $C \subset \mathbb{P}^r$ is a 1-dimensional subscheme. Let $S = H^0_*(\mathcal{O}_{\mathbb{P}^r})$ be the homogeneous coordinate ring of $\mathbb{P}^r$, and let $R_C = S/I_C$ be the homogeneous coordinate ring of C. The following conditions are equivalent:*

(1) *The natural injective map $R_C \to H^0_*(\mathcal{O}_C(1))$ is surjective.*

(2) $H^1_*(\mathcal{J}_{C/\mathbb{P}^r}) = 0$.

(3) *The homogeneous coordinate ring R_C of C is a Cohen–Macaulay ring; that is, there are linear forms h, h' on $\mathbb{P}^r$ whose images in R_C form a regular sequence. In particular, R_C has no 0-dimensional primary components, so C is purely 1-dimensional and thus Cohen–Macaulay as a scheme.*

(4) *For every hyperplane $H \subset \mathbb{P}^r$ that does not contain any component of C, the homogeneous ideal of $H \cap C$ is equal to $I_C + (h)$, where h is a linear form defining H.*

Proof. (1) $\iff$ (2): We may assume that $r \geq 2$, so $H^1_*(\mathcal{O}_{\mathbb{P}^r}(n)) = 0$ for all n. Using this and the exact sequence $0 \to \mathcal{J}_{C/\mathbb{P}^r} \to \mathcal{O}_{\mathbb{P}^r} \to \mathcal{O}_C \to 0$ we see that $H^0(\mathcal{O}_{\mathbb{P}^r}(n)) \to H^0(\mathcal{O}_C(n))$ is surjective for all n if and only if $H^1(\mathcal{J}_{C/\mathbb{P}^r}(n)) = 0$ for all n, proving the equivalence of (1) and (2).

(1) $\iff$ (3): First let $C \subset \mathbb{P}^r$ be an arbitrary 1-dimensional subscheme, and let $R = H^0_*(\mathcal{O}_C) := \bigoplus_{n \in \mathbb{Z}} H^0(\mathcal{O}_C(n))$. If H is a general hyperplane, with equation $h = 0$, then h does not vanish on any primary component of C, and thus the sequence

$$0 \to \mathcal{O}_C(-1) \xrightarrow{h} \mathcal{O}_C \to \mathcal{O}_{C \cap H} \to 0$$

is exact. Applying H^0_* and using (2), we see that h is a nonzerodivisor on R, and that R/hR is a subring of $H^0_*(\mathcal{O}_{C \cap H})$. A general linear form h' doesn't vanish on any point of $C \cap H$, so h' is a unit on $H^0_*(\mathcal{O}_{C \cap H})$ and thus a nonzerodivisor on R/hR.

The ring R_C is the image of the natural map $S \to R$, and by definition C is ACM if and only if this map is surjective, so that $R_C = R$. This shows that if C is arithmetically Cohen–Macaulay then R_C is a Cohen–Macaulay ring, and is in particular unmixed; that is, C has no 0-dimensional primary components (see for example [Eisenbud 1995, Chapter 18] for general information about Cohen–Macaulay rings). This proves the equivalence of conditions (1) and (3).

(3) $\iff$ (4): If h does not vanish on any component of C then h is a nonzerodivisor on R_C. The ideal $I_{C \cap H}$ is in any case the saturation of $I_C + (h)$. If $I_{C \cap H} = I_C + (h)$, then any linear form h' not containing a point of $C \cap H$ is a nonzerodivisor on $I_C + (h)$, so C satisfies condition (2). Conversely, in a 2-dimensional positively graded Cohen–Macaulay ring, any series of parameters is a regular sequence [Eisenbud 1995, Section 18.2], so $I_C + (h)$ is unmixed, and in particular, saturated. $\square$

Cheerful Fact 3.19. The Cohen–Macaulay property is hard to interpret geometrically; the definition is justified by its usefulness. Here are two results that help our intuition:

(1) A scheme X is Cohen–Macaulay if some (equivalently every) finite map $f : X \to P$ to a smooth scheme P of the same dimension is flat, or equivalently the pushforward $f_*(\mathcal{O}_X)$ is locally free. (This follows from the Auslander–Buchsbaum formula [Eisenbud 1995, Section 19.3].)

(2) (Hartshorne) If a scheme X is Cohen–Macaulay then X is connected in codimension 1 (that is, X remains connected after removing any closed subset of codimension ≥ 2). See [Eisenbud 1995, Theorem 18.12] for a proof.

3.4. Exercises

Exercise 3.1. Let $\mathcal{V} = (\mathcal{O}_{\mathbb{P}^1}(d), V)$ be the linear series of degree d on $\mathbb{P}^1$ defined by the vector space $V = \langle s^d, s^{d-1}t, st^{d-1}, t^d \rangle$, where s, t are coordinates on $\mathbb{P}^1$. Show that $\mathcal{V}$ defines an isomorphism from $\mathbb{P}^1$ onto a smooth curve of type $(1, d-1)$ on a quadric surface. ◆

Exercise 3.2. With notation as in Section 3.1, show that the sheaf associated to the graded module coker M, that is, the cokernel of the map $\mathcal{O}_{\mathbb{P}^d}^d(-1) \to \mathcal{O}_{\mathbb{P}^d}^2$ defined by M, is the unique invertible sheaf of degree 1 on the rational normal curve C, and that thus the associated complete linear series defines the isomorphism $C \to \mathbb{P}^1$ inverse to the Veronese map.

Exercise 3.3. Considering $\mathbb{P}^n$ as $\operatorname{Proj} \mathbb{C}[x_0, \ldots, x_n]$, we may index the variables of $\mathbb{P}^{\binom{n+d}{d}-1}$ by monomials p of degree d in the x_i. Let $M_{n,d}$ be an $(n+1) \times \binom{n+d-1}{n}$ matrix whose rows are indexed by the variables x_i, whose columns are indexed by the monomials m of degree $d-1$ in the x_i and whose (i, m) entry is the variable corresponding to the monomial $x_i m$. (The matrix M of Section 3.1 is $M_{1,d}$.) Show that the 2×2 minors of $M_{n,d}$ generate the ideal of the image $V_{n,d}$ of the Veronese map $\mathbb{P}^n \to \mathbb{P}^{\binom{n+d}{d}-1}$, and that the cokernel of $M_{n,d}$ is the unique invertible sheaf of degree 1 supported on $V_{n,d}$.

Exercise 3.4. Let $\nu_d : \mathbb{P}^r \to \mathbb{P}^{\binom{r+d}{r}-1}$ be the d-Veronese map, and let $C \subset \mathbb{P}^r$ be the rational normal curve of degree r. Is $\nu_d(C)$ nondegenerate? If not, what is the dimension of its linear span (that is, of the smallest linear space that contains it)? ◆

Exercise 3.5. Let C_1 be the union of two skew lines in $\mathbb{P}^3$, and let C_2 be the double line on a smooth quadric in $\mathbb{P}^3$. Show that a general hyperplane section of C_1 and of C_2 violates the conclusion of Proposition 3.8.

Exercise 3.6. Suppose that $X \subset \mathbb{P}^r$ is a subscheme and Z is the hypersurface in $\mathbb{P}^r$ defined by a polynomial $F \in H^0(\mathcal{O}_{\mathbb{P}^r}(m))$. If the restriction of F is a nonzerodivisor in $\mathcal{O}_X(m)$ then there is a short exact sequence

$$\mathcal{I}_X(d-m) \xrightarrow{F} \mathcal{I}_X(m) \to \mathcal{I}_{X \cap Z}(m) \to 0.$$

Exercise 3.7. Let $C \subset \mathbb{P}^3$ be the smooth rational quartic (or any smooth curve embedded by an incomplete linear series), and let h be a linear form defining a hyperplane H. Show that the irrelevant ideal is associated to the homogeneous ideal $I_C + (h)$, and thus $I_C(1)/hI_c$ is not the saturated homogeneous ideal of the finite set $C \cap H$.

Exercise 3.8. Show that the twisted cubic is the unique irreducible, nondegenerate space curve lying on three quadrics by considering the possible intersections of two of the quadrics.   ◆

Exercise 3.9. As a consequence of our description of rational quartic rational quartic curves on a smooth quadric in Proposition 3.14, show that a general g_4^3 on $\mathbb{P}^1$ is uniquely expressible as a sum of the g_1^1 and a g_3^1 (in other words, a general 4-dimensional vector space of quartic polynomials on $\mathbb{P}^1$ is uniquely expressible as the product of a 2-dimensional vector space of cubics and the 2-dimensional space of linear forms.   ◆

Exercise 3.10. Show that, up to projective equivalence, there is a 1-parameter family of embeddings of $\mathbb{P}^1$ as a smooth quartic curve in $\mathbb{P}^3$ by constructing an invariant that distinguishes them.   ◆

Exercise 3.11. Complete the proof of Proposition 3.11 by showing that if C, C' are two rational normal curves in $\subset \mathbb{P}^n$ meeting in at least $n + 3$ distinct points, then $C = C'$.   ◆

Exercise 3.12. Let $V = \mathbb{C} \cdot e_1 \oplus \mathbb{C} \cdot e_2$ be a 2-dimensional vector space.

The group $\mathrm{SL}_2 = \mathrm{SL}(V)$ acts on the rational normal curve of degree d through automorphisms induced from its action on the ambient space $\mathbb{P}^d$ of the rational normal curve, which may be identified with $\mathbb{P}(\mathrm{Sym}^d(V))$.

In [Fulton and Harris 1991, pp. 146–150] it is shown that every finite-dimensional rational representation of $\mathrm{SL}(V)$ is a direct sum of representations of the form $\mathrm{Sym}^e(V)$ for various $e \geq 0$. There it is explained that to understand how a given representation decomposes one should look at the action of the torus generator

$$\alpha := \begin{pmatrix} t & 0 \\ 0 & t^{-1} \end{pmatrix} \in \mathrm{SL}(V).$$

The eigenvectors of α are called the *weight vectors* of the representation. Note that $\mathrm{Sym}^e(V)$ is spanned by the weight vectors $w_s := e_1^{e-s} e_2^s$ that satisfy $\alpha w_s = t^{e-2s}$ for $s = 0, \dots e$. To decompose an arbitrary representation W, knowing that W is a direct sum of $\mathrm{Sym}^{e_i} V$, it is enough to know the eigenvalues for the action of α: We begin by finding an element $w \in W$ that is an eigenvector of α and transforms by α as $\alpha w = t^m w$ with the highest possible m (this is called a "highest weight vector"). Such a w must be contained in a summand $\mathrm{Sym}^m(V)$, and after removing the eigenvalues of the action of SL_2 on $\mathrm{Sym}^m(V)$, we continue.

(1) Use this method to show that

$$\mathrm{Sym}^d(V) \otimes \mathrm{Sym}^d(V) = \mathrm{Sym}^{2d}(V) \oplus \mathrm{Sym}^{2d-2}(V) \oplus \mathrm{Sym}^{2d-4}(V) \otimes \cdots ,$$
$$\mathrm{Sym}^2(\mathrm{Sym}^d(V)) = \mathrm{Sym}^{2d}(V) \oplus \mathrm{Sym}^{2d-4}(V) \oplus \mathrm{Sym}^{2d-8}(V) \otimes \cdots ,$$
$$\textstyle\bigwedge^2(\mathrm{Sym}^d(V)) = \mathrm{Sym}^{2d-2}(V) \oplus \mathrm{Sym}^{2d-6}(V) \oplus \mathrm{Sym}^{2d-10}(V) \otimes \cdots ,$$

where we take $\mathrm{Sym}^m(V) = 0$ when $m < 0$.

(2) Show that the space of quadrics containing the rational normal curve is a representation of SL_2 of the form

$$\mathrm{Sym}^{2d-4}(V) \oplus \mathrm{Sym}^{2d-8}(V) \cdots$$

(3) Show there is a distinguished nonsingular skew-symmetric form (up to scalars) on the ambient space of the twisted cubic. Thus a twisted cubic in $\mathbb{P}^3$ determines, for each point of $\mathbb{P}^3$, a distinguished plane containing that point.

(4) Show that if d is divisible by 4 there is a distinguished quadric in the ideal of the rational normal curve.

Exercise 3.13. Let $\mathbb{P}^1 \hookrightarrow C \subset \mathbb{P}^3$ be a twisted cubic. Show that the normal bundle $\mathcal{N}_{C/\mathbb{P}^3}$ (defined to be the quotient of the restriction $T_{\mathbb{P}^3}|_C$ to C of the tangent bundle of $\mathbb{P}^3$ by the tangent bundle T_C) is $\mathcal{N}_{C/\mathbb{P}^3} \cong \mathcal{O}_{\mathbb{P}^1}(5) \oplus \mathcal{O}_{\mathbb{P}^1}(5)$. ♦

Exercise 3.14. Let $\mathbb{P}^1 \hookrightarrow C \subset \mathbb{P}^d$ be a rational normal curve. Show that the normal bundle $\mathcal{N}_{C/\mathbb{P}^d}$ is

$$\mathcal{N}_{C/\mathbb{P}^d} \cong \bigoplus_{i=1}^{d-1} \mathcal{O}_{\mathbb{P}^1}(d+2). \quad \blacklozenge$$

Exercise 3.15. In the situation of Exercise 3.14, the set of direct summands of $\mathcal{N}_{C/\mathbb{P}^d}$ is a projective space $\mathbb{P}^{d-2}$. How does the group of automorphisms of $\mathbb{P}^d$ carrying C to itself act on this $\mathbb{P}^{d-2}$? ♦

See [Coskun and Riedl 2018], for example, for more on normal bundles of rational curves.

Exercise 3.16. If $C = \bigcap_{i=1}^{r-1} X_i \subset \mathbb{P}^r$ is a complete intersection of hypersurfaces, then C is arithmetically Cohen–Macaulay. ♦

Exercise 3.17. Give a proof of Corollary 3.9 without using Bertini's theorem, by projecting X from a general point and using induction on $\operatorname{codim} X$. ♦

Chapter 4

Smooth plane curves and curves of genus 1

If C is a curve of genus 1, then by Corollary 2.19, any complete linear series D of degree $3 = 2g + 1$ on C is very ample. By the Riemann–Roch theorem any invertible sheaf of degree $d > 0$ on C has d sections, so the morphism associated to D embeds C as a plane curve of degree 3, on which D is the intersection of C with a line.

We therefore begin this chapter with a general explanation of sheaves of differentials and linear series on smooth plane curves, and use that theory to say a little about other embeddings of curves of genus 1. Though most curves cannot be realized as smooth plane curves, we shall see in Proposition 10.11 that every curve can be projected birationally to a plane *nodal curve* (that is, one having only ordinary nodes as singularities), and in Chapter 15 we will explain the analogous treatment of differentials and linear series on such nodal curves, and more generally — but with less specificity — on all reduced plane curves.

Using the plane model of a curve of genus 1, we shall see that the theory is only a little more complicated than for genus 0. (But curves of genus 1 over number fields occupy many modern number theorists!)

4.1. Riemann, Clebsch, Brill and Noether

For a long time, plane curves were the only algebraic curves that were studied. Originally these were curves in the affine plane over the real numbers, but by the second half of the nineteenth century the complex projective plane was well understood, and curves in $\mathbb{P}^2 = \mathbb{P}^2_{\mathbb{C}}$, corresponding to irreducible forms in 3 variables, were recognized as the natural objects of study (see Chapter 19.12.)

The work of Bernhard Riemann dramatically changed the focus of the theory to branched coverings of what we know as the Riemann sphere, or $\mathbb{P}^1_{\mathbb{C}}$. The Riemann–Roch theorem, in particular, gave information about the existence of meromorphic functions on such coverings, well beyond what could be done in the earlier theory. However, Riemann's work, depending as it did on the then-obscure "Dirichlet principle", was not universally accepted. In the 1860s Alfred Clebsch and, after the death of Clebsch in 1872, Alexander Brill and Max Noether (Emmy Noether's father) undertook the ambitious program of redoing the Riemann–Roch theorem entirely in terms of plane curves. They went beyond Riemann in certain directions, too: the Brill–Noether theorem treated in our Chapter 12 was formulated by Brill and Noether, and "proved" by them through an unsupported general position assumption.

A central difficulty in the Brill–Noether attempt on the Riemann–Roch theorem was that, although any smooth curve can be embedded in $\mathbb{P}^r$ for any $r \geq 3$, most curves cannot be embedded in the plane. However, as is shown in Section 10.3, we can embed C as a curve $C \subset \mathbb{P}^r$ in a higher-dimensional projective space and find a projection $\mathbb{P}^r \to \mathbb{P}^2$ that carries C birationally onto its image C_0, called a plane model of C. The curve C_0 typically has singularities, and C is the normalization of C_0. Brill and Noether wanted to prove the Riemann–Roch theorem for C by formulating and proving a related theorem for C_0. In particular, they tried to characterize linear equivalence of divisors on C in terms of certain "clusters" of points — we would say 0-dimensional subschemes — of C_0.

To carry out this program, a key step was to show that if $D \subset C_0$ is contained in the intersection D' of C_0 and some other plane curve C'_0, then a divisor $E := D' - D$ can be defined with properties such as that $D' - E = D$; Brill and Noether seem simply to have assumed that this is so. A bit later Frances Sowerby Macaulay proved that this is in fact possible and also understood that it would not generally be possible if D' were the intersection of three or more curves. (In modern terms, the intersection of two curves is Gorenstein; the intersection of 3 is generally not.)

Macaulay exploited this theory of residuation to prove what he called the generalized Riemann–Roch theorem (Theorem 4.5). This early work of Macaulay led directly to his definitions of "perfection" (a homogeneous ideal $I \subset S := k[x_0, \ldots, x_n]$ is perfect if S/I is Cohen–Macaulay) and "super-perfection" (the case when S/I is Gorenstein). For all this, see [Eisenbud and Gray 2023].

In this section we will take the point of view of Clebsch, Brill and Noether, and explain how to understand the differential forms and, given a (possibly ineffective) divisor D on C, how to find all the effective divisors on C that are linearly equivalent to D, in terms of a smooth plane curve. In Chapter 15 we will return to this point of view and treat the case of nodal curves and the case of arbitrary reduced plane curves. We will give effective algorithms for two constructions:

(1) Given the equation $F(X, Y, Z)$ of a smooth d plane curve C of degree d, we can construct a basis for $H^0(K_C)$.

(2) Given a possibly ineffective divisor $D = D_+ - D_-$ on C we can construct the complete linear series $|D|$ on C. In particular we can:
   (a) determine whether D is equivalent to any effective divisor on C; and if so,
   (b) find all effective divisors E on C with $E \sim D$;
   (c) find a basis of $H^0(\mathcal{O}_C(D))$, expressed in terms of curves of high degree with base locus;
   (d) find the homogeneous coordinate ring of the morphism defined by $|D|$ or a subseries.

4.2. Smooth plane curves

4.2.1. Differentials on a smooth plane curve. Let $C \subset \mathbb{P}^2$ be a smooth plane curve, given as the zero locus of a homogeneous polynomial $F(X, Y, Z)$ of degree d. By the adjunction formula (Proposition 2.8) the canonical divisors on C are the intersections of C with curves of degree $d - 3$. In the spirit of Brill and Noether we will make this explicit by constructing all the regular differential forms on C in terms of forms of degree $d - 3$.

For this purpose we introduce coordinates $x = X/Z$ and $y = Y/Z$ on the affine open subset $U \cong \mathbb{A}^2$ given by $Z \neq 0$, and let $f(x, y) = F(x, y, 1)$ be the inhomogeneous form of F, so that $C^\circ = C \cap U$ is the zero locus $V(f) \subset \mathbb{A}^2$.

Since an automorphism of $\mathbb{P}^2$ can carry any point in $\mathbb{P}^2$ to any other point, we may assume that the point $[0, 1, 0]$ (that is, the point at infinity in the vertical direction) does not lie on C, so that the projection $\pi : C \to \mathbb{P}^1$ from $(0, 1, 0)$, which is given by $[X, Y, Z] \mapsto [X, Z]$ (or, in affine coordinates, $(x, y) \mapsto x$) has degree d. Let D be the divisor defined by the intersection of C with the line $Z = 0$ at infinity.

Consider the regular 1-form dx on $\mathbb{A}^2$, which we may regard as the pullback of the form dx on $\mathbb{A}^1$. Since the form dx on $\mathbb{P}^1$ has a double pole at infinity the form $dx|_C$ has polar locus $2D$.

How do we get rid of the poles of dx? The extension to $\mathbb{P}^2$ of a polynomial $h(x, y)$ of degree m on $\mathbb{A}^2$ has a pole of order m along the line L at infinity. Thus if h has degree at least 2 then dx/h has no poles at infinity. However, $h(x, y)$ may well vanish at points of C°, and this may create new poles of dx/h. Of course if h vanishes only at points of C° where dx has a zero, the zeroes of h may cancel the zeroes of dx rather than creating new poles.

To avoid producing new poles in this way we may take

$$h(x, y) = f_y := \frac{\partial f}{\partial y}(x, y).$$

We claim that

$$\varphi_0 = \frac{dx}{f_y}$$

is everywhere regular and nowhere 0 in C°.

Note that df vanishes identically when restricted to C°, so

$$0 \equiv df|_{C^\circ} = f_x dx|_{C^\circ} + f_y dy|_{C^\circ}.$$

Clearly φ_0 is regular at points p where $f_y(p) \neq 0$. At such a point, if $dx|_{C^\circ}$ were 0, then since C° is smooth we would have $dy|_{C^\circ} \neq 0$, contradicting the equation above. Thus φ_0 is both regular and nonzero at such points. On the other hand, if $f_y(p) = 0$, then since C° is smooth at p we also have $f_x(p) \neq 0$, so $dx|_{C^\circ}$ and f_y vanish to the same order, whence, again, $\varphi_0 = (dx/f_y)|_{C^\circ}$ is regular and nonzero, proving the claim.

Put differently, if L is the line at infinity, so that $U = \mathbb{P}^2 \setminus L \cong \mathbb{A}^2$, then the cotangent bundle on U is $\Omega_U = \mathcal{O}_U dx \oplus \mathcal{O}_U dy$, so the cotangent bundle Ω_{C° on C°, which is the canonical sheaf ω_{C°, is the cokernel of the map from the normal sheaf $\mathcal{O}_C(-d)|_U$ to the restriction of Ω_U to C°. This map sends the local generator f of the normal sheaf to $f_x\, dx + f_y\, dy \in \Omega_U$, and because f_x and f_y have no common zeros, the generator $d_y \in \Omega_{C^\circ}$ is a multiple of dx/f_y. Thus the free $\mathcal{O}_{C^\circ}$-module $\omega_C|_{C^\circ}$ is generated by dx/f_y.

Since f_y has degree $d-1$, the rational function $1/f_y$ vanishes to order $d-1$ on the line at infinity, and thus in particular on the divisor D. Thus φ_0 vanishes to order $d-3$ on D; in other words, as divisors,

$$(\varphi_0) = (d-3)D.$$

In particular, if $d \geq 3$ then φ_0 is a globally on C. The divisor of zeros of this differential has degree $d(d-3)$. If g is the genus of C, then $2g - 2 = d(d-3)$, whence

$$g = \frac{d(d-3)}{2} + 1 = \binom{d-1}{2}.$$

We can produce a vector space of $\binom{d-1}{2}$ regular differentials by multiplying φ_0 by polynomials $e(x, y)$ of degree $d-3$, since this does not introduce any poles. This proves:

Theorem 4.1. *The space of regular differentials on a smooth plane curve C with affine equation $f = 0$ is*

$$\left\{ \frac{e(x, y)\, dx}{f_y} \,\middle|\, e(x,y) \text{ is a polynomial degree at most } d-3 \right\}. \qquad \square$$

4.2.2. Linear series on a smooth plane curve. Any divisor on a smooth plane curve C may be expressed as the difference of two effective divisors, $D = D_+ - D_-$. We would like to find all the *effective* divisors linearly equivalent to D, that is, of the form $D + (H/G)$, where G, H are forms of the same degree m. We begin by choosing an integer m, large enough so there is a form G of degree m vanishing on D_+ but not on all of C, so that the divisor of zeros of G is $A + D_+$ for some divisor A on C.

Theorem 4.2. *Let $D = D_+ - D_-$ be a divisor on a smooth plane curve C. If there is a form G of degree m vanishing on D_+ but not on all of C then we may write $(G) = D_+ + A$, and then the effective divisors equivalent to D, if any, are precisely those of the form $(H) - A - D_-$, where H is a form of degree m vanishing on $D_- + A$, but not on all of C.*

See Figure 4.1 for an illustration with $D_+ = r + s + t$, $D_- = u$. In particular, if no homogeneous polynomial H of degree m vanishes on $A + D_-$ but not on C, then D is not linearly equivalent to any effective divisor. The existence of such an H is thus independent of the choices of m and G, as we shall see in the proof.

The simplest special case of the theorem is the completeness of the linear series defined by intersections of C with curves of a given degree, or, equivalently, the fact that the restriction maps

$$H^0(\mathcal{O}_{\mathbb{P}^n}(m)) \to H^0(\mathcal{O}_C(m))$$

are surjective for all m.

Proposition 4.3. *If C is a smooth plane curve, then any Cartier divisor on D that is linearly equivalent to the divisor of a form of degree m is itself the divisor of a form of degree m; that is, plane curves are arithmetically Cohen–Macaulay.*

The results for singular curves explained in Chapter 15 depend on strengthenings of this condition.

Proof. Let C be the plane curve defined by $F = 0$, and let D be the divisor on C defined by a form L of degree m, not vanishing on (any component of) the curve C. If G and H are forms of the same degree t, not vanishing on any component of C, and $D + (G/H)$ is effective, then the divisor (LG) on C must contain the divisor (H) on C. This means that the subscheme of $\mathbb{P}^2$ defined by (LG, F) contains the scheme defined by (H, F). Since H and F have no components in common, Theorem 0.1 implies that $LG = AH + BF$ for some forms A, B with $\deg A = \deg LG - \deg H = m$, whence $D + (G/H) = (A)$ as required.   □

Example 4.4. Suppose that C has degree 3 and thus genus 1. If we choose as origin on the curve C a point o, then to add two points p and $q \in C$ means to find the (unique) effective divisor of degree 1 linearly equivalent to $p + q - o$. In

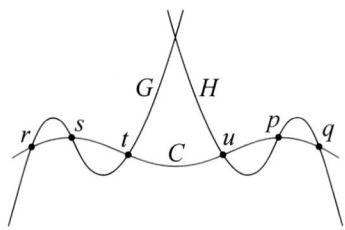

Figure 4.1. If G and H have the same degree, then $r + s + t - u \sim p + q$ on C.

this situation, Theorem 4.2 applies with $m = 1$: there is a line L containing $p+q$ defined by a linear form G. If $r \in C$ be the remaining point of intersection of L with C we can choose a linear form H vanishing on $o + r$, and the line it defines meets C in one additional point s. This is the classical construction of the group law on the points of C (or, for curves over a field that is not algebraically closed, on the rational points of C). See Figure 4.2.

Proof of Theorem 4.2. First, suppose that we can find a form H of degree m as in the theorem. Setting $D' = (H) - (D_- + A)$ we have

$$D' = D + (H/G) = D_+ - D_- - (D_+ + A) + (D_- + A + D')$$

so D' is linearly equivalent to D.

We claim that we find in this way all effective divisors $D' \sim D$. To see this, suppose D' is any effective divisor with $D' \sim D$, so that

$$\mathcal{O}_C(A + D_- + D') = \mathcal{O}_C(A + D_- + D) = \mathcal{O}_C(m),$$

that is, $A + D_- + D' \equiv (G)$ for some form G of degree m. By Proposition 4.3 this implies that $A + D_+ = (H)$ for some form of degree m, as required. $\qquad\square$

The argument given in Proposition 4.2 can be stated more generally thus: if curves $F(X, Y, Z) = 0$ and $Q(X, Y, Z) = 0$ meet only in a finite set of points Γ in $\mathbb{P}^2$, and $E(X, Y, Z) = 0$ is a curve containing the intersection in an appropriate sense, then $E = QH + LF$ for some forms H and L, and this statement applies to arbitrarily singular curves. Recognizing its importance for the argument above and the generalizations to come, Max Noether in [Noether 1873] dubbed it the *fundamental theorem*, noting that it had often been used by geometers but not proven. After successive attempts and criticisms involving many mathematicians, he and Brill gave a complete proof in [Brill and Nöther 1874]. For more of this story see the account in [Eisenbud and Gray 2023].

4.2.3. The Cayley–Bacharach–Macaulay theorem.

The following result was proven by Macaulay [Macaulay 1900, p. 424], as a version of the Riemann–Roch theorem. It is now widely referred to as the Cayley–Bacharach theorem, named for an incorrect version asserted by Cayley and a correct special case later proved by Bacharach [1886]; see [Eisenbud and Gray 2023, Section 2.3] for more on this history, and [Eisenbud et al. 1996] (where the result is incorrectly attributed to Bacharach) for generalizations and related conjectures. Here is the version for divisors on a smooth plane curve:

Theorem 4.5 (Cayley–Bacharach–Macaulay). *Let C be a smooth plane curve of degree d, and suppose that E', E'' are effective divisors on C such that $E := E' + E'' = C \cap C'$, the complete intersection of C with a curve C' of degree d'. For any integer $0 \leq k \leq d + d' - 3$, the difference in the number of conditions imposed on forms of degree k by E'' and by E is equal to the degree of E' minus the number of conditions E' imposes on forms of degree $d'' := d + d' - 3 - k$ — that is, the failure of E' to impose independent conditions on forms of degree d''.*

Writing H for the divisor class on C of the intersection of C with a line, and setting $s := d + d' - 3 - k$, this is the equality

$$h^0(kH - E) - h^0(kH - E') = \deg E'' - \left(h^0(sH) - h^0(sH - E'') \right).$$

Proof. Set $e' := \deg E'$, $e'' := \deg E''$ and $e = e' + e'' = \deg E$. By the adjunction formula, the divisor class of the canonical sheaf on C is $K = (d - 3)H$. Using the Riemann–Roch theorem, the left side of the equality is

$$kd - e - h^0(K - (kH - E)) - \left(kd - e' - h^0(K - (kH - E')) \right).$$

Since $K - (kH - E) = K - (kH - d'H) = sH$ and $K - (kH - E') = sH + E''$, we see that the left side is equal to $e'' - h^0(sH) + h^0(sH + E'')$ as required. $\square$

As noted in [Eisenbud and Gray 2023], the converse, proving the Riemann–Roch theorem for C from Theorem 4.5, is also easy.

Corollary 4.6. *A divisor E' on a smooth plane curve $C \subset \mathbb{P}^2$ of degree d moves in a linear series of dimension r if and only if E' fails by r to impose independent conditions on curves of degree $d - 3$.*

Proof. For sufficiently large d' we can choose a curve C' of degree d' containing E and meeting C in $E = E' + E''$, where E'' is disjoint from E'. Since $C \cap C'$ is a complete intersection, every form vanishing on $E + E'$ is a linear combination of the forms defining C and C', and thus the dimension of the space of forms of degree d' vanishing on E modulo those vanishing on C is 1. By Theorem 4.2 the dimension r of the linear series $|E'|$ is the dimension of the space of forms of degree d' modulo those vanishing on $E + E'$, and by Theorem 4.5 this is the failure of E' to impose independent conditions on forms of degree $d'' = d + d' - d' - 3 = d - 3$. $\square$

To apply Theorem 4.5 it is helpful to know when points impose independent conditions on forms of a certain degree. Here is a first result of this kind:

Proposition 4.7. *Suppose that $1 \leq n$. Any set Γ of $k \leq n$ distinct points in $\mathbb{P}^2$ imposes independent conditions on forms of degree $n-1$; and if Γ is not contained in a line then Γ imposes independent conditions on forms of degree $n - 2$.*

Proof. To show that Γ imposes independent conditions on forms of degree d we must produce, for each $p \in \Gamma$, a form of degree d vanishing on

$$\Gamma_p := \Gamma \setminus \{p\}$$

but not p. If Γ imposes independent conditions on forms of degree d then Γ automatically imposes independent conditions on forms of degree $d + 1$, so we may assume that $k = n$.

The product of linear forms vanishing on general lines through the points of Γ_p does not vanish at p, proving the first statement.

Now assume that Γ is not contained in a line and $p \in \Gamma$. It suffices to show that for each $p \in \Gamma$ there is a form of degree $n - 2$ vanishing on Γ_p but not on p. Since Γ spans $\mathbb{P}^2$, there is a spanning set of three points p, q, r of Γ. The union of the line spanned by q, r with general lines through the $n - 3$ points of $\Gamma_p \setminus \{q, r\}$ is defined by a form of degree $n - 2$ containing Γ_p but not p, as required. $\quad\square$

Corollary 4.8. *Suppose that $C \subset \mathbb{P}^2$ is a smooth plane curve of degree $d \geq 3$.*

(1) *If $\mathcal{V}$ is a g_e^1 on C with $e \leq d-1$ then $e = d-1$ and $\mathcal{V}$ corresponds to projection from a point of C.*

(2) *If $\mathcal{V}$ is a g_e^2 on C with $e \leq d \geq 4$ then $e = d$. Furthermore, the embedding of C in $\mathbb{P}^2$ is unique up to automorphisms of $\mathbb{P}^2$.*

Proof. (1) If E is a divisor in the linear series $\mathcal{V}$ then by Corollary 4.6 the points of E fail to impose independent conditions on forms of degree $d - 3$. By Proposition 4.7 the degree of E is $d - 1$ and the points lie in a line L, which must meet C in an additional point p. By Theorem 4.2 the linear series $|E|$ is residual to p in curves of degree 1; that is, it is the linear series corresponding to projection from p.

(2) In this case a divisor in the series $\mathcal{V}$ fails by 2 to impose independent conditions on forms of degree $d - 3$, and thus fails (by at least 1) to impose independent conditions on forms of degree $d-2$. By Proposition 4.7 the degree of E is d. Moreover, the points lie in a line L; thus $\mathcal{V}$ is a subseries of the series by which C is embedded. $\quad\square$

4.3. Curves of genus 1 and the group law of an elliptic curve

We will describe the maps of a curve of genus 1 given by the complete linear series in the lowest degree cases of interest: $d = 2, 3, 4$ and 5. Along the way we will see several ways of parametrizing the family of curves of genus 1 by one-dimensional varieties, forerunners of the moduli spaces that we will introduce in Chapters 7 and 8.

On a smooth, irreducible curve E of genus 1 the canonical sheaf has degree 0; and since it has a global section, it must be $\mathcal{O}_C$. Since invertible sheaves of negative degree cannot have nonzero sections, the Riemann–Roch theorem shows that $h^0(\mathcal{L}) = \deg \mathcal{L}$ for any $\mathcal{L}$ of positive degree. Among the surprising consequences is that, given a point $o \in E$, there is a natural structure of abelian algebraic group on the points of E for which o is the zero element. A curve of genus 1 with a chosen point o is called an *elliptic curve*.

Proposition 4.9. *Let E be a curve of genus 1 and let o be a chosen point $o \in E$. If we set $p + q = r$, where r is the unique effective divisor linearly equivalent to $p+q-o$, then E becomes a commutative algebraic group. Moreover, the group of divisor classes is divisible, in the sense that for any divisor D of degree $n > 0$ there is a point p such that $D \sim np$.*

Proof. The group operation is easy to describe: Let E be an elliptic curve with the point $o \in E$ chosen arbitrarily. If p, q are points of E then $\mathcal{O}_E(p + q - o)$ has degree 1, and thus has a unique global section. This vanishes at a unique point r, which may also be described as the unique effective divisor linearly equivalent to $p + q - o$, and thus $p + q = r$ in the group operation, which is thus obviously commutative. For the inverse, if r is the unique point linearly equivalent to $2o - p$ then $p + r - o \sim o$, so that $r = -p$. For any divisor D we define ΣD to be the unique point linearly equivalent to $\Sigma D - (\deg D - 1)o$. In this group operation any two linearly equivalent divisors have the same sum $\Sigma D = \Sigma D'$.

We can determine the point $r \sim p + q - o$ and show that the group law is given by regular functions as follows. By Corollary 2.19 we may embed E in $\mathbb{P}^2$ by any linear series of degree 3. Let L be the line through $p + q$, and suppose that L meets C in an additional point s. Since s is unique, its coordinates are functions polynomial functions (on a suitable affine chart) of the coordinates of p and q. The line through s and o meets C additionally in a point r, and by Theorem 4.2 we have $r \sim p + q - o$, as in Figure 4.2. and similarly the point r is a polynomial function of the coordinates of s and o.

To see that E is a divisible group, consider the maps $n : E \to E$ given by multiplication by integers n. Since the group law is given by polynomial functions, the number of solutions p of the equation $np = q$ is finite, and thus for $n \neq 0$ this map is nonconstant. This implies that it is surjective. $\square$

It is convenient for some purposes to suppose that the linear series used to embed C in $\mathbb{P}^2$ is $|3o|$; this has the advantage that $p, q, r \in C$ are collinear if and only if $p + q + r = o$ in the group law.

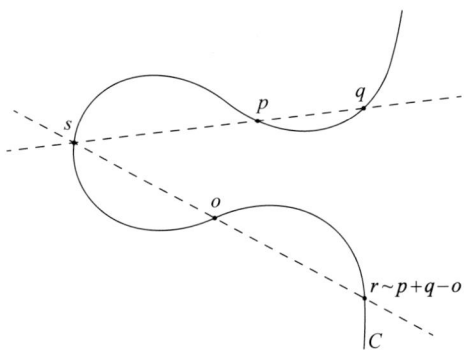

Figure 4.2. Adding points p, q on a plane cubic with origin o.

Remark 4.10. From the definition it is obvious that the map $E \to \mathrm{Pic}_0(E)$ sending p to $\mathcal{O}_E(p - o)$ is an isomorphism of groups, and adding multiples of o induces an isomorphism with each $\mathrm{Pic}_d(E)$ as well. This provides a natural sense in which the family of invertible sheaves on C can be treated as a smooth curve. In Chapter 5 we will see a general construction: the Picard group $\mathrm{Pic}_0(C)$ can be made into a variety, and for a curve C of genus g the effective divisors of degree g form a variety that surjects birationally to $\mathrm{Pic}_g(C)$.

Corollary 4.11. *Given two invertible sheaves $\mathcal{L}, \mathcal{L}'$ of the same degree on a curve E genus 1, there is an automorphism $\sigma : E \to E$ such that $\sigma^* \mathcal{L} = \mathcal{L}'$.*

Proof. By Proposition 4.9 we may write $\mathcal{L} \cong \mathcal{O}_E(np)$ and $\mathcal{L}' \cong \mathcal{O}_E(np')$ for some points p, p'; and translation by $p - p'$ is an automorphism of E carrying one into the other. $\qquad\square$

4.4. Low degree divisors on curves of genus 1

In this section we will describe the complete linear series of degrees 2, 3, 4, and 5 on a curve of genus 1, and say something about the geometry of each case. Several of these descriptions lead to natural guesses answering the apparently silly question: How many different curves of genus 1 are there?

The dimension of families. To make sense of this question, we observe a fundamental fact of algebraic curve theory that the set of isomorphism classes of smooth, projective curves of a given genus g is naturally parametrized by the points of an irreducible quasiprojective variety M_g, called the *moduli space* of curves of genus g. We will have a great deal more to say about moduli spaces in general, and M_g in particular, in Chapters 7 and 8.

We will use the low-degree embeddings to describe the family of isomorphism classes of curves of genus 1 in several ways, giving what seem to be natural estimates of its dimension. We leave to Chapter 8 the reasoning that leads to a map from the base of such a family to M_1, validating the estimates we give here.

Double covers of $\mathbb{P}^1$. Let E be a smooth projective curve of genus 1 and let $\mathcal{L}$ be an invertible sheaf of degree 2 on E. By the Riemann–Roch theorem, $h^0(\mathcal{L}) = 2$ and the linear series $|\mathcal{L}|$ is basepoint free, so we get a map ϕ : $E \to \mathbb{P}^1$ of degree 2. By the Riemann–Hurwitz theorem the map ϕ will have 4 branch points, which must be distinct because in a degree 2 map only simple branching is possible. By Corollary 4.11, this set of four points are determined, up to automorphisms of $\mathbb{P}^1$, by the curve E, and are independent of the choice of $\mathcal{L}$.

We will see in Chapter 6 that a double cover of $\mathbb{P}^1$ is determined by its branch points, so the family $\mathcal{H}$ of double covers of $\mathbb{P}^1$ having genus 1 is four-dimensional. There is a map $\mathcal{H} \to M_1$, sending such a double cover to the isomorphism class of the cover, and by what we have said, the fibers of this map PGL_2 are isomorphic to the automorphism group PGL_2 of $\mathbb{P}^1$; thus we expect that M_1 has dimension $4 - 3 = 1$.

Plane cubics. We can also represent an arbitrary smooth curve of genus 1 as a plane cubic: Let $\mathcal{L}$ be an invertible sheaf of degree 3 on E. As in the proof of Proposition 4.9 the linear series $|\mathcal{L}|$ gives an embedding of E as a smooth plane cubic curve of degree 3; conversely, the adjunction formula implies that any smooth plane cubic curve has genus 1.

The space of plane cubic curves is parametrized by the space of cubic forms in 3 variables up to scalars, a $\mathbb{P}^9$. The locus of forms defining smooth curves is a Zariski open subset. If two plane cubics are abstractly isomorphic, that is, if we have two different degree 3 linear series $|\mathcal{L}|, |\mathcal{L}'|$ mapping a given genus 1 curve E to the plane, then by Proposition 4.11 we may precompose one of the maps with an automorphism of E and suppose that $\mathcal{L} = \mathcal{L}'$. Thus the two curves differ by an element of PGL_3 of automorphisms of $\mathbb{P}^2$. Since the group PGL_3 has dimension 8, one should expect that the family of such curves up to isomorphism has dimension 1, which accords with the dimension computed in the previous section.

Corollary 4.8 does not determine the degree 3 maps of E to $\mathbb{P}^2$, but Theorem 4.2 can be applied directly, given one embedding $E \subset \mathbb{P}^2$. If D is a divisor of degree 3 on $E \subset \mathbb{P}^2$, we can find a conic containing D. The conic will meet E in D plus another divisor D' of degree 3, and the complete linear series D is then cut out by conics containing D' — that is, the divisors equivalent to D are those residual to D' in the intersection of E with conics containing D'. Since D fails by 2 to impose independent conditions on constants, the linear series has dimension 2 as expected. In fact, since any n points of E fail by $n - 1$ to impose independent conditions on constants, this shows again that $\dim |D| = \deg D - 1$, and we can find the divisors in D using Theorem 4.2

4.5. Genus 1 quartics in $\mathbb{P}^3$

By Corollary 2.19, any divisor of degree 4 on E embeds E in $\mathbb{P}^3$. Representing E as a plane cubic, we see that a conic containing a divisor D of degree 4 on E meets E in $D + D'$, where D' has degree 2. Following the prescription of Theorem 4.2, we see that the linear series $|D|$ may be represented as the residual to D' in the series of conics containing D'. We can also regard this space of conics as a linear series on $\mathbb{P}^2$, with base locus D'. As such it maps $\mathbb{P}^2$ rationally to $\mathbb{P}^3$. The pullbacks of planes in $\mathbb{P}^3$ are the conics through D', and the fact that a general pair of them meet, away from D', in 2 points, means that the intersection of the image surface with a line in $\mathbb{P}^3$ consists of two points — the surface is a quadric. Assuming for simplicity that $D' = p + q$, the sum of two distinct points, the linear series is well-defined on the blowup of the plane at p and q. However, a conic meeting the line L spanned by p and q in any additional point r must contain L, and thus L is contracted by the linear series: the smooth quadric in $\mathbb{P}^3$ can be described in this way.

The quadric we have constructed is not distinguished in the ideal of E:

Proposition 4.12. *The image of a genus 1 curve E by a linear series of degree 4 is the complete intersection of two quadrics in $\mathbb{P}^3$, and conversely any smooth complete intersection of two quadric in $\mathbb{P}^3$ is a curve of genus 1.*

Proof. Consider the restriction map

$$\rho_2 : H^0(\mathcal{O}_{\mathbb{P}^3}(2)) \to H^0(\mathcal{O}_E(2)) = H^0(\mathcal{L}^2).$$

The space on the left — the space of homogeneous polynomials of degree 2 in four variables — has dimension 10, while by the Riemann–Roch theorem the space $H^0(\mathcal{L}^2)$ has dimension 8. It follows that E lies on at least two linearly independent quadrics Q and Q'. Since E does not lie in any plane, neither Q nor Q' can be reducible, so they have no component in common. Thus by Bézout's theorem $Q \cap Q'$ has degree 4, so

$$E = Q \cap Q'.$$

Moreover, Q, Q' form a regular sequence, so the ideal (Q, Q') is unmixed, and thus the homogeneous ideal $I(E)$ is generated by these two quadrics.

Conversely, if $E := Q \cap Q'$ is a smooth complete intersection of two quadrics then every quadric in the pencil of quadrics spanned by Q and Q' is nonsingular along the base locus E, and thus by Bertini's theorem the general member Q_0 of this pencil is nonsingular. Since E is the intersection of Q_0 with another quadric, we see that E has class $(2, 2)$ on Q_0, and thus has genus 1 by the adjunction formula. (In fact the same argument works for complete intersection of any two quadrics, showing that it has arithmetic genus 1, as we will see in Chapter 16. □

From Proposition 4.12 we see that an elliptic quartic in $\mathbb{P}^3$ determines a point in the Grassmannian $G(2, H^0(\mathcal{O}_{\mathbb{P}^3}(2))) = G(2, 10)$ of pencils of quadrics; and by Bertini's theorem, a Zariski open subset of that Grassmannian corresponds to smooth quartic curves of genus 1. The Grassmannian $G(2, 10)$ has dimension 16, while the group PGL_4 of automorphisms of $\mathbb{P}^3$ has dimension 15, so this suggests again that the family of curves of genus 1 up to isomorphism has dimension 1.

There is a direct way to go back and forth between the representation of the smooth genus 1 curve E as the intersection of two quadrics in $\mathbb{P}^3$ and the representation of E as a double cover of $\mathbb{P}^1$ branched at 4 distinct points. First, by Bertini's theorem, we may take the two quadrics to be nonsingular, since they must meet transversely along E, and elsewhere the pencil of quadrics they span has no basepoints. Representing the quadrics as symmetric matrices A, B, the pencil of all quadrics containing E can be written as $sA + tB$. A quadric in the pencil is singular at the points (s, t) such that the quartic polynomial $det(sA + tB)$ vanishes; thus at 4 points.

A smooth quadric has two rulings by lines; a cone has one. Thus the family

$$\Phi := \{(\lambda, \mathcal{L}) \mid \mathcal{L} \in \mathrm{Pic}(Q_\lambda) \text{ is the class of a ruling of } Q_\lambda\}$$

is — at least set-theoretically — a 2-sheeted cover of $\mathbb{P}^1$, branched over the four values of λ corresponding to singular quadrics in the pencil. In fact, we claim:

Proposition 4.13. *There is an isomorphism of Φ with E, and thus the branch points of Φ over $\mathbb{P}^1$ — that is, the singular elements of the pencil of quadrics — are the same, up to automorphisms of $\mathbb{P}^1$, as the four points over which a double cover of $\mathbb{P}^1$ by E are ramified.*

Proof. First, choose a basepoint $o \in E$. We will construct inverse maps $E \to \Phi$ and $\Phi \to E$ as follows:

(1) Suppose $q \in E$ is any point other than o, and let $M \subset \mathbb{P}^3$ be the line $\overline{oq}$. Every quadric Q_λ contains the two points $o, q \in M$. It is one linear condition for a quadric Q_λ to contain a given point so if $r \in M$ is any third point, there will be a unique λ such that r, and hence all of M, lies in Q_λ. Thus the choice of q determines both one of the quadrics Q_λ of the pencil, and a ruling of that quadric, giving us a map $E \to \Phi$.

(2) Given a quadric Q_λ and a choice of ruling of Q_λ, there is a unique line $M \subset Q_\lambda$ of that ruling passing through o, and that line M will meet the curve E in one other point q; this gives us the inverse map $\Phi \to E$. □

Cheerful Fact 4.14. There is a beautiful extension of this result to pencils of quadrics in any odd-dimensional projective space. Briefly: a smooth quadric $Q \subset \mathbb{P}^{2g+1}$ has two rulings by g-planes, which merge into one family when the quadric specializes to quadric of rank $2g + 1$, that is, a cone over a smooth quadric in $\mathbb{P}^{2g}$. If $\{Q_\lambda\}_{\lambda\in\mathbb{P}^1}$ is a pencil with smooth base locus $X = \bigcap_{\lambda\in\mathbb{P}^1} Q_\lambda$, then exactly $2g + 2$ of the quadrics will be singular, and they will all be of rank $2g + 1$. The space Φ of rulings of the quadrics Q_λ is thus a double cover of $\mathbb{P}^1$ branched at $2g + 2$ points. We shall see in Chapter 6 that this double cover is a hyperelliptic curve of genus g. For a proof see for example [Harris 1992, Proposition 22.34]. This shows in particular that the polynomial $\det(sA + tB)$ has $2g + 2$ *distinct* roots.

There is a remarkable analogue of Proposition 4.13 for any g: the variety $F_{g-1}(X)$ of $(g-1)$-planes in the base locus X of the pencil is isomorphic to the Jacobian of the curve Φ. (We will discuss Jacobians in Chapter 5.) A proof of this in case $g = 2$ can be found in [Griffiths and Harris 1978]; for all g it is done in [Donagi 1980]. For a further study of the equivalence, see [Eisenbud and Schreyer 2022].

4.6. Genus 1 quintics in $\mathbb{P}^4$

Now suppose that D is a divisor of degree 5 on the smooth cubic $E \subset \mathbb{P}^2$, and consider the embedding $\phi_D : E \hookrightarrow \mathbb{P}^4$, the embedding by the complete linear series $|D|$. We first observe that since D is contained in a cubic, at most three points of D can lie on a line. Thus D imposes independent conditions on conics: that is, there is a unique conic C containing D, and thus we may write the divisor $C \cap E$ as $D + p$ for some point $p \in E$. By Theorem 4.2 the linear series $|D|$ consists of the divisors residual to p in the intersection of E with conics through p.

This description again determines an interesting surface containing $\phi_D(E)$: the image X of $\mathbb{P}^2$ by the rational map defined by the conics containing p. This map is well-defined on the blowup of $\mathbb{P}^2$ at p, and it follows that the degree of X is the number of intersections of two conics away from p, that is, $\deg X = 3$. We can also see from this that X is ruled by lines, since the images of lines in $\mathbb{P}^2$ through p become disjoint in the blowup. The image of a line through p meets $\phi_D(E)$ in the images of the two points of intersection $E \cap L$ away from p; that is, the lines are the spans of the points in the g_2^1 on E given by projection from p. We shall study this construction in a much more general setting in Chapter 17.

Another interesting surface containing $\phi_D(E)$ can be constructed in a similar way. Since D imposes independent conditions on conics, it also imposes independent conditions on cubics in $\mathbb{P}^2$. If we choose a cubic E' containing D and distinct from E, then we may write $E \cap E' = D + D'$ where $D' \subset E$ is a divisor of degree $9 - 5 = 4$. By Theorem 4.2 the series $|D|$ is also the linear series of intersections of E with cubics containing D'.

The linear series of cubics in $\mathbb{P}^2$ containing D', without restricting it to E, is a series of dimension 6, and maps $\mathbb{P}^2$ rationally to a surface $Y \subset \mathbb{P}^5$, called a *del Pezzo surface of degree 5* (see Section 12.3.2 for more on del Pezzo surfaces). Since E itself is one of the cubics containing D', the embedded curve $\phi_D(E)$ is a hyperplane section of Y, which must have degree 5; the quadrics containing Y in $\mathbb{P}^5$ pull back to sextic curves in $\mathbb{P}^2$ vanishing doubly on D'. Vanishing doubly at a point imposes 3 linear conditions on a form (one to vanish, and two more for the derivatives to vanish), so the linear series of quadrics in $\mathbb{P}^5$, restricted to Y, has dimension at least $\binom{6+2}{2} - 3*4 = 16$ and one can show (Exercise 4.2) that in this case the conditions are independent. Thus the family of quadrics containing Y, modulo the ideal of Y, has vector space dimension 16, and we see that Y lies on $5 = \binom{5+2}{2} - 16$ quadrics.

On the other hand, we can determine the space of quadrics containing the quintic curve $\phi_D(E)$ as the kernel of the map

$$H^0(\mathcal{O}_{\mathbb{P}^4}(2)) \to H^0(\mathcal{O}_{\phi_D(E)}(2)) = H^0(\mathcal{O}_E(5)).$$

The left-hand term has dimension 15, and the right-hand term dimension 5. As we shall see in Corollary 10.7, the curve $\phi_D(E)$ is arithmetically Cohen–Macaulay, so the map above is surjective and the kernel has dimension exactly 5; it is thus the restriction to the hyperplane section of the family of quadrics containing the del Pezzo surface Y.

Much more can be said about the surface Y and the equations of $\phi_D(E)$, which we will only summarize:

Cheerful Fact 4.15 (quintic curves of genus 1 in $\mathbb{P}^4$). Recall that if A is a skew-symmetric matrix of even size, then the determinant of A is the square of a polynomial in the entries of A called the Pfaffian of A. For example, if

$$M = \begin{pmatrix} 0 & x_{1,1} & x_{1,2} & x_{1,3} \\ -x_{1,1} & 0 & x_{2,2} & x_{2,3} \\ -x_{1,2} & -x_{2,2} & 0 & x_{3,3} \\ -x_{1,3} & -x_{2,3} & -x_{3,3} & 0 \end{pmatrix}$$

then the Pfaffian of M is by definition the quadric $x_{1,1}x_{3,3} - x_{1,2}x_{2,3} + x_{1,3}x_{2,2}$.

Let $\tilde{A}$ be a 5×5 generic alternating matrix (that is, a skew-symmetric matrix with 0 on the diagonal and independent variables in the 10 entries above the diagonal), and let I be the ideal generated by the 4×4 Pfaffians of the submatrices of A leaving out one row and the corresponding column.

As a consequence of the main theorem of [Buchsbaum and Eisenbud 1977a] (see also [Eisenbud 1995, Theorem 11]) we have:

Proposition 4.16. *The scheme defined by the ideal I introduced just above is a smooth irreducible Cohen–Macaulay variety $\tilde{Y}$ of codimension 3 in $\mathbb{P}^9$, and both the quintic del Pezzo surface Y and the quintic curve $\phi_D(E)$ are plane sections* *of $\tilde{Y}$. With a suitable choice of bases and variables, the presentation matrix of $I(\phi_D(E))$ is a 5×5 matrix of linear forms*

$$B = \begin{pmatrix} 0 & 0 & x_0 & x_1 & x_2 \\ 0 & 0 & x_1 & x_2 & x_3 \\ -x_0 & -x_1 & 0 & \ell_1 & \ell_2 \\ -x_1 & -x_2 & -\ell_1 & 0 & \ell_3 \\ -x_2 & -x_3 & -\ell_2 & -\ell_3 & 0 \end{pmatrix}$$

and the homogeneous ideal of $\phi_D(E)$ is generated by the 4×4 Pfaffians of this matrix. The ideal of the scroll X is defined by the 2×2 minors of the 2×3 matrix in the upper right corner of A, which are equal to the Pfaffians of the matrices leaving out the third, fourth and fifth rows and corresponding columns of B.

4.7. Exercises

Exercise 4.1. Let C be a smooth plane curve of degree d. Show that C admits a one-parameter family of maps $C \to \mathbb{P}^1$ of degree $d - 1$. Using the Riemann–Roch theorem and Proposition 4.7, show that C does not admit a map $C \to \mathbb{P}^1$ of degree $d - 2$ or less. ◆

Exercise 4.2. Show that the vector space of sextic curves in $\mathbb{P}^2$ vanishing doubly at 4 points, no three of which are collinear, has dimension exactly 16.

◆

Exercise 4.3. Let $p \in C$ be a smooth point of a plane curve with equation $F(x_0, x_1, x_2) = 0$ of degree $d > 1$. Show that the tangent line to C at p meets C in a scheme of order $m + 2$ at p if and only if the determinant of the Hessian matrix vanishes along C to order m as is done in [Kunz 2005, pp. 84–85]:

(1) Assume that $p = (0,0) \in \mathbb{A}^2 \subset \mathbb{P}^2$, where $\mathbb{A}^2$ is the locus $x_0 \neq 0$, and suppose that $f(x, y) = 0$ is the affine equation of C, with $x = x_1/x_0$, $y = x_2/x_0$. Reduce to the affine case by showing that

$$\det \text{Hess}(C) = x_0^2 \det \begin{pmatrix} d(d-1)F & (d-1)\partial F/\partial x_1 & (d-1)\partial F/\partial x_2 \\ (d-1)\partial F/\partial x_1 & \partial^2 F/\partial x_1 \partial x_1 & \partial^2 F/\partial x_1 \partial x_2 \\ (d-1)\partial F/\partial x_2 & \partial^2 F/\partial x_2 \partial x_1 & \partial^2 F/\partial x_2 \partial x_2 \end{pmatrix}.$$

Writing the partial derivatives as subscripts, this becomes

$$\det \begin{pmatrix} d(d-1)f & (d-1)f_x & (d-1)f_y \\ (d-1)f_x & f_{xx} & f_{xy} \\ (d-1)f_y & f_{xy} & f_{yy} \end{pmatrix}$$

when restricted to $\mathbb{A}^2$.

(2) Assume that the tangent line to C at p is $y = 0$. Show that f can be written as

$$f = x^{m+2}\phi(x) + y\psi(x, y)$$

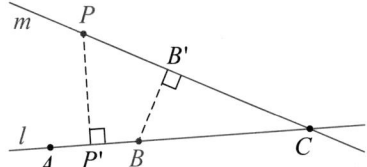

Figure 4.3. Construction for the Sylvester-Gallai theorem.

where $\phi(0) \neq 0$ and $\psi(0,0) \neq 0$, and thus, modulo f, the Hessian determinant, up to a constant factor, has the form

$$f_x^2 f_{yy} + f_y^2 f_{xx} - 2f_x f_y f_{xy}.$$

Note that x is a local parameter at p. Using the form of f above, show that there is a unique term vanishing to order m at p, and no term vanishing to lower order there.

Exercise 4.4. As we have seen, each line through two flexes of a smooth cubic curve in $\mathbb{P}_{\mathbb{C}}^2$ passes through a third flex; but such a configuration is not possible in $\mathbb{P}_{\mathbb{R}}^2$. Prove:

Theorem 4.17 (Sylvester–Gallai theorem). *In any finite set $\Gamma \subset \mathbb{P}_{\mathbb{R}}^2$ there are three noncollinear points unless Γ is contained in a line.*

Hint (following Leroy Milton Kelly): We may assume that $\Gamma \subset \mathbb{R}^2$. Choose a pair P, L consisting of a point $P \in \Gamma$ and line ℓ containing at least 2 of the points of Γ, but not P, such that the distance from L to p is minimal among such pairs. If there were at least 3 points of Γ on L then, considering Figure 4.3, show that there is a point of Γ and a line L' violating the minimal distance hypothesis.

Chapter 5

Jacobians

We have seen that the points of an elliptic curve naturally form an algebraic group. This is not true for curves of higher genus, but there is a substitute: the Jacobian. An essential construction in studying a curve C is the association of an invertible sheaf to a divisor — in other words, the map

$$\{\text{effective divisors of degree } d \text{ on } C\} \xrightarrow{\mu} \{\text{invertible sheaves of degree } d \text{ on } C\}$$

sending D to $\mathcal{O}_C(D)$.

A priori, this is a map of sets. But it is a fundamental fact that both source and target may be given the structure of an algebraic variety in a natural way, and that in terms of this structure the map between them is a morphism. In many ways the geometry of the map governs the geometry of the curve. In this chapter we will describe the source and target of μ, and give references to proofs of their properties.

We start with the effective divisors. Since C is smooth, an effective divisor of degree d on C is the same thing as a subscheme $D \subset C$ of dimension 0 and degree d, and thus the family C_d of effective divisors of degree d on C is a Hilbert scheme; see Section 7.3. This Hilbert scheme may be identified with the d-th *symmetric power C_d* of C, described in Section 5.1.

The parametrization of the set of invertible sheaves on C of a given degree d by the variety $\text{Pic}_d(C)$, called the *Picard variety* of C, requires different techniques. We will define it by a universal property in the category of schemes, and exhibit its construction as an analytic variety, actually a complex torus $\text{Jac}(C)$, whose group structure reflects the tensor product of sheaves in $\text{Pic}_0(C)$. Historically, the algebraic construction was a major milestone, first reached in the work of André Weil in the middle of the twentieth century, and then reshaped by Grothendieck and his school. The interested reader will find a beautiful, detailed account both of the history and the modern theory of the scheme of divisors and the Picard scheme in [Kleiman 2005], which has extensive references to the original literature.

A consequence of the modern theory is that we can extend the construction of the Picard variety to singular curves as well as smooth ones. Also, given a family $\pi : \mathcal{C} \to B$ of curves there is a corresponding family of Picard varieties $\mathrm{Pic}_d(\mathcal{C}/B) \to B$, called the *relative Picard variety*, whose fiber over a closed point $b \in B$ is the Picard scheme $\mathrm{Pic}_d(C_b)$ of the corresponding curve C_b in our family. In Chapter 14 we'll have occasion to describe the Picard varieties of g-nodal and g-cuspidal curves, and to see how these fit into families with Picard varieties of smooth curves.

As an application of the existence of the spaces C_d and Pic_d, we show in Section 5.5 that a general divisor of degree $g + 3$ on any curve of genus g gives rise to an embedding in $\mathbb{P}^3$ as a curve of degree $g + 3$, and there are related results for general divisors of degree $g + 2$ and $g + 1$.

5.1. Symmetric products and the universal divisor

Let C be a smooth curve. The space of effective divisors on C can be characterized by a universal property:

Definition 5.1. Let B be any scheme. A *family* of effective divisors of degree d on C, parametrized by the scheme B, is a Cartier divisor $X \subset B \times C$ whose intersections with fibers $\{b\} \times C \cong C$ over points of B are divisors of degree d on C.

Given this, we have a contravariant functor

$$F : (\text{schemes}) \to (\text{sets}),$$

taking a scheme B to the set of families of divisors of degree d on C over B; if $\pi : B' \to B$ is any morphism, the induced map $F(B) \to F(B')$ is defined by taking a family $\mathcal{D} \subset B \times C$ to the fiber product $\mathcal{D}' := B' \times_B \mathcal{D} \subset B' \times C$. We say that a scheme C_d is a fine moduli space for divisors of degree d on C if there is an isomorphism of functors

$$F \cong \mathrm{Hom}_{\text{Schemes}}(-, C_d).$$

By Yoneda's lemma (Exercise 5.4), this is equivalent to the existence of a *universal family* $\mathcal{D} \subset C_d \times C$, with the property that for any family $X \subset B \times C$ of divisors on C over any scheme B, there is a unique map $\phi : B \to C_d$ such that $X = (\phi \times id_C)^{-1}(\mathcal{D})$.

From the universal property it is clear that a fine moduli space for divisors of degree d on C is unique if it exists. Indeed, it does exist, and we'll sketch the construction, using symmetric products. This construction relies on the existence of quotients of schemes by finite groups, and we pause to discuss such quotients.

Finite group quotients. If G is a finite group acting by automorphisms on an affine scheme $X := \operatorname{Spec} A$ then X/G is by definition $\operatorname{Spec}(A^G)$, the spectrum of the ring A^G of invariant elements of A. Since the functions in A^G are constant on orbits, every fiber is a union of orbits, but something much better is true:

Theorem 5.2. *The map $\pi : X \to X/G$ induced by the inclusion of rings is finite. If X is a normal variety, then each fiber of π is a single orbit of G.*

For a proof, see for example [Eisenbud 1995, Proposition 13.10].

Since the map $X \to X/G$ is finite, we have $\dim X/G = \dim X$.

The construction commutes with the passage to G-invariant open affine sets, and thus passes to quasiprojective schemes as well (Exercise 5.1), though not to arbitrary schemes (see [Olsson 2016, Example 5.3.2]).

When the group G is positive-dimensional, the situation becomes much more complex, and is the subject of *geometric invariant theory* — see Chapter 8 for some idea of what can and cannot be done in a special case.

If X is any quasiprojective scheme we define the *d-th symmetric power $X^{(d)}$ of X* to be the quotient of the Cartesian product X^d of d copies of X by the action of the symmetric group of permutations of the factors, and let $\pi_d : X^d \to X^{(d)}$ be the projection. There is a natural subvariety

$$\mathcal{D} = \{(x, y) \in X^{(d)} \times X \mid y \text{ is in the projection of } \pi_d^{-1}(x) \text{ to the first factor}\};$$

in other words, if we think of a point $x \in X^{(d)}$ as an unordered d-tuple of points on X, the fiber of $\mathcal{D}$ over x is the collection of points in x.

If $X = \mathbb{A}^1$ then $X^d = \mathbb{A}^d = \operatorname{Spec} \mathbb{C}[x_1, \dots, x_d]$. The ring of invariants of the symmetric group acting on $\mathbb{C}[x_1, \dots, x_d]$ by permuting the variables is generated by the d elementary symmetric functions, which generate a polynomial subring, and $X^{(d)}$ is isomorphic to $\mathbb{A}^d$. Since the symmetric functions of the roots of a polynomial in one variable are the coefficients of that polynomial, we may identify $X^{(d)}$ with the affine space of monic polynomials of degree d in $\mathbb{C}[z]$. (See [Eisenbud 1995, Exercises 1.6, 13.2, 13.3, 13.4].)

Next consider $X = \mathbb{P}^1$ and the product $(\mathbb{P}^1)^d$. Taking the homogeneous coordinates of the i-th copy of $\mathbb{P}^1$ to be (s_i, t_i), the $d+1$ multilinear symmetric functions of degree d,

$$t_1 t_2 \cdots t_d, \quad \sum_i s_i t_1 \cdots \hat{t}_i \cdots t_d, \quad \dots \quad s_1 \cdots s_d$$

are the coefficients of the polynomial

$$(s_1 a + t_1 b)(s_2 a + t_2 b) \cdots (s_d a + t_d b)$$

defining a general subscheme of $\mathbb{P}^1$ of degree d. These forms define an isomorphism $(\mathbb{P}^1)^{(d)} \to \mathbb{P}^d$. Again, we may think of this map as taking a d-tuple of points to the unique-up-to-scalars homogeneous form of degree d vanishing on it. Identifying $(\mathbb{P}^1)^{(d)}$ with the space of forms F of degree d, the fiber of the subscheme $\mathcal{D} \subset \mathbb{P}^1 \times \mathbb{P}^d \to \mathbb{P}^d$ over a form F is the subscheme defined by F.

We shall see that when C is a smooth curve of higher genus, the global geometry of $C^{(d)}$ is nontrivial, but at least the local geometry is simple:

Proposition 5.3. *If C is a smooth curve then each symmetric power $C^{(d)}$ is smooth, and the subscheme $\mathcal{D} \subset C \times C^{(d)}$ is a family of divisors.*

Proof. It suffices to show that the quotient of an invariant formal neighborhood of the preimage $\{p_1, \dots, p_s\}$ of a point $\bar{p} \in C^{(d)}$ in C^d is smooth. After completing the local rings, we get an action of the symmetric group G on the product of the completions of X at the p_i, and this depends only on the orbit structure of G acting on $\{p_1, \dots, p_s\}$. Thus it would be the same for some orbit of points on $\mathbb{A}^1$, so the general case follows from the smoothness of $(\mathbb{A}^1)^{(d)}$. $\square$

By contrast, if $\dim X \geq 2$, the symmetric powers $X^{(d)}$ are singular for all $d \geq 2$. See Exercise 5.3 for the case of $(\mathbb{A}^2)^{(2)}$ and Exercise 5.2 for a well-behaved case.

Cheerful Fact 5.4. It is clear from Theorem 5.2 that the points of $C^{(d)}$ are in one-to-one correspondence with the effective divisors of degree d on C, but much more is true:

If C is a smooth projective curve, then the d-th symmetric power $C^{(d)}$ of C is the fine moduli space C_d of effective divisors of degree d on C with universal family $\mathcal{D} \subset C \times C_d \to C_d$.

For a proof in the analytic category, see [Arbarello et al. 1985, p. 164]; for the algebraic fact, see [Kleiman 2005, Remark 9.3.9]. The result remains true for families of curves, that is, smooth curves in the category of schemes over a base scheme B.

Henceforward we will write C_d in place of $C^{(d)}$.

5.2. The Picard varieties

As with C_d, we may define $\mathrm{Pic}_d(C)$ by a universal property. We start by saying what we mean by a family of invertible sheaves on a smooth curve C:

Definition 5.5. For any scheme B, a *family of invertible sheaves on C over B* is an equivalence class of invertible sheaves $\mathcal{L}$ on $B \times C$, where two such families

$\mathcal{L}$ and $\mathcal{L}'$ are equivalent if they differ by an invertible sheaf pulled back along the projection $\pi_1 : B \times C \to B$, that is, if

$$\mathcal{L}' \cong \mathcal{L} \otimes \pi_1^* \mathcal{F}$$

for some invertible sheaf $\mathcal{F}$ on B. The family $\mathcal{L}$ is a *family of invertible sheaves of degree d* if, moreover, the restriction of $\mathcal{L}$ to $\{b\} \times C$ has degree d for each point $b \in B$.

If $p \in C$ and $\mathcal{F}$ is a sheaf on B then $\pi_1^*(\mathcal{F})\,|_{B \times p} = \mathcal{F}$, so we could have eliminated the equivalence relation at the expense of choosing a point by insisting that the restriction of $\mathcal{L}$ to $B \times \{p\}$ be trivial.

We define the functor

$$\mathrm{Pic}_d : (\text{schemes}) \to (\text{sets})$$

by associating to any scheme B the set of invertible sheaves of degree d on $B \times C$, modulo tensoring with pullbacks of invertible sheaves on B. Since the tensor products and inverses of invertible sheaves of degree 0 again have degree 0, Pic_0 factors through the category of abelian groups. These functors too are representable by schemes:

Cheerful Fact 5.6 [Kleiman 2005, Theorem 9.4.8]. Kleiman's cited paper gives the proof (in a still more general setting) that if $f : X \to S$ is projective over S with reduced irreducible fibers, then there exists a fine moduli space $\mathrm{Pic}_d(C)$ for invertible sheaves of degree d on C; that is, a scheme $\mathrm{Pic}_d(C)$ and an invertible sheaf $\mathcal{P}$ of degree d on $\mathrm{Pic}_d(C) \times C$, such that for any scheme B there is, in the complex analytic category, a natural bijection between families of invertible sheaves of degree d over B, modulo invertible sheaves on B, and morphisms $B \to \mathrm{Pic}_d(C)$, defined by pulling back $\mathcal{P}$. The tensor product of invertible sheaves makes $\mathrm{Pic}_0(C)$ an algebraic group, which acts on each $\mathrm{Pic}_d(C)$. The sheaf $\mathcal{P}$ is called a *Poincaré sheaf* for C.

The choice of $\mathcal{P}$ makes the variety $\mathrm{Pic}_0(C)$ into a projective algebraic group with group law given by tensor product of invertible sheaves of degree 0, and origin the point of $\mathrm{Pic}_0(C)$ that is the image of $\mathrm{Spec}\,\mathbb{C}$ under the map corresponding to the structure sheaf of C regarded as family of sheaves over $\mathrm{Spec}\,\mathbb{C}$; and for each d, $\mathrm{Pic}_d(C)$ is a for $\mathrm{Pic}_0(C)$.

The construction works for families of curves as long as there is a section (the Poincaré sheaf on $\mathrm{Pic}_d(C) \times C$ relies on the existence of a point on C, so that it does not apply to families of curves unless the family has a section). That is, given a family of curves $\mathcal{C} \to S$ with a section $\sigma : S \to \mathcal{C}$ there is a scheme $\mathrm{Pic}_d(\mathcal{C}/S) \to S$ that represents the functor of families of invertible sheaves of degree d over S-schemes $B \to S$. We will use this important refinement in the proof of the Brill–Noether theorem given in Chapter 14.

If $\mathcal{L}$ is any invertible sheaf of degree e on C, we can define a bijection between families of invertible sheaves of degree d over B and families of invertible sheaves of degree $d + e$ over B, uniformly for all B, by tensoring with the pullback $\pi_2^* \mathcal{L}$. Thus $\mathrm{Pic}_d(C) \cong \mathrm{Pic}_{d+e}(C)$ (but not canonically, since the isomorphism depends on the choice of $\mathcal{L}$).

The description of the functors represented by the symmetric product C_d and the Picard scheme Pic_d implies that the set-theoretic map

$$C_d \to \mathrm{Pic}_d, \quad D \mapsto \mathcal{O}_c(D),$$

is the underlying set map of a morphism of schemes: on the level of functors the map takes a family of divisors $\mathcal{D} \subset B \times C$ to the family of sheaves $\mathcal{O}_{B \times C}(\mathcal{D})$.

The characterization of $\mathrm{Pic}_d(C)$ given above does not shed much light on the geometry of $\mathrm{Pic}_d(C)$: whether it's irreducible, for example, or what its dimension is. But we can get a good picture from the classical construction of the Jacobian, $\mathrm{Jac}(C)$, which has roots in the nineteenth century.

5.3. Jacobians

The history leading to the analytic construction of the Jacobian starts from a calculus problem. A goal of the nineteenth century mathematicians was to make sense of integrals of algebraic functions. In the early development of calculus, mathematicians figured out how to evaluate integrals such as

$$\int_{t_0}^{t} \frac{dx}{\sqrt{x^2 + 1}}.$$

From a modern point of view such integrals can be thought of as path integrals of meromorphic differentials on the Riemann surface associated to the equation $y^2 = x^2 + 1$. This surface is isomorphic to $\mathbb{P}^1$, meaning that x and y can be expressed as rational functions of a single variable z; making the corresponding change of variables transformed the integral into one of the form

$$\int_{s_0}^{s} R(z)\, dz,$$

with R a rational function, and such integrals can be evaluated by the technique of partial fractions.

When they tried to extend this to similar-looking integrals, such as

$$\int_{t_0}^{t} \frac{dx}{\sqrt{x^3 + 1}},$$

which arises when one studies the length of an arc of an ellipse, they were stymied. The reason gradually emerged: the problem is that the Riemann surface associated to the equation $y^2 = x^3 + 1$ is not $\mathbb{P}^1$, but rather a curve

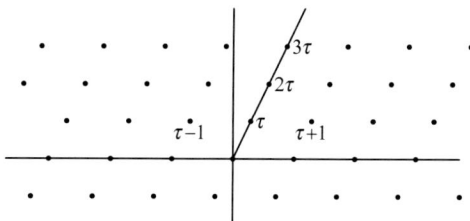

Figure 5.1. An elliptic curve is the quotient of $\mathbb{C}$ by a lattice $\mathbb{Z} \oplus \mathbb{Z}\tau$.

of genus 1, and so has nontrivial homology group $H_1(C, \mathbb{Z}) \cong \mathbb{Z}^2$. In particular, if one expresses this "function" of t as a path integral, then the value depends on a choice of path; it is defined only modulo a lattice $\mathbb{Z}^2 \subset \mathbb{C}$ (Figure 5.1). This implies that the inverse function is a doubly periodic meromorphic function on $\mathbb{C}$, and not an elementary function. Many new special functions, such as the Weierstrass $\mathcal{P}$-function, were studied as a result. The name "elliptic curve" arose from these considerations.

Once this case was understood, the next step was to extend the theory to path integrals of holomorphic differentials on curves of arbitrary genus. One problem is that the dependence of the integral on the choice of path is much worse (Figure 5.2); the set of homology classes of paths between two points $p_0, p \in C$ is identified with $H_1(C, \mathbb{Z}) \cong \mathbb{Z}^{2g}$ rather than $\mathbb{Z}^2$, rendering the expression $\int_p^q \omega$ for a given 1-form ω virtually meaningless.

To express the solution to this problems in relatively modern terms, let C be a smooth projective curve of genus g over $\mathbb{C}$, and let ω_C be the sheaf of differential forms on C. We will consider C as a complex manifold. Every meromorphic differential form is in fact algebraic [Serre 1955/56], and we consider ω_C as a sheaf in the analytic topology.

Integration over a closed loop in C defines a linear function on 1-forms, so that we have a map

$$\iota : \mathbb{Z}^{2g} = H_1(C, \mathbb{Z}) \rightarrow H^0(\omega_C)^* \cong H^1(\mathcal{O}_C) = \mathbb{C}^g.$$

By Hodge theory,

$$H^1(C, \mathbb{C}) \cong H^1(C, \mathcal{O}_C) \oplus \overline{H^1(C, \mathcal{O}_C)},$$

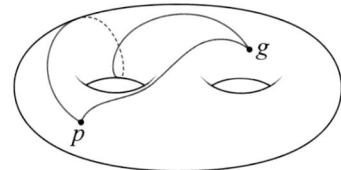

Figure 5.2. Two paths from p to q might give different values of a path integral.

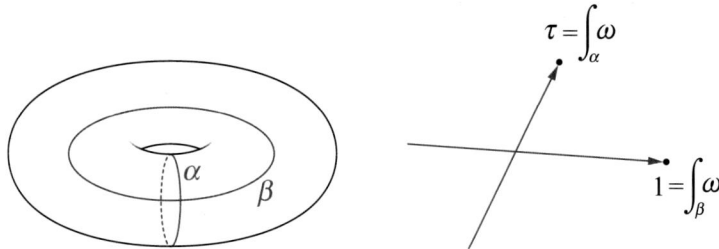

Figure 5.3. The map μ maps a curve of genus 1 to the quotient of the plane by the lattice $\Lambda = \mathbb{Z}\langle 1, \tau \rangle$.

where the bar denotes complex conjugation in $H^1(C, \mathbb{C})$, and the map ι is the composition of the natural inclusion with the projection to the first summand. Now $H_1(C, \mathbb{C}) = \mathbb{C} \otimes_{\mathbb{Z}} H_1(C, \mathbb{Z})$, so any basis of $H_1(C, \mathbb{Z})$ maps to a basis of $H^1(C, \mathbb{C})$ invariant under conjugation in $H^1(C, \mathbb{C})$ — see [Voisin 2007] or [Griffiths and Harris 1978, p. 116].

One can show that the image of ι is a lattice in $H^0(\omega_C)^*$, and thus the quotient is a torus of real dimension $2g$. Moreover, the complex structure on $H^0(\omega_C)^*$ yields a complex analytic structure on the quotient $\mathbb{C}^g/\iota(\mathbb{Z}^{2g})$, which is thus a complex torus of dimension g.

Definition 5.7. The complex torus $J(C) := \mathbb{C}^g/\iota(\mathbb{Z}^{2g})$ is called the *Jacobian* of C.

The point of this construction is that for any pair of points $p, q \in C$, the expression $\int_q^p$ describes a linear functional on $H^0(\omega_C)$, defined up to functionals obtained by integration over closed loops, and thus a point of $J(C)$. We can think of $p - q$ as a divisor of degree 0, and the map $p - q \mapsto \int_q^p$ extends to a well defined map μ_0 from divisors of degree 0 to $J(C)$ because if p', q' are two more points, then

$$\int_q^p + \int_{q'}^{p'} - \left(\int_q^{p'} + \int_{q'}^p \right)$$

is the integral around a closed path $q \to p \to q' \to p' \to q$.

Further, if we choose a basepoint $p \in C$, we can define the holomorphic map (Figure 5.3)

$$\mu : C \to J(C), \quad q \mapsto \int_p^q,$$

and more generally maps

$$\mu_d : C_d \to J(C), \quad (q_1, \ldots, q_d) \mapsto \sum_i \int_p^{q_i}.$$

These are called the *Abel–Jacobi* maps. When there is no ambiguity about d, we will denote all these maps by μ, and we extend the definition by defining

$\mu(-D)$ to be $-\mu(D)$. The map μ is a group homomorphism in the sense that if D, E are divisors, then $\mu(D + E) = \mu(D) + \mu(E)$; this is immediate when the divisors are effective, and follows in general because the group of divisors is a free abelian group.

5.4. Abel's theorem

The connection between the discussion above and the geometry of linear series is made by one of the landmark theorems of the nineteenth century:

Theorem 5.8 (Abel's theorem). *Two effective divisors of degree d on C are linearly equivalent if and only if their images under the Abel–Jacobi map are equal; equivalently, the fibers of μ_d are the complete linear series of degree d on C. Thus μ_0 factors through a canonical isomorphism $\mathrm{Pic}_0(C) \to J(C)$ and, after choosing a basepoint p, an isomorphism $\mathrm{Pic}_d(C) \to \mathrm{Jac}(C)$, factoring through the isomorphism*

$$\mathrm{Pic}_d(C) \xrightarrow{-\otimes \mathcal{O}_C(-dp)} \mathrm{Pic}_0(C) \to J(C).$$

We will prove only one implication, the part proved by Abel; the converse, which is substantially more subtle, was proved by Clebsch. For a complete proof, see [Griffiths and Harris 1978, Section 2.2].

Proof of "only if". If D and D' are linearly equivalent, then they are the zero loci of sections σ, σ' of an invertible sheaf $\mathcal{L}$. Taking linear combinations of σ and σ', we get a pencil $\{D_\lambda\}_{\lambda \in \mathbb{P}^1}$ of divisors on C, with

$$D_\lambda = V(\lambda_0 \sigma + \lambda_1 \sigma'),$$

and by the universal property of the symmetric product (Cheerful Fact 5.4) this corresponds to a regular map $\alpha : \mathbb{P}^1 \to C_d$.

Consider the composition

$$\phi = \mu \circ \alpha : \mathbb{P}^1 \to J(C).$$

If z is any linear functional on $V = H^0(\omega_C)^*$ then the differential dz on V descends to a global holomorphic 1-form on $J(C)$, since this is the quotient of V by a discrete lattice. Thus the regular one-forms on $J(C)$ generate the cotangent space to $J(C)$ at every point. But for any 1-form ω on $J(C)$, the pullback $\phi^*\omega$ is a global holomorphic 1-form on $\mathbb{P}^1$, and hence identically zero. It follows that the differential $d\phi$ vanishes identically, and hence that ϕ is constant, proving that $\mu(D) = \mu(D')$. $\square$

Corollary 5.9. *Let C be a curve of genus g. The number of isomorphism classes of invertible sheaves whose n-th power is $\mathcal{O}_C$ is n^{2g}.*

Proof. This is the number of n-torsion points in the group

$$\mathbb{C}^g / \mathbb{Z}^{2g} \cong (S^1)^{2g}.$$ $\square$

André Weil applied Abel's theorem to give an algebraic description of the structure of the Jacobian, using the following fact:

Corollary 5.10 (Jacobi inversion theorem). *If C is a smooth curve of genus g then the Abel–Jacobi map $\mu_g : C_d \to J(C)$ is surjective for $d \geq g$ and generically injective for $d \leq g$. In particular, $\mu_g : C_g \to J(C)$ is birational.*

Proof. For $d \leq g = \dim H^0(\omega_C)$, a divisor D that is the sum of d general points $p_1, \ldots, p_d \in C$ will impose independent vanishing conditions on the sections of ω_C, and thus

$$h^0(\omega_C(-D)) = g - d.$$

Using this, the Riemann–Roch formula gives $h^0(\mathcal{O}_C(D)) = 1$, so the fiber of μ_d consists of a single point, proving generic injectivity.

If $d \geq g$ then $h^0(\omega_C(-D)) = 0$ and hence $r(D) = d - g = \dim C_d - \dim J(C)$, and it follows that the image of C_d has dimension g. Since C_d is compact and $J(C)$ is irreducible, μ_d is surjective. $\square$

Here we are using the fact that the analytic construction can be made algebraic, one of the "miracles" mentioned by Barry Mazur:

> The useful generalization [of the group structure on the points of an elliptic curve] ... has the advantage that by its very definition it is seen to be intrinsic (it depends only on the birational equivalence class of the curve and not upon its imbedding in the projective plane) and, moreover, three felicitous miracles occur (analytic, algebraic, and arithmetic) which make it an extraordinarily powerful tool.
>
> — [Mazur 1986]

The differential of the Abel–Jacobi map.

Theorem 5.11. *The Abel–Jacobi map $\mu_1 : C \to \mathrm{Pic}_1(C)$ is a closed embedding. If $p \in C$ is a point on a smooth projective curve of genus $g > 0$ then the cotangent space of C at p is $\kappa(p) \otimes \omega_C$ and the cotangent space to Pic_1 at $\mu_1(p)$ is $H^0(\omega_C)$ where $\kappa(p) \cong \mathbb{C}$ is the residue class field at p. With these identifications the cotangent map*

$$T^*(\mu_1) : H^0(\omega_C) \to \kappa(p) \otimes \omega_C$$

is evaluation at p, which is surjective, with kernel $H^0(\omega_C(-p))$.

Proof. Since $g(C)$ is positive distinct points $p, q \in C$ cannot be linearly equivalent, so the Abel–Jacobi map μ_1 is set-theoretically injective. To prove that it is an embedding, we must show that its differential is everywhere nonzero.

To compute the differential of μ_1 at a point $p \in C$ we choose a local parameter z at p, so that any differential form on a neighborhood of p has the form $f(z)dz$. We may thus write the cotangent space to C at p as $\kappa(p) \otimes \omega_C \cong \mathbb{C}dz$.

The tangent space to $J(C) = H^0(\omega_C)^*/H_1(C, \mathbb{Z})$ at any point is $H^0(\omega_C)^*$, so the cotangent space is $H^0(\omega_C)$.

The map μ_1 sends each point q to

$$\mu_1 : C \to J(C) := H^0(\omega_C)^*/H_1(C, \mathbb{Z}), \quad q \mapsto \int_p^q .$$

The fundamental theorem of calculus tells us the derivative of an integral: for any $q \in C$ close to p we have

$$\frac{\partial}{\partial q} \int_p^q f(z)dz = f(q).$$

It follows that the codifferential $T^*(\mu_1)$ of μ_1 is the natural restriction map $H^0(\omega_C) \to \kappa(p) \otimes \omega_C$ sending a differential form to its value at p. Thus the kernel of $T^*(\mu_1)$ evaluated at p is the space of differential forms vanishing at p, that is $H^0(\omega_C(-p))$. Since ω_C is basepoint free, there is a differential form not vanishing at p, so $T^*(\mu_1)$ is surjective, proving that the differential of μ_1 is injective as required. $\qquad\qquad\qquad\qquad\qquad\qquad\qquad\qquad\qquad\qquad\qquad$ □

Corollary 5.12. *The composition of the map $\mu_1 : C \to \mathrm{Pic}_1(C)$ with the Gauss map that sends a point $\mu_1(p)$ to the tangent line to $\mu_1(C)$ at p is the canonical embedding.*

Proof. Since the kernel of $T^*(\mu_1)(p)$ is the hyperplane $H^0\omega_C(-p)) \subset H^0(\omega_C)$, the differential of μ_1 carries the tangent space to C at p to the line in $H^0(\omega_C)^*$ corresponding to the hyperplane $H^0(\omega_C(-p)) \subset H^0(\omega_C)$. This is the image of p in $\mathbb{P}(H^0(\omega_C)) = \mathbb{P}^{g-1}$ under the canonical map. $\qquad\qquad$ □

5.5. The $g + 3$ theorem

The Picard varieties are mysterious objects. But even the existence of a moduli space for invertible sheaves of degree d on C, and the fact that this space is an irreducible algebraic variety of dimension g, suffice to prove nontrivial results.

We proved in Corollary 2.19 that every divisor of degree $\geq 2g + 1$ is very ample, and this is sharp as stated: on any curve C there exist divisors of every degree $d \leq 2g$ that are not very ample. But if we look at a *general* divisor D of degree d on C, we can do much better:

Theorem 5.13 ($g + 3$ theorem). *Let C be a smooth projective curve of genus g. If $D \in C_{g+3}$ is a general effective divisor of degree $g + 3$ on C, then D is very ample. Thus every curve of genus g can be embedded in $\mathbb{P}^3$ as a curve of degree $g + 3$.*

Moreover, this is sharp: if $g \geq 2$ and C is hyperelliptic, then no invertible sheaf of degree $\leq g + 2$ is very ample on C.

This is not the final word on low-degree embeddings of curves: if we look at a general curve C, Theorem 12.9 says that C can in fact be embedded in $\mathbb{P}^3$ as a curve of degree $d = \lceil 3g/4 \rceil + 3$, and can be birationally embedded in $\mathbb{P}^2$ as a nodal curve of degree $d = \lceil 2g/3 \rceil + 2$.

Proof of Theorem 5.13. If D is general of degree $g + 3$ then D is nonspecial, so $h^0(\mathcal{O}_C(D)) = 4$. By Proposition 1.11 the divisor D is very ample if and only if $h^0(\mathcal{O}_C(D - F)) = 2$ for every effective divisor $F = p + q$ of degree 2 on C.

If, on the contrary, $h^0(\mathcal{O}_C(D-F)) \geq 3$ then by the Riemann–Roch theorem $h^0(\omega_C(-D+F)) \geq 1$ so $K_C - D + F$ is linearly equivalent to an effective divisor E of degree $g - 3$.

Now, consider the map $\nu : C_{g-3} \times C_2 \to J(C)$ given by

$$\nu : (E, F) \mapsto \mu_{g-3}(K_C - E + F) = \mu_{2g-2}(K_C) - \mu_{g-3}(E) + \mu_2(F),$$

where the $+$ and $-$ on the right refer to the group law on $J(C)$.

By what we have just said and Abel's theorem, the divisor D is very ample if and only if $\mu(D)$ is not in the image of ν. However the source of ν has dimension $g - 3 + 2 = g - 1$, so its image in $J(C)$ is a proper subvariety. Since μ_{g+3} is surjective, the image of a general divisor $D \in C_{g-3}$ is a general point of $J(C)$ and thus will not lie in the image of ν.

To prove that $g + 3$ is sharp for hyperelliptic curves, recall that a smooth plane curve C of genus at least 2 has degree $d \geq 4$ and canonical class $\omega_C = \mathcal{O}_C(d - 3)$, which is very ample. This shows that no hyperelliptic curve can be embedded in the plane.

Thus every very ample divisor D on a hyperelliptic curve has $h^0(D) \geq 3$. If also $\deg D \leq g + 2$ then D is a special divisor. By Corollary 2.32, D would have the form rD_0, where D_0 is the g_2^1, the unique divisor of degree 2 with 2 independent sections, and $r \leq (g + 2)/2$. But the smallest multiple of the D_0 that is very ample is $(g + 1)D_0$, and since $g \geq 2$ we have $(g + 2)/2 < g + 1$, so D is not very ample. $\qquad \square$

It would be natural to speculate that a general divisor of degree $g+2$ defines a map to $\mathbb{P}^2$ whose image has only nodes as singularities, but this is not quite true. Here is the correct result, whose proof we give in Theorem 10.15 because it depends on a general position result that we will prove in Section 10.1.

Theorem 5.14 ($g + 2$ theorem). *Let C be any smooth projective curve of genus g, and let D be a general divisor of degree $g + 2$ on C.*

(1) *If C is not hyperelliptic then the map $\phi_D : C \to \mathbb{P}^2$ is birational onto its image C_0, and C_0 is a plane curve of degree $g + 2$ with exactly $\binom{g}{2}$ nodes and no other singularities.*

(2) *If C is hyperelliptic then the map $\phi_D : C \to \mathbb{P}^2$ is birational onto its image C_0, and C_0 is a plane curve of degree $g + 2$ with one ordinary g-fold point and no other singularities.*

5.6. The schemes $W_d^r(C)$

Now that we have a parameter space $\mathrm{Pic}_d(C)$ for invertible sheaves of degree d, we can ask about the geometry of the set of invertible sheaves defining linear series of dimension $\geq r$, that is,

$$W_d^r(C) := \{\mathcal{L} \in \mathrm{Pic}_d(C) \mid h^0(\mathcal{L}) \geq r + 1\}.$$

Thus, for example, $W_d^0(C)$ is the locus of effective divisor classes, the image of the map $\mu : C_d \to \mathrm{Pic}_d(C)$. We often omit the 0 in this case and write this simply as $W_d(C)$.

$W_d^r(C)$ can also be characterized as the locus

$$W_d^r(C) = \{\mathcal{L} \in \mathrm{Pic}_d(C) \mid \dim \mu^{-1}(\mathcal{L}) \geq r\},$$

where $\mu : C_d \to \mathrm{Pic}_d(C)$ is the Abel–Jacobi map. By the upper semicontinuity of fiber dimension, the support of $W_d^r(C)$ is a Zariski-closed subset, so we can talk about its dimension, irreducibility, smoothness or singularity and so on. The preimage $\mu^{-1}(W_d^r(C)) \subset C_d$ is the set of effective divisors D with $r(D) \geq r$, and is denoted C_d^r.

The set $W_d^r(C)$ can be given the structure of a scheme; see [Arbarello et al. 1985, Section IV.3]. With this structure, it is a fine moduli space representing the functor of families of invertible sheaves with $r + 1$ or more sections (that is, the functor that associates to a scheme B the set of invertible sheaves $\mathcal{L}$ on $B \times C$ such that the pushforward $(\pi_2)_* \mathcal{L}$ has a locally free subsheaf of rank $r + 1$ over every curvilinear[1] subscheme of B, as always modulo tensoring with pullbacks of invertible sheaves on B). In Chapter 9 we will see examples of curves C of genus 4 for which the scheme $W_3^1(C)$ is reducible, and in Chapter 12 we will see curves C of genus 6 for which the scheme $W_4^1(C)$ is nonreduced — in fact, in one case isomorphic to $\mathrm{Spec}\, \mathbb{C}[\epsilon]/(\epsilon^5)$.

5.7. Examples in low genus

Genus 1. If C has genus 1, then since the Abel–Jacobi map $\mu : C \to \mathrm{Pic}_1(C)$ is always an embedding, it is an isomorphism, and also $\mathrm{Pic}_d(C) \cong C$ for all d. If we fix any point $p \in C$, we get a map $C \to \mathrm{Pic}_0(C)$ sending $q \in C$ to the invertible sheaf $\mathcal{O}_C(q - p)$, which is again an isomorphism.

The fact that a curve of genus 1 over $\mathbb{C}$ is the quotient of the complex plane by a lattice $\Lambda \subset \mathbb{C}$ has an immediate consequence:

---

[1]This word means the Zariski tangent space has dimension at most 1 everywhere.

Corollary 5.15. *If E is a smooth projective curve of genus 1 and $p \in E$ any point, there are only finitely many automorphisms $\phi : E \to E$ with $\phi(p) = p$.*

Proof. The choice of p allows us to identify C with $\mathbb{C}/\Lambda$ for some lattice Λ. Any automorphism ϕ lifts to a linear map $\tilde{\phi} : \mathbb{C} \to \mathbb{C}$ carrying Λ to itself. ☐

Genus 2. The map $\mu_2 : C_2 \to J(C)$ is an isomorphism except along the locus $\Gamma \subset C_2$ of divisors of the unique g_2^1 on C. In fact, the symmetric square C_2 of C is the blowup of $J(C)$ at the point corresponding to the invertible sheaf corresponding to the g_2^1; see Exercise 5.10.

Genus 3. Let C be a curve of genus 3. The geometry of the map μ_2 depends on whether or not C is hyperelliptic. If it is, μ_2 will collapse the locus in C_2 of divisors of the hyperelliptic g_2^1 to a point, but is otherwise one-to-one, and $W_2(C)$ is singular. If C is not hyperelliptic, by contrast, μ_2 is an embedding, and $W_2(C)$ is smooth.

The birational surjection $\mu_3 = \mu_g$ is the blowup of $J(C) \cong \mathrm{Pic}_3(C)$ along the locus $W_3^1(C)$. At the same time, we know that

$$W_3^1(C) = K - W_1(C),$$

in the sense that any invertible sheaf $\mathcal{L}$ defining a g_3^1 has, by the Riemann–Roch theorem, the form $\omega_C \otimes \mathcal{L}'$, where $\mathcal{L}'$ has degree 1 and $h^0(\mathcal{L}') = 1$, so W_3^1 is isomorphic to C. Thus C_3 is the blowup of $J(C)$ along the curve C. See [Mumford 1975, pp. 53–54] for a discussion.

5.8. Martens' theorem

If C is a smooth projective curve of genus g and $r \leq d - g$, then by the Riemann–Roch theorem $W_d^r = \mathrm{Pic}_d$. On the other hand Clifford's theorem says that if C is a smooth projective curve of genus g and $d \leq 2g - 2$ then the set $W_d^r(C)$ is empty when $d - 2r < 0$. Martens' theorem extends Clifford's theorem to the statement $\dim W_d^r \leq d - 2r$ whenever $r > d - g$.

Theorem 5.16 [Martens 1967]. *Let C be a smooth projective curve of genus g. If $r > d - g$, then*

$$\dim(W_d^r(C)) \leq d - 2r;$$

moreover, equality holds for some r in this range if and only if $d = 2g - 2$ and $r = g - 1$ (in which case $W_d^r(C) = \{\omega_C\}$) or $d = r = 0$ (in which case $W_d^r(C) = \{\mathcal{O}_C\}$) or C is hyperelliptic. In the last case, equality holds for all d and r with $d \geq 2r > 2(d - g)$.

Given a smooth projective curve $C \subset \mathbb{P}^r$, we write $\Sigma_d^r \subset C_d$ for the locus of effective divisors D of degree d on C with $\dim \overline{D} \leq d - r - 1$. This is a closed condition, so Σ_d^r is an algebraic subset. Using the geometric Riemann–Roch theorem we will reduce the inequality in Martens' theorem to the following.

Lemma 5.17 (elementary secant plane lemma). *Let $C \subset \mathbb{P}^r$ be a smooth, non-degenerate curve. If $d \leq r$ and $r \geq 0$ then*

$$\dim \Sigma_d^r \leq d - r.$$

Proof. Since $\Sigma_d^0 = C^d$, the conclusion is immediate for $r = 0$.

Suppose that $D \in \Sigma_d^r$ with $r > 0$. Since $\dim \overline{D} \leq d - r - 1 \leq d - 2$ the plane Λ spanned by D is already spanned by some divisor $D' := D - p$ of degree $d - 1$, obtained by subtracting one point in the support of D. It follows that for points $q \notin \Lambda \cap C$, the divisor $q + D'$ spans a plane of dimension $\leq d - r$, so $q + D' \in \Sigma_d^{r-1}$. Furthermore, D is in the closure of the set of divisors

$$\{q + D' \mid q \notin \Lambda \cap C\}.$$

From this we see that $\Sigma_d^r \setminus \Sigma_d^{r-1}$ is in the closure of Σ_d^{r-1}. It follows that

$$\dim \Sigma_d^0 > \dim \Sigma_d^1 > \cdots > \dim \Sigma_d^r,$$

and thus $\dim \Sigma_d^r \leq d - r$ as claimed. $\qquad\square$

Proof of Martens' inequality. The listed cases of equality are easy to verify: for example on a hyperelliptic curve Corollary 2.32 shows that a special invertible sheaf with $r(\mathcal{L}) = s$ can be written $\mathcal{L}_0^{\otimes s}(p_1 + \cdots + p_{d-2s})$, where $\mathcal{L}_0$ is the unique g_2^1 on C and the p_i are basepoints — a family of dimension $d - 2s$. We postpone the proof that these are the only cases of equality to Theorem 10.12, which we prove using a stronger version of Lemma 5.17 that is a corollary of Theorem 11.3.

Since we have assumed $r > d - g$, the set $W_d^r(C)$ contains only sheaves $\mathcal{L}$ with $h^1(\mathcal{L}) \neq 0$. For such sheaves there is a one-to-one correspondence $\mathcal{L} \mapsto \omega_C \otimes \mathcal{L}^{-1}$ inducing an isomorphism

$$\{\mathcal{L} \in W_d^r(C) \mid h^1(\mathcal{L}) \neq 0\} = \{\mathcal{L} \in W_{2g-2-d}^{g-1-d+r} \mid h^1(\mathcal{L}) \neq 0\}.$$

Since

$$2g - 2 - d - 2(g - 1 - d + r) = d - 2r,$$

it will suffice to prove the theorem in case $d \leq g - 1$.

Let $C_d^r \subset C_d$ be the preimage of $W_d^r(C)$. Since the fibers of C_d^r over points of $W_d^r(C)$ have dimension at least r, Marten's inequality will follow from the inequality

$$\dim C_d^r \leq d - r.$$

Let $\phi_K : C \to \mathbb{P}^g - 1$ be the canonical map. By the geometric Riemann–Roch theorem (Theorem 2.29), $r(D) = \deg D - 1 - \dim \overline{\phi_K(D)}$, so $D \in C_d^r$ if and only if $\dim \overline{\phi_K(D)} \leq d - r - 1$. Since $d \leq g - 1$ we may apply Lemma 5.17, showing that the set of such D has dimension $\leq d - r$, as required. $\qquad\square$

If C is hyperelliptic with $g_2^1 = |D|$ then for $r \le g - 1$ we have $\mu(rD) \in W_{2r}^r$ and

$$W_d^r(C) \supset W_{d-2r}(C) + \mu(rD).$$

In fact this is an equality by Corollary 2.32. Since $W_{d-2r}(C)$ has dimension $d - 2r$, we see that Martens' theorem is sharp.

The theorem can be extended to show that $\dim W_d^r(C) < d - 2r$ in certain cases; see for example [Mumford 1974; Keem 1990; Coppens 1983].

5.9. Exercises

Exercise 5.1. Let G be a finite group acting on a quasiprojective scheme X. Show that there is a finite covering of X by invariant open affine sets.   ◆

Exercise 5.2. We say that a group G acts freely on X if $gx = gy$ only when g is the identity or $x = y$. Show that if G is a finite group acting freely on a smooth affine variety X then the quotient X/G is smooth.

Exercise 5.3. Let $X = (\mathbb{A}^2)^2$ and let $G := \mathbb{Z}/2$ act on X by permuting the two copies of $\mathbb{A}^2$; algebraically, $(\mathbb{A}^2)^2 = \operatorname{Spec} S$, with $S = k[x_1, x_2, y_1, y_2]$ and the nontrivial element $\sigma \in G$ acts by $\sigma(x_i) = y_i$.

(1) Show that G acts freely on the complement of the diagonal, but fixes the diagonal pointwise.

(2) Supposing that the ground field has characteristic $\neq 2$, show that the algebra S^G is the affine coordinate ring of the product of A^2 with the affine cone over a conic in the projective plane.   ◆

(3) Conclude that the symmetric square has singular locus equal to the image of the diagonal in $(\mathbb{A}^2)^2$.

Exercise 5.4. It is not an accident that we can characterize a fine moduli space M in terms of the maps into it. Let X be a category, and F, G two functors from X to the category of sets. A morphism $\eta : F \to G$ in the category of functors is called a *natural transformation*: for every object $a \in X$ there is a morphism $\eta_a : F(a) \to G(a)$ such that for every morphism $f : a \to b$ in X the compositions $G(f) \circ \eta_a$ and $\eta_b \circ F(f)$ are equal.

(1) Prove Yoneda's lemma: If X is any category, and F is a contravariant functor from X to the category of sets, then

$$\operatorname{Hom}_{\text{Functors on } X}(\operatorname{Hom}_X(-, Z), F) = F(Z).$$

(2) Conclude that if the functors $\operatorname{Hom}_X(-, Z)$ and $\operatorname{Hom}_X(-, Z')$ are isomorphic in the functor category, then $Z \cong Z'$ in X; in other words, the functor $\operatorname{Hom}_X(-, Z)$ determines the object Z up to isomorphism.

Exercise 5.5. Show that if $r \geq d - g$, then $W_d^r(C) \setminus W_d^{r+1}(C)$ is dense in $W_d^r(C)$ (that is, $W_d^{r+1}(C)$ does not contain any irreducible component of $W_d^r(C)$). ◆

Exercise 5.6. Let C be a curve of genus 2, and let $C \subset J(C)$ be the image of the Abel–Jacobi map μ_1. Show that the self-intersection of the curve C in $J(C)$ is 2 in two different ways:

(1) by applying the adjunction formula to $C \subset J(C)$; and

(2) by calculating the self-intersection of its preimage $C + p \subset C_2$ and using the geometry of the Abel–Jacobi map $\mu_2 : C_2 \to J(C)$. ◆

Exercise 5.7. Let C be a curve of genus 2, and consider the map $\nu : C \times C \to$ $\mathrm{Pic}_0(C)$ defined by sending $(p, q) \in C \times C$ to the invertible sheaf $\mathcal{O}_C(p - q)$. Show that this map is generically finite, and compute its degree. ◆

Exercise 5.8. Let C_d^r be the preimage of W_d^r, the set of divisors moving in at least an r-dimensional linear series. Show that the image of the differential of the Abel–Jacobi map $C_d \to J(C)$ at a point of $C_d \setminus C_d^1$ corresponding to a reduced divisor is the plane in $\mathbb{P}^{g-1}$ spanned by the divisor D on the canonical curve.

Exercise 5.9 ($g + 1$ theorem). Let C be a smooth projective curve of genus g. If $D \in C_{g+1}$ is a general effective divisor of degree $g + 1$ on C, then D is basepoint free and defines a map to $\mathbb{P}^1$ that has only simple ramification, with $2g + 2$ distinct branch points. ◆

Exercise 5.10. Let C be a curve of genus 2. The canonical map $\phi_K : C \to \mathbb{P}^1$ expresses C as a 2-sheeted cover of $\mathbb{P}^1$, and we have correspondingly an involution $\tau : C \to C$ exchanging points in the fibers of ϕ_K (equivalently, for any $p \in C$, we have $h^0(K_C(-p)) = 1$; τ will send p to the unique zero of the unique section $\sigma \in H^0(K_C(-p))$). Let $\Gamma \subset C \times C$ be the graph of τ.

(1) Using the fact that a birational morphism of smooth surfaces is the inverse of a sequence of blowups of reduced points [Hartshorne 1977, Chapter V, Theorem 5.5], show the self-intersection of the image C_2^1 of Γ in C_2 is -1.

(2) Find the self-intersection of Γ in $C \times C$.

Chapter 6

Hyperelliptic curves and curves of genus 2 and 3

6.1. Hyperelliptic curves

Recall that a hyperelliptic curve C is a curve of genus ≥ 2 admitting a map $\pi : C \to \mathbb{P}^1$ of degree 2. We met hyperelliptic curves in Chapter 2 and proved that the canonical map from C is the composition of π with the embedding of $\mathbb{P}^1$ in $\mathbb{P}^{g-1}$ as a rational normal curve, showing in particular that π is unique up to automorphisms of $\mathbb{P}^1$.

We used this to show that every special linear series on a hyperelliptic curve is a sum of a multiple of the unique g_2^1 plus basepoints. We will begin this chapter with an explicit construction of hyperelliptic curves and use it to give a concrete computation of the canonical series, reproving what we did in Chapter 2. Then we will consider the projective embeddings of curves of genus 2 (which are all hyperelliptic) and genus 3.

There will be a further discussion of hyperelliptic curves in Chapter 17.

The equation of a hyperelliptic curve. Because the degree of the canonical map is 2, each point in $\mathbb{P}^1$ has either two distinct preimages, or only one; in the latter case, this point is a ramification point with ramification index 1; that is, the map is given in terms of local analytic coordinates on C and $\mathbb{P}^1$ by $z \mapsto z^2$. In particular, both the ramification divisor and the branch divisor (as defined in Chapter 2) are reduced. By Hurwitz's formula there are exactly $2g+2$ branch points in $\mathbb{P}^1$. These points determine the curve:

Theorem 6.1. *There is a unique smooth projective hyperelliptic curve C express-ible as a 2-sheeted cover of $\mathbb{P}^1$ branched over any given set of $2g + 2$ distinct points* $\{q_1, \ldots, q_{2g+2}\}$.

Proof. We will exhibit such a curve, leaving the proof of uniqueness to Section 6.2. If the coordinate of the point $q_i \in \mathbb{P}^1$ is λ_i, we take for C the smooth projective model of the affine curve

$$C^\circ = \left\{(x, y) \in \mathbb{A}^2 \,\middle|\, y^2 = \prod_{i=1}^{2g+2} (x - \lambda_i)\right\}.$$

Note that we're choosing a coordinate x on $\mathbb{P}^1$ with the point $x = \infty$ at infinity not among the q_i, so that the preimage of $\infty \in \mathbb{P}^1$ is two points $r, s \in C$. Concretely, we see that as $x \to \infty$, the ratio y^2/x^{2g+2} approaches 1, so that

$$\lim_{x \to \infty} \frac{y}{x^{g+1}} = \pm 1.$$

The two possible values of this limit correspond to the two points $r, s \in C$. $\qquad\square$

The curve C thus constructed is *not* simply the closure of the affine curve $C^\circ \subset \mathbb{A}^2$ in either $\mathbb{P}^2$ or $\mathbb{P}^1 \times \mathbb{P}^1$: as you can see from a direct examination of the equation, each of these closures will be singular at the (unique) point at infinity.

To give a smooth projective model of a hyperelliptic curve C with given branch divisor, we divide the $2g + 2$ branch points into two sets of the same size, $\{q_1, \ldots, q_{g+1}\}$ and $\{q_{g+2}, \ldots, q_{2g+2}\}$. We can then take C to be the closure in $\mathbb{P}^1 \times \mathbb{P}^1$ of the locus

$$\left\{(x, y) \in \mathbb{A}^2 \,\middle|\, y^2 \prod_{i=1}^{g+1} (x - \lambda_i) = \prod_{i=g+2}^{2g+2} (x - \lambda_i)\right\};$$

in projective coordinates, this is

$$C = \left\{((X_0, X_1), (Y_0, Y_1)) \in \mathbb{P}^1 \times \mathbb{P}^1 \,\middle|\, Y_1^2 \prod_{i=1}^{g+1} (X_1 - \lambda_i X_0) = Y_0^2 \prod_{i=g+2}^{2g+2} (X_1 - \lambda_i X_0)\right\}.$$

To see that $C \subset \mathbb{P}^1 \times \mathbb{P}^1$ is smooth we note that it is a curve of bidegree $(2, g+1)$ in $\mathbb{P}^1 \times \mathbb{P}^1$, and the formula for the genus of a curve in $\mathbb{P}^1 \times \mathbb{P}^1$ derived in Example 2.7 tells us that such a curve has arithmetic genus g, and thus no singular points.

From this model, we deduce:

Corollary 6.2. *If C is a hyperelliptic curve and $p_1, \ldots, p_{2g+2} \in C$ are the ramification points of the unique degree 2 map $C \to \mathbb{P}^1$, then for any division of* $\{1, \ldots, 2g + 2\}$ *into two sets A, B of cardinality $g + 1$,*

$$\sum_{i \in A} p_i \sim \sum_{i \in B} p_i.$$

Proof. The abstract curve $C \subset \mathbb{P}^1 \times \mathbb{P}^1$ above is independent of the choice of A and B, since in any case the projection to the first factor is ramified at the same set $p_1, \ldots, p_{2g+2}$. Given the representation above, the sets $\{p_i \mid i \in A\}$ and $\{p_i \mid i \in B\}$ are preimages of $(0, 1)$ and $(1, 0)$ in the second factor. □

The map $\iota : C \to C$ that exchanges the two points in each reduced fiber of the map $C \to \mathbb{P}^1$ and fixes the ramification points is algebraic: in terms of the last representation of C, it is given by $((X_0, X_1), (Y_0, Y_1)) \mapsto ((X_0, X_1), (Y_0, -Y_1))$. The map ι is called the *hyperelliptic involution* on C.

Differentials on a hyperelliptic curve. We can give a pleasantly concrete description of the differentials, and thus the canonical linear series, on a hyperelliptic curve C by working with the affine model $C^\circ = V(f) \subset \mathbb{A}^2$, where

$$f(x, y) = y^2 - \prod_{i=1}^{2g+2} (x - \lambda_i).$$

We will again denote the two points at infinity (that is, the two points of $C \setminus C^\circ$) by r and s; for convenience, we'll denote the divisor $r + s$ by D. We write $\pi : C \to \mathbb{P}^1$ for the morphism that, on C°, sends $(x, y) \in C$ to x.

We can construct a differential form on C by following the proof of Hurwitz's theorem in Chapter 2. Let dx denote the usual differential on $\mathbb{P}^1$ having a double pole at infinity, and consider $\pi^* dx$ on C. The function x is regular on C°, and is a local parameter over points other than the λ_i; from the local description of the map π, we see that $\pi^* dx$ is regular on C° with simple zeros at the ramification points $q_i = (\lambda_i, 0)$. Since dx has a double pole at the point at $\infty \in \mathbb{P}^1$ and π is a local isomorphism near r and s, the differential $\pi^* dx$ has double poles at the points r and s. Thus the canonical divisor of C is

$$K_C \sim (dx) \sim R - 2D,$$

where R denotes the ramification divisor, in this case the sum of the ramification points.

How can we find differentials that are regular everywhere on C? If we divide dx by x^2 (or any quadratic polynomial in x) to kill the poles we introduce new poles in the finite part C° of C.

Instead, we want to multiply dx by a rational function with zeros at r and s, but whose poles occur only at the points where dx has zeroes — that is, the points λ_i. A natural choice is the reciprocal of the partial derivative $f_y := \partial f / \partial y = 2y$, which vanishes at the points q_i, and has a pole of order $g + 1$ at each of the points r and s (reason: y/x^{g+1} approaches ± 1 as x goes to infinity, and x has a pole of order 1 at $\infty \in \mathbb{P}^1$ and thus also at each of r, s). In other words, as long as $g \geq 1$, the differential

$$\omega = \pi^* \left(\frac{dx}{f_y} \right)$$

is regular, with divisor

$$(\omega) = (g-1)r + (g-1)s = (g-1)D.$$

The remaining regular differentials on C are now easy to find: Since x has only a simple pole at the two points at infinity we can multiply ω by any x^k with $k = 0, 1, \ldots, g-1$. This gives us g differentials

$$\omega, x\omega, \ldots, x^{g-1}\omega$$

that are independent, and so form a basis for $H^0(K_C)$.

With this description of the differentials, we can see clearly why the canonical map of a hyperelliptic curves is degree 2 onto a rational normal curve, as proved in Chapter 2: the relations on $\omega, x\omega, \ldots, x^{g-1}\omega$ are the relations on x^i, and we see that the canonical image is the rational normal curve of degree $g-1$.

6.2. Branched covers with specified branching

Given a curve B and points $p_1, \ldots, p_b$ in B, what are the branched covers $\pi : C \to B$ of degree d with specified branching over each of the points p_i, up to isomorphism over B? We will reduce this question to the classification of topological covering spaces of the complement $U = B \backslash \Delta$; we will then use properties of the fundamental group of U to enumerate such covering spaces. We will prove the uniqueness of hyperelliptic curves with specified branch points at the end of this section as a special case of a general analysis of branched covers.

Theorem 6.3. *Let B be a smooth curve, let $\Delta \subset B$ be a finite set of points, and let $U := B \backslash \Delta$. If $\pi^\circ : V \to U$ is a topological covering space then V may be given the structure of a Riemann surface in a unique way so that the map π° is holomorphic; and V may be compactified to a compact Riemann surface C in a unique way such that the map π° extends to a holomorphic map $\pi : C \to B$.*

Proof. The space V inherits the structure of a complex manifold from U because if $D \subset U$ is any simply connected coordinate chart, then the preimage $(\pi^\circ)^{-1}(D)$ is a disjoint union of d copies of D, and we may use them as coordinate charts on V.

To compactify V we observe that if $D^* = \{z \in \mathbb{C} \mid 0 < |z| < 1\}$ is a punctured disc, then the map $z \mapsto z^n$ on the unit disk restricts to a connected n-fold covering space $D^* \to D^*$. Since $\pi_1(D^*) = \mathbb{Z}$, any connected covering space E of degree n is homeomorphic to this one by a homeomorphism inducing the identity on the target of π. If we define a holomorphic structure on E by pulling back the one on D, then this homeomorphism is biholomorphic.

Thus if D_i is a small neighborhood of the point $p_i \in B$ biholomorphic to a disc, then the preimage in V of the punctured disc $D_i^* := D_i \cap U$ is a disjoint union of punctured discs $E_{i,j}^*$. The maps $E_{i,j}^* \to D_i^*$ are homeomorphic to the maps $z \mapsto z^{n_{i,j}}$ of the punctured unit disc for some $n_{i,j}$. Because of the way the holomorphic structure of V is defined, the maps $E_{i,j}^* \to D_i^*$ are actually holomorphic. Thus they extend holomorphically to maps of the full disks $E_{i,j} \to D_i$ and $V \cup \bigcup E_{i,j}$ is a compact Riemann surface in a unique way. $\qquad\square$

The problem of classifying smooth curves C that have a map $\pi : C \to B$ of degree d thus becomes one of classifying covering spaces of U.

Branched covers of $\mathbb{P}^1$. We continue with the notation $U = B \setminus \Delta$, now supposing that $B = \mathbb{P}^1_{\mathbb{C}}$, the Riemann sphere. Again, let $\pi : V \to U$ be a covering space.

Choose a basepoint $p_0 \in U$, and draw simple, nonintersecting arcs γ_i joining p_0 to p_i in U. If Σ is the complement of the union of these arcs in the sphere, then the preimage of Σ in V will be the disjoint union of d copies of Σ, called the *sheets* of the cover; label these $\Sigma_1, \ldots, \Sigma_d$.

Given U, elementary homotopy theory asserts the existence of a bijection between coverings $V \to U$ of U of degree d (up to homeomorphisms of V fixing U) and group homomorphisms

$$M : \pi_1(U, p_0) \to S_d,$$

to the symmetric group on d letters, up to inner automorphisms of S_d. The map M is called the *monodromy* of the covering: given $V \to U$ and a labeling of the d sheets of V over the point p_0, the value of M at a loop β in U based at p_0 is the permutation of the points of $\pi^{-1}(p_0)$ given by sending a point $q \in \pi^{-1}(p_0)$ to the endpoint of the unique lift of β starting at q. A permutation σ of the labels of the sheets leads to a map M' equal to the composition of M with conjugation by σ.

A convenient set of generators of $\pi_1(U, p_0)$ is the set of paths β_i indicated in Figure 6.1: starting at p_0, going out along the arc γ_i until just short of p_i, going once around p_i and then going back to p_0 along the same path γ_i. The fundamental group of U is the free group generated by the paths $\beta_1, \ldots \beta_b$ modulo the relation $\prod_{i=1}^{b} \beta_i = 1$ which comes from the fact that the sphere minus the part enclosed by the paths β_i is contractible.

Given a degree d covering space V and a labeling of the d sheets over the point p_0, let τ_i be the permutation of $\{1, 2, \ldots, d\}$ corresponding to the path β_i.

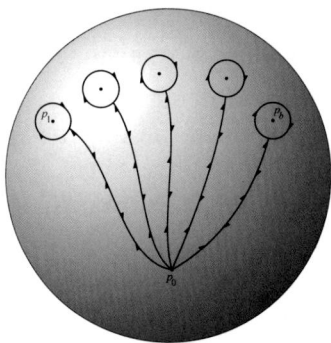

Figure 6.1. Generators for the fundamental group of a multiply punctured sphere.

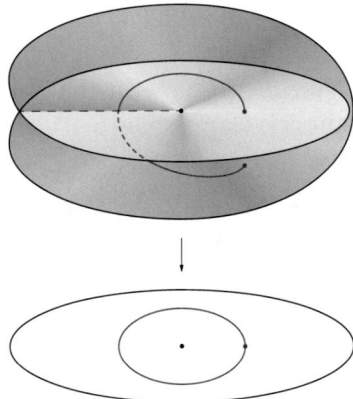

Figure 6.2. Local picture of a simple branch point $z \mapsto z^2$.

The space V is connected if and only if the τ_i generate a transitive subgroup of S_d. The case $d = 2$ is illustrated in Figure 6.2.

If we start from a map of Riemann surfaces $C \to \mathbb{P}^1$ that is simply branched, and take V to be the complement of the set of branch points, then each τ_i is a transposition.

Summarizing we have proven:

Lemma 6.4. *Let* $p_1, \dots, p_b \in \mathbb{P}^1$ *be any* b *distinct points. There is a natural bijection between*

(1) *the set of simply branched covers* $\pi : C \to \mathbb{P}^1$ *of degree* d, *branched over the points* p_i, *up to isomorphism over* $\mathbb{P}^1$; *and*

(2) *the set of* b-*tuples of transpositions* $\tau_1, \dots, \tau_b \in S_d$ *such that*

    (a) $\prod \tau_i$ *is the identity, and*

    (b) $\tau_1, \dots, \tau_b$ *generate a transitive subgroup of* S_d,

modulo simultaneous conjugation by S_d.   □

Proof of the uniqueness statement in Theorem 6.1. In the case of double covers of $\mathbb{P}^1$ that is relevant to hyperelliptic curves, we note that there is only one transposition in S_2. Thus there is a unique double cover of $\mathbb{P}^1$ with given branch points $p_1, \ldots, p_b$. (The product condition shows again that the number of branch points must be even.) This completes the proof of Theorem 6.1. $\qquad\square$

Example 6.5. In contrast to the situation of double covers of $\mathbb{P}^1$, there are generally many branched covers of specified degree greater than 2 or with given branch points and given conjugacy classes of the local monodromy. The number of these is called the *Hurwitz number* of the configuration, and its computation in general is the subject of a large and active literature; see for example [Ekedahl et al. 2001].

To illustrate this, we can use Lemma 6.4 to count the number of degree 3 branched covers $C \to \mathbb{P}^1$ with given simple branch points, using that fact that every odd permutation $\tau \in S_3$ is a transposition. Thus if b is even and $\tau_1, \ldots, \tau_{b-1} \in S_3$ are arbitrary transpositions, then the product $\tau_1 \cdots \tau_{b-1}$ is also a transposition. It follows that the number of ordered b-tuples of transpositions $\tau_1, \ldots, \tau_b \in S_3$ with $\prod \tau_i$ equal to the identity is 3^{b-1}. The requirement that the group generated by the τ_i is transitive eliminates just the three cases where all the τ_i are equal. The group S_3 acts on the set of b-tuples of permutations without stabilizing any b-tuple, so every cover corresponds to exactly 6 sequences $\tau_1, \ldots, \tau_b$. In sum, the number of simply branched 3-sheeted covers of $\mathbb{P}^1$ with specified branch points $q_1, \ldots, q_b \in \mathbb{P}^1$ is

$$\frac{3^{b-1} - 3}{6} = \frac{3^{b-2} - 1}{2}.$$

One can use a similar strategy to count covers in other cases, when the target has higher genus and/or the degree of the covering is larger, but the combinatorics becomes more complicated.

6.3. Curves of genus 2

Since curves of genus 2 are hyperelliptic, everything we said above applies to them; in particular, the canonical map $\phi_K : C \to \mathbb{P}^1$ on a curve of genus 2 is the expression of C as a double cover of $\mathbb{P}^1$, simply branched over 6 points in $\mathbb{P}^1$, which are unique up to automorphisms of $\mathbb{P}^1$.

In this section, we'll consider other maps from hyperelliptic curves C to projective space, starting with maps $C \to \mathbb{P}^1$. See for example [Eisenbud 1980] for a treatment of certain embeddings of hyperelliptic curves of all genera.

Maps of C to $\mathbb{P}^1$. The curve C has a unique degree 2 morphism to $\mathbb{P}^1$ associated to the canonical system $|K_C|$. But there are many other morphisms to $\mathbb{P}^1$. For example, there is a 2-parameter family of maps of degree 3:

Let $\mathcal{L}$ be an invertible sheaf of degree 3 on C. Since $3 > 2g - 2$, the Riemann–Roch theorem tells us immediately that $h^0(\mathcal{L}) = 2$, and there are two possibilities:

(1) If the linear series $|\mathcal{L}|$ has a basepoint $p \in C$, then $h^0(\mathcal{L}(-p)) = 2$, and hence $\mathcal{L}$ must be of the form $\mathcal{L} = K_C(p)$. Conversely, if $\mathcal{L} = K_C(p)$, then $h^0(\mathcal{L}(-p)) = h^0(\mathcal{L})$, which is to say p is a basepoint of $|\mathcal{L}|$. There is a 1-parameter family of such $\mathcal{L}$.

(2) If $\mathcal{L}$ is not of the form $\mathcal{L} = K_C(p)$, then $|\mathcal{L}|$ does not have a basepoint, and so defines a degree 3 map $\phi_{\mathcal{L}} : C \to \mathbb{P}^1$.

Since the variety $\mathrm{Pic}_3(C)$ has dimension $g = 2$ the general invertible sheaf of degree 3 is of the second kind, and this gives a 2-parameter family of such maps.

There are plenty of higher-degree maps as well: an invertible sheaf of degree $d \geq 4 = 2g$ is basepoint free, and gives a map to $\mathbb{P}^{d-2}$, from which we can project in many ways to $\mathbb{P}^1$.

Maps of C to $\mathbb{P}^2$. Next consider maps of a curve C of genus 2 to the plane. By the Riemann–Roch theorem, an invertible sheaf $\mathcal{L}$ of degree 4 on C has $h^0(\mathcal{L}) = 3$ and is basepoint free by Corollary 2.19, so the linear series $|\mathcal{L}|$ gives a morphism $\phi_{\mathcal{L}} : C \to \mathbb{P}^2$. The invertible sheaf $\mathcal{L} \otimes \omega_C^{-1}$ is either ω_C or nonspecial; in either case, by the Riemann–Roch theorem, it has at least one section, so we may write $\mathcal{L} \otimes \omega_C^{-1} = \mathcal{O}_C(p + q)$ for some points p, q. There are two possibilities:

(1) If $p + q = K_C$, then $\mathcal{L} = \omega_C^2$. Since the elements of $H^0(\omega_C)$ may be written as $\omega, x\omega$, the map
$$\mathrm{Sym}^2 H^0(\omega_C) \to H^0(\mathcal{L})$$
is injective, and since both sides are 3-dimensional vector spaces, they are equal. In other words, every divisor $D \sim 2K_C$ is the sum of two divisors $D_1, D_2 \in |K_C|$. We conclude that the map $\phi_{\mathcal{L}}$ is the composition of the canonical map $\phi_K : C \to \mathbb{P}^1$ with the Veronese embedding $\nu_2 : \mathbb{P}^1 \to \mathbb{P}^2$ of $\mathbb{P}^1$ as a conic in the plane and the map $\phi_{\mathcal{L}}$ is generically 2-to-1 onto the conic.

(2) If $p + q \neq K_C$ then $h^0(p + q) = 1$, so the pair p, q is unique. Furthermore, $h^0(\mathcal{L} - p) = 2 = h^0(\mathcal{L}(-p-q))$ so $H^0(\mathcal{L}(-p)) = H^0(\mathcal{L}(-q))$ and $\phi_{\mathcal{L}}(p) = \phi_{\mathcal{L}}(q)$. By the genus formula, the δ invariant of this point must be 1. By Exercise 2.11 this is a node (if $p \neq q$) or cusp (if $p = q$).

Thus for $\mathcal{L}$ in an open subset of $\mathrm{Pic}_4(C)$ the image is a quartic with a node; for a one-dimensional locus in $\mathrm{Pic}_4(C)$, the image is a quartic with a cusp; and for one point in $\mathrm{Pic}_4(C)$ the image is a conic.

Embeddings in $\mathbb{P}^3$. By Corollary 2.19 any invertible sheaf $\mathcal{L}$ of degree 5 is very ample. Write $\phi_{\mathcal{L}} : C \to \mathbb{P}^3$ for the map given by the complete linear series $|\mathcal{L}|$. Since $\phi_{\mathcal{L}}$ is an embedding, we'll also denote the image $\phi_{\mathcal{L}}(C) \subset \mathbb{P}^3$ by C and write $\mathcal{O}_C(1)$ for $\mathcal{L}$.

What degree surfaces in $\mathbb{P}^3$ contain the curve C? We start with degree 2, and consider the restriction map

$$H^0(\mathcal{O}_{\mathbb{P}^3}(2)) \to H^0(\mathcal{O}_C(2)) = H^0(\mathcal{L}^2).$$

The space on the left has dimension 10; by the Riemann–Roch theorem we have $h^0(\mathcal{L}^2) = 2 \cdot 5 - 2 + 1 = 9$. It follows that C lies on a quadric surface Q. Since C is not contained in a plane or a union of planes, any quadric containing C is irreducible; if there were more than one such, Bézout's theorem would imply that $\deg C \leq 4$. Thus Q is unique.

We might ask at this point: is Q smooth or a quadric cone? The answer depends on the choice of invertible sheaf $\mathcal{L}$.

Proposition 6.6. *Let $C \subset \mathbb{P}^3$ be a smooth curve of degree 5 and genus 2, and set $\mathcal{L} = \mathcal{O}_C(1)$. The unique quadric Q containing C is singular if and only if*

$$\mathcal{L} \cong K^2(p)$$

for some point $p \in C$; in this case, the point p maps to the vertex of Q.

Proof. Suppose first that $\mathcal{L} \cong K^2(p)$ for some $p \in C$. Then $\mathcal{L}(-p) \cong K^2$, so that the map $\pi : C \to \mathbb{P}^2$ given by projection from p is the map $\phi_{K^2} : C \to \mathbb{P}^2$ given by the square of the canonical sheaf. As we have seen, the map ϕ_{K^2} is two-to-one onto a conic $E \subset \mathbb{P}^2$, so that the curve C lies on the cone Q over E with vertex p, and this is the unique quadric surface containing C.

On the other hand, if $\mathcal{L}$ is not of the form $K^2(p)$, then we can write

$$\mathcal{L} = K \otimes \mathcal{M},$$

where by hypothesis $\mathcal{M}$ is not of the form $K(p)$. We are in case (2) at the bottom of the previous page; that is, the pencil $|\mathcal{M}|$ gives a degree 3 map $C \to \mathbb{P}^1$.

This gives us a way of factoring the map $\phi_{\mathcal{L}} : C \to \mathbb{P}^3$: we have maps $\phi_K : C \to \mathbb{P}^1$ of degree 2 and $\phi_{\mathcal{M}} : C \to \mathbb{P}^1$ of degree 3, and we can compose their product with the Segre embedding $\sigma : \mathbb{P}^1 \times \mathbb{P}^1 \to \mathbb{P}^3$:

$$C \xrightarrow{\phi_K \times \phi_{\mathcal{M}}} \mathbb{P}^1 \times \mathbb{P}^1 \xrightarrow{\sigma} \mathbb{P}^3.$$

This description of the map $\phi_{\mathcal{L}}$ shows that C is a curve of type $(2,3)$ on a smooth quadric $Q \subset \mathbb{P}^3$, completing the proof of Proposition 6.6. $\square$

The variety $\mathrm{Pic}_5(C)$ has dimension 2, while the sheaves $K^2(p)$ form a one-dimensional subfamily. Thus for a general invertible sheaf $\mathcal{L} \in \mathrm{Pic}_5(C)$ the unique quadric Q containing $\phi_{\mathcal{L}}(C)$ is smooth.

The ideal of a quintic space curve of genus 2. Continuing the discussion above, let $C \subset \mathbb{P}^3$ be a smooth quintic curve of genus 2. To describe a minimal set of generators of the homogeneous ideal $I(C) \subset \mathbb{C}[x_0, x_1, x_2, x_3]$ we look at the restriction map

$$H^0(\mathcal{O}_{\mathbb{P}^3}(3)) \to H^0(\mathcal{O}_C(3)).$$

Since the dimensions of these spaces are 20 and $15 - 2 + 1 = 14$ respectively, we see that the vector space of cubics vanishing on C has dimension at least 6. The subspace of cubics divisible by Q has dimension 4. It follows that there are at least two cubics vanishing on C that are linearly independent modulo those vanishing on Q.

We can identify these cubics geometrically. Suppose first that Q is smooth, so that C is a curve of type $(2, 3)$ on Q. In that case, if $L \subset Q$ is any line of the first ruling, the sum $C + L$ is the complete intersection of Q with a cubic S_L, unique modulo the ideal of Q; conversely, if S is any cubic containing C but not containing Q, the intersection $S \cap Q$ will be the union of C and a line L of the first ruling; thus $S = S_L$ modulo $I(Q)$. A similar argument applies in case Q is a cone, and L is any line of the (unique) ruling of Q. In Exercise 6.6 you may show that there are no more cubics containing C.

The dimension of the family of genus 2 curves. Each of the types of maps that we described from a curve C of genus 2 to projective space suggests a way to compute the dimension of the family of genus 2 curves, and indeed, as we will explain in Chapter 8, there is a moduli space of this dimension.

First, every curve of genus 2 is uniquely expressible as a double cover of $\mathbb{P}^1$ branched at six points, modulo the group PGL_2 of automorphisms of $\mathbb{P}^1$. The space of such double covers has dimension 6, and $\dim \mathrm{PGL}_2 = 3$, and since the group acts with finite stabilizers this gives a family of dimension $6 - 3 = 3$.

Also, each curve of genus 2 is expressible as a 3-sheeted cover of $\mathbb{P}^1$ (with eight branch points) in a 2-dimensional family of ways. As we saw in Section 6.2, such a triple cover is determined up to a finite number of choices by its branch divisor, so the space of such triple covers has dimension 8; modulo PGL_2 it has dimension 5, and since every curve is expressible as a triple cover in a two-dimensional family of ways, we arrive again at a family of dimension $5 - 2 = 3$.

We've also seen that each curve of genus 2 can be realized as (the normalization of) a plane quartic with a node in a 2-dimensional family of ways. The space of plane quartics has dimension 14; the family of those with a node has codimension one (Proposition 8.14) and hence dimension 13. Since the automorphism group PGL_3 of $\mathbb{P}^2$ has dimension 8, this suggests that the family of nodal plane quartics modulo PGL_3 has dimension 5. Finally, since every curve of genus 2 corresponds to a 2-parameter family of such curves, this again suggests a family of dimension $5 - 2 = 3$.

Finally, a curve of genus 2 may be realized as a quintic curve in $\mathbb{P}^3$ in a two-parameter family of ways. To count the dimension of the family of such curves, note that each one lies on a unique quadric Q. We can assume for this purpose that Q is smooth, since the singular quadrics and curves on them occur in codimension 1. The curve C is of type $(2,3)$ on Q. Thus to specify such a curve we have to specify Q (9 parameters) and then a bihomogeneous polynomial of bidegree $(2,3)$ on $Q \cong \mathbb{P}^1 \times \mathbb{P}^1$ up to scalars; these have $3 \cdot 4 - 1 = 11$ parameters. Thus there is a 20-dimensional family of such divisors; modulo the automorphism group PGL_4 of $\mathbb{P}^3$, this is a 5-dimensional family. Again, every abstract curve C of genus 2 corresponds to a 2-parameter family of these curves modulo PGL_4, so once more this suggests a family of dimension $5 - 2 = 3$.

6.4. Curves of genus 3

Let C be a smooth projective curve of genus 3. Since we have already discussed hyperelliptic curves, we will assume that C is not hyperelliptic. By Theorem 2.27, the canonical map $\phi_K : C \to \mathbb{P}^2$ embeds C as a smooth plane quartic curve. Conversely, by Proposition 2.8 any smooth plane curve of degree 4 has genus 3 and is embedded by the complete canonical series.

Since the space of plane quartic curves is 14-dimensional and $\mathrm{PGL}(3)$ has dimension 8, this suggests that there is a 6-dimensional family of curves of genus 3, and in Chapter 8 we will see that this is indeed the case.

Other representations of a curve of genus 3. Since we have assumed that C is not hyperelliptic there is no degree 2 cover of $\mathbb{P}^1$. On the other hand, there are degree 3 covers: if $\mathcal{L} \in \mathrm{Pic}_3(C)$ is an invertible sheaf of degree 3 then, by the Riemann–Roch theorem, we have

$$h^0(\mathcal{L}) = \begin{cases} 2 & \text{if } \mathcal{L} \cong K - p \text{ for some point } p \in C, \\ 1 & \text{otherwise.} \end{cases}$$

There is thus a 1-dimensional family of representations of C as a 3-sheeted cover of $\mathbb{P}^1$. These are visible directly from the canonical model: a degree 3 map $\phi_{K-p} : C \to \mathbb{P}^1$ is the composition of the canonical embedding $\phi_K : C \to \mathbb{P}^2$ with a projection from p, as illustrated in Figure 6.3.

There are other representations of C as the normalization of a plane curve. By Clifford's theorem C has no g_3^2, and the canonical system is the only g_4^2, but there are plenty of models as plane quintic curves: by Proposition 1.11, if $\mathcal{L}$ is any invertible sheaf of degree 5, the linear series $|\mathcal{L}|$ will be a basepoint free g_5^2 as long as L is not of the form $K + p$, so that $\phi_{\mathcal{L}}$ maps C birationally onto a plane quintic curve $C_0 \subset \mathbb{P}^2$. These can also be described geometrically in terms of the canonical model: any such invertible sheaf $\mathcal{L}$ is of the form $2K - p - q - r$ for some trio of points $p, q, r \in C$ that are not collinear in the canonical model, and we see that C_0 is obtained from the canonical model of C by applying a

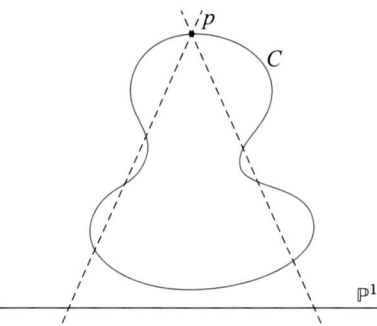

Figure 6.3. Expression of a plane quartic C of genus 3 as a 3-sheeted cover of $\mathbb{P}^1$ by projecting the canonical model from a point on it.

Cremona transform with respect to the points p, q and r, that is, by applying the birational transformation of the plane defined by the linear series of conics through p, q, r.

Proposition 1.11 implies that a divisor D of degree 6 is very ample if and only if it is not of the form $K + p + q$ for any $p, q \in C$ and since the family of invertible sheaves on C has dimension 3, we see that a general invertible sheaf of degree 6 is very ample (indeed, this is a simple case of Theorem 5.13).

If $C \subset \mathbb{P}^3$ is a curve of genus 3 embedded as a curve of degree 6, then C cannot lie on a singular quadric since by Example 2.7 it would have to be a complete intersection of the quadric with a cubic, and then such a curve has genus 4. If C lies on a smooth quadric in class (a, b) then a or b would be 2, so C would be hyperelliptic, and conversely any curve in class $(2, 4)$ is a hyperelliptic curve of genus 3, degree 6.

Thus if C is not hyperelliptic, then C does not lie on a quadric surface. We have $h^0(\mathcal{O}_{\mathbb{P}^3}(3)) = 20$ while, by the Riemann–Roch formula, $h^0(\mathcal{O}_C(3)) = 18 - 3 + 1 = 16$, so C lies on (at least) 4 independent cubics. Each of these cubics must be irreducible, so any two of them intersect in a curve of degree 9 containing C and another component or components D of degree totaling 3. By Bertini's theorem if we choose two *general* cubics containing C, then each of the components of D will be smooth. We shall see in Theorem 16.1 that the arithmetic genus of D must be 0; thus D must be a twisted cubic curve. The ideal of the twisted cubic is generated by the 2×2 minors of a matrix of the form

$$\begin{pmatrix} \ell_0 & \ell_1 & \ell_2 \\ \ell_1 & \ell_2 & \ell_3 \end{pmatrix}$$

where the ℓ_i are linear forms, and it follows that the two cubics can be written as the two 3×3 minors involving the first two rows of a matrix of the form

$$\begin{pmatrix} \ell_0 & \ell_1 & \ell_2 \\ \ell_1 & \ell_2 & \ell_3 \\ \ell_4 & \ell_5 & \ell_6 \\ \ell_7 & \ell_8 & \ell_9 \end{pmatrix}$$

where $\ell_4, \ldots, \ell_9$ are linear forms as well. From the Hilbert–Burch theorem (Corollary 18.15) one can show that the ideal of C is generated by the four 3×3 minors of this matrix, whose columns generate the syzygies of the ideal of the curve.

6.5. Theta characteristics

In this section we sketch the algebraic theory of theta characteristics, starting with the case of curves of genus 3.

Suppose that $C \subset \mathbb{P}^2$ is a smooth plane curve. A *bitangent* to C is a line $L \subset \mathbb{P}^2$ that is either tangent to C at two distinct points, or has contact of order ≥ 4 with C at a point. Alternatively, we can say that a bitangent corresponds to an effective divisor of degree 2 on C such that $2D$ is contained in the intersection of C with a line $L \subset \mathbb{P}^2$.

A naive dimension count suggests that a smooth plane curve should have a finite number of bitangents (it's one condition on a line $L \in (\mathbb{P}^2)^*$ to be tangent to C, so it should be two conditions for it to be bitangent). Indeed, this is the case; by Bézout's theorem a conic or cubic curve cannot have any bitangents, but as we will show in Section 13.1 every smooth curve of degree $d \geq 4$ has

$$12 \binom{d+1}{4} - 4d(d-2),$$

counted with appropriate multiplicities — for example, a line simply tangent to C at 3 points counts as three bitangents. Accordingly, a smooth plane quartic has 28 bitangents. (Figure 6.4 illustrates a special case in which all these bitangents are realized over $\mathbb{R}$. The first explicit such example was published by Plücker; see his drawing on page 366.)

The bitangents to a smooth plane quartic C (a canonical curve of genus 3) have a special significance: since $4 = 2 \times 2$, if $D = p + q$ is a bitangent, then the divisor $2D$ comprises the complete intersection of C with a line; in other words, we have a linear equivalence

$$2D \sim K_C$$

or equivalently the invertible sheaf $\mathcal{O}_C(D)$ is a square root of the canonical sheaf of C. Because of their appearance in the theory of theta functions, Riemann named the square roots of the canonical sheaf *theta characteristics*.

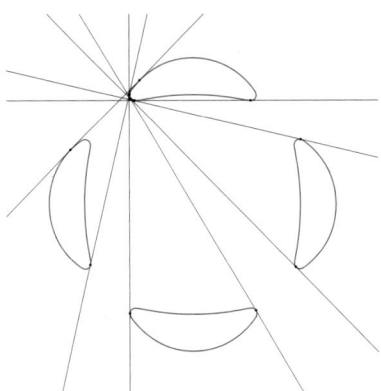

Figure 6.4. The smooth plane quartic $144(x^4 + y^4) - 225(x^2 + y^2) + 350x^2y^2 + 81 = 0$, known as the Trott curve [1997], has all 28 bitangents real. (Only seven are shown for neatness. The involution $x \leftrightarrow -x$ gives $7 - 1 = 6$ more, then $y \leftrightarrow -y$ gives $13 - 2 = 11$ more, and $x \leftrightarrow y$ the remaining four.)

How many such square roots are there? If $\mathcal{L}$ and $\mathcal{M}$ are invertible sheaves with $\mathcal{L}^2 = \mathcal{M}^2 = K$, then $\mathcal{L}$ and $\mathcal{M}$ differ by an invertible sheaf of order 2:

$$\mathcal{M} = \mathcal{L} \otimes \mathcal{F}, \quad \text{where} \quad \mathcal{F} \otimes \mathcal{F} \sim \mathcal{O}_C.$$

In other words, $\mathcal{F}$ is an invertible sheaf of degree 0 and, having fixed $\mathcal{L}$, the other sheaf, $\mathcal{M}$, corresponds to a point of order 2 in the Picard group $\mathrm{Pic}_0(C)$. Since we've seen that $\mathrm{Pic}_0(C) = \mathrm{Jac}(C)$ is a complex torus of dimension $g = 3$ — the quotient of $\mathbb{C}^3$ by a lattice $\Lambda \cong \mathbb{Z}^6$ — we see that there are $2^6 = 64$ such invertible sheaves, and thus, given that there is some invertible sheaf $\mathcal{L}$ satisfying $\mathcal{L}^2 \cong K_C$, there are exactly $64 = 2^{2g}$ of them.

The reader will have noticed that the number 64 of theta characteristics does not agree with the number 28 of bitangents. The reason is that bitangents correspond to *effective* divisors D with $2D \sim K$, while a theta characteristic $\mathcal{L}$ may have $h^0(\mathcal{L}) = 0$, that is, may not correspond to an effective divisor. This situation also occurs in other genera. What can we say about the dimensions $h^0(\mathcal{L})$ of the space of sections of the theta characteristics on C?

There is a beautiful partial answer to this question, which can be deduced from a remarkable fact: the dimension $h^0(\mathcal{L})$ of the space of sections of a theta characteristic mod 2 is invariant in families. We will now sketch the necessary results; see [Mumford 1971] and [Harris 1982b] for a full treatment.

Theorem 6.7. *Let $\mathcal{C} \to B$ be a family of smooth curves, and $\mathcal{L}_b$ a family of theta characteristics on the curves in this family — in other words, an invertible sheaf $\mathcal{L}$ on $\mathcal{C}$ such that $(\mathcal{L}|_{C_b})^2 \cong K_{C_b}$ for each $b \in B$. If $f : B \to \mathbb{Z}/2$ is defined by*

$$f(b) = h^0(\mathcal{L}|_{C_b}) \pmod 2,$$

then f is locally constant.

We say that a theta characteristic $\mathcal{L}$ is *even* or *odd* according to the parity of $h^0(\mathcal{L})$. Given the irreducibility of the space of smooth irreducible curves of genus g (which we'll discuss in Chapter 8), Theorem 6.7 suggests that all curves of genus g have the same number of even (equivalently, of odd) theta characteristics, and this is in fact the case.

Theorem 6.8. *If C is a curve of genus g, then of the 2^{2g} theta characteristics on C there are $2^{g-1}(2^g + 1)$ even theta characteristics and $2^{g-1}(2^g - 1)$ odd theta characteristics.*

Using Theorem 6.7 and the connectedness of the moduli space of curves, Theorem 6.8 is reduced to the case when C is hyperelliptic. We will compute the number of theta characteristics in the hyperelliptic case later in this section (page 127).

For a nonhyperelliptic curve C of genus 3, the dimension $h^0(\mathcal{L})$ of a theta characteristic $\mathcal{L}$ cannot be ≥ 2, so the odd theta characteristics are exactly the effective theta characteristics, and this says that there are $2^{g-1}(2^g - 1) = 28$ effective theta characteristics corresponding to the 28 bitangents.

We will present a proof of Theorem 6.7 using an ingenious construction of Mumford's, after explaining the necessary facts about quadratic forms in an even number of variables.

Cheerful Fact 6.9. Suppose that V is a $2n$-dimensional complex vector space with a nondegenerate bilinear form Q. An *isotropic subspace* for Q is a subspace $\Lambda \subset V$ such that $Q(\Lambda, \Lambda) = 0$.

(1) The maximal isotropic subspaces for Q have dimension n.

(2) The set of maximal isotropic subspaces for Q is a subvariety of the Grassmannian $G(n, V)$, of dimension $\binom{n}{2}$, that has exactly two connected components.

(3) If $\Lambda, \Lambda' \subset V$ are any two maximal isotropic subspaces, then

$$\dim(\Lambda \cap \Lambda') \equiv n \,(\text{mod } 2) \quad \Longleftrightarrow \quad \Lambda, \Lambda' \text{ belong to the same ruling.}$$

A proof is given in [Griffiths and Harris 1978, pp. 735–740].

Remark 6.10. The first assertion in Cheerful Fact 6.9 is elementary: since the map $\widetilde{Q} : V \xrightarrow{\cong} V^*$ associated to the form Q carries an isotropic subspace to its annihilator, there can't be an isotropic subspace of dimension $> n$; and similarly if $\Lambda \subset V$ is any isotropic subspace of dimension $< n$ we can include Λ in a larger isotropic subspace by adding any vector v with $\overline{Q}(v, v) = 0$ for the induced bilinear form $\overline{Q}$ on $\text{ann}(\Lambda)/\Lambda$.

The second and third assertions are less elementary, but the reader may already have seen the first two nontrivial cases of each:

Example 6.11. When $n = 2$ the form Q corresponds to a smooth quadric surface in $\mathbb{P}^3$, and the lines on this surface correspond to the isotropic 2-planes in $\mathbb{C}^4$. There are two rulings by lines, and lines of opposite rulings meet in a point, while lines of the same ruling are either disjoint or equal.

Example 6.12. When $n = 3$, the Grassmannian $\mathbb{G}(1, 3)$, in its Plücker embedding, is a smooth quadric in $\mathbb{P}^5$. The isotropic subspaces in the two distinct components are easy to describe: in one component they are the projective 2-plane of lines containing a given point $p \in \mathbb{P}^3$. In the other component they are the planes corresponding to the lines contained in a given plane $H \subset \mathbb{P}^3$. These families visibly satisfy property (3) above. See Exercise 6.9.

Proof of Theorem 6.7. Suppose that C is a smooth curve of genus g, and let $\mathcal{L}$ be an invertible sheaf on C with $\mathcal{L}^2 \cong K_C$ — that is, a theta characteristic. Choose a divisor $D = p_1 + \cdots + p_n$ of degree $n > g - 1$ consisting of distinct points, and let V be the $2n$-dimensional vector space

$$V := H^0(\mathcal{L}(D)/\mathcal{L}(-D)).$$

From the exact sequence

$$0 \to \mathcal{L}(-D) \to \mathcal{L}(D) \to \mathcal{L}(D)/\mathcal{L}(-D) \to 0$$

we see that the sheaf $\mathcal{L}(D)/\mathcal{L}(-D)$ is supported on D, with stalk isomorphic to $\mathcal{O}_p/\mathfrak{m}_{C,p}^2$ of dimension 2 at each $p \in D$. We can define a bilinear form on V by setting

$$Q(\sigma, \tau) := \sum_i \operatorname{Res}_{p_i}(\sigma\tau)$$

where we use the isomorphism $\mathcal{L}^2 \cong K_C$ to identify the product $\sigma\tau$ with a rational differential.

We now introduce two isotropic subspaces for Q. The first is

$$\Lambda := H^0(\mathcal{L}/\mathcal{L}(-D)),$$

which is isotropic because the product of two of its elements corresponds to a regular differential, and so has no residues. Second, we set

$$\Lambda' := \operatorname{im}\bigl(H^0(\mathcal{L}(D)) \to H^0(\mathcal{L}(D)/\mathcal{L}(-D))\bigr).$$

Since $H^0(\mathcal{L}(-D)) = 0$, the map is injective and according to the Riemann–Roch theorem we have $h^0(\mathcal{L}(D)) = n$, so this is again an n-dimensional subspace of V; it's isotropic because the sum of the residues of a global rational differential on C is 0. Finally,

$$H^0(\mathcal{L}) \cong \Lambda \cap \Lambda',$$

and Theorem 6.7 follows. $\square$

Counting theta characteristics (proof of Theorem 6.8). One way to count the number of odd and even theta characteristics on a curve of genus g is to describe them explicitly in the case of a hyperelliptic curve and use Theorem 6.7 to deduce the corresponding statements for any smooth curve of genus g. The reader may wish to try a relatively simple case in Exercise 6.7 before looking at the general case below. We start with some preliminary calculations:

Lemma 6.13. *For any positive integer n, we have*

(1)
$$\sum_{k=0}^{n} \binom{2n}{2k} = \sum_{k=0}^{n-1} \binom{2n}{2k+1} = 2^{2n-1},$$

(2) $\quad \sum_{k=0}^{n} \binom{4n}{4k} = 2^{4n-2} + (-1)^n 2^{2n-1}, \quad \sum_{k=0}^{n-1} \binom{4n}{4k+2} = 2^{4n-2} - (-1)^n 2^{2n-1},$

(3) $\quad \sum_{k=0}^{n} \binom{4n+2}{4k+1} = 2^{4n} + (-1)^n 2^{2n}, \quad \sum_{k=0}^{n-1} \binom{4n}{4k+3} = 2^{4n} - (-1)^n 2^{2n}.$

Proof. Equality (1) is elementary; by the binomial theorem, we have

$$2^{2n} = (1+1)^{2n} = \sum_{l=0}^{2n} \binom{2n}{l} \quad \text{and} \quad 0 = (1-1)^{2n} = \sum_{l=0}^{2n} (-1)^l \binom{2n}{l},$$

and taking the sum and the difference of these two equations yields the result.

The equalities in (2) follow similarly by applying the binomial theorem to the expression $(1+i)^{4n} = (-1)^n 2^{2n}$. Equating the real parts, we have

$$\sum_{k=0}^{n} \binom{4n}{4k} - \sum_{k=0}^{n-1} \binom{4n}{4k+2} = (-1)^n 2^{2n},$$

while by (1) we have

$$\sum_{k=0}^{n} \binom{4n}{4k} + \sum_{k=0}^{n-1} \binom{4n}{4k+2} = 2^{4n-1}.$$

Taking the sum and difference of these equations yields the desired formulas.

For (3) we apply the binomial theorem to the expression $(1+i)^{4n+2} = (-1)^n 2^{2n+1} i$. Equating the imaginary parts, this gives

$$\sum_{k=0}^{n} \binom{4n+2}{4k+1} - \sum_{k=0}^{n-1} \binom{4n+2}{4k+3} = (-1)^n 2^{2n+1},$$

whereas by (1),

$$\sum_{k=0}^{n} \binom{4n+2}{4k+1} + \sum_{k=0}^{n-1} \binom{4n+2}{4k+3} = (-1)^n 2^{4n+1},$$

and as before taking the sum and difference yields the result. $\square$

We will count the number of theta characteristics on a hyperelliptic curve in terms of sums of subsets of the ramification points, so we need to know what linear equivalences exist among sums of these subsets:

Lemma 6.14. *Let C be the hyperelliptic curve of genus g expressed as a 2-sheeted cover of* $\mathbb{P}^1$ *with ramification points* $p_1, \ldots, p_{2g+2}$. *The divisor class of any half of the ramification points is equal to the divisor class of the other half, but there are no smaller relations. More precisely, let* I_1, I_2 *be subsets of* $\{1, \ldots 2g+2\}$ *and set*

$$D_i = \sum_{j \in I_i} p_j.$$

The divisors D_1, D_2 *are linearly equivalent if and only if* $I_1 = I_2$ *or they have the same cardinality* $g+1$ *and* $I_1 \cup I_2 = \{1, \ldots, 2g+2\}$.

Proof. The "if" part is simply Corollary 6.2 above.

For the "only if" part, subtracting whatever points D_1 and D_2 have in common we may suppose that $I_1 \cap I_2 = \emptyset$. If $D_1 \sim D_2$, it follows at once that they have the same degree, $d \leq g+1$, and we must show that either $d = 0$ or $d = g+1$.

We have $D_1 \sim D_2$ if and only if $D_1 + D_2 \equiv 2D_1$. If $d \leq g$ we have $r(2D_1) = d$: for $d < g$ this is the extremal case of Clifford's theorem, while for $d = g$ this follows simply from the Riemann–Roch formula. Thus in case $d \leq g$ every divisor in $|2D_1|$ is a sum of d fibers of the 2 to 1 map of C to $\mathbb{P}^1$, and for such a divisor to be a sum of distinct points p_i the degree d must be 0, concluding the argument. $\square$

Returning now to the counting, let C be the hyperelliptic curve of genus g, expressed as a 2-sheeted cover of $\mathbb{P}^1$, with ramification points $p_1, \ldots, p_{2g+2}$.

First of all, if we denote the class of the unique g_2^1 on C by E, and D is any theta characteristic, then $D + E$ will be effective, and so we can write

$$D \sim mE + F$$

with $-1 \leq m \leq (g-1)/2$ and F the sum of $g - 1 - 2m$ distinct points p_i. This representation is unique unless $m = -1$; in that case, we note that the sum of $g + 1$ of the branch points of C is linearly equivalent to the sum of the other $g + 1$ by Corollary 6.2. Thus the total number of theta characteristics is a sum of binomial coefficients; if g is odd, it is

$$\binom{2g+2}{0} + \binom{2g+2}{2} + \binom{2g+2}{4} + \cdots + \binom{2g+2}{g-1} + \frac{1}{2}\binom{2g+2}{g+1}$$

and similarly if g is even it is

$$\binom{2g+2}{1} + \binom{2g+2}{3} + \binom{2g+2}{5} + \cdots + \binom{2g+2}{g-1} + \frac{1}{2}\binom{2g+2}{g+1}.$$

In either case, we are adding up every other entry in the $(2g+2)$-nd row of Pascal's triangle, starting from the left and ending up with one half of the middle term. This sum is exactly one half of the sum of every other entry in the whole row; by the first part of Lemma 6.13 this equals $\frac{1}{4} \cdot 2^{2g+2} = 2^{2g}$.

Finally, we can add up the number of even and odd theta characteristics separately simply by taking every other term in the sums above; using equalities (2) and (3) in Lemma 6.13 (in case g is odd and even, respectively) we can conclude that C has $2^{g-1}(2^g - 1)$ odd theta characteristics and $2^{g-1}(2^g + 1)$ even theta characteristics. By Theorem 6.7 and the connectedness of the space of smooth irreducible curves of genus g, this count then holds for all curves of genus g, establishing Theorem 6.8. □

It is also possible to describe the configurations of odd and even theta characteristics as subsets of the set S of all theta characteristics, which as we've seen is a principal homogeneous space for the group $\mathrm{Jac}(C)_2 \cong (\mathbb{Z}/2\mathbb{Z})^{2g}$ of points of order 2 on the Jacobian. This leads to an alternative proof of Theorem 6.8 as in [Harris 1982b].

Cheerful Fact 6.15. There is more to say about the configuration of theta characteristics. As noted, if we choose any theta characteristic on a curve C, we may identify the set S^- of odd theta characteristics with a subset of the group $\mathrm{Jac}(C)_2$ of points of order 2 on the Jacobian of C. We might expect that some 4-tuples of these points will add up to 0 in $\mathrm{Jac}(C)$; in other words, there should exist some 4-tuples $\mathcal{L}_1, \ldots, \mathcal{L}_4 \in S^-$ such that

$$\mathcal{L}_1 \otimes \cdots \otimes \mathcal{L}_4 = 2K_C.$$

What this means in the case of genus $g = 3$ is that among the 28 bitangents to a smooth plane quartic curve C, there are some subsets of 4 whose eight points of tangency form the intersection of C with a plane conic. From the more detailed knowledge of the configuration S^- we can say how many. Indeed, the number was first found by Salmon [1879]; it is 315.

6.6. Exercises

Exercise 6.1. We have seen that a curve C of genus $g = 1$ is expressible as a 2-sheeted cover of $\mathbb{P}^1$ branched over four points; that is, as the smooth projective curve associated to the affine curve $C^\circ \subset \mathbb{A}^2$ given by $y^2 - \prod_{i=1}^4 (x - \lambda_i)$. Show that the closure $\overline{C^\circ}$ of $C^\circ \subset \mathbb{A}^2$ in either $\mathbb{P}^2$ or $\mathbb{P}^1 \times \mathbb{P}^1$ consists of the union of C° with one additional point, with that point a tacnode of $\overline{C^\circ}$ in either case. ◆

Exercise 6.2. Find the number of 3-sheeted covers $C \to \mathbb{P}^1$ of genus g with simple branching except for one point of total ramification (that is, one point with just a single preimage point.) ◆

Exercise 6.3. Let C be a curve of genus g. How many unramified double covers of C are there? ◆

Exercise 6.4. Show that unramified double covers of a smooth curve C are in one-to-one correspondence with invertible sheaves $\mathcal{L}$ on C such that $\mathcal{L}^2 \cong \mathcal{O}_C$, that is with the 2-torsion points of $\mathrm{Jac}(C)$.   ◆

Exercise 6.5. Let E be a curve of genus 1, and $q_1, \ldots, q_b \in E$. How many double covers $C \to E$ are there branched over the q_i?   ◆

Exercise 6.6. In this exercise, we ask you to complete the earlier description of the ideal of a quintic space curve of genus 2, keeping the notation of page 120.

Show that for any pair of lines L, L' of the appropriate ruling of Q, the three polynomials Q, S_L and $S_{L'}$ generate the homogeneous ideal $I(C)$. Find relations among them. Write out the minimal resolution of $I(C)$.   ◆

Exercise 6.7. Let C be a curve of genus 2, expressed as a 2-sheeted cover of $\mathbb{P}^1$ with ramification points $p_1, \ldots, p_6$. In this exercise we will count the number of even and odd theta characteristics. The text contains the count for a hyperelliptic curve of any genus; we offer the case of genus 2 as a warmup.

(1) Show that the theta characteristics on C are either of the form $\mathcal{L} = \mathcal{O}_C(p_i)$ or of the form $\mathcal{L} = \mathcal{O}_C(p_i + p_j - p_k)$ with i, j, k distinct.

(2) Show that in the first case we have $h^0(\mathcal{L}) = 1$, and in the second case we have $h^0(\mathcal{L}) = 0$.

(3) Finally, show that there are six of the former kind, and 10 of the latter, making $2^4 = 16$ in all.   ◆

Exercise 6.8. Let C be a curve of genus 2 and let $\mathcal{L} \in \mathrm{Pic}_4(C)$ be an invertible sheaf of the form $\mathcal{L} = K_C(p+q)$ with $p \neq q$ and $p + q \nsim K_C$ as in 2. Show that

(1) $h^0(\mathcal{L}(-2r)) = 1$ for any point $r \in C$, and

(2) $h^0(\mathcal{L}(-2p - 2q)) = 0$.

Deduce from this that the map $\phi_\mathcal{L}$ is an immersion, and that the tangent lines to the two branches of $\phi_\mathcal{L}(C)$ at the point $\phi_\mathcal{L}(p) = \phi_\mathcal{L}(q)$ are distinct, meaning the point $\phi_\mathcal{L}(p) = \phi_\mathcal{L}(q)$ is a node of $\phi_\mathcal{L}(C)$.   ◆

Exercise 6.9. We can represent any line in $\mathbb{P}^3$ as the row space of a 2×4 matrix by choosing 2 points on the line and using their coordinates as the rows. The *Plücker coordinates* of the line are the six 2×2 minors

$$\{p_{i,j}\}_{0 \leq i < j \leq 3}$$

of this matrix. They are independent, up to a common scalar multiple, of the two points chosen, and define the *Plücker embedding* of the Grassmannian $\mathbb{G}(1, 3)$ in $\mathbb{P}^5$.

The minors $p_{i,j}$ satisfy a nonsingular quadratic equation: if we stack two copies of the 2×2 matrix to produce a 4×4 matrix, its determinant is zero, and the Laplace expansion of this determinant is the *Plücker equation*

$$p_{0,1}p_{2,3} - p_{0,2}p_{1,3} + p_{0,3}p_{1,2} = 0.$$

(1) Show that the quadratic form $Q = p_{0,1}p_{2,3} - p_{0,2}p_{1,3} + p_{0,3}p_{1,2}$ is non-singular, and deduce that it generates the ideal of $\mathbb{G}(1,3)$ in $\mathbb{P}^5$.

(2) Write the bilinear form corresponding to Q as the determinant of a matrix, and deduce that two points in $\mathbb{G}(1,3)$ correspond to vectors that pair to 0 if and only if they correspond to lines that intersect.

(3) Deduce that a maximal isotropic subspace for Q corresponds either to the set of lines containing a given point or the set of lines contained in a given plane; and that two such sets of lines of the same type meet in a single point or coincide.

Chapter 7

Fine moduli spaces

7.1. What is a moduli problem?

Algebraic geometry is special among geometric theories in that the objects in-volved — varieties, schemes or maps between them — can be parametrized by other varieties or schemes. The set of submanifolds of a given manifold, or more generally of maps between two given manifolds, seems too large to be given the structure of a finite-dimensional manifold itself. By contrast, any algebraic variety is specified by a finite collection of polynomials, which in turn have a finite number of coefficients, so it's not too far-fetched that the set of all varieties with specified numerical invariants, or morphisms between two given varieties, could be given the structure of a "moduli space" that is a variety (or scheme or...) in its own right.

For an example — perhaps the original one — projective plane curves of degree d are in natural one-to-one correspondence with the forms of degree d modulo the group of nonzero scalars — that is, with the points of the dual of the projective space $\mathbb{P}(H^0(\mathcal{O}_{\mathbb{P}^2}(d))) = \mathbb{P}^{\binom{d+2}{2}-1}$. Thus, for example, plane cubics are parametrized by $\mathbb{P}^9$, and a family $\mathcal{C} \to B$ of cubics corresponds to a map $\phi : B \to \mathbb{P}^9$ (Figure 7.1). In this chapter, we'll give a general framework for the notion of moduli space, introducing the main examples that we will treat.

There are several ways in which the possibility of making moduli spaces has been useful in algebraic geometry. First, the existence of a moduli space

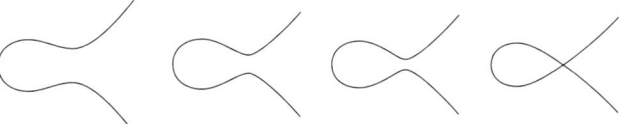

Figure 7.1. A one-parameter family of plane cubics.

that parametrizes objects of a certain type allows us to speak of the "general object," meaning that we allow ourselves to avoid the properties of "special objects" parametrized by closed subvarieties of the moduli space. For example, we can make precise sense of the statement "The general plane curve of degree d is smooth." It means that the set of smooth curves corresponds to an open subset of the projective space $\mathbb{P}^{\binom{d+2}{2}-1}$. We have already used this possibility in many places in this book.

Second, it allows us to speak coherently about families of objects. Some moduli spaces carry *universal families*, and every nice family of the sort of objects they parametrize is pulled back from this one by a unique map.

This idea was already exploited informally in the nineteenth century in the guise of "preservation of number," used to count configurations of points or curves with a given property by specializing the data, and we have also exploited this idea in Chapter 5 to explain the count of odd and even theta characteristics on a general curve by appealing to the existence of a specialization to a hyperelliptic curve. In a related manner, we have already seen how the fact that invertible sheaves of degree d on a curve C are parametrized by a g-dimensional variety allows us to prove the $g + 3$ theorem (Theorem 5.13).

Third, to the extent that we can describe the intersection theory of a moduli space, it opens up the possibility of doing enumerative geometry on it to count solutions of geometric problems — and in particular to prove the existence of solutions. For example, knowing that the parameter space for lines in $\mathbb{P}^3$ is the Grassmannian $\mathbb{G}(1, 3)$, a projective variety of dimension 4, and that the condition that the set of lines meeting a given curve of degree d is a divisor linearly equivalent to d times the hyperplane section (Figure 7.2), we can conclude that there exists a line in $\mathbb{P}^3$ meeting any four given curves [Eisenbud and Harris 2016, Section 3.4.1]. We will use the same idea to prove the much deeper existence of certain linear series on all curves (the existence half of the Brill–Noether theorem, discussed in Chapter 12).

In modern terms, a *moduli problem* consists of a class of algebraic geometric

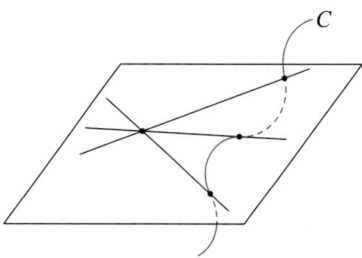

Figure 7.2. A general one-parameter linear family of lines in $\mathbb{P}^3$ — that is, the family of lines contained in a general plane and passing through a general point in that plane — meets a space curve C in $\deg C$ points.

objects — schemes, subschemes of a given scheme, maps of schemes, sheaves on schemes, typically defined by some common attributes — and a notion of what it means to have a *family* of these objects parametrized by a scheme B. The notion is formalized in the idea of a *moduli functor*, which associates to each scheme B the set of families over B of the given sort. Examples will make this vague notion more concrete.

Example 7.1 (effective divisors on a given curve). The objects are effective divisors of given degree on a given smooth, projective curve C. A family of such divisors is a subscheme $\mathcal{D} \subset B \times C$, flat of degree d over B. Here we are using the equivalence between divisors of degree d on a smooth curve and degree d subschemes of the curve. The moduli space is the d-th symmetric power C_d of C, discussed in Section 5.1.

Example 7.2 (invertible sheaves on a given curve). The objects are invertible sheaves on a given smooth projective curve C. A family over a scheme B is an equivalence class of invertible sheaves $\mathcal{L}$ on $B \times C$ whose restriction to each fiber of $B \times C$ over B has degree d, where two families $\mathcal{L}$ and $\mathcal{L}'$ on $B \times C$ are equivalent if $\mathcal{L} \cong \mathcal{L}' \otimes \mathcal{M}$, where $\mathcal{M}$ is an invertible sheaf pulled back from B. Usually one restricts attention to invertible sheaves whose restrictions to each fiber $b \times C$ have a given degree d. The moduli spaces are the Jacobian and Picard varieties, discussed in Section 5.2.

Example 7.3 (moduli of smooth curves). The objects are isomorphism classes of smooth, projective curves of genus g. A family over B is an equivalence class of smooth, projective morphisms $f : \mathcal{C} \to B$ whose fibers are curves of genus g, where two such families f, f' are equivalent if there is an isomorphism from the source of f to the source of f' making the following diagram commute:

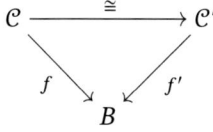

The moduli spaces M_g of curves are harder to construct, and we will have a separate discussion of them in Chapter 8. Nonetheless, it will be useful at several points in this chapter to assume their existence.

Example 7.4 (Hurwitz spaces). An object is a smooth projective curve C of a given genus, together with a map $f : C \to \mathbb{P}^1$ of given degree, up to isomorphisms of the curves that commute with the map, as in this diagram:

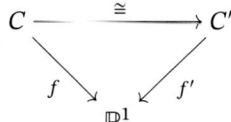

A family over B is a family $\mathcal{C} \to B$ of smooth projective curves of a given genus, together with flat map $\mathcal{C} \to B \times \mathbb{P}^1$ of degree d. Often the allowable ramification indices are specified. We will discuss the simplest Hurwitz scheme in Chapter 8.

Example 7.5 (Severi varieties). The objects are plane curves of given degree d and geometric genus g. If $g \neq \binom{d-1}{2}$ the curves will necessarily be singular, but are often constrained to have only mild singularities, usually only nodes. We will discuss Severi varieties in Chapter 8.

Example 7.6 (Hilbert schemes). The objects are subschemes of a given projective space; a family over B is a subscheme $\mathcal{C} \subset B \times \mathbb{P}^r$, flat over B. Since the Hilbert polynomials of the fibers of a flat family are all equal [Hartshorne 1977, Chapter III, §9], the Hilbert scheme is the disjoint union of subschemes corresponding to particular Hilbert polynomials, and these subschemes are of finite type. If B is reduced then the flatness condition is equivalent to the constancy of the Hilbert polynomial ([Hartshorne 1977, Theorem 9.9] or [Eisenbud and Harris 2000, Chapter III, §3.2]).

In Chapter 19 we study the open set consisting of smooth curves $C \subset \mathbb{P}^r$ of given degree and genus.

7.2. What is a solution to a moduli problem?

Typically what we want from a solution to a moduli problem is to understand all possible families of the objects in question. In particular, the individual objects should be in natural one-to-one correspondence with the closed points of the moduli space. The word *natural* is the key. Most of the time, the set of objects we are interested in has the cardinality of the continuum, as do all positive-dimensional varieties M over $\mathbb{C}$, so a mere bijection between the points of M and the objects to be parametrized is meaningless.

In the nicest situations there is a *universal* family $\phi : \mathcal{X} \to M$ of these objects over M, such that any family over a scheme B is pulled back from the one on M via a unique morphism $B \to M$. Such a space M with its universal family ϕ, if it exists, is called a *fine moduli space*. This can be expressed more abstractly but more succinctly by saying that the moduli scheme *represents the moduli functor*, which means there is an isomorphism of functors

$$\{B \mapsto \text{families of } X \text{ over } B\} \cong \{B \mapsto \text{Mor}_{\text{Schemes}}(B, M)\}.$$

If M is a fine moduli space then the identity map $M \to M$ corresponds to the "universal family" $\phi : \mathcal{X} \to M$. The Hilbert scheme, as well as $\text{Div}_d(C)$ and $\text{Pic}_d(C)$ are fine moduli spaces but the moduli space of curves and the Hurwitz schemes are not.

If a fine moduli space and its universal family exist, then it is unique up to unique isomorphism: given two avatars M and M' the universal family on M corresponds to a map $M \to M'$, and we similarly produce a map $M' \to M$. The pullback of the universal family on M by the composition of these two maps is again the universal family, so the composition is the identity map.

Although the moduli space of curves and the Hurwitz spaces are not fine moduli spaces, they are still defined in a way that makes them unique, as we shall see below.

One of the useful features of a fine moduli space is that it makes the computation of tangent spaces relatively easy. Recall that if $(R, \mathfrak{m})$ is the local ring at a point m on a variety $\mathcal{M}$ then the (Zariski) tangent space to M at m is the vector space of linear functionals $\mathrm{Hom}_{R/\mathfrak{m}}(\mathfrak{m}/\mathfrak{m}^2, R/\mathfrak{m})$. Assuming that R contains its residue field $k := R/\mathfrak{m}$, such functionals are precisely the restrictions to $\mathfrak{m}/\mathfrak{m}^2$ of the ring homomorphisms $R \to k[\epsilon]/(\epsilon^2)$ inducing the identity on k. If $\mathcal{M}$ is a fine moduli space for some functor F, then such maps are in one-to-one correspondence with the set $F(E)$ of families over $E := \mathrm{Spec}(k[\epsilon]/(\epsilon^2))$.

The vector space structure on the tangent space is also accessible from this description: the sum of tangent vectors corresponding to families $X_i \to E$ is the restriction to the diagonal $E \subset E \times E$ of the product family $X_1 \times X_2 \to E \times E$. Often one can to compute the tangent space to a moduli space before even knowing that the moduli space exists!

7.3. Hilbert schemes

The Hilbert scheme is a fine moduli space representing the functor of flat families of subschemes of $\mathbb{P}^r$, that is, the functor that takes a scheme B over $\mathbb{C}$ to the set of subschemes $\mathcal{X} \subset B \times \mathbb{P}^r$ that are flat over B; a morphism $f : B' \to B$ induces a map carrying a family to the pullback of the family by f.

Example 7.7 (the Hilbert scheme of plane curves). Let $p(m) := dm + 1 - \binom{d-1}{2}$ be the Hilbert polynomial of a plane curve of degree d, and consider the family $\mathrm{Hilb}_{p(m)}(\mathbb{P}^2)$ of schemes $X \subset \mathbb{P}^2$ having Hilbert polynomial p. Since $\dim X = \deg p = 1$, X has at least one component that is 1-dimensional. The union of such components must have degree equal to d, the leading coefficient of $p(m)$. Since the Hilbert polynomial of this union is also equal to $p(m)$ we see that X is in fact a (possibly nonreduced and/or reducible) plane curve of degree d. The space $\mathrm{Hilb}_{p(m)}(\mathbb{P}^2)$ is the projective space $\mathbb{P}^{\binom{d+2}{2}-1}$ of forms of degree d, and the universal family is the projection universal family

$$\{(x, F) \in \mathbb{P}^2 \times \mathbb{P}^{\binom{d+2}{2}-1} \mid F(x) = 0\} \to \mathbb{P}^{\binom{d+2}{2}-1}.$$

More typically, the set of curves $C \subset \mathbb{P}^r$ of degree d and genus g corresponds to a subset of the Hilbert scheme parametrizing subschemes of $\mathbb{P}^r$ with

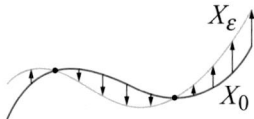

Figure 7.3. Infinitesimal perturbation of a curve by a normal vector field.

Hilbert polynomial $p(m) = dm - g + 1$, though not all schemes with this Hilbert polynomial are purely one-dimensional subschemes, as we shall see.

In this section we compute the Zariski tangent space to the Hilbert scheme at a point and sketch the construction of the scheme itself. For a rigorous treatment including many generalizations, see [Álvarez et al. 2008] or [Nitsure 2005]. In Chapter 19, we'll describe the open subsets of smooth curves in the Hilbert schemes of curves of low degree and genus in $\mathbb{P}^3$ in more detail.

7.3.1. The tangent space to the Hilbert scheme. Following the general recipe for tangent spaces to fine moduli spaces, we need to understand flat families of projective schemes over $E := \operatorname{Spec} \mathbb{C}[\epsilon]/(\epsilon^2)$. Recall from [Hartshorne 1977, p. 182], for example, that if $X \subset \mathbb{P}^r$ is a smooth subscheme then the normal bundle $\mathcal{N}_{X/\mathbb{P}^r}$ of X in $\mathbb{P}^r$ is defined in terms of the tangent bundles of X and $\mathbb{P}^r$ by the exact sequence:

$$0 \to \mathcal{T}_X \xrightarrow{\phi} \mathcal{T}_{\mathbb{P}^r}|_X \to \mathcal{N}_{X/\mathbb{P}^r} \to 0.$$

Thus a global section of $\mathcal{N}_{X/\mathbb{P}^r}$ can be thought of as an infinitesimal motion of X (Figure 7.3).

To define the normal sheaf for locally complete intersection subschemes X we first define the *conormal sheaf* to be the kernel of the dual of ϕ, which is the natural map $\Omega_{\mathbb{P}^r}|_X \to \Omega_X$, and since X is locally a complete intersection, this corresponds to the exact sequence:

$$0 \to \mathcal{I}_{X/\mathbb{P}^r}/\mathcal{I}_{X/\mathbb{P}^r}^2 \to \Omega_{\mathbb{P}^r}|_X \to \Omega_X \to 0.$$

We thus define $\mathcal{N}_{X/\mathbb{P}^r} := \mathcal{H}om(\mathcal{I}_{X/\mathbb{P}^r}/\mathcal{I}_{X/\mathbb{P}^r}^2, \mathcal{O}_X)$.

By analogy we define the *normal sheaf* of any subscheme $X \subset \mathbb{P}^r$

$$\mathcal{N}_{X/\mathbb{P}^r} := \mathcal{H}om(\mathcal{I}_{X/\mathbb{P}^r}/\mathcal{I}_{X/\mathbb{P}^r}^2, \mathcal{O}_X).$$

The following theorem shows that this has the same connection with infinitesimal motions for arbitrary subschemes X.

Theorem 7.8. *Let $X \subset \mathbb{P}^r$ be a subscheme with Hilbert polynomial $p(m)$ and let $E = \operatorname{Spec} \mathbb{C}[\epsilon]/(\epsilon)^2$. The flat families $\mathcal{X} \subset \mathbb{P}^r \times E$ specializing to X at the closed point defined by (ϵ) are in natural one-to-one correspondence with the vector space $\operatorname{Hom}_{\mathbb{P}^r}(\mathcal{I}_{X/\mathbb{P}^r}, \mathcal{O}_X) = H^0(\mathcal{N}_{X/\mathbb{P}^r})$, which is thus the tangent space to the Hilbert scheme $\operatorname{Hilb}_p(\mathbb{P}^r)$ at $[X]$.*

Proof. We will actually treat the analogous result for an affine subscheme $X \subset \mathbb{A}^r = \operatorname{Spec} S$, where $S = \mathbb{C}[x_1, \ldots, x_r]$; since our construction is natural, it will patch on an affine cover to give the projective case stated above.

An E-module M is flat over E if and only if $\operatorname{Tor}_1^E(\mathbb{C}, M) = 0$. Since the free resolution of $\mathbb{C}$ as an E-module has the form

$$\cdots \xrightarrow{\epsilon} E \xrightarrow{\epsilon} E \xrightarrow{\epsilon} E \longrightarrow \mathbb{C} \longrightarrow 0,$$

we have $\operatorname{Tor}_1^E(\mathbb{C}, M) = 0$ if and only if the submodule of M annihilated by ϵ is ϵM.

We first construct a flat family from a homomorphism: Let

$$I = (g_1, \ldots, g_t) \subset S = \mathbb{C}[x_1, \ldots, x_r]$$

be the ideal defining X. Note that $\operatorname{Hom}_S(I/I^2, S/I) = \operatorname{Hom}_S(I, S/I)$. Let $\phi : I \to S/I$ be a homomorphism, and let $h_i \in S$ be any element reducing to $\phi(g_i)$ modulo I. The ideal

$$I' := (g_1 + \epsilon h_1, \ldots, g_t + \epsilon h_t) \subset S[\epsilon]/(\epsilon^2) =: S'$$

defines a scheme $\operatorname{Spec} S'/I'$ over E which restricts to X modulo (ϵ); that is, $I' + (\epsilon) = I$.

To see that I' is independent of the lifting chosen, consider $k_1, \ldots, k_t \in I$ such that the elements $g_i + \epsilon(h_i + k_i)$ are a different lifting, generating an ideal I''. Writing $k_i = \sum r_{i,j} g_j$ we have

$$g_i + \epsilon(h_i + k_i) = g_i + \epsilon h_i + \sum \epsilon r_{i,j}(g_j + \epsilon h_j),$$

so $I'' \subset I'$, and symmetrically $I' \subset I''$.

To prove flatness, suppose that $a \in S'$ and $\epsilon a \in I'$, so that we can write $\epsilon a = \sum (r_i + \epsilon s_i)(g_i + \epsilon h_i)$ for some $r_i, s_i \in S$. It follows that $\sum r_i g_i = 0$. Since ϕ is a homomorphism this implies $\sum r_i h_i \in I$. Thus $\epsilon a \in \epsilon I \subset S'$, whence $a \in I + (\epsilon)$. Writing $a = \sum p_i g_i + \epsilon b'$ and using the relations $g_i \equiv -\epsilon h_i \bmod I'$ we get $a \equiv \epsilon(-\sum p_i h_i + b') \bmod I'$, as required.

Finally, starting from a flat family S'/I' over E with $I' + (\epsilon) = I + \epsilon$, let J be the image of I' in $S'/(\epsilon I) = S \oplus (\epsilon S)/(\epsilon I)$. We claim that J is the graph of a homomorphism $\phi : I \to (\epsilon S)/(\epsilon I) \cong S/I$.

Since $I' + \epsilon S = I + \epsilon S$, the projection to the summand $S \subset S'$ maps J onto I. To prove that J is the graph of a homomorphism, it suffices to show that this projection is an isomorphism. The kernel is the intersection of J with $(\epsilon S)/(\epsilon I)$, so we must show that if $r \in S$ and $\epsilon r \in I'$ then $\epsilon r \in \epsilon I$.

Since $\epsilon r \in I'$, the condition of flatness implies that the image of r in S'/I' is in $\epsilon(S'/I')$, which is to say that $r \in I' + \epsilon S = I + \epsilon S$. Thus $r \in I + \epsilon S$, whence $\epsilon r \in \epsilon I$ as required.

This shows that J is the graph of a homomorphism ϕ such that I' is generated by $\{g + \phi g \mid g \in I\}$, so the two constructions are inverse to each other. $\square$

Example 7.9. If C is the plane curve of degree d defined by a form F, then the ideal sheaf of C is $\mathcal{O}_{\mathbb{P}^2}(-d)$, and thus

$$\mathcal{H}om(\mathcal{I}_{C/\mathbb{P}^2}/\mathcal{I}^2_{C/\mathbb{P}^2}, \mathcal{O}_C) = \mathcal{O}_{\mathbb{P}^2}(d)|_C = \mathcal{O}_C(d).$$

From the exact sequence

$$0 \to \mathcal{O}_{\mathbb{P}^2} \xrightarrow{F} \mathcal{O}_{\mathbb{P}^2}(d) \to \mathcal{O}_C(d) \to 0$$

we deduce that the dimension of the tangent space to $\mathrm{Hilb}_{p(m)}(\mathbb{P}^2) = \mathbb{P}^{\binom{d+2}{2}-1}$ at C with $p(m) = dm+1-\binom{d-1}{2}$, is $h^0(\mathcal{O}_C(d)) = h^0(\mathcal{O}_{\mathbb{P}^2}(d))-1 = \dim \mathbb{P}^{\binom{d+2}{2}-1}$, as expected.

7.3.2. Parametrizing twisted cubics.
By Lemma 7.10 below, the set of twisted cubics — that is, smooth, irreducible, nondegenerate curves of degree 3 in $\mathbb{P}^3$ — is an open subset of the Hilbert scheme $\mathrm{Hilb}_{3m+1}(\mathbb{P}^3)$. As we've seen, a twisted cubic curve $C \subset \mathbb{P}^3$ can be described as the zero locus of three homogeneous quadratic polynomials Q_1, Q_2 and Q_3 in the homogeneous coordinates on $\mathbb{P}^3$; to specify the twisted cubic we could just list the $3 \times 10 = 30$ coefficients of these. But of course we could replace the three quadrics Q_i with any three independent linear combinations of them; what matters — and what is naturally associated to C — is the vector space $V = \langle Q_1, Q_2, Q_3 \rangle \subset H^0(\mathcal{O}_{\mathbb{P}^3}(2))$. This suggests that we consider the map of sets

$$h : \{\text{twisted cubic curves } C \subset \mathbb{P}^3\} \to G = G(3, H^0(\mathcal{O}_{\mathbb{P}^3}(2)))$$

obtained by associating to a twisted cubic C the second graded piece of its homogeneous ideal.

This differs significantly from the example of plane curves introduced in the second paragraph of this chapter (page 133; see also Examples 7.7 and 7.9): there, the objects to be parametrized were the zero locus of a single polynomial, and we could vary those coefficients arbitrarily and still have a plane curve; thus, the image of the analogous map was open in the projective space $\mathbb{P}^{\binom{d+2}{2}-1}$. But if we generically perturb the coefficients of the three quadratic polynomials Q_i defining the twisted cubic the resulting quadrics will be a complete intersection, generating the ideal of a set of eight points. Thus the image of the map h does not contain an open set of G. We will give equations defining the image in Section 7.3.3, and we can consider it the Hilbert scheme of twisted cubics.

As in this case, we will mainly be interested in the subsets of the Hilbert schemes corresponding to smooth irreducible curves:

Lemma 7.10. *Suppose that $X \to B$ is a flat family of projective schemes (that is, the projection to the second factor of $X \subset B \times \mathbb{P}^n$ is flat). The points $b \in B$ such that the fiber X_b is smooth and irreducible form an open set.*

Proof. To prove the lemma we may assume that $B = \operatorname{Spec} A$ is affine and irreducible. Let $x_0 \ldots, x_r$ be homogeneous coordinates on $\mathbb{P}^r$, and suppose that $X \subset \mathbb{P}^r_\mathbb{C}$ is defined by the ideal $I_X = (F_1, \ldots, F_n) \in A[x_0, \ldots x_r]$. Since X is flat over B the fiber dimension d is constant, and the singular locus of a given fiber X_b is the subscheme of X_b defined by the ideal J of $(r-d) \times (r-d)$ minors of the Jacobian matrix $(\partial F_i / \partial x_j)$.

Let $Y \subset B \times \mathbb{P}^r$ be the closed subscheme defined by J alone. It follows that $Y \cap X \subset X$ defines a scheme of singular points of fibers of the family, so this set is closed in X. Since the map $X \to B$ is projective, the image of $X \cap Y$ is closed, and its complement is open. Finally, a smooth fiber X_b is irreducible if and only if it is connected. This holds if and only if $h^0(\mathcal{O}_{X_b}) < 2$ and by the base change theorem [Hartshorne 1977, Theorem 12.11] this set is open as well. $\square$

Proposition 7.11. *The open subset $\mathcal{H}^\circ$ of the Hilbert scheme $\operatorname{Hilb}_{3m+1}(\mathbb{P}^3)$ parametrizing twisted cubics is irreducible of dimension* 12.

Proof. Let $C_0 \subset \mathbb{P}^3$ be a twisted cubic, and consider the family of translates of C_0 by automorphisms $A \in \operatorname{PGL}_4$ of $\mathbb{P}^3$: that is, the family

$$\mathcal{C} = \big\{ (A, p) \in \operatorname{PGL}_4 \times \mathbb{P}^3 \mid p \in A(C_0) \big\}.$$

Via the projection $\pi : \mathcal{C} \to \operatorname{PGL}_4$, this is a family of twisted cubics, and so it induces a map

$$\phi : \operatorname{PGL}_4 \to \mathcal{H}^\circ.$$

Since every twisted cubic is a translate of C_0, this is surjective, with fibers isomorphic to the stabilizer of C_0, that is, the subgroup of PGL_4 of automorphisms of $\mathbb{P}^3$ carrying C_0 to itself. By Exercise 1.5, every automorphism of C_0 is induced by an automorphism of $\mathbb{P}^3$, so the stabilizer is isomorphic to PGL_2 and thus has dimension 3. Since PGL_4 is irreducible of dimension 15, we conclude that $\mathcal{H}^\circ$ is irreducible of dimension 12. $\square$

7.3.3. Construction of the Hilbert scheme in general. The Hilbert scheme is more complicated than would appear from the examples above, and this is even true for the Hilbert polynomial $3m+1$. There are many subschemes of $\mathbb{P}^3$ that have the same Hilbert polynomial $3m+1$ as a twisted cubic — for example, the union of a plane cubic and a point — and are not the intersection of the quadrics containing them. (See Exercises 2.1 and 7.1.) In Chapter 19 we will discuss many more components of Hilbert schemes.

A fundamental result of Matsusaka provides a place to start:

Theorem 7.12 [Matsusaka 1972]. *Let $p(m) \in \mathbb{Q}[m]$ be a polynomial. There exists an integer m_0 such that:*

(1) *For any subscheme $X \subset \mathbb{P}^r$ with Hilbert polynomial $p_X = p$ we have*

$$h^0(\mathcal{J}_{X/\mathbb{P}^r}(m)) = \binom{m+r}{r} - p(m) \quad \textit{for all } m \geq m_0;$$

in other words, the Hilbert function of X agrees with the Hilbert polynomial $p_X = p$ for all $m \geq m_0$.

(2) *For any subscheme $X \subset \mathbb{P}^r$ with Hilbert polynomial $p_X = p$ and the saturated ideal of X is defined by forms of degree $\leq m$.*

Note that for any given X the existence of an m_0 satisfying the statement of the lemma is immediate by Theorem 2.1. The point of the lemma is that we can find one value of m_0 that works for all X with Hilbert polynomial p. The following result of Gotzmann [1978] provides a method for determining m_0.

Theorem 7.13. *The Hilbert polynomial of the homogeneous coordinate ring of any scheme $X \subset \mathbb{P}^r$ can be written uniquely in the form*

$$\chi(\mathcal{O}_X(m)) = \binom{m+a_1}{a_1} + \binom{m+a_2-1}{a_2} + \cdots + \binom{m+a_s-(s-1)}{a_s}$$

with

$$a_1 \geq \cdots \geq a_s \geq 0,$$

where the binomial coefficients are interpreted as polynomials in m. Moreover, the saturated homogeneous ideal of X is generated in degrees $\leq s$, and one can take $m_0 = s$ in the construction of the Hilbert scheme above.

See [Green 1989] for an exposition and a proof. From the coefficients a_j one can read off uniform vanishing theorems for $H^i(\mathcal{I}_X(m))$ as well.

For example, the Hilbert polynomial $3m + 1$ of the twisted cubic may be written as

$$3m + 1 = \binom{m+1}{1} + \binom{m+1-1}{1} + \binom{m+1-2}{1} + \binom{m+0-3}{0},$$

Here $s = 4$, and indeed the homogeneous ideal of the union of a plane cubic with a point, also in the plane, requires equations of degree 4.

7.3.4. Grassmannians. The simplest and most fundamental Hilbert schemes are the Grassmannians — including the projective spaces themselves. They parametrize the families of linear subspaces of given dimension in vector spaces or in projective spaces. For $0 \leq k \leq r$ we write $\mathbb{G}(k, r)$ for the set of k-planes in $\mathbb{P}^r$ and identify it with $G(k + 1, r + 1)$, the set of $(k + 1)$-dimensional vector subspaces of an $(r + 1)$-dimensional vector space. When we want to make the $(r + 1)$-dimensional vector space V explicit, we write $G(k + 1, V)$ instead.

We embed $G(k + 1, V)$ in $\mathbb{P}(\bigwedge^{r-k} V) = \mathbb{P}^{\binom{r+1}{r-k}-1}$ by sending a subspace $W \subset V$ to the 1-quotient $\bigwedge^{r-k} V \to \bigwedge^{r-k}(V/W)$. This map is a monomorphism called the *Plücker embedding*, and its image is an algebraic subvariety, which we take to be the algebraic structure of the Grassmannian.

Concretely, choose a basis of V/W so that the projection map $V \to V/W$ is given by an $(r - k) \times (r + 1)$ matrix A. The coordinates of the image of W are the $(r - k) \times (r - k)$ minors of A, called *Plücker coordinates* of W. On the open set where the first $(r - k) \times (r - k)$ minor is nonzero, we may multiply by its inverse, and it is not hard to check that the minors become the entries of the complementary $k \times (r - k)$ submatrix; thus $G(k, V)$ is covered by open sets isomorphic to affine $k(r - k)$-space, so that $G(k + 1, r + 1) = \mathbb{G}(k, r)$ is smooth and irreducible of dimension $k(r - k)$.

Example 7.14. The Grassmannian $\mathbb{G}(1, 3)$ of lines in $\mathbb{P}^3$ has Plücker coordinates of a line L that is the span of points $q, r \in \mathbb{P}^3$ the 2×2 minors of the 2×4 matrix whose rows are the coordinates of the points:

$$\begin{pmatrix} q_0 & q_1 & q_2 & q_3 \\ r_0 & r_1 & r_2 & r_3 \end{pmatrix}.$$

Indexing the minors $p_{i,j}$ by pairs of distinct column indices one can easily prove the *Plücker relation*

$$p_{0,1}p_{2,3} - p_{0,2}p_{1,3} + p_{0,3}p_{1,2} = 0,$$

and this defines $\mathbb{G}(1, 3) \subset \mathbb{P}^5$ as a quadric hypersurface.

The *universal subbundle* $W \subset V \times G(k+1, r+1)$, also called the *tautological subbundle*, is the vector bundle with fiber $W \subset V$ over the point of $G(k + 1, V)$ corresponding to W. It is universal in the sense that given any scheme X and a $(k + 1)$-dimensional subbundle W' of the trivial bundle $V \times X$ there is a unique morphism $X \to G(k + 1, V)$ such that the pullback of W is W'.

For a thorough introduction to the Grassmannian, the reader may consult Chapters 3–4 of [Eisenbud and Harris 2016].

7.3.5. Equations defining the Hilbert scheme. Matsusaka's theorem allows us to define an injective map of sets

$$h : \{\text{subschemes } X \subset \mathbb{P}^r \text{ with } p_X = p\} \to G\left(\binom{m_0+r}{r} - p(m_0), \binom{m_0+r}{r}\right)$$

by sending X to $H^0(\mathcal{J}_{X/\mathbb{P}^r}(m_0))$, and its image is the set of closed points of the Hilbert scheme. It remains to describe the scheme structure.

We observed above that though there are vector spaces V of 3 quadrics in $\mathbb{P}^3$ that define twisted cubics, a general such vector space would generate the ideal of 8 points, not a twisted cubic. What we want to know is how to tell these cases apart algebraically. Consider the multiplication map

$$V \otimes H^0(\mathcal{O}_{\mathbb{P}^3}(1)) \to H^0(\mathcal{O}_{\mathbb{P}^3}(3)).$$

We saw in Chapter 3 that the cokernel of this map is the 10-dimensional space $H^0(\mathcal{O}_{\mathbb{P}^1}(9))$, so the image of this map is 10-dimensional, whereas 3 general quadrics form a complete intersection and would have only Koszul syzygies, so in the case of general quadrics this map would have 12-dimensional image. This is a map from a 12-dimensional vector space to a 20-dimensional one, and what we've seen is that if V is the net of quadrics containing a twisted cubic, it has a 2-dimensional kernel; that is, it has rank 10.

Thus if $\mathcal{E}$ is the universal subbundle on

$$G = G(3, H^0(\mathcal{O}_{\mathbb{P}^3}(2))),$$

and $H^0(\mathcal{O}_{\mathbb{P}^3}(d)) \otimes \mathcal{O}_G$ is the trivial bundle, then the multiplication map above gives a map of vector bundles

$$\mu : \mathcal{E} \otimes H^0(\mathcal{O}_{\mathbb{P}^3}(1)) \to H^0(\mathcal{O}_{\mathbb{P}^3}(3)).$$

We can represent this locally as a matrix of functions, and the 11×11 minors of this matrix define the rank 10 locus, and thus vanish on the points of the Hilbert scheme: in a neighborhood of a point in G corresponding to a twisted cubic, the common zero locus of these minors is the locus of nets of quadrics containing a twisted cubic.

In fact, the construction of the Hilbert scheme in general is no more structurally complicated than this special case. Given a polynomial $p(m)$, we find a value of m_0 that satisfies the statement of Theorem 7.12; we let

$$G = G\left(\binom{m_0+r}{r} - p(m_0), \binom{m_0+r}{r}\right)$$

be the Grassmannian, and let h be the map from the set of subschemes of $\mathbb{P}^r$ with Hilbert polynomial p to G sending X to $H^0(\mathcal{J}_{X/\mathbb{P}^r}(m_0))$. We then get a map of vector bundles on G:

$$\mathcal{E} \otimes H^0(\mathcal{O}_{\mathbb{P}^r}(1)) \to H^0(\mathcal{O}_{\mathbb{P}^r}(m_0 + 1)).$$

In a neighborhood of a point of G in the image of h, the common zero locus of the minors of size $\binom{r+m_0+1}{r} - p(m_0 + 1)$ of a matrix representative of this map is the image of h. Thus these functions define the Hilbert scheme.

7.4. Bounding the number of maps between curves

A priori, the Hilbert scheme parametrizes subschemes of projective space. But the construction is adaptable to many other situations. In this section we'll sketch a proof of such an application. As we have seen, there can be infinitely many maps from a given curve to $\mathbb{P}^1$, and this is also the case for maps to a curve of genus 1, even modulo the automorphisms of the target. But this is not the case in higher genus: given two smooth projective curves C and D of genera $g, h \geq 2$, we'll show that there are at most a finite number of nonconstant morphisms $C \to D$. In fact, the number is bounded purely in terms of g and h:

Theorem 7.15. *Given integers $g, h \geq 2$ there is a bound $N(g, h)$ on the number of distinct nonconstant morphisms from a smooth projective curve C of genus g to a smooth projective curve D of genus h.*

A special case of this result is a bound on the size of the group of auto-morphisms of a curve of genus $g \geq 2$. We will give a second proof of that result, which doesn't rely on the Hilbert scheme, in Theorem 13.9.

Proof. Hurwitz's theorem (Theorem 2.12) implies a bound on the degree d of a morphism $f : C \to D$, so it suffices to bound the number of morphisms of a fixed degree d.

We will use the Hilbert scheme in the relative setting, as Grothendieck orig-inally defined it: Given a base scheme S, the scheme $\mathrm{Hilb}_{p(m)}(\mathbb{P}^r_S)$ represents the functor on S-schemes $X \to S$ that associates to X the set of subschemes of $\mathbb{P}^r_X$, flat over X, whose fibers have Hilbert polynomial p.

We can construct a family of products of pairs of curves of genera g and h embedded in projective space as follows (the details don't matter — only the existence): Let $S_g \to S$ be the Hilbert scheme of smooth curves of genus g embedded by invertible sheaves of degree $2g + 1$ in $\mathbb{P}^{g+1}_S$ and similarly for S_h and curves of genus h, and write $([C], [D])$ for the corresponding point in the fiber of $S_g \times_S S_h$. From the universal families of curves over S_g and S_h we may construct a family $\mathcal{C} \to S = S_g \times_S S_h$ whose fiber over s includes all products of pairs of smooth curves of genera g and h, embedded in $\mathbb{P}^{g+1}_s \times_S \mathbb{P}^{h+1}_s$. Finally we embed this product of projective spaces by the Segre embedding in $\mathbb{P}^N_S$, with

$$N = (g + 2)(h + 2) - 1.$$

If $\Gamma \subset C \times D \subset \mathbb{P}^N$ is the graph of a morphism $f : C \to D$ of degree d, then Γ has genus g and degree $2g + 1 + d(2h + 1)$, so we know its Hilbert polynomial p. The set of subschemes in this Hilbert scheme that project isomorphically to C and d-to-1 to D correspond to the points of a locally closed subset of this Hilbert scheme, and therefore are a scheme of finite type. The fibers of this scheme over S are the sets of morphisms of degree d from curves of genus g to curves of genus h.

We first prove that each fiber is finite — that is, there can only be finitely many maps of degree d from one fixed curve to another. By Theorem 7.8 it suffices for this to show that if we fix the curves C, D and a map ϕ with graph Γ_ϕ then the normal bundle

$$\mathcal{N}_{\Gamma_\phi/C \times D} = \mathcal{I}_{C \times D}/\mathcal{I}_{\Gamma_\phi}$$

has no sections. The projection onto the first factor identifies $\mathcal{I}_{\Sigma_\phi}$ with $\mathcal{I}_C$, and the quotient is thus $\phi^* \mathcal{I}_D$, which has degree $2 - 2h < 0$. Thus, fiber by fiber, the scheme of morphisms is finite.

Since the base S of the family is also a scheme of finite type and the degree of the fiber is semicontinuous, there is an absolute bound $N(g, h)$ as claimed. $\square$

A small variation of this construction, essentially the case $g = h$ and $d = 1$, parametrizes families of isomorphisms of curves: given two families of projective curves $\mathcal{X} \subset B \times \mathbb{P}_S^r$ and $\mathcal{Y} \subset B \times \mathbb{P}_S^s$, the result is a scheme $\mathrm{Isom}(\mathcal{X}, \mathcal{Y}) \to S$ whose fiber over a point of S is the set of isomorphisms of $X_b \to Y_b$ for b in the fiber over s. This turns out to be useful in describing the properties of the moduli space M_g treated in the next chapter.

7.5. Exercises

Exercise 7.1. Suppose that a scheme $X \subset \mathbb{P}^n$ is the disjoint union of subschemes Y, Z. Show that the Hilbert polynomial of X is the sum of the Hilbert polynomials of Y and Z. What statement can you make about the Hilbert functions? ◆

Exercise 7.2. More generally, suppose that a scheme $X \subset \mathbb{P}^n$ is the union of subschemes Y, Z. Show that the Hilbert polynomial of X is the sum of the Hilbert polynomials of Y and Z minus the Hilbert polynomial of $Y \cap Z$. ◆

Exercise 7.3. Let $H \subset \mathbb{P}^3$ be a 2-plane; let $C \subset H$ be a plane cubic curve and $p \in H \setminus C$ and point in H not on C; let $X = C \cup \{p\}$.

(1) Show that the Hilbert polynomial of X is $p_X(m) = 3m + 1$.

(2) Show that the smallest value of m_0 satisfying the statement of Matsusaka's theorem (Theorem 7.12) is 4. ◆

Exercise 7.4. Use an argument like that of Proposition 7.11 to show that the restricted Hilbert scheme $\mathcal{H}^\circ$ of rational normal curves $C \subset \mathbb{P}^r$ is irreducible of dimension $r^2 + 2r - 3$. ◆

Exercise 7.5. If $C = X \cap Y \subset \mathbb{P}^3$ is a complete intersection of surfaces of degrees d, e, then Hilb is smooth at the point $[C]$, of dimension $2\binom{3+d}{3} - 4$ if $d = e$ or $\binom{3+d}{3} + \binom{3+e}{3} - \binom{3+e-d}{3} - 2$ if $d < e$. ◆

Chapter 8

Moduli of curves

In the preceding chapter, we described the *Hilbert scheme,* a fine moduli space for curves in projective space. In this chapter we will discuss the second moduli space central to the theory of algebraic curves: M_g, which parametrizes isomorphism classes of smooth projective curves of genus g. As we'll see, M_g is not a fine moduli space, but it comes close.

To describe the situation, we will start with the case of curves of genus 1, where everything can be made explicit.

8.1. Curves of genus 1

Let C be a smooth curve of genus 1. Any invertible sheaf of degree 2 on C can be written as $\mathcal{O}_C(2p)$, and defines a morphism to $\mathbb{P}^1$ with 4 distinct branch points. Since the automorphism group of C is transitive, these 4 points in $\mathbb{P}^1$ are independent of the choice of p, and are well-defined up to an automorphism of $\mathbb{P}^1$. As explained in Section 6.2, this means that every such curve C can be realized as the completion of an affine curve

$$y^2 = f(x)$$

where f is a quartic polynomial with distinct roots:

$$f(x) = \prod_{i=1}^{4}(x - \lambda_i).$$

Thus we would like to define M_1 to be the set of 4-tuples of distinct points $\{\lambda_1, \dots, \lambda_4\}$ of $\mathbb{P}^1$ modulo the action of $\operatorname{Aut} \mathbb{P}^1 = \operatorname{PGL}_2$.

As we will explain in the next sections, quotients by infinite groups can behave badly, but in this case we can compute the quotient in a much simpler way: There is a unique automorphism of $\mathbb{P}^1$ carrying the three points $\lambda_1, \lambda_2, \lambda_3$

to the points $0, 1$ and $\infty \in \mathbb{P}^1$ respectively, so that we can write C as the zero locus of

$$y^2 = x(x-1)(x-\lambda)$$

for some complex number $\lambda \in \mathbb{P}^1 \setminus \{0, 1, \infty\}$; we'll call this curve C_λ. This expression is not unique, since if we reordered the original four points λ_i, we might arrive at a different value of λ; for example, if we exchanged 0 and ∞ and fixed 1, λ would be replaced by $1/\lambda$. Thus the symmetric group S_4 acts on the set $\mathbb{A}^1 \setminus \{0, 1, \infty\}$ and one can show that the orbit of λ under this action is

$$\left\{ \lambda, 1 - \lambda, \frac{1}{\lambda}, \frac{1}{1-\lambda}, \frac{\lambda-1}{\lambda}, \frac{\lambda}{\lambda-1} \right\}.$$

There are 6 points in the orbit rather than 24 because the Klein 4-group $K = \mathbb{Z}/2 \times \mathbb{Z}/2 \subset S_4$ of fixed-point-free involutions acts trivially, so what we really have is an action of $S_4/K \cong S_3$.

Since S_3 is finite and $\mathbb{P}^1 \setminus \{0, 1, \infty\}$ is a normal affine curve, the quotient space by the action is again a normal affine curve whose points are in one-to-one correspondence with the orbits, and thus with the set of curves of genus 1. By Lüroth's theorem (Theorem 3.2), the quotient is rational, meaning that the field of rational functions on the quotient — that is, the subfield of $\mathbb{C}(\lambda)$ invariant under the action of S_3 — is of the form $\mathbb{C}(j)$ for some rational function $j(\lambda)$ of degree 6. Of course, there are many possible generators of the field of rational functions on the quotient; one that works is

$$(*) \qquad\qquad j(\lambda) := 256 \frac{(\lambda^2 - \lambda + 1)^3}{\lambda^2(\lambda - 1)^2},$$

known as the *j-function*. As λ varies in $\mathbb{P}^1 \setminus \{0, 1, \infty\}$, the values of $j(\lambda)$ range over all of $\mathbb{A}^1$.

Summarizing, we have proven:

Theorem 8.1. *The set of isomorphism classes of smooth projective curves of genus 1 is in bijection with the points of the affine line $M_1 \cong \mathbb{A}^1$. The bijection maps the curve defined by $y^2 = x(x-1)(x-\lambda)$ to $j(\lambda) \in \mathbb{A}^1$.*

M_1 is a coarse moduli space. As we will see in Exercises 8.3 and 8.4, M_1 is not a fine moduli space, but it comes close in two senses.

Proposition 8.2. *To any family $\pi : \mathcal{C} \to B$ of smooth projective curves of genus 1 over a reduced base B we can associate a natural morphism of schemes*

$$\phi : B \to M_1$$

whose value at any point $b \in B$ is the j-invariant of the corresponding fiber C_b.

Proof. To start, we will work locally in B: for a given $b_0 \in B$, we will choose a suitably small neighborhood U of $b_0 \in B$ and restrict ourselves to the preimage $\mathcal{C}_U = \pi^{-1}(U)$. The first thing to do is to express the curves C_b in our family as

2-sheeted covers of $\mathbb{P}^1$, which is to say we want to choose an invertible sheaf on $\mathcal{C}_U$ having degree 2 on each fiber C_b. Since we're working locally in B, we can find a section $\rho : U \to \mathcal{C}_U$ of $\pi : \mathcal{C} \to B$. If we let $D = \rho(U) \subset \mathcal{C}_U$ be the image, then we can take our invertible sheaf to be $\mathcal{L} := \mathcal{O}_{\mathcal{C}_U}(2D)$.

Next, we use the following result, which is a special case of the theorem on cohomology and base change (see [Eisenbud and Harris 2016, Appendix, Theorems B.5 and B.9] or [Hartshorne 1977, Theorem 12.11].)

Theorem 8.3 (cohomology and base change). *If $f : X \to Y$ is a morphism and $\mathcal{F}$ is a coherent sheaf on X such that $H^1(\mathcal{F}|_{f^{-1}(y)}) = 0$ for all $y \in Y$, then $h^0(\mathcal{F}|_{f^{-1}(y)})$ is a constant function of y, and $f_*(\mathcal{F})$ is a vector bundle of this rank.*
□

This result implies that the direct image $\mathcal{E} := \pi_*(\mathcal{O}_{\mathcal{C}_U}(2D))$ is locally free of rank 2, and we get a morphism $\mathcal{C}_U \to \mathbb{P}(\mathcal{E})$ expressing each curve C_b as a 2-sheeted cover of the corresponding fiber $\mathbb{P}(\mathcal{E}_b)$. Again, since we are working locally in B, we can trivialize the bundle $\mathcal{E}$, so that we get a diagram

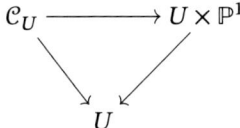

Once more restricting to a smaller neighborhood U if necessary, we can write the family $\mathcal{C}_U \to U$ as the locus

$$y^2 = \prod_1^4 (x - \lambda_i),$$

where the λ_i are regular functions on U. The j-function of the λ_i yields a map $U \to M_1$; since the value of the j-function at a point is determined by the isomorphism type of the fiber over this point, these maps agree on overlaps to give the desired morphism $B \to M_1$.
□

The good news. The second way in which M_1 comes close to being a fine moduli space is seen in the next result:

Proposition 8.4. *Let B be a reduced scheme.*

(1) *If $j : B \to \mathbb{A}^1$ is any regular function on B, then there exists a finite cover $\alpha : B' \to B$ such that $j \circ \alpha$ is the j-function of a family of curves of genus 1 on B'.*

(2) *If $\pi : \mathcal{C} \to B$ and $\eta : \mathcal{D} \to B$ are two families of curves of genus 1 with the same associated j-function, then there exists a finite cover $\alpha : B' \to B$ and*

an isomorphism $\mathcal{C} \times_B B' \cong \mathcal{D} \times_B B'$ *such that the diagram*

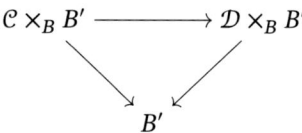

commutes.

Proof. For the first of these assertions, let

$$B' := \big\{ (b, \lambda) \in B \times (\mathbb{A}^1 \setminus \{0, 1\}) \mid j(b) = j(\lambda) \big\},$$

where $j(\lambda)$ is as given by formula (∗) on page 148. We have already described a family of curves of genus 1 over the λ-*line* $\mathbb{A}^1 \setminus \{0, 1\}$; the pullback to B' is the desired family.

For the second half, we want to do something similar. Specifically, we want to choose sections $\sigma : B \to \mathcal{C}$ and $\tau : B \to \mathcal{D}$ and take

$$B' := \big\{ (b, \phi) \mid b \in B, \ \phi : C_b \xrightarrow{\cong} D_b \text{ and } \phi(\sigma(b)) = \tau(b) \big\};$$

as a set, B' is the set of isomorphisms between corresponding fibers in the two families. By Corollary 5.15, B' is a finite cover of B and when we pull back the two families to B' we have a tautological isomorphism between them. The only issue is how to give B' an appropriate scheme structure, and for this we can use the Isom scheme described at the end of Section 7.4. $\square$

Thus, M_1 is not a fine moduli space for smooth curves of genus 1, but it is the next best thing: we don't get a bijection between families of curves of genus 1 over a given base B and maps $j : B \to M_1$; but we do get a map from the former to the latter with "finite kernel and cokernel".

Compactifying M_1. A natural question to ask is, if every value of $j \in \mathbb{A}^1$ corresponds to an isomorphism class of curves C_j of genus 1, what happens to the curves C_j as j goes to ∞? Equivalently, what happens to the curve C_λ given as the double cover

$$y^2 = x(x - 1)(x - \lambda)$$

when λ approaches 0, 1 or ∞ — the other branch points of the double cover? The answer is seen from the equation: when two branch points of a double cover of smooth curves coalesce the limiting curve has a node (Figure 8.1). In fact, there is a unique isomorphism class of irreducible curves of arithmetic genus 1 having a node; it's represented by the curve defined by $y^2 = x^2(x - 1)$.

The upshot is that if we enlarge the original class of curves parametrized by M_1 — smooth projective curves of genus 1 — to the slightly larger class of irreducible nodal projective curves of arithmetic genus 1, we still have a coarse moduli space $\overline{M}_1$ for this slightly larger class of objects. This enlarged moduli

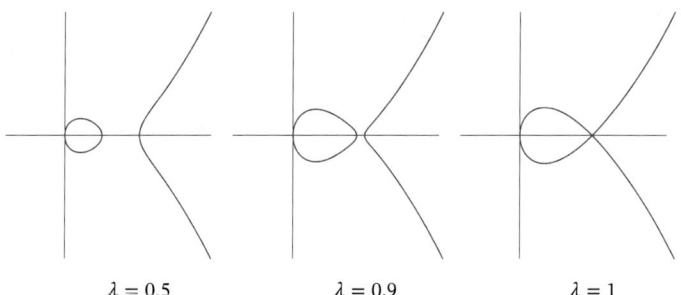

$\lambda = 0.5$ $\lambda = 0.9$ $\lambda = 1$

Figure 8.1. A curve of genus 1 degenerating to a rational curve with a node in the family $y^2 = x(x-1)(x-\lambda)$.

space is obtained by adding one point "at ∞" to the existing space $M_1 \cong \mathbb{A}^1$ to form $\overline{M}_1 \cong \mathbb{P}^1$.

This is an example of what is called a *modular compactification*. There is no precise definition, but if we have a class of objects parametrized by a (noncompact) moduli space M we may be able enlarge the class of objects to be parametrized, with the result that the moduli space $\overline{M}$ of the larger class is compact.

Modular compactifications of a given moduli problem may or may not exist. It's sometimes a tricky problem to find a suitable class of objects to parametrize: if we don't add enough additional isomorphism classes, not every 1-parameter family of objects in our original class will have a limit in the larger class, meaning the enlarged moduli space will still not be compact; if we add too many, 1-parameter families may have more than one possible limit, meaning the enlarged space won't be separated. For example, in the family of curves C_t given as

$$C_t = V(y^2 - x^3 - t^2 x - t^3),$$

the *j*-function is constant when $t \neq 0$, but the limiting curve C_0 has a cusp (Figure 8.2). This shows that we could not have added cuspidal curves to M_1.

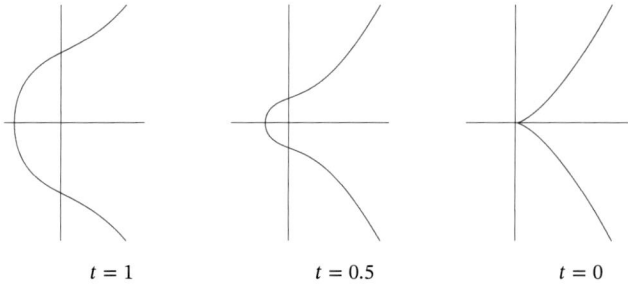

$t = 1$ $t = 0.5$ $t = 0$

Figure 8.2. A curve of genus 1 degenerating to a cuspidal curve in the family $C_t = V(y^2 - x^3 - t^2 x - t^3)$.

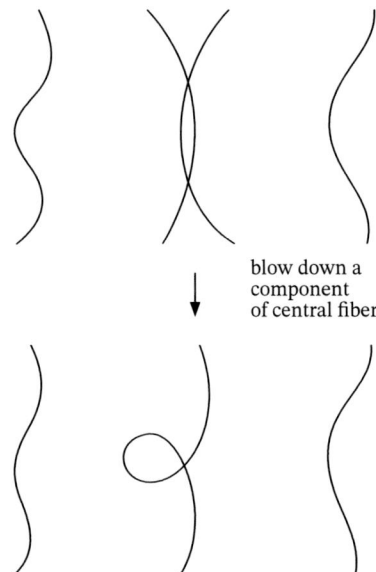

Figure 8.3. In this case a birational modification of the total space of the family changes the unstable reducible curve to a stable curve.

When modular compactifications do exist, they are extremely valuable for the study of both the space M and of the objects parametrized by M: compactness allows us to apply the techniques of modern algebraic geometry to the space $\overline{M}$, while the fact that it is still a moduli space gives us a handle on its geometry. In the following section, we will describe a modular compactification of M_g. The objects parametrized are called *stable curves*.

Getting back to the moduli space $\overline{M}_1$, if we have a family where $j(\lambda)$ has a pole, we would like to say that the limit of the curves in the family is an irreducible nodal curve, but this is not necessarily true! For example, the limit of the curves

$$y^2 = x(x-t)(tx-1)$$

as $t \to 0$ is reducible, with two components meeting in two points, 0 and ∞. What is true is that a process called *semistable reduction* shows that after a base change and a birational modification of the family around the pole we can replace the family with one where the singular fiber is indeed an irreducible nodal curve (Figure 8.3). See [Harris and Morrison 1998] for a description of this process in general, and several explicit examples.

8.2. Higher genus

The idea is analogous to the one used for genus 1 curves: to construct a moduli space, first parametrize curves with a choice of some additional structure, such as a map to projective space, and then mod out by the choices made. For any smooth projective curve C of genus $g \geq 2$, the tricanonical linear series $|3K_C|$ is very ample; it embeds C as a curve of degree $6g - 6$ in $\mathbb{P}^{5g-6}$. Thus we have a way of realizing a given abstract curve C as a curve in projective space, unique up to automorphisms of $\mathbb{P}^{5g-6}$.

We claim next that the set of smooth, tricanonically embedded curves is a locally closed subset X of the Hilbert scheme $\mathrm{Hilb}_{(6g-6)m+1-g}(\mathbb{P}^{5g-6})$ parametrizing curves of genus g and degree $6g - 6$ in $\mathbb{P}^{5g-6}$. By Lemma 7.10, the set of points in the base over which the curves are smooth is open. Let

$$\mathrm{Hilb}^{\circ} = \mathrm{Hilb}^{\circ}_{(6g-6)m+1-g}(\mathbb{P}^{5g-6}) \subset \mathrm{Hilb}_{(6g-6)m+1-g}(\mathbb{P}^{5g-6})$$

be this open set.

Next, on the universal family $\mathcal{C} \subset \mathrm{Hilb}^{\circ} \times \mathbb{P}^{5g-6}$, we have two families of invertible sheaves: we have the pullback of $\mathcal{O}_{\mathbb{P}^{5g-6}}(1)$; and we have the cube K^3 of the dualizing sheaf. Each gives rise to a section of the relative Picard variety over Hilb°, and the locus where they agree is thus a closed subset $X \subset \mathrm{Hilb}^{\circ}$.

The group PGL_{5g-5} of automorphisms of $\mathbb{P}^{5g-6}$ acts on the variety X and its orbits are the isomorphism classes of smooth curves of genus g; thus, we might hope to realize the moduli space M_g as the quotient of X by PGL_{5g-5}. But here things go awry in a hurry: unlike the case of an action of a finite group on a variety, the orbit spaces of infinite groups are often not algebraic varieties. (Think of the action of $\mathbb{C}^*$ on $\mathbb{C}$ by multiplication.) What is needed is a tool to extract the "best possible approximation" to a quotient. Happily, David Mumford created a tool that does exactly this: *geometric invariant theory* (GIT). To see how GIT can be used in this setting to produce the space M_g, see the wonderful introduction in [Mumford and Suominen 1972] or the more technical version in [Mumford 1977].

Theorem 8.5 (Mumford). *The space of orbits of* PGL_{5g-5} *acting on the subset of the Hilbert scheme representing tricanonical curves has the structure of an algebraic variety* M_g *which is a* coarse moduli space *in the following sense:*

(1) *Given any flat family* $Y \to B$ *of smooth curves of genus g there is a morphism of schemes* $B \to M_g$ *sending each closed point $p \in B$ to the point of M_g representing the fiber Y_b.*

(2) *These maps form a natural transformation from the functor $G(-)$ of families of smooth curves to the functor* $\mathrm{Mor}_{\mathrm{schemes}}(-, M_g)$ *through which any natural transformation* $G \to \mathrm{Mor}_{\mathrm{schemes}}(-, M')$ *factors.*

The power of the theory of the moduli space of curves was greatly increased when compactifications of the space (there are many interesting ones) were introduced. One of these, the compactification of $M_1 = \mathbb{A}^1$ to $\overline{M}_1 = \mathbb{P}^1$ by adding a nodal curve, has already been mentioned. This has the desirable properties that the subset added to M_1 is a divisor; and the compactification is *modular* in the sense that the point added corresponds to a curve almost of the same type as the curves in M_1.

There are two reasons why a compactification is important:

First, the great majority of the techniques that algebraic geometers have developed for dealing with varieties apply directly only to projective varieties. For example, the Satake compactification is a projective variety containing M_g in such a way that the complement — usually referred to as the boundary — has codimension 2. Taking successive hyperplane sections that pass through a given point but don't meet the boundary, we see that for $g \geq 2$ there is a complete one-dimensional family of *smooth* curves containing any smooth curve of genus ≥ 2.

Often, though, we can learn the most from a compactification where the added boundary is a divisor, and this is the case for the Deligne–Mumford compactification $\overline{M}_g$, described below, introduced in the groundbreaking 1969 paper [Deligne and Mumford 1969]. A central example of how this is used is given in Section 8.4, where we take up the question, "can we write down a general curve of genus g?"

To describe this compactification, we first explain some of the language and results of geometric invariant theory.

Stable, semistable, unstable. Given a quasiprojective variety $X \subset \mathbb{P}^N$ and a group $G \subset \mathrm{PGL}_{N+1}$ that carries X into itself, we wish to construct as good a map as possible from the set of orbits to a projective space. Whatever map we take, the closure of the image will correspond to a graded ring. We want to preserve as much of the structure of the orbit space as possible, and on an open affine cover this means finding as many functions as possible that are invariant on the orbits. Thus it is natural to take the ring of invariants of the homogeneous coordinate ring A of the closure of X as the homogeneous coordinate ring of the closure of the image of X.

The first difficulty is that the elements of A are not functions on X, so G may not even act on A. However, it is possible to lift the action of G to an action on A of the slightly larger group, SL_{N+1}, a process called *linearization*. The kernel of the map $\mathrm{SL}_{N+1} \to \mathrm{PGL}_{N+1}$ consists of diagonal matrices of finite order dividing $N+1$, and the choice of a linearization amounts to a choice of a character of this abelian group. However, the choice doesn't matter, since the kernel acts trivially on forms of degree a multiple of $N+1$, and thus the action of PGL_{N+1} itself extends to an action on the homogeneous coordinate ring of the $(N+1)$-st Veronese embedding. Another way to say this is to introduce the cone $\overline{X} \subset \mathbb{A}^{N+1}$ over X; a linearization amounts to an action of SL_{N+1} on $\overline{X}$.

The second difficulty in this program is that the ring of invariants of an infinite group may not be finitely generated, so it may not correspond to a projective variety. Hilbert showed that if $G = \mathrm{SL}_{N+1}$, then the ring of invariants is finitely generated. Since Hilbert's time this result has been extended to the class of *linearly reductive* groups — see [Haboush 1975]. Thus the subring $A^G \subset A$ of invariant elements is finitely generated over the ground field.

The third difficulty is that the points of $\mathrm{Proj}(A^G)$, usually denoted $X /\!\!/ G$, are generally not in one-to-one correspondence with the orbits of G on X!

Geometric invariant theory explains the relationship of $X /\!\!/ G$ to the set of orbits. To do this, it performs a sort of triage on the points of X (or their orbits), dividing them into three classes: stable, semistable and unstable. The theory also provides tools for determining this stratification.

(1) *Stable points.* These are the points whose orbits in $\mathbb{A}^{N+1}$ are closed. They comprise an open subset $X^{\mathrm{stable}} \subset X$, and the points of an open subset of $X /\!\!/ G$ correspond one-to-one to the stable orbits, that is, an open subset that is set-theoretically X^{stable}/G. In general, this set may be empty, but in the case of the action of PGL_3 on the $\mathbb{P}^9$ of plane cubics, the stable points are the smooth plane cubics, and the quotient is the affine j-line.

(2) *Strictly semistable points.* These are the points p such that there exists an invariant form not vanishing at p. Together with the stable points, comprise a larger open subset $X^{\mathrm{semistable}} \subset X$, called the *semistable* locus. Two semistable points p, q map to the same point in $X /\!\!/ G$ if and only if $\overline{Gp} \cap \overline{Gq} \cap X^{\mathrm{semistable}} \neq \emptyset$. In the example of the action of PGL_3 on $\mathbb{P}^9$, the semistable locus contains the orbits of smooth and nodal plane cubics; that is, smooth cubics together with the three orbits consisting of irreducible cubics with a node, unions of lines and conics meeting transversely, and triangles. In the quotient, these last three orbits correspond to just one additional point, and this quotient is the compactification of the affine line to the projective line obtained by adding one point.

(3) *Unstable orbits.* These are the points p on which all invariant polynomials vanish, so that the induced map $\mathrm{Proj}\, A \to \mathrm{Proj}(A^G)$ is not even defined at p. Thus unstable points do not correspond to any points of $X /\!\!/ G$; in fact, they cannot be included in any topologically separated quotient of an open subset of X defined in this way, though there may be other compactifications, coming from other representations of M_g as $X' /\!\!/ G'$; see [Smyth 2013].

8.3. Stable curves

The compactification $\overline{M}_g$ is also a *modular* compactification in the sense that the points of the boundary correspond to slightly more general objects of the same type as the points of M_g.

Definition 8.6. A reduced irreducible connected curve is *stable* if it has at most nodes as singularities and if every smooth rational component meets the other components at least three times (Figure 8.4).

The last phrase of the definition could be replaced by the equivalent condition that the automorphism group of C is finite.

These are stable points in the Hilbert scheme of tricanonical embeddings in the sense of geometric invariant theory, and the result is that M_g has a modular compactification that is a projective variety:

Theorem 8.7 (properties of $\overline{M}_g$ [Deligne and Mumford 1969; Knudsen 1983]).

(1) $\overline{M}_g$ *is a projective variety.*

(2) *The points of $\overline{M}_g$ correspond one-to-one to isomorphism classes of smooth curves.*

(3) *For every family $\mathcal{C} \to B$ of stable curves there is a morphism of schemes $B \to M_g$ carrying each closed point $b \in B$ to the point representing the isomorphism class of the fiber of $\mathcal{C}$ over b. These maps form a natural transformation from the functor $G(-)$ of families of stable curves to the functor*

$$\mathrm{Mor}_{\mathrm{schemes}}(-, \overline{M}_g)$$

through which any natural transformation $G \to \mathrm{Mor}_{\mathrm{schemes}}(-, M')$ factors.

The deepest theorems about M_g have been proven using the divisor class group of $\overline{M}_g$, and many of the divisors that play a role are actually supported on the complement $\overline{M}_g \setminus M_g$, often called the *boundary*.

Cheerful Fact 8.8. *For $g \geq 1$ the boundary $\overline{M}_g \setminus M_g$ is the union of $1 + \lfloor g/2 \rfloor$ divisors whose generic points are*

(1) *irreducible nodal curves of geometric genus $g - 1$ and*

(2) *for $i = 1, \ldots, \lfloor g/2 \rfloor$ the union of two smooth curves $C_i \cup C_{g-i}$ of genera i and $g - i$ meeting in a point.*

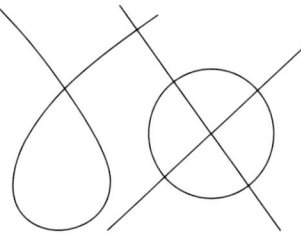

Figure 8.4. A stable curve.

We will not prove either of Theorems 8.5 and 8.7. For an introduction to the proofs, with references, see [Harris and Morrison 1998].

How we deal with the fact that $\overline{M}_g$ is not fine. The fact that $\overline{M}_g$ is not a fine moduli space — and that correspondingly there does not exist a universal family of curves over it — is unquestionably a nuisance. Nonetheless, there are ways of dealing with the situation. The first step is to identify the cause of the problem, which is that some curves have nontrivial automorphisms. There are three ways to proceed:

(1) *Kill the automorphisms.* The idea here is to add additional structure to the objects parametrized, so as to eliminate automorphisms. Here is an example of such a construction. We saw in Chapter 5 that on a smooth projective curve C of genus g, the collection of invertible sheaves $\mathcal{L}$ with $\mathcal{L}^m \cong \mathcal{O}_C$ forms a group isomorphic to $(\mathbb{Z}/m)^{2g}$. We define a *curve with level m structure* to be such a curve, together with a choice of 2g generators $\mathcal{L}_1, \ldots, \mathcal{L}_{2g}$ for this group. On every curve C of genus ≥ 2 an automorphism fixing all line bundles of order $m \geq 3$ is trivial, and there does exist a fine moduli space $M_g[m]$ for curves with level m structure; this space is a finite cover of M_g. Thus, while a universal family does not exist over M_g, one does exist over a finite cover of M_g, and this is sufficient for many purposes.

(2) *Ignore the automorphisms.* Here we use a basic fact, which we'll establish in Section 13.3: in M_g, the locus $A \subset M_g$ of curves that do have automorphisms other than the identity has codimension $g - 2$. If we restrict to the complement $M_g^\circ = M_g \setminus A$, accordingly, there does exist a universal family, and again this is sufficient for many purposes; for example, if $g \geq 4$ then a divisor class on M_g is determined by its restriction to M_g°, so we can just work over that open set.

(3) *Embrace the automorphisms.* We mentioned above that there does not exist a fine moduli space for curves of genus g in the category of schemes. But there is a larger category, called *stacks*, in which a fine moduli space does exist. This solution to the problem, pioneered by Deligne and Mumford, has many advantages but involves a substantial investment in mastering the technical issues; readers who wish to pursue this avenue may consult [Deligne and Mumford 1969], [Olsson 2016], or the forthcoming book *Stacks and moduli* by Jarod Alper.

8.4. Can one write down a general curve of genus g?

We have made a fuss over the value of compactifying M_g to a projective variety. To see an example of the usefulness of $\overline{M}_g$, we'll take up a question we've raised before: Can one write down a general curve of genus g? More precisely, does there exist a family of curves depending freely on parameters that includes all the curves in an open subset of M_g, as the equation $y^2 = x(x-1)(x-\lambda)$ represents general curves of genus 1? Still more precisely, we say that a variety is *unirational* if it admits a dominant morphism from an open subset of $\mathbb{A}^n$. Our question is: Is M_g unirational?

We have produced families with free parameters in genera 2 and 3. Essentially the same approach works in genera 4 and 5; in each case a general canonical curve is a complete intersection, so that if we take the coefficients of its defining polynomials to be general scalars we have a general curve.

This method breaks down when we get to genus 6, where a canonical curve is not a complete intersection. But it's close enough: as discussed in Chapter 12, a general canonical curve of genus 6 is the intersection of a smooth del Pezzo surface $S \subset \mathbb{P}^5$ with a quadric hypersurface Q; since all smooth del Pezzo surfaces in $\mathbb{P}^5$ are isomorphic, we can just fix one such surface S and let Q be a general quadric.

It gets harder as the genus increases. Already genus 7 calls for a different approach. Here we want to argue that, by Brill–Noether theory, a general curve of genus 7 can be realized as (the normalization of) a plane septic curve with 8 nodes $p_1, \ldots, p_8 \in \mathbb{P}^2$. Conversely, if $p_1, \ldots, p_8 \in \mathbb{P}^2$ are general points then having nodes at the points p_i imposes $3 \times 8 = 24$ independent conditions on the $\mathbb{P}^{35}$ of curves of degree 7, so that we would expect that the septic curves double at the p_i form a linear series, parametrized by a projective space $\mathbb{P}^{11}$.

This suggests that we consider the space

$$\Sigma := \left\{ (p_1, \ldots, p_8, C) \in (\mathbb{P}^2)^8 \times \mathbb{P}^{35} \mid C \text{ is singular at } p_1, \ldots, p_8 \right\}$$

With a little work, we can see that there is a unique component Σ° of Σ dominating $(\mathbb{P}^2)^8$, which is a $\mathbb{P}^{11}$-bundle over an open subset of $(\mathbb{P}^2)^8$ and hence rational; this component dominates M_7, showing that M_7 is unirational.

A similar approach works through genus 10, and Severi conjectured that it would be possible to do something similar for all genera. The approach through plane curves, however, fails in genus 11: by the Brill–Noether theorem, the smallest degree of a planar embedding of a general curve of genus 11 is 10; by Theorem 12.9 (itself a consequence of the Brill–Noether theorem), such a curve has $\binom{9}{2} - 11 = 25$ nodes. But $3 \times 25 > 65$, the dimension of the space of plane curves of degree 10. Thus, if we introduce the analogue of the incidence correspondence we used in the case of genus 7 — that is,

$$\Sigma := \left\{ (p_1, \ldots, p_{25}, C) \in (\mathbb{P}^2)^{25} \times \mathbb{P}^{65} \mid C \text{ is singular at } p_1, \ldots, p_{25} \right\}$$

— then the projection $\Sigma \to (\mathbb{P}^2)^{25}$ is not dominant, and we have no idea if Σ is rational. Ad hoc (and much more difficult) arguments have been given in genera 11, 12, 13 and 14, but so far no-one can go further in producing general curves; in genus 15 it is only known that any two general curves can be connected by a chain of rational curves that passes through the locus of irreducible nodal curves in $\overline{M}_g$ [Bruno and Verra 2005]. In genera 15 and 16 Chang and Ran showed the weaker statement that $\overline{M}_g$ has no pluricanonical divisors.

However the issue is resolved for all genera ≥ 22. Surprisingly, this depends (in the current state of our knowledge) on an understanding of the complement $\overline{M}_g \setminus M_g$ and its image in the divisor class group of $\overline{M}_g$. The starting point is the fact that a smooth n-dimensional projective variety X with an effective pluricanonical canonical divisor — that is, a nonzero section of the sheaf $\omega_X^{\otimes p}$ for some $p > 0$ — cannot be unirational: if there were a dominant rational map $\mathbb{P}^n \to X$, we could pull this section back to get an effective pluricanonical divisor on $\mathbb{P}^n$, which doesn't exist because the canonical divisor on $\mathbb{P}^n$ has negative degree. At the same time, we can analyze the divisor class theory of the space $\overline{M}_g$ and for large g exhibit an effective pluricanonical divisor on M_g by using components of $\overline{M}_g \setminus M_g$. The upshot is this:

Theorem 8.9 [Harris and Mumford 1982; Harris 1984; Eisenbud and Harris 1987c; Farkas et al. 2023]. *For all $g \geq 22$, M_g is not unirational.*

In each case, what is actually proven is the stronger but more technical statement that $\overline{M}_g$ has *general type*. This line of argument requires a great deal of work; the interested reader can find more details, plus a guide to the literature, in [Harris and Morrison 1998].

8.5. Hurwitz spaces

Hurwitz spaces are spaces parametrizing branched covers. They are fascinating objects; we know quite a bit about their geometry but there is much that is unknown as well. In this discussion, we'll stick to the simplest case, that of the *small Hurwitz spaces*, parametrizing simply branched covers of $\mathbb{P}^1$.

To start with the definition: the small Hurwitz space $\mathrm{Hur}^\circ_{g,d}$ parametrizes pairs (C, f) where C is a smooth curve of genus g and $f : C \to \mathbb{P}^1$ a map of degree d with simple branching; that is,

$$\mathrm{Hur}^\circ_{g,d} = \{(C, f) \mid C \in M_g \text{ and } f : C \to \mathbb{P}^1 \text{ simply branched of degree } d\}.$$

There are two natural maps from the Hurwitz space to other spaces. First, we can "project on the first factor;" that is, simply forget the map f to arrive at a map $\pi : \mathrm{Hur}^\circ_{g,d} \to M_g$. Secondly, we can associate to a point $(C, f) \in \mathrm{Hur}^\circ_{g,d}$ the branch divisor $B \subset \mathbb{P}^1$, which is an unordered b-tuple of distinct points in $\mathbb{P}^1$, which we can think of as a point in the b-th symmetric product $(\mathbb{P}^1)_b \cong \mathbb{P}^b$. We thus have a diagram

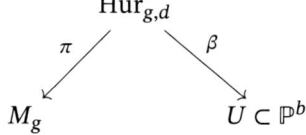

where $U \subset \mathbb{P}^b$ is the complement of the hypersurface in $\mathbb{P}^b$ where at least 2 of the b points are equal, called the discriminant hypersurface. Thus the Hurwitz space is positioned between an object U we understand relatively well, and an object M_g about which we would like to know more; this accounts for the historical importance of Hurwitz spaces. We'll now illustrate how this can be exploited.

To begin with, by the analysis in Section 6.2, we see that *the map β is a covering space*: for any reduced divisor $B \subset \mathbb{P}^1$ there are a finite number of simply branched covers of $\mathbb{P}^1$ with branch divisor B; and as we vary the points of B locally we can deform the cover along with them. This allows us to give the Hurwitz space $\mathrm{Hur}^\circ_{g,d}$ the structure of a smooth variety, and also tells us that

$$\dim(\mathrm{Hur}^\circ_{g,d}) = b = 2d + 2g - 2.$$

The dimension of M_g. Next, we look at the projection $\pi : \mathrm{Hur}^\circ_{g,d} \to M_g$. To start, let's assume d is large relative to g; $d \geq g + 1$ suffices, but you can take d as large as you like; taking $d > 2g$ may make the argument simpler.

Proposition 8.10. *If $d \geq g + 1$, the map $\pi : \mathrm{Hur}^\circ_{g,d} \to M_g$ is surjective, with fibers of dimension $2d - g + 1$.*

Proof. The question is, how many simply branched maps $f : C \to \mathbb{P}^1$ of degree d are there for a given curve C? To begin with, the $g + 1$ theorem (Theorem 5.9) tells us that there are some, whence we see that π is surjective.

We can compute the dimension of the fibers, too. To specify a map $f : C \to \mathbb{P}^1$, we can start by choosing a divisor $D \in C_d$, which will be the divisor $f^{-1}(\infty)$; this can be a general divisor of degree d on C. Second, we choose a divisor E which will be $f^{-1}(0)$; this can be a general member of the linear series $|D|$, which has dimension $d - g$. Finally, specifying $f^{-1}(\infty)$ and $f^{-1}(0)$ determines the map f up to scalar multiplication on $\mathbb{P}^1$; adding up the degrees of freedom, we see that the fibers of π have dimension

$$d + (d - g) + 1 = 2d - g + 1. \quad \square$$

Finally, we conclude that if $g \geq 2$ then

$$\dim(M_g) = (2d + 2g - 2) - (2d - g + 1) = 3g - 3.$$

We can use this in turn to analyze the cases of smaller d. As a basic application, note that the group PGL_2 of automorphisms of $\mathbb{P}^1$ acts on the Hurwitz space: given $\varphi \in \mathrm{PGL}_2$, we can send (C, f) to $(C, \varphi \circ f)$. Moreover, the orbits of this action lie in fibers of the projection $\pi : \mathrm{Hur}^\circ_{g,d} \to M_g$, meaning that the fibers of π have dimension at least 3.

Corollary 8.11. *If $d < \lceil \frac{g}{2} \rceil + 1$, then a general curve C of genus g does not admit a map of degree d to $\mathbb{P}^1$.*

This is one-half of the case $r = 1$ of the Brill–Noether theorem, about which we will say much more later.

Irreducibility of M_g. Another important application is the original proof of the irreducibility of M_g. Hurwitz [1891] analyzed the monodromy of the map $\beta : \mathrm{Hur}^{\circ}_{g,d} \to U \subset \mathbb{P}^b$, which describes what happens when you let the branch points of a cover wander around in U before coming back to their original locations. He proved that the monodromy is transitive, and hence that the Hurwitz space $\mathrm{Hur}^{\circ}_{g,d}$ is irreducible; since $\mathrm{Hur}^{\circ}_{g,d}$ dominates M_g for d large, he deduced that M_g must be irreducible as well.

Hurwitz's argument illustrates a fundamental point: in practice, moduli spaces of curves "with extra structure," such as a map to projective space, are often easier to work with, and provide a useful tool for understanding the geometry of abstract moduli spaces. Given an abstract curve C of genus g, it's hard without developing a fair amount of deformation theory, to show that C varies in a nontrivial family. But if C is expressed as a branched cover, we can find such families just by varying the branch points.

There are many open problems connected with the Hurwitz scheme; here are a few:

(1) A compactification of the Hurwitz scheme by *admissible covers* (allowing both source and target of the covering to be reducible in a controlled way) is known [Harris and Morrison 1998], but the boundary is very complicated, and it would be interesting to find a simpler one.

(2) It is conjectured that the Picard group of the Hurwitz scheme is torsion; see [Deopurkar and Patel 2015], where the conjecture is proved for $g \leq 5$, and [Mullane 2023] for the case $d > g - 1$.

(3) There is active work and many open problems around computing the *Hurwitz numbers*, that is, the number of curves having maps to $\mathbb{P}^1$ with specified degree and branching; see for example [Hurwitz 1901] and [Ekedahl et al. 2001].

8.6. The Severi variety

Despite having been studied for so long, many questions about plane curves remain open — for example: which ones degenerate into which others, and in what way. All plane curves of degree d have the same Hilbert function, and thus the same arithmetic genus $\binom{d-1}{2}$, but since curves of degree d can have different sorts and numbers of singularities, they can have geometric genera from 0 to $\binom{d-1}{2}$. In this section we will explore the subset of (reduced, irreducible) curves of degree d with a fixed geometric genus. We will focus on the open set consisting of nodal curves (those with only nodes as singularities), and compute its dimension.

Let $\mathbb{P}^N := \mathbb{P}^{\binom{d+2}{2}-1}$ be the projective space parametrizing plane curves of degree d. Within $\mathbb{P}^N$ the set of reduced irreducible curves is open — it is the complement of the union of the images of the maps

$$\mathbb{P}^{\binom{d_1+2}{2}-1} \times \mathbb{P}^{\binom{d_2+2}{2}-1} \to \mathbb{P}^N$$

with $d_1 + d_2 = d$ given by multiplication of forms.

Definition 8.12. The *Severi variety* $V_{d,g} \subset \overline{V}_{d,g}$ is the locus of irreducible plane curves of degree d with $\delta = \binom{d-1}{2} - g$ nodes and no other singularities. This is a locally closed subset of $\mathbb{P}^N$. (Reason: having only nodes as singularities is an open condition; having at least a certain number of them is a closed condition.) It is sometimes called the *small Severi variety*, since we are excluding curves with more complicated singularities.

We will see that the closure $\overline{V}_{d,g}$ is well behaved in a neighborhood of $V_{d,g}$; but away from this, even the singularities of $\overline{V}_{d,g}$ are not well understood. It is an interesting open problem to find a simpler partial compactification of $V_{d,g}$.

Cheerful Fact 8.13. Corollary 8.17 says that the variety $V_{d,g}$ is smooth. In 1921 F. Severi gave an incorrect proof that $V_{d,g}$ is connected, and thus irreducible. A correct proof was finally given in [Harris 1986].

Local geometry of the Severi variety. We first consider the universal singular point

$$\Phi := \{(C, p) \in \mathbb{P}^N \times \mathbb{P}^2 \mid p \in C_{\text{sing}}\}$$

and its image $\Delta \subset \mathbb{P}^N$, the *discriminant variety*.

Proposition 8.14. Φ *is smooth and irreducible of dimension* $N - 1$, *and the discriminant* Δ *is a hypersurface in* $\mathbb{P}^N$.

Proof. Projection on the second factor expresses Φ as a $\mathbb{P}^{N-3}$-bundle over $\mathbb{P}^2$. Explicitly, if $[X, Y, Z]$ are homogeneous coordinates on $\mathbb{P}^2$, and $\{a_{i,j,k} \mid i + j + k = d\}$ are homogeneous coordinates on $\mathbb{P}^N$, then the universal curve

$$\mathbb{C} := \{(C, p) \in \mathbb{P}^N \times \mathbb{P}^2 \mid p \in C\}$$

is given as the zero locus of the single bihomogeneous polynomial

$$F([a_{i,j,k}], [X, Y, Z]) = \sum a_{i,j,k} X^i Y^j Z^k$$

of bidegree $(1, d)$; and the universal singular point is the common zero locus of the three partial derivatives $\partial F/\partial X$, $\partial F/\partial Y$ and $\partial F/\partial Z$.

The set of forms F that define curves singular at a given point is defined by 3 independent linear conditions, and since the set of points is 2-dimensional, the set Δ of singular forms has dimension $N - 1$. $\square$

We next compute the differential of the map $\pi : \Phi \to \mathbb{P}^N$:

Lemma 8.15. *Suppose that* $(C, p) \in \Phi$, *with* p *a node of* C. *The differential*

$$d\pi : T_{(C,p)}\Phi \to T_C\mathbb{P}^N$$

is injective, with image the hyperplane $H_p \subset \mathbb{P}^N$ *of plane curves containing the point* p.

Thus, if p is a node of C and the only singularity of C, then Δ is smooth at C; and more generally the image of a small analytic neighborhood of $(C, p) \in \Phi$ is smooth, and we can identify its tangent space at p with the hyperplane H_p.

Proof. We will prove this using affine coordinates on $\mathbb{P}^2$ and $\mathbb{P}^N$. Changing coordinates if necessary, we may assume that the point $[1, 0, 0]$ is not in C, and that the point p is $[0, 0, 1]$. Let $x = X/Z$ and $y = Y/Z$ be coordinates on the affine plane $Z \neq 0$ and write the polynomial $F(x, y, 1)$ above as

$$f(x, y) = \sum_{i+j \leq d} a_{i,j} x^i y^j,$$

with $a_{d,0}$ normalized to 1.

Let g, h be the two partial derivatives of f:

$$g(x, y) := \frac{\partial f}{\partial x} = \sum_{i+j \leq d} i a_{i,j} x^{i-1} y^j$$

$$h(x, y) := \frac{\partial f}{\partial y} = \sum_{i+j \leq d} j a_{i,j} x^i y^{j-1}.$$

The functions f, g and h are local defining equations for Φ; we consider their partial derivatives with respect to x, y and $a_{0,0}$, evaluated at the point (C, p), as in the table:

	f	g	h
$\partial/\partial x$	0	$a_{2,0}$	$a_{1,1}$
$\partial/\partial y$	0	$a_{1,1}$	$a_{0,2}$
$\partial/\partial a_{0,0}$	1	0	0

The fact that p is a node of C (and not a more complicated singularity) implies that the upper right 2×2 submatrix is nonsingular, which shows that the differential $d\pi$ is injective, and its image is the hyperplane $a_{0,0} = 0$ in $\mathbb{P}^N$, which is exactly the hyperplane of curves containing p. $\square$

Lemma 8.16. *The nodes* q_i *of an irreducible nodal plane curve* C *of degree* d *impose independent conditions on curves of degree* $d - 3$, *and hence on curves of any degree* $m \geq d - 3$.

Proof. We will prove in Chapter 15 that the g sections of the canonical sheaf on the normalization $\widetilde{C}$ of C are the preimages of the sections of $\mathcal{O}_C(d-3)$ that vanish at the nodes. On the other hand, $h^0(\mathcal{O}_C(d-3)) = \binom{d-1}{2}$, and the difference is exactly the number of nodes. $\qquad\square$

Corollary 8.17. *If C is a nodal curve of degree d and geometric genus $g = \binom{d-1}{2}-\delta$, then in a neighborhood of $C \in \mathbb{P}^N$ the discriminant hypersurface of all singular curves consists of δ smooth sheets, meeting transversely, and hence $V_{d,g}$ is smooth.*

In a neighborhood of $C \in \mathbb{P}^N$ the variety $\overline{V}_{d,g'}$ with $g' = \binom{d-1}{2} - \delta' > g$ is the union of $\binom{\delta}{\delta'}$ smooth branches, each of dimension $N - \delta'$, corresponding bijectively with subsets of $\{p_1, \dots, p_\delta\}$ of cardinality δ'.

Figure 8.5 shows the case $\delta = 2$, $\delta' = 1$.

Proof. Lemma 8.15 shows that in an analytic neighborhood of $C \in \mathbb{P}^N$ the discriminant hypersurface Δ consists of δ smooth sheets, each corresponding to one node, and Lemma 8.16 implies that the tangent spaces to these sheets are linearly independent. $\qquad\square$

Corollary 8.18. *The Severi variety $V_{d,g}$ has pure dimension $N - \delta$, where*

$$\delta = \binom{d-1}{2} - g.$$

In Section 19.11, we give a heuristic calculation of the "expected dimension" $h(g,r,d)$ of the variety parametrizing curves of degree d and genus g in $\mathbb{P}^r$:

$$h(g,r,d) := 4g - 3 + (r+1)(d-g+1) - 1.$$

The actual dimension of the restricted Hilbert scheme may be quite different. But Corollary 8.18 shows that in case $r = 2$ (as in the case of $r = 1$), the actual dimension is always the expected.

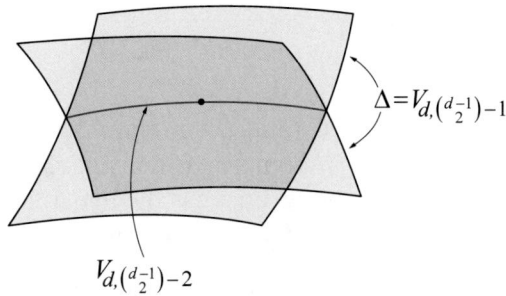

Figure 8.5. Near the point corresponding to a plane curve with 2 nodes, $V_{d,\binom{d-1}{2}-2}$ is the transverse intersection of two smooth hypersurfaces.

8.7. Exercises

Exercise 8.1. Consider the action of G_m on $\mathbb{P}^3$ given in coordinates by

$$t : (x_0, x_1, x_2, x_3) \mapsto (tx_0, \ tx_1, \ t^{-1}x_2, \ t^{-1}x_3)$$

for $t \in G_m = \mathbb{C}^*$.

(1) Show that the ring of forms in $\mathbb{C}[x_0, \dots, x_3]$ that are invariant is generated by

$$x_0 x_3, \ x_0 x_2, \ x_1 x_3, \ x_1 x_2$$

and thus $\mathbb{P}^3 \; /\!\!/ \; G_m \cong \mathbb{P}^1 \times \mathbb{P}^1$.

(2) Show that the unstable locus for this action is the union of the two lines $x_0 = x_1 = 0$ and $x_2 = x_3 = 0$.

(3) Show that the orbits of G_m are the points on the unstable lines and, for each point p not on an unstable line, a copy of $\mathbb{P}^1 \setminus \{0, \infty\} \cong G_m$ whose closure is the unique line containing p and meeting both unstable lines.

Exercise 8.2. Consider the action of G_m on $\mathbb{P}^3$ given in coordinates by

$$t : (x_0, x_1, x_2, x_3) \mapsto (tx_0, \ tx_1, \ tx_2, \ t^{-1}x_3)$$

for $t \in G_m = \mathbb{C}^*$.

(1) Show that the ring of forms in $\mathbb{C}[x_0, \dots, x_3]$ that are invariant is generated by forms

$$F(x_0, x_1, x_2)x_3$$

where F is a cubic form on $\mathbb{P}^2$, and thus $\mathbb{P}^3 \; /\!\!/ \; G_m \cong \mathbb{P}^2$, with the embedding given by the third Veronese map.

(2) Show that the unstable locus for this action is the union of the point $x_0 = x_1 = x_2 = 0$ and the plane $x_3 = 0$.

(3) Show that the orbits of G_m are the points on the components of the unstable locus and, for each point p that is not unstable, a copy of $\mathbb{P}^1 \setminus \{0, \infty\} \cong G_m$ whose closure is the unique line containing p and the unstable point. Thus the quotient map is the composition of the linear projection from the unstable point with the 3-uple embedding.

Exercise 8.3. Show from the explicit formula for the j-function on page 148 that if $j : B \to M_1 = \mathbb{A}^1$ is a map associated to a family $\mathcal{C} \to B$ of curves of genus 1, then every zero of the j-function has multiplicity divisible by 3, and conclude that some maps $B \to M_1$ do not correspond to families of curves; in particular there is no universal family over M_1, and thus M_1 is not a fine moduli space for curves of genus 1. There is a similar problem at $j(\lambda) = 1728$.

Exercise 8.4. In Exercise 8.3 we saw a local obstruction to the existence of a universal family over M_1. There is also a global obstruction, coming from the fact that some genus 1 curves have extra automorphisms. Show that there is a "tautological" family over the punctured j-line $L := \mathbb{A}^1 \setminus \{0, 1728\}$ — that is, a family $\mathcal{X} \to L$ whose fiber over t has j-invariant t; but show that this family is not universal as follows:

Let B be any curve of genus 1 and $\tau : B \to B$ a translation of order 2, and let E be a fixed elliptic curve (that is, a curve of genus 1 with a chosen point, so that we may identify the points of E with an abelian group). Let $\mathcal{X} \to L$ be the family $E \times B$ modulo the equivalence relation $(e, b) \sim (-e, \tau(b))$. The projection to B/τ has all fibers isomorphic to $E/(\pm) \cong E$. But the family is not isomorphic to the trivial family $E \times B/\tau \to B/\tau$.   ♦

Chapter 9

Curves of genus 4 and 5

In this chapter we study the linear series that exist on curves of genus 4 and 5, and what this says about maps to $\mathbb{P}^r$, focusing as usual on the nonhyperelliptic case.

9.1. Curves of genus 4

In genus 4 we face a question that the elementary theory based on the Riemann–Roch formula cannot answer: are nonhyperelliptic curves of genus 4 *trigonal*, that is, expressible as 3-sheeted covers of $\mathbb{P}^1$? The answer will emerge from our analysis in Proposition 9.2.

The canonical model. Let C be a nonhyperelliptic curve of genus 4. The canonical map $\phi_K : C \hookrightarrow \mathbb{P}^3$ embeds C as a curve of degree 6 in $\mathbb{P}^3$, and we identify C with the image. As in previous cases we may describe the homogeneous ideal I of C by considering the restriction maps

$$\rho_m : H^0(\mathcal{O}_{\mathbb{P}^3}(m)) \;\to\; H^0(\mathcal{O}_C(m)) = H^0(mK_C).$$

If $m = 2$ we have $h^0(\mathcal{O}_{\mathbb{P}^3}(2)) = \binom{5}{3} = 10$, while by the Riemann–Roch theorem

$$h^0(\mathcal{O}_C(2)) = 12 - 4 + 1 = 9.$$

Thus $C \subset \mathbb{P}^3$ lies on at least one quadric surface Q. The quadric Q must be irreducible, since any reducible and/or nonreduced quadric must be a union of planes, and thus cannot contain an irreducible nondegenerate curve. If $Q' \subset \mathbb{P}^3$ is another quadric then, by Bézout's theorem, $Q \cap Q'$ is a curve of degree 4 and thus cannot contain C. From this we see that Q is unique, and it follows that ρ_2 is surjective.

What about cubics? Again we consider the restriction map

$$\rho_3 : H^0(\mathcal{O}_{\mathbb{P}^3}(3)) \;\to\; H^0(\mathcal{O}_C(3)) = H^0(3K_C).$$

The space $H^0(\mathcal{O}_{\mathbb{P}^3}(3))$ has dimension $\binom{6}{3} = 20$, while the Riemann–Roch formula shows that

$$h^0(\mathcal{O}_C(3)) = 18 - 4 + 1 = 15.$$

It follows that the ideal of C contains at least a 5-dimensional vector space of cubic polynomials. We can get a 4-dimensional subspace as products of the unique quadratic polynomial F vanishing on C with linear forms — these define the cubic surfaces containing Q. Since $5 > 4$ we conclude that the curve C lies on at least one cubic surface S not containing Q. Bézout's theorem shows that the curve $Q \cap S$ has degree 6; thus it must be equal to C.

Let $G = 0$ be the cubic form defining the surface S. By Lasker's theorem the ideal (F, G) is unmixed, and thus is equal to the homogeneous ideal of C.

Conversely, let $C = Q \cap S$ with Q a quadric and S a cubic. By Corollary 2.10 the canonical sheaf of C is

$$\omega_C = (\omega_Q \otimes \mathcal{O}_Q(3))|_C = \mathcal{O}_C(-2 + 3) = \mathcal{O}_C(1),$$

so C is a canonical curve. Since C is smooth, the quadric and the cubic must meet transversely at points that are nonsingular on each of them. Thus:

Theorem 9.1. *The canonical model of a nonhyperelliptic curve of genus 4 is a complete intersection of a quadric $Q = V(F)$ and a cubic surface $S = V(G)$ meeting transversely along nonsingular points of each. Conversely, any smooth curve that is the complete intersection of a quadric and a cubic surface in $\mathbb{P}^3$ is the canonical model of a nonhyperelliptic curve of genus 4.* $\qquad\square$

Since the quadric surface Q containing the canonical curve C is unique, its rank is an invariant of C. Since C is irreducible and nondegenerate, the quadric cannot be a double plane or the union of two planes, but it can be singular (rank 3) or smooth (rank 4). On the other hand, the singularities of a cubic S such that $S \cap Q = C$ play no role. Of course S must be nonsingular along C, since else C would be singular. We can vary S by adding a multiple of the equation of Q to the equation of S, and since this linear series of cubics has base locus only along C, Bertini's theorem shows that the general such cubic is nonsingular everywhere.

Maps to projective space.

Maps to $\mathbb{P}^1$. We can now answer the question we asked at the outset, whether a nonhyperelliptic curve of genus 4 can be expressed as a 3-sheeted cover of $\mathbb{P}^1$. This amounts to asking if there are any divisors D on C of degree 3 with $r(D) \geq 1$; since we can take D to be a general fiber of a map $\pi : C \to \mathbb{P}^1$, we can for simplicity assume $D = p + q + r$ is the sum of three distinct points.

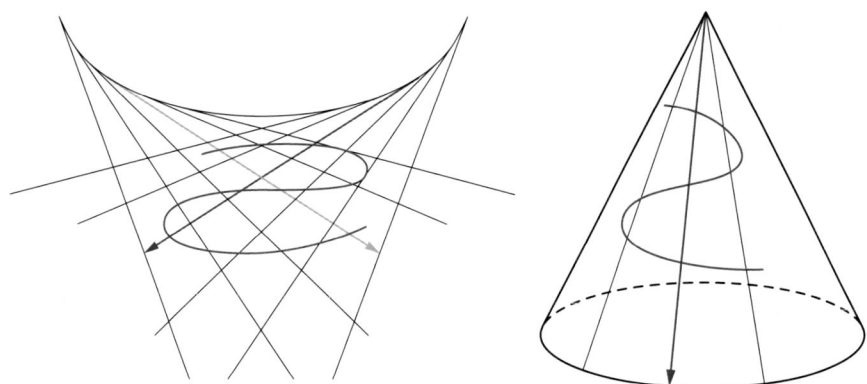

Figure 9.1. Left: canonical curve of genus 4 on a smooth quadric, showing two g_3^1s. Right: canonical curve of genus 4 on a quadric cone, with just one g_3^1.

By the geometric Riemann–Roch theorem, a divisor $D = p + q + r$ on a canonical curve $C \subset \mathbb{P}^{g-1}$ has $r(D) \geq 1$ if and only if the three points $p, q, r \in C$ are collinear. If three points $p, q, r \in C \subset Q$ lie on a line $L \subset \mathbb{P}^3$ then the quadric Q will meet L in at least three points, and hence will contain L. Conversely, if L is a line contained in Q, then the divisor $D = C \cap L = S \cap L$ on C has degree 3. Thus we can answer our question in terms of the family of lines contained in Q.

Any smooth quadric is isomorphic to $\mathbb{P}^1 \times \mathbb{P}^1$, and contains two families of lines, or *rulings*. On the other hand, any quadric of rank 3 is a cone over a smooth plane conic, and thus has just one ruling. By the argument above, the pencils of divisors on C cut out by the lines of these rulings are the g_3^1s on C. This proves:

Proposition 9.2. *A nonhyperelliptic curve of genus* 4 *may be expressed as a 3-sheeted cover of* $\mathbb{P}^1$ *in either one or two ways, depending on whether the unique quadric containing the canonical model of the curve is singular or smooth.*

See Figure 9.1 for an idea how this might look in each case.

Putting together the analyses of the preceding sections, we have shown that *every curve of genus g with* $2 \leq g \leq 4$ *is either hyperelliptic or trigonal.*

Maps to $\mathbb{P}^2$. We can also describe the lowest degree plane models of nonhyperelliptic curves C of genus 4. We can get a plane model of degree 5 by projecting C from a point p of the canonical model of C. Moreover, the Riemann–Roch theorem shows that if D is a divisor of degree 5 with $r(D) = 2$ then $h^0(K - D) = 1$. Thus D is of the form $K - p$ for some point $p \in C$, and the map to $\mathbb{P}^2$ corresponding to D is π_p. These maps $\pi_p : C \to \mathbb{P}^2$ have the lowest possible degree (except for those whose image is contained in a line) because by Clifford's theorem a nonhyperelliptic curve of genus 4 cannot have a g_4^2.

We now consider the singularities of the plane quintic $\pi_p(C)$. Suppose as above that $C = Q \cap S$, with Q a quadric. If a line L through p meets C in p plus a divisor of degree ≥ 2 then, as we have seen, L must lie in Q. Any line through p that is not contained in Q meets C in at most a single reduced point, whose image is thus a nonsingular point of $\pi(C)$. Moreover, a line that met C in a divisor of degree 4 or more would have to lie in both the quadric and the cubic containing C, and therefore would be contained in C. Since C is irreducible there can be no such line; thus the image $\pi_p(C)$ has at most double points.

We distinguish two cases, depending on whether the quadric Q is smooth or singular. We will make use of the Gauss map of the quadric, described by the next lemma.

Lemma 9.3. *Let $L \subset S \subset \mathbb{P}^3$ be a line on a surface $S \subset \mathbb{P}^3$ of degree $d \geq 2$, and write S_{sing} for the singular locus of S. The Gauss map $\mathcal{G} : S \to (\mathbb{P}^3)^*$ sending each point $p \in S \setminus S_{\mathrm{sing}}$ to the tangent plane $\mathbb{T}_p(S)$ maps L into the dual line in $\mathbb{P}^3$ (that is, the locus of planes containing L); if S is smooth along L then $\mathcal{G}$ has degree $d - 1$, and if S is singular anywhere along L it has strictly lower degree.*

The geometric idea behind this result is easy to understand: If S is smooth along L and $H \cong \mathbb{P}^2$ is a plane containing L, then $H \cap S = L \cup D$ for a plane curve D of degree $d - 1$. The plane H is tangent to S at a point p if and only if $H \cap S$ is singular at p; and this occurs at each of the $d - 1$ points (counted with multiplicity) at which D meets L, suggesting that the restriction of the Gauss map is $(d-1)$-to-one in this case.

If S is singular somewhere along L, then every plane through that point would be among the tangent planes in the sense above, and the image of L would be a component of a curve of degree $d - 1$ containing a line; thus of lower degree.

To see that the multiplicities count correctly in this argument, we will give an algebraic version:

Proof. Suppose that in terms of homogeneous coordinates $[X, Y, Z, W]$ on $\mathbb{P}^3$ the line L is given by $X = Y = 0$. Then the defining equation F of S can be written

$$F(X, Y, Z, W) = X \cdot G(Z, W) + Y \cdot H(Z, W) + J(X, Y, Z, W)$$

where J vanishes to order 2 along L; that is, $J \in (X, Y)^2$. The Gauss map $\mathcal{G}|_L$ restricted to L is then given by

$$[0, 0, Z, W] \mapsto [G(Z, W), H(Z, W), 0, 0].$$

The polynomials G and H have degree $d - 1$, and have a common zero if and only if S is singular somewhere along L; the lemma follows. $\square$

Example 9.4 (Gauss map of a quadric). Let $Q \subset \mathbb{P}^3$ be a smooth quadric, and let $L \subset Q$ be the line $X = Y = 0$. Since we may write the equation of Q as $XZ + YW = 0$, the Gauss map of Q, restricted to L, maps L one-to-one onto the dual line. Indeed, the Gauss map takes Q isomorphically onto its dual, which is also a smooth quadric.

We can also see this geometrically: if $H \subset \mathbb{P}^3$ is any plane containing the line $L \subset Q$, then H intersects Q in the union of L and a line M; the hyperplane section $Q \cap H = L \cup M$ is then singular at a unique point $p \in L$. Thus the Gauss map gives a bijection between points on L and planes containing L.

Given this, we can analyze the geometry of projections $\pi_p(C)$ of our canonical curve $C = Q \cap S$ as follows:

(1) Q is nonsingular: In this case there are two lines L_1, L_2 on Q that pass through p; they meet C in p plus divisors E_1 and E_2 of degree 2. If each E_i consists of distinct points then, since the tangent planes to the quadric along L_i are all distinct by Example 9.4, the plane curve $\pi(C)$ has two nodes, one at the image of each E_i.

On the other hand, if E_i consists of a double point $2q$ (that is, L_i is tangent to C at $q \neq p$, or meets C three times at $q = p$), then $\pi(C)$ has a cusp at the corresponding image point.

In either case, $\pi(C)$ has two distinct singular points, each either a node or a cusp. The two g_3^1s on C correspond to the projections from these singular points. These possibilities are illustrated in Figure 9.2.

(2) Q is a cone: In this case, since the curve cannot pass through the singular point of Q there is a unique line $L \subset Q$ that passes through p. Let $p + E$ be the divisor on C in which this line meets C. The tangent planes to Q along L are all the same. Thus if $E = q_1 + q_2$ consists of two distinct points,

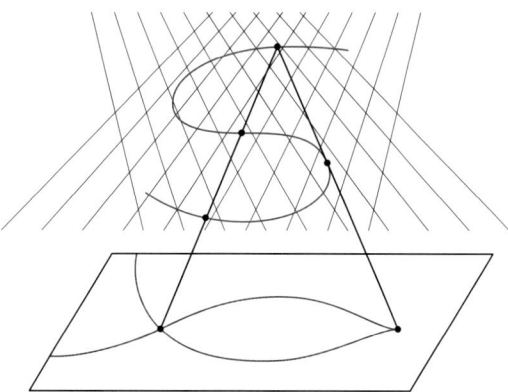

Figure 9.2. If a genus 4 canonical curve C lies on a smooth quadric, the image of the projection of C from a point $p \in C$ is a plane quintic with two singular points.

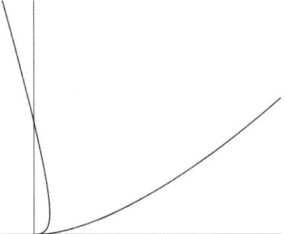

Figure 9.3. A ramphoid cusp on a quartic curve with affine parametrization (t^2+t^3, t^4). Note that the real points of this affine curve lie entirely in one of the half-planes determined by the tangent line at the cusp. This was first recorded in a 1744 letter from Leonhard Euler to Gabriel Cramer, who had conjectured that for an algebraic curve, the tangent at a cusp always lies between the two branches.

the image $\pi_p(C)$ has two smooth branches sharing a common tangent line at $\pi_p(q_1) = \pi_p(q_2)$. Such a point is called a *tacnode* (Figure 2.4) of $\pi_p(C)$. On the other hand, if $E = 2q$, that is, if L meets C tangentially at one point $q \neq p$ (or meets C three times at p) then the image curve has a higher-order cusp, called a *ramphoid* ("beak-like") *cusp* (Figure 9.3). In either case, the one g_3^1 on C is the projection from the unique singular point of $\pi(C)$.

9.2. Curves of genus 5

We next consider a nonhyperelliptic curve C of genus 5. There are now two questions that cannot be answered by simple application of the Riemann–Roch theorem:

(1) Is C expressible as a 3-sheeted cover of $\mathbb{P}^1$? Equivalently, does C have a g_3^1?

(2) Is C expressible as a 4-sheeted cover!of $\mathbb{P}^1$ of $\mathbb{P}^1$? In other words, does C have a basepoint free g_4^1?

The first question is answered in the negative by the dimension count of Section 8.5, but the analysis below will give us a way of characterizing those curves of genus 5 that are trigonal. Note also that if a curve C of genus 5 is trigonal, the g_3^1 on C is unique: if there were two distinct g_3^1s on C, with associated maps $\alpha, \beta : C \to \mathbb{P}^1$, the product map

$$\alpha \times \beta : C \to \mathbb{P}^1 \times \mathbb{P}^1$$

would give a birational embedding of C as a curve of bidegree $(3, 3)$ in $\mathbb{P}^1 \times \mathbb{P}^1$, and it would follow from the genus formula for curves on $\mathbb{P}^1 \times \mathbb{P}^1$ that the genus of C would be at most 4.

The argument of the preceding paragraph can be applied more broadly, and it's worth stating the results here:

Proposition 9.5. *Let d and e be relatively prime integers. If C is a curve of genus g with a basepoint free g_d^1 and a basepoint free g_e^1, then*

$$g \le (d-1)(e-1).$$

The same conclusion holds when C has two distinct $g_d^1 s$ with d prime.

Proof. As before, we let α and $\beta : C \to \mathbb{P}^1$ be the maps associated to the g_d^1 and the g_e^1. In either case the product map $\alpha \times \beta : C \to \mathbb{P}^1 \times \mathbb{P}^1$ is a birational embedding of C as a curve of bidegree (d,e) in $\mathbb{P}^1 \times \mathbb{P}^1$, and the result follows.   □

The condition of (relative) primeness is necessary: if $d = ma$ and $e = mb$, the map $\alpha \times \beta : C \to \mathbb{P}^1 \times \mathbb{P}^1$ could express C as an m-sheeted cover of a curve $C_0 \subset \mathbb{P}^1 \times \mathbb{P}^1$ of bidegree (a, b), and the genus of C could be arbitrarily large.

See Exercise 13.3 for the corresponding inequality where the target curves have higher genus.

9.3. Canonical curves of genus 5

As in the case of genus 4, the answers to the basic questions above about linear series on a curve C of genus 5 can be found through an investigation of the geometry of the canonical model $C \subset \mathbb{P}^4$ of C. This is an octic curve in $\mathbb{P}^4$, and as before the first question to ask is what sort of polynomial equations define C. We start with quadrics, by considering the restriction map

$$\rho_2 : H^0(\mathcal{O}_{\mathbb{P}^4}(2)) \to H^0(\mathcal{O}_C(2)).$$

On the left, we have the space of homogeneous quadratic polynomials on $\mathbb{P}^4$, which has dimension $\binom{6}{4} = 15$, while by the Riemann–Roch theorem the target is a vector space of dimension

$$2 \cdot 8 - 5 + 1 = 12.$$

We deduce that C lies on at least 3 independent quadrics. (We will see in the course of the following analysis that it is exactly 3; that is, ρ_2 is surjective.) Since C is irreducible and, by construction, does not lie on a hyperplane, each of the quadrics containing C is irreducible, and thus the intersection of any two is a surface of degree 4. There are now two possibilities: The intersection of (some) three independent quadrics $Q_1 \cap Q_2 \cap Q_3$ containing the curve is 1-dimensional; or every such intersection is 2-dimensional.

First case: the intersection of the quadrics has dimension 1. By Bézout's theorem the intersection is a curve of degree 8, and since C also has degree 8 we must have $C = Q_1 \cap Q_2 \cap Q_3$; that is, the canonical curve is a complete intersection. Lasker's theorem then shows that the three quadrics generate the

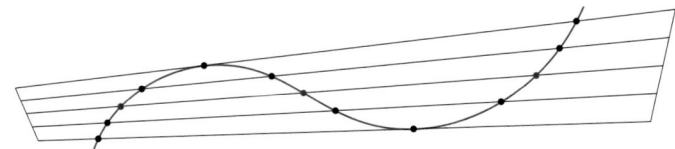

Figure 9.4. If C has a g^1_3 then, in the canonical embedding, the three points of each divisor are collinear, and the lines they span sweep out a surface.

whole homogeneous ideal of C; in particular, there are no additional quadrics containing C.

We can now answer the first of our two questions for curves of this type. As in the genus 4 case the geometric Riemann–Roch theorem implies that C has a g^1_3 if and only if the canonical model of C contains 3 collinear points or, more generally, meets a line L in a divisor of degree 3 (Figure 9.4). When C is the intersection of quadrics, this cannot happen, since the line L would have to be contained in all the quadrics that contain C. Thus, in this case, C has no g^1_3.

What about g^1_4s? Again invoking the geometric Riemann–Roch theorem, a divisor of degree 4 moving in a pencil lies in a 2-plane; so the question is, does $C \subset \mathbb{P}^4$ contain a divisor of degree 4, say $D = p_1 + \cdots + p_4 \subset C$, that lies in a plane Λ? Supposing this is so, we consider the restriction map

$$H^0(\mathcal{J}_{C/\mathbb{P}^4}(2)) \; \to \; H^0(\mathcal{J}_{D/\Lambda}(2)).$$

By what we have said, the left-hand space is 3-dimensional. We will show that the right-hand space is 2-dimensional, so that one of the quadrics vanishes identically on Λ.

Lemma 9.6. *Let $\Gamma \subset \mathbb{P}^2$ be any scheme of dimension 0 and degree 4. Either Γ is contained in a line $L \subset \mathbb{P}^2$, or Γ imposes independent conditions on quadrics, that is, $h^0(\mathcal{J}_{\Gamma/\mathbb{P}^2}(2)) = 2$.*

Proof. We will do this in case Γ is reduced, that is, consists of four distinct points; the reader is asked to supply the analogous argument in the general case in Exercise 9.8. Suppose to begin with that Γ fails to impose independent conditions on quadrics, and let $q \in \mathbb{P}^2$ be a general point. Since we are assuming that $h^0(\mathcal{J}_{\Gamma/\mathbb{P}^2}(2)) \geq 3$, we see that there are at least two conics $C', C'' \subset \mathbb{P}^2$ containing $\Gamma \cup \{q\}$. By Bézout's theorem, these two conics have a component in common, which can only be a line L; thus we can write $C' = L \cup L'$ and $C'' = L \cup L''$ for some pair of distinct lines $L', L'' \subset \mathbb{P}^2$. The intersection $C' \cap C''$ thus consists of the line L and the single point $L' \cap L''$. Since this must contain $\Gamma \cup \{q\}$, and q does not lie on the line joining any two points of Γ, we conclude that $L' \cap L'' = \{q\}$ and hence $\Gamma \subset L$. □

Cheerful Fact 9.7. Lemma 9.6 is the first case of a more general statement: If $n \le 2d + 1$ points in the plane fail to impose independent conditions on forms of degree d, then $d + 2$ of the points lie on a line. See [Eisenbud et al. 1996, p. 302] for a proof.

It follows that the 2-plane Λ spanned by D must be contained in one of the quadrics $Q \subset \mathbb{P}^4$ containing C. This implies in particular that the quadric is singular: If $V = \mathbb{C}^3 \subset \mathbb{C}^5$ is a 3-dimensional subspace of a 5-dimensional inner-product space, then V meets its orthogonal space in a line, which is a singular point of the corresponding quadric.

Thus Q is a cone over a quadric in $\mathbb{P}^3$, and it is ruled by the (one or two) families of 2-planes it contains, which are the cones over the (one or two) rulings of the quadric in $\mathbb{P}^3$. The argument above shows that the existence of a g^1_4s on C in this case implies the existence of a singular quadric containing C.

Conversely, suppose that $Q \subset \mathbb{P}^4$ is a singular quadric containing $C = Q_1 \cap Q_2 \cap Q_3$. Now say $\Lambda \subset Q$ is a 2-plane. If Q' and Q'' are quadrics that with Q span the space of quadrics containing C, then we can write

$$\Lambda \cap C = \Lambda \cap Q' \cap Q'',$$

from which we see that $D = \Lambda \cap C$ is a divisor of degree 4 on C, and so has $r(D) = 1$ by the geometric Riemann–Roch theorem. Thus, the rulings of singular quadrics containing C cut out on C pencils of degree 4; and every pencil of degree 4 on C arises in this way.

Does C lie on singular quadrics? There is a $\mathbb{P}^2$ of quadrics containing C—a 2-plane in the space $\mathbb{P}^{14}$ of quadrics in $\mathbb{P}^4$—and the family of singular quadrics consists of a hypersurface of degree 5 in $\mathbb{P}^{14}$, called the *discriminant* hypersurface. By Bertini's theorem, not every quadric containing C is singular. Thus the set of singular quadrics containing C is a plane curve B cut out by a quintic equation. So C does indeed have a g^1_4, and is expressible as a 4-sheeted cover of $\mathbb{P}^1$. Each singular quadric contributes either one or two g^1_4s, depending on whether it has rank 3 or 4. In sum, we have proven:

Proposition 9.8. *Let $C \subset \mathbb{P}^4$ be a canonical curve, and assume C is the complete intersection of three quadrics in $\mathbb{P}^4$. Then C may be expressed as a 4-sheeted cover of $\mathbb{P}^1$ in a one-dimensional family of ways, and there is a map from the set $W^1_4(C)$ of g^1_4s on C to a plane quintic curve B, whose fibers have cardinality 1 or 2.*

One can go further and ask about the geometry of the plane curve B and how it relates to the geometry of C. The list of possibilities is given in [Arbarello et al. 1985, p. 274].

Second case: the intersection of the quadrics is a surface. We will show that the intersection must contain an irreducible, nondegenerate surface. This follows from Fulton's *elementary Bézout theorem*:

Theorem 9.9 [Fulton 1984]. *Let $Z_1, \ldots, Z_k \subset \mathbb{P}^n$ be hypersurfaces of degrees $d_1, \ldots, d_k$. If $\Gamma_1, \ldots, \Gamma_m$ are the irreducible components of the intersection $\bigcap_1^k Z_j$, then*

$$\sum_{\alpha=1}^m \deg \Gamma_\alpha \leq \prod_{i=1}^k d_i.$$

Proof. We use induction on k, the result being trivial for $k = 1$. Assuming that the result is true for $k - 1$, we consider the irreducible components V_i of $\bigcap_1^{k-1} Z_j$. If Z_k contains V_i, then V_i is again a component of $\bigcap_1^k Z_j$. Otherwise, $V_i \cap Z_k$ is a union of components whose degrees sum to $d_k \deg V_i$. Thus the sum of the degrees of the components of $\bigcap_1^k Z_j$ is at most d_i times the sum of the degrees of components of $\bigcap_1^{k-1} Z_j$, as required. $\qquad \square$

Returning to the canonical curve $C \subset \mathbb{P}^4$, suppose that the intersection $X = Q_1 \cap Q_2 \cap Q_3$ of the three quadrics containing C has dimension 2. If C were a component of X, then the sum of the degrees of the irreducible components of X would be strictly greater than 8, which Fulton's theorem doesn't allow. Thus C must be contained in a 2-dimensional irreducible component S of X, and this surface S is necessarily nondegenerate.

We will return to this surface in Chapter 17, where we develop the theory of rational normal scrolls. Here is the result:

Theorem 9.10. *Let $C \subset \mathbb{P}^4$ be a canonical curve of genus 5. Then C lies on exactly three quadrics, and either*

(1) *C is the intersection of these quadrics; in which case C is not trigonal, and the variety $W_4^1(C)$ of expressions of C as a 4-sheeted cover of $\mathbb{P}^1$ is a 2-sheeted cover of a plane quintic curve; or*

(2) *the quadrics containing C intersect in a cubic surface that is a nonsingular rational normal scroll; in this case, C has a unique g_3^1, and the variety $W_4^1(C)$ consists of the union of two copies of C meeting at two points.*

Cheerful Fact 9.11. In the second case of the theorem, the ideal of the curve has the form $I_C = (Q_1, Q_2, Q_3, F_1, F_2)$ where the Q_i are quadrics and the F_i are cubic forms. The quadrics Q_i cut out a surface scroll, which must be smooth by Corollary 17.25.

There are 2 linear relations among the Q_i, and we shall see in Chapter 18 that this is a special case of the Eagon–Northcott complex. The generators of

I_C above may be written as the 4×4 Pfaffians of a skew-symmetric 5×5 matrix of the form

$$\begin{pmatrix} 0 & g_1 & g_2 & \ell_0 & \ell_1 \\ -g_1 & 0 & g_3 & \ell_2 & \ell_3 \\ -g_2 & -g_3 & 0 & \ell_3 & \ell_4 \\ -\ell_0 & -\ell_2 & -\ell_3 & 0 & 0 \\ -\ell_1 & -\ell_3 & -\ell_4 & 0 & 0 \end{pmatrix}$$

where $\ell_0, \dots, \ell_3$ are linear forms and g_1, g_2, g_3 are quadrics. The 2×2 minors of the matrix of linear forms in the last two columns are the three quadrics Q_i contained in the ideal of C, and those two columns are the linear relations on the Q_i mentioned above. The columns of the whole 5×5 matrix generate the syzygies of I_C. Moreover, the 4×4 Pfaffians of any sufficiently general matrix of this form define a trigonal canonical curve; see [Buchsbaum and Eisenbud 1977a].

9.4. Exercises

Exercise 9.1. Let C be a smooth projective nonhyperelliptic curve of genus 4, and $|D|$ a g_3^1 on C. Show that the following are equivalent:

(1) $D \sim K - D$.

(2) The multiplication map $\mu : H^0(D) \otimes H^0(K - D) \to H^0(K)$ fails to be surjective.

(3) The unique quadric Q containing the canonical curve of C is singular.

(4) $|D|$ is the unique g_3^1 on C.   ◆

Exercise 9.2. Let C again be a smooth projective nonhyperelliptic curve of genus 4 whose canonical model lies on a smooth quadric. We have seen that C is birational to a quintic plane curve C_0 with two nodes $p, q \in C_0$. Show that the canonical series of C is cut out by the system of plane conic curves passing through p and q in the sense that if D is a curve meeting C at each node and transverse to each branch at the nodes, then the sum of the points of $D \cap C$ *other than the nodes* is a canonical divisor.   ◆

Exercise 9.3. Let C be a smooth projective nonhyperelliptic curve of genus 4 and let D be a general divisor of degree 7 on C. By the $g + 3$ theorem (Theorem 5.13), $h^0(D) = 4$ and the map $\phi_D : C \to \mathbb{P}^3$ is an embedding. Show that the image $C \subset \mathbb{P}^3$ does not lie on any quadric surfaces, but does lie on two cubic surfaces S and T; describe the intersection $S \cap T$.   ◆

Exercise 9.4. We keep the setting of the preceding exercise, but now suppose that C *is* hyperelliptic. Show that in this case the image of C under the map $\phi_D : C \to \mathbb{P}^3$ does lie on a quadric surface Q, and in fact is a curve of type $(2, 5)$ on Q. Show also that if D is of the form $D \sim 2g_2^1 + p + q + r$ then the quadric surface Q is singular, and the image curve $\phi_D(C)$ has a triple point at the vertex of Q.   ◆

Exercise 9.5. Consider a space $M^r_{d,g}$ parametrizing g^r_ds on curves of genus g; that is,

$$M^r_{d,g} = \left\{ (C,L) \mid C \text{ a smooth curve of genus } g, L \in \mathrm{Pic}^d(C) \text{ and } h^0(L) \geq r+1 \right\}.$$

The analysis of this chapter shows that $M^1_{3,4}$ is a 2-sheeted cover of M_4. Show that it is in fact irreducible.   ◆

Exercise 9.6. The arguments in the chapter show that the canonical model of a nonhyperelliptic trigonal curve of genus 5 lies on an irreducible, nondegenerate cubic surface $S \subset \mathbb{P}^4$. Chapter 17, we'll see that such a surface is either smooth or a cone over a twisted cubic curve. Show that the latter case cannot occur.   ◆

Exercise 9.7. Let C be a smooth projective curve of genus 5. The $g+3$ theorem (Theorem 5.13) says that C admits an embedding in $\mathbb{P}^3$ as a curve of degree 8. Does it admit an embedding of degree 7?   ◆

Exercise 9.8. Complete the proof of Lemma 9.6.   ◆

Chapter 10

Hyperplane sections of a curve

One way to study a curve in $\mathbb{P}^r$ is to use properties of its hyperplane sections. In this chapter, we will take up the question: if $C \subset \mathbb{P}^r$ is a reduced, irreducible and nondegenerate curve, what can we say about the geometry of the points in the hyperplane section $\Gamma = C \cap H$ of C, and what more can we say about a general hyperplane section?

We prove and apply a result originally due to Castelnuovo, that every subset of r points in a general hyperplane section of C spans the hyperplane; that is, any r points of Γ are linearly independent. Recall (from just above Proposition 3.11) that in this situation the points are said to be in *linearly general position*. Castelnuovo's result holds for smooth curves in any characteristic; Exercise 10.2 has counterexamples involving singular curves in positive characteristic.

This is followed by a series of applications, including a famous theorem of Castelnuovo bounding the genus of a curve in terms of its degree, and, through that result, a proof that canonical curves and linearly normal curves "of high degree" are arithmetically Cohen–Macaulay (Theorem 10.7).

In Chapter 11, we will explain and prove a still stronger general position result, which requires characteristic 0, and give applications to the irreducibility of fiber products, to the Hilbert functions of subsets of the general hyperplane section, and to sums of linear series.

10.1. Linearly general position

In this section we allow our algebraically closed ground field to have arbitrary characteristic.

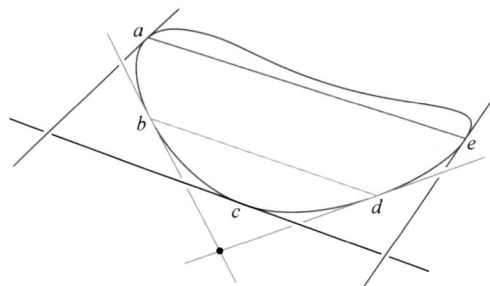

Figure 10.1. The lines bd and ae are stationary secants to a degree four space curve in red, because the tangents at b and d have a nonempty intersection, as do (at infinity) the tangents at a and e.

An irreducible and reduced curve $C \subset \mathbb{P}^r$ over an algebraically closed field, other than a line, is called *strange* if all the tangent lines at smooth points of C meet in a point $p \in \mathbb{P}^r$. If this condition seems somewhat... strange, that may be because no such curves exist in characteristic 0: if all the tangent lines at smooth points of C met in a point, then the projection of C from that point would have derivative 0 everywhere and hence be constant. But strange curves do exist in positive characteristic, albeit rarely: Pierre Samuel showed that the only *smooth* curve that is strange (in any characteristic) is the conic in characteristic 2; a proof is given in [Hartshorne 1977, Theorem IV.3.9]. Without the smoothness hypothesis there are more examples; see Exercise 10.2.

Theorem 10.1 [Rathmann 1987, Lemma 1.1]. *Suppose that $n \geq 3$. If $C \subset \mathbb{P}^r$ is a nondegenerate, irreducible, reduced curve that is not strange, and H is a general hyperplane, then any r points of $H \cap C$ are linearly independent.*

We postpone the proof to introduce some related definitions.

For example, the theorem says that for $C \subset \mathbb{P}^3$, the general hyperplane section does not contain a line that meets C three or more times — a *multisecant*. Another special configuration which is not present in a general hyperplane section, as it turns out, is any pair of points $p \neq q$ on C such that the tangent lines to C at p and q intersect — what was classically called a *stationary secant* (Figure 10.1).

More formally we define two subsets of the symmetric square C_2 of C:

$$\operatorname{mulsec} C := \{(p, q) \in C_2 \mid \overline{pq} \text{ meets } C \text{ in a scheme of length} \geq 3\},$$

$$\operatorname{stat} C := \{(p, q) \in C_2 \setminus \Delta \mid \text{the tangent lines to } C \text{ at } p, q \text{ meet}\}.$$

Here $\overline{pq}$ denotes the secant line through p, q (or the tangent line if $p = q$) and $\Delta \subset C_2$ is the diagonal.

Clearly if C is strange or planar then $\operatorname{stat} C = C_2 \setminus \Delta$, and if C is planar with $\deg C > 2$ then $\operatorname{mulsec} C = C_2$. These are the only such cases:

Proposition 10.2. *If $C \subset \mathbb{P}^r$ is a reduced, irreducible, curve that is neither planar nor strange, then* mulsec $C \subset C_2$ *and* stat $C \subset C_2 \setminus \Delta$ *are closed subsets of dimension at most 1.*

Proof. To show that mulsec C is closed, consider the projection map

$$\{((p,q),r) \in C_2 \times \mathbb{P}^r \mid r \in \overline{pq}\} \to C_2$$

sending $((p,q),r)$ to (p,q). By the semicontinuity of the degree in proper families of constant fiber dimension, the set of points lying in fibers of length ≥ 3 is closed, and mulsec C is the image of this set. Because the map is finite, mulsec C is closed as well.

Similarly, since $\{((p,q),r) \in (C_2 \setminus \Delta) \times \mathbb{P}^r \mid r \in \mathbb{T}_p(C) \cup \mathbb{T}_q(C)\}$ is closed in $(C_2 \setminus \Delta) \times \mathbb{P}^r$, its image under the projection to the first factor, which is stat C, is closed in $C_2 \setminus \Delta$.

Suppose, contrary to the proposition, that dim stat $C > 1$. Since stat C is a closed subset of the irreducible variety $C_2 \setminus \Delta$ it follows that stat $C = C_2 \setminus \Delta$, so every pair of tangent lines meet. Any three lines in projective space that meet pairwise must be either coplanar or meet in a point. Thus if C is not strange, then all the tangents to C lie in a common plane, and C is planar, contradicting our hypothesis and proving that dim stat $C \leq 1$.

Finally suppose, contrary to the proposition, that dim mulsec $C > 1$. As in the previous argument, this implies that every secant is a multisecant. We will show in this case that dim stat $C = 2$, a contradiction.

To this end, consider the projections $\pi_p : C \to \mathbb{P}^{r-1}$ and the set

$$I = \left\{ (p,r,r') \in C^3 \,\middle|\, \begin{array}{l} p,r,r' \text{ are distinct and} \\ \pi_p(r) = \pi_p(r') \text{ is a smooth point of } \pi_p(C) \end{array} \right\}.$$

The tangent lines $\mathbb{T}_r(C)$ and $\mathbb{T}_{r'}(C)$ both project from p to the tangent line $\mathbb{T}_{\pi_p(r)}(\pi_C)$, and thus they lie in the 2-plane spanned by p and this line; it follows that $(r,r') \in$ stat C. The projection $I \to C_2 : (p,r,r') \mapsto (r,r')$ is finite since the secant line $\overline{rr'}$ meets C in only finitely many points, so dim $I \leq$ dim stat C.

If the projection from some point p were everywhere ramified, then all tangents to C would pass through p, and C would be strange. Since also every secant to C is a multisecant it follows that for every smooth point $p \in C$ there is an open set of points $q \in \pi_p(C)$ such that the fiber $\pi_p^{-1}(q)$ contains two distinct points of C. Thus the projection $I \to C : (p,r,r') \mapsto p$ maps I onto an open set of C, with 1-dimensional fibers, so $2 = $ dim $I \leq$ dim stat C as claimed. $\qquad\square$

Proof of Theorem 10.1. If $C \subset \mathbb{P}^n$ is a reduced curve then, by Bertini's theorem, a general hyperplane meets C transversely. First suppose that $n = 3$, and consider the incidence correspondence

$$I := \{((p,q),H) \in \text{mulsec } C \times (\mathbb{P}^3)^* \mid \overline{pq} \in H\}.$$

Since dim $\overline{pq} = 1$, the fibers of the projection of I to mulsec C have dimension 1, so dim $I = 1 + \dim \operatorname{mulsec} C \leq 2$. Consequently the image of the projection of I to $(\mathbb{P}^3)^*$ has dimension ≤ 2, so there is an open set of planes that contain no multisecants, proving the theorem in this case.

Cheerful Fact 10.3. Since a secant line to a curve C that is not itself a line is determined by its intersections with C, every curve other than a line has a 2-parameter family of secant lines. If $C \subset \mathbb{P}^3$ is nondegenerate, this means that it is 2 conditions for a line in $\mathbb{P}^3$ to be a secant to C. Since a pencil of lines meets C in codimension 1, one might expect that, correspondingly, there would be a 1-parameter family of trisecant lines and a finite number of 4-secant lines. One can compute the expected number enumeratively [Griffiths and Harris 1978, p. 296]. For example a general curve of degree ≤ 4 can have no 4-secant line (it would meet planes containing that line in at least 5 points), but a general rational curve of degree 5 in $\mathbb{P}^3$ has exactly 1 [Eisenbud and Harris 2016, Section 12.4.4]. See Figure 10.2 for a construction in which the 4-secant line is visible.

This fact entered history in a peculiar way: After Kronecker [1882] proved that every algebraic set in $\mathbb{P}^n$ is set-theoretically the intersection of $n+1$ hypersurfaces, Vahlen [1891] claimed that Kronecker's result was sharp "because" a general rational quintic with one 4-secant line could not be the intersection of 3 hypersurfaces. Vahlen played a role in the rise of the Nazi party, and perhaps for that reason Perron [1941] produced 3 hypersurfaces intersecting precisely in a given general rational quintic. It is now known that every algebraic set in $\mathbb{P}^n$ over an infinite field can be written set-theoretically as the intersection of just n hypersurfaces [Eisenbud and Evans 1973].

We next do induction on $n \geq 4$. We will show first that the image C' of C under the projection $\pi_p : C \to C'$ from a general point of C is not strange.

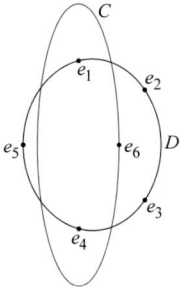

Figure 10.2. A smooth rational quintic in $\mathbb{P}^3$ with a unique 4-fold secant line can be obtained as the image of a plane conic C by the rational map of $\mathbb{P}^2$ to $\mathbb{P}^3$ induced by the 4 cubics through 6 points $e_1, \ldots e_6$, exactly one of which lies on C; the image of the unique conic D through the other 5 points becomes the 4-secant line to the image of C.

Since $C \subset \mathbb{P}^r$ is neither planar nor strange Proposition 10.2 shows that we may choose two smooth points $r, r' \in C \subset \mathbb{P}^r$ such that $\mathbb{T}_r(C) \cap \mathbb{T}_{r'}(C) = \emptyset$, and thus $\mathbb{T}_r(C)$ and $\mathbb{T}_{r'}(C)$ together span a 3-plane L. Since C is nondegenerate we may choose a point $p \in C$ outside L, and it follows that π_p restricted to L is an isomorphism. Thus $\mathbb{T}_{\pi_p(r)}(C') \cap \mathbb{T}_{\pi_p(r')}(C') = \emptyset$, so C' is not strange, and since $n \geq 4$ the curve C' is not planar either.

Arguing by contradiction, suppose that the general hyperplane section of C contains a set of r linearly dependent points. Consider the closed subsets

$$I_1 \subset I \subset \{(p, H) \in C \times (\mathbb{P}^n)^*\},$$

where I is defined by the condition that $p \in H$ and I_1 is defined by the condition that, in addition, there exist $p_2, \ldots, p_r \in H$ such that $p, p_2, \ldots, p_r$ are dependent. Both I_1 and I are closed subsets.

By hypothesis, both I_1 and I project onto open sets of $(\mathbb{P}^n)^*$, so they have the same dimension. But I is irreducible: it projects onto an open subset of C with fibers isomorphic to $\mathbb{P}^{r-1}$. Thus $I_1 = I$, and p is part of a dependent set $\{p, p_2, \ldots, p_r\}$ in a general hyperplane containing p.

Let H' be a general hyperplane in $\mathbb{P}^{n-1}$ and let $H = \pi_p^{-1}(H')$, which is a general hyperplane containing p. The intersection $H \cap C$ contains r dependent points $\{p, p_2, \ldots, p_r\}$, and it follows that $\pi_p(p_2), \ldots, \pi_p(p_n)$ are dependent points of $H' \cap C'$. This contradicts the induction hypothesis, and proves that the general hyperplane section of C is in linearly general position. $\square$

10.2. Castelnuovo's theorem

Clifford's theorem gives a complete and sharp answer to the question, "what linear series can exist on a curve of genus g?" But maybe that wasn't the question we meant to ask! After all, we're interested in describing curves in projective space as images of abstract curves C under maps given by linear series on C. Observing that the linear series that achieve equality in Clifford's theorem give maps of degree 2 onto a rational curve, we might hope that we would have a different and more relevant answer if we restrict our attention to linear series $\mathcal{D} = (\mathcal{L}, V)$ for which the associated map $\phi_{\mathcal{D}}$ is at least birational onto its image.

A classical result of Castelnuovo does exactly that: it gives a sharp bound on the arithmetic genus of a reduced, irreducible, nondegenerate curve of degree d in $\mathbb{P}^r$. To state it, for positive integers d and r, let

$$M := M(d, r) := \left\lfloor \frac{d-1}{r-1} \right\rfloor,$$

so that

$$d - 1 = M(r - 1) + \epsilon$$

for some $\epsilon = \epsilon(d, r)$ with $0 \leq \epsilon \leq r - 2$.

Theorem 10.4 (Castelnuovo bound). *Let $C \subset \mathbb{P}^r$ be a reduced, irreducible, nondegenerate curve of degree d. With M and ϵ defined as above,*

$$p_a(C) \leq \pi(d,r) := \frac{M(M-1)}{2}(r-1) + M\epsilon.$$

Moreover, if $p_a(C) = \pi(d,r)$ then C is arithmetically Cohen–Macaulay and every hyperplane section $H \cap C$ has Hilbert function $\dim(R_{H\cap C})_m = \min\{m(r-1), d\}$.

We will say that a curve achieving the bound is a *Castelnuovo curve*.

Proof of Castelnuovo's bound. The idea of Castelnuovo's proof is simple and beautiful. To start with, the hypothesis that C is nondegenerate in $\mathbb{P}^r$ says that $h^0(\mathcal{O}_C(1)) \geq r + 1$. We'd like to use this and the Riemann–Roch formula, in the form

$$g = d + 1 - h^0(\mathcal{O}_C(1)) + h^1(\mathcal{O}_C(1)),$$

but this doesn't work because we have a priori no way to estimate $h^1(\mathcal{O}_C(1))$.

Castelnuovo's solution is to derive lower bounds not just on $h^0(\mathcal{O}_C(1))$ but on $h^0(\mathcal{O}_C(m))$ for all m; since we know that $h^1(\mathcal{O}_C(m)) = 0$ for large m, a lower bound on $h^0(\mathcal{O}_C(m))$ for large m will translate directly into an upper bound on g.

How can we estimate $h^0(\mathcal{O}_C(m))$? The answer is to derive lower bounds on the successive differences $h^0(\mathcal{O}_C(m)) - h^0(\mathcal{O}_C(m-1))$, by letting $\Gamma = C \cap H$ be a general hyperplane section of C, and viewing $H^0(\mathcal{O}_C(m-1))$ as the subspace of $H^0(\mathcal{O}_C(m))$ consisting of sections vanishing on Γ. The estimates we get may seem crude — not surprisingly, considering how little we know about Γ — but remarkably, the bound we ultimately derive on the genus of C turns out to be sharp!

For the proof, the following definition will be convenient:

Definition 10.5. If $\mathcal{V} = (V, \mathcal{L})$ is a linear series on a variety X and Γ is a subscheme then the *number of conditions imposed by Γ on $\mathcal{V}$* (that is, the number of independent linear conditions for a section in V to vanish identically on Γ) is the dimension of the image of V in $H^0(\mathcal{L}|_\Gamma) = H^0(\mathcal{L} \otimes \mathcal{O}_\Gamma)$; or, numerically,

$$\dim V - \dim\left(V \cap H^0(\mathcal{L} \otimes \mathcal{I}_{\Gamma/X})\right).$$

Thus, for example, if $\Gamma \subset \mathbb{P}^r$, then the number of conditions imposed by Γ on $H^0(\mathcal{O}_{\mathbb{P}^r}(m))$ is the value $h_\Gamma(m)$ of the Hilbert function of Γ at m. Note that in case Γ is zero-dimensional, the number of conditions imposed by Γ on a linear series V is necessarily less than or equal to the degree d of Γ; if it is equal we say that Γ *imposes independent conditions on V*.

Proof of Theorem 10.4. Suppose that $C \subset \mathbb{P}^r$ is an irreducible, nondegenerate curve. For large m we have $h^0(\mathcal{O}_C(m)) = md - p_a(C) + 1$, so to bound the genus from above we must bound $h^0(\mathcal{O}_C(m))$ from below.

Let $\Gamma = C \cap H$ be a general hyperplane section of C, and let $V_m \subset H^0(\mathcal{O}_C(m))$ be the linear series cut on C by hypersurfaces of degree m in $\mathbb{P}^r$, that is, the image of the restriction map

$$H^0(\mathcal{O}_{\mathbb{P}^r}(m)) \to H^0(\mathcal{O}_C(m)).$$

The number of conditions imposed by Γ on V_m is the rank of the restriction map $\rho_m : V_m \to \mathcal{O}_\Gamma(m)$. The number of conditions imposed by Γ on $H^0(\mathcal{O}_C(m))$ is the rank of the restriction map from this potentially larger space, and thus is at least as large. This accounts for the inequality in the following:

$$
\begin{aligned}
h^0(\mathcal{O}_C(m)) - h^0(\mathcal{O}_C(m-1)) &= \text{\# of conditions imposed by } \Gamma \text{ on } H^0(\mathcal{O}_C(m)) \\
&\geq \text{\# of conditions imposed by } \Gamma \text{ on } V_m \\
&= \text{\# of conditions imposed by } \Gamma \text{ on } H^0(\mathcal{O}_{\mathbb{P}^r}(m)) \\
&= h_\Gamma(m).
\end{aligned}
$$

Replacing m by k in the display above and summing over k, we obtain the lower bound

$$h^0(\mathcal{O}_C(m)) \geq \sum_{k=0}^{m} h_\Gamma(k).$$

In order to bound the genus C from above, we will bound the Hilbert function of its hyperplane section Γ from below; and for this, we need to know something about the geometry of Γ. In fact, all we need to know is Theorem 10.1, which says that the points of Γ are in linearly general position!

Proposition 10.6. *If $\Gamma \subset \mathbb{P}^r$ is a collection of d points in linearly general position that span $\mathbb{P}^r$, then*

$$
h_\Gamma(m) \geq \begin{cases} mr + 1 & \text{if } m \leq M(d, r+1), \\ d & \text{otherwise.} \end{cases}
$$

One way to understand the bound $mr + 1$ is to realize that if Γ is any finite subscheme of a rational normal curve $C \subset \mathbb{P}^r$ of degree r, then $h_\Gamma(m) = \min\{\deg \Gamma, mr + 1\}$ for every m (Exercise 10.5). Thus the bound in Proposition 10.6 is best possible. On the other hand, sets of points on a rational normal curve are almost the only sets for which the bound is sharp. See Figure 10.3 for an illustration.

Proof. Suppose first that $d \geq mr+1$, and let $p_1, \dots, p_{mr+1} \in \Gamma$ be any subset of $mr+1$ points. It suffices to show that $\Gamma' = \{p_1, \dots, p_{mr+1}\}$ imposes independent conditions on $H^0(\mathcal{O}_{\mathbb{P}^r}(m))$, that is, for any $p_i \in \Gamma'$ there is a hypersurface $X \subset \mathbb{P}^r$ of degree m containing all the points $p_1, \dots, \hat{p}_i, \dots, p_{mr+1}$ but not containing p_i.

To construct such an X, group the mr points of $\Gamma' \setminus \{p_i\}$ into m subsets Γ_k of cardinality r; each set Γ_k will span a hyperplane $H_k \subset \mathbb{P}^r$, and we can take $X = H_1 \cup \cdots \cup H_m$.

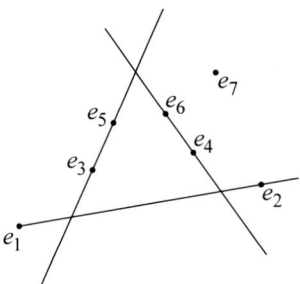

Figure 10.3. The case $d = 3 \cdot 2 + 1$: the point e_7 imposes an additional condition on the cubics (such as the union of the three lines shown) passing through $e_1, \dots, e_6$.

In the case where $d < mr + 1$, we add $mr + 1 - d$ general points; each one imposes exactly one additional condition on hypersurfaces of degree m.   □

To complete the proof of Theorem 10.4 we add up the lower bounds in the proposition. To this end, let $C \subset \mathbb{P}^r$ be an irreducible, nondegenerate curve of degree d, and set $M = M(d,r) = \lfloor \frac{d-1}{r-1} \rfloor$ as on page 183. We have

$$h^0(\mathcal{O}_C(M)) = \sum_{k=0}^{M} h^0(\mathcal{O}_C(k)) - h^0(\mathcal{O}_C(k-1))$$

$$\geq \sum_{k=0}^{M} (k(r-1)+1) = \frac{M(M+1)}{2}(r-1) + M + 1,$$

and similarly

$$h^0(\mathcal{O}_C(M+m)) \geq \frac{M(M+1)}{2}(r-1) + M + md + 1.$$

For sufficiently large m, the line bundle $\mathcal{O}_C(M+m)$ is nonspecial, so by the Riemann–Roch theorem,

$$g = (M+m)d - h^0(\mathcal{O}_C(M+m)) + 1$$

$$\leq (M+m)d - \left(\frac{M(M+1)}{2}(r-1) + M + 1 + md\right) + 1$$

$$= M(M(r-1)+1+\epsilon) - \left(\frac{M(M+1)}{2}(r-1) + M\right)$$

$$= \frac{M(M-1)}{2}(r-1) + M\epsilon.$$

In the case of equality, to show that C is arithmetically Cohen–Macaulay we must show that $V_m = H^0(\mathcal{O}_C(m))$ for all m. Consider the diagram

$$
\begin{array}{ccccccc}
0 & \longrightarrow & H^0(\mathcal{O}_{\mathbb{P}^r}(m-1)) & \longrightarrow & H^0(\mathcal{O}_{\mathbb{P}^r}(m)) & \longrightarrow & H^0(\mathcal{O}_\Gamma(m)) \\
& & \downarrow{\scriptstyle\text{surjection}} & & \downarrow{\scriptstyle\text{surjection}} & & \downarrow{\scriptstyle =} \\
0 & \longrightarrow & V_{m-1} & \longrightarrow & V_m & \longrightarrow & H^0(\mathcal{O}_\Gamma(m)) \\
& & \downarrow{\scriptstyle\text{injection}} & & \downarrow{\scriptstyle\text{injection}} & & \downarrow{\scriptstyle =} \\
0 & \longrightarrow & H^0(\mathcal{O}_C(m-1)) & \longrightarrow & H^0(\mathcal{O}_C(m)) & \longrightarrow & H^0(\mathcal{O}_\Gamma(m)).
\end{array}
$$

The top and bottom rows are left exact. From the left exactness of the bottom row it follows that the map $V_{m-1} \to V_m$ is an injection. From the left exactness of the top row it follows that this map is the kernel of the restriction map $V_m \to H^0(\mathcal{O}_\Gamma(m))$. Thus the middle row is also left exact, and we see that the number of conditions that Γ imposes on V_m is equal to $\dim V_m - \dim V_{m-1}$.

From the first part of the proof we have inequalities

# of conditions imposed by Γ on $H^0(\mathcal{O}_C(m))$

$$\geq \text{\# of conditions imposed by } \Gamma \text{ on } V_m,$$

that is,

$$h^0(\mathcal{O}_C(m)) - h^0(\mathcal{O}_C(m-1)) \geq \dim V_m - \dim V_{m-1}.$$

If $g = \pi(d,r)$ then these must both be equalities. Thus for large m the restriction map $H^0(\mathcal{O}_{\mathbb{P}^r}(m)) \to H^0(\mathcal{O}_C(m))$ is surjective, so

$$\sum_{k=0}^{m} (h^0(\mathcal{O}_C(k)) - h^0(\mathcal{O}_C(k-1))) = h^0(\mathcal{O}_C(m)) = \dim V_m$$

$$= \sum_{k=0}^{m} (\dim V_k - \dim V_{k-1}).$$

However, for each k we have

$$\dim V_k - \dim V_{k-1} \geq h^0(\mathcal{O}_C(k)) - h^0(\mathcal{O}_C(k-1)),$$

and thus we must have equality for all k, whence $\dim V_k = h^0(\mathcal{O}_C(k))$ for all k; that is, C is arithmetically Cohen–Macaulay.

Always supposing that $p_a(C) = \pi(d,r)$, the inequalities in Proposition 10.6 must be equalities, so for a general hyperplane section $\Gamma = H \cap C$ we have $h_\Gamma(m) = \min\{m(r-1)+1, d\}$ for all $m \geq 0$. However, since C is arithmetically Cohen–Macaulay, $h_\Gamma(m) = h^0(\mathcal{O}_C(m)) - h^0(\mathcal{O}_C(m-1))$ for every hyperplane section Γ, completing the proof. $\square$

Consequences and special cases. First of all, if $r = 2$ we have $\pi(d,r) = \binom{d-1}{2}$, so every plane curve (smooth or not) is a Castelnuovo curve.

In case $r = 3$, we have

$$\pi(d,r) = \begin{cases} \left(\frac{d-2}{2}\right)^2 & \text{if } d \text{ is even,} \\ \left(\frac{d-1}{2}\right)\left(\frac{d-3}{2}\right) & \text{if } d \text{ is odd,} \end{cases}$$

which we can recognize as the genus of curves of type $(d/2, d/2)$ on a quadric surface in case d is even, and the genus of curves of type $\big((d+1)/2, (d-1)/2\big)$ on a quadric when d is odd. Thus the Castelnuovo bound is sharp when $r = 3$ as well.

Indeed, the Castelnuovo bound is sharp for all d and r; we'll prove this, and describe explicitly the curves that achieve it, in Chapter 17.

Corollary 10.7. *Suppose that $C \subset \mathbb{P}^r$ is a smooth curve of arithmetic genus p_a and degree d embedded by a complete linear series. The curve C is Castelnuovo (and thus arithmetically Cohen–Macaulay) if and only if one of the following conditions holds:*

(1) *$d < 2r$ and $d \geq 2p_a + 1$.*

(2) *$d = 2r$ and C is a canonical curve.*

(3) *$r = 3, d > 6$, and C is a divisor on a (smooth or singular) irreducible quadric in the class mH or $mH + L$, where H is the class of a hyperplane and L is the class of a line.*

(4) *More generally, $d > 2r$ and C lies on a rational normal scroll in a certain range of divisor classes (Theorem 17.20).*

Cheerful Fact 10.8. Corollary 10.7 remains true with appropriate definitions also for reduced irreducible curves. This can be proven with the techniques developed in Chapters 16 and 17.

Proof. Simple arithmetic shows that if $d \geq 2p_a + 1$ then $d < 2r$ and $p_a = \pi(d,r)$. Moreover if $p_a = \pi(d,r)$ and $d < 2r$, then indeed $d \geq 2p_a + 1$, proving (1).

If $d = 2r$ then $\pi(d,r) = r + 1$, and Clifford's theorem shows that C is a canonical curve, proving (2).

Now suppose that $r = 3, d > 6$ and let Γ be a hyperplane section of C. Since $h_\Gamma(2) = \min\{2 \cdot 2 + 1, d\} = 5$, Γ lies on a conic, and since C is arithmetically Cohen–Macaulay, C must lie on a quadric. If C is smooth then arithmetic plus the result of Section 2.7 establishes the claim. For the more general case (4), and the case $r > 3$, see Theorem 17.20 in Chapter 17, where we define rational normal scrolls. $\square$

The most important special case is that of canonical curves:

Corollary 10.9. *If $C \subset \mathbb{P}^{g-1}$ is a canonical curve then the Hilbert function of the homogeneous coordinate ring S_C of C depends only on g, and is given by*

$$\dim(S_C)_n = h^0(\mathcal{O}_C(n)) = \begin{cases} 0 & \text{if } d < 0, \\ 1 & \text{if } d = 0, \\ g & \text{if } d = 1, \\ (2g-2)n + 1 - g & \text{if } d > 1. \end{cases}$$

Proof. Theorem 10.4 implies that the homogeneous coordinate ring of C can be identified with $\bigoplus_{n\in\mathbb{Z}} H^0(\mathcal{O}_C(n))$, and the dimensions of these spaces are determined by the Riemann–Roch theorem. □

Cheerful Fact 10.10. A famous theorem of Gruson and Peskine [1982] (see [Hartshorne 1982] for an exposition and also the case of positive characteristic) completes the picture of the possibilities for the degree d and genus g of a smooth curve in $\mathbb{P}^3$. If the curve does not lie on a plane or a quadric, then the genus satisfies the stronger inequality

$$g \leq \pi_1(d, 3) := \frac{d^2 - 3d}{6} + 1$$

and smooth curves with all such degree and genus exist; they can all be realized as curves on cubic or quartic surfaces.

Note that there can be gaps in the possible genera of curves a given degree: for example $\pi_1(9, 3) = 10 < \pi(9, 3) = 12$ but there is no curve of degree 9 and genus 11. The full range of possible degrees and genera for curves in $\mathbb{P}^r$ remains open for larger r. See [Hartshorne 1980] for a summary and a number of conjectures.

10.3. Other applications of linearly general position

Existence of good projections. We can use Theorem 10.1 to show that every smooth curve C is birational to a nodal plane curve $C_0 \subset \mathbb{P}^2$, in many ways.

Proposition 10.11. *If $C \subset \mathbb{P}^n$ is a smooth nondegenerate curve in projective space, let $\Lambda \cong \mathbb{P}^k \subset \mathbb{P}^n$ be a general k-plane, and let $\pi_\Lambda : C \to \mathbb{P}^{n-k-1}$ be the projection from Λ, restricted to C. If $n-k-1 \geq 3$ then $\pi_\Lambda : C \to \mathbb{P}^{n-k-1}$ defines an isomorphism of C onto its image, while if $n - k - 1 = 2$ then π_Λ is birational onto its image, which is a curve with only nodes.*

Proof. Recall that the secant variety of C consists of the union of the lines $\overline{qr}$ joining pairs of distinct points $q, r \in C$, plus the tangent lines $\mathbb{T}_q(C)$; altogether, these lines form a family, parametrized by the symmetric square C_2 of C. More precisely every subscheme λ of length 2 in $\mathbb{P}^n$ spans a line $\overline{\lambda}$. The incidence variety

$$I := \{(\lambda, p) \mid \lambda \in C_2 \text{ is a divisor of degree 2 on } C, \ p \in \overline{\lambda} \subset \mathbb{P}^n\}$$

projects to C_2 with 1-dimensional fibers isomorphic to $\mathbb{P}^1$, and thus is irreducible of dimension 3. Its image in $\mathbb{P}^n$ under the second projection is the secant variety of C, which is thus irreducible of dimension ≤ 3. It follows that a general $(n-4)$-plane does not meet the secant variety, and the first statement of the proposition follows.

If $n > 3$ then by first projecting from a general $(n-4)$-plane inside Λ we may reduce to the case $n = 3$, and assume that Λ is a general point of $\mathbb{P}^3$. By a variant of the argument above, the union of the tangent lines to C is a surface, and thus does not contain Λ. It follows that π_Λ is locally in the source an analytic isomorphism.

To show that the fibers of π_Λ are subschemes of length at most 2, we need to show that Λ does not lie on any multisecant line.

By Theorem 10.1 the family of multisecant lines to C is a proper subscheme of the irreducible two-dimensional family of secant lines, so the union of the trisecant lines is at most 2 dimensional, and we see that the fibers of π_Λ all have degree ≤ 2. Furthermore, the general fiber of the projection from the incidence correspondence I to $\mathbb{P}^3$ is empty or finite, so only a finite number of secant lines contain Λ, and we see that π_Λ is birational.

We have shown that the map π_Λ is an immersion, and at most two-to-one everywhere; thus the image curve $C_0 \subset \mathbb{P}^2$ has at most double points, and an analytic neighborhood of each double point consists of two smooth branches. To complete the proof of Proposition 10.11 we have to show that those two branches have distinct tangent lines; that is, that if $q, r \in C$ are any two points collinear with Λ, then the images of the tangent lines $\mathbb{T}_q(C)$ and $\mathbb{T}_r(C)$ in $\mathbb{P}^2$ are distinct. But if $\pi_p(\mathbb{T}_q(C)) = \pi_p(\mathbb{T}_r(C))$ then $\mathbb{T}_q(C)$ and $\mathbb{T}_r(C)$ lie in a plane, and thus intersect.

Since the family of all secant lines is irreducible of dimension 2, it will suffice to show that not every secant line to C is a stationary secant or, equivalently, that not every pair of tangent lines to C meet. We saw in Proposition 10.2 that in the contrary case C would be either strange or planar, a contradiction. $\qquad\square$

The case of equality in Martens' theorem. We can now revisit the case of equality in Martens' theorem bounding the dimension of the variety $W_d^r(C)$ that parametrizes divisor classes of degree d on a curve C with $r(D) \geq r$ (Theorem 5.16). To start, recall the statement:

Theorem 10.12 (Martens). *If C is any smooth projective curve of genus g, then for any $r > d - g$ we have*

$$\dim W_d^r(C) \leq d - 2r.$$

Equality holds if and only if either C is hyperelliptic or $d = r = 0$ or $d = 2g - 2$, $r = g - 1$; in either of the last two cases W_d^r is a single point.

Note that the inequality $\dim W_d^r(C) \leq d - 2r$ is equivalent to the inequality $\dim C_d^r \leq d - r$, which is what we actually showed in Chapter 5. There we combined the geometric Riemann–Roch theorem with an elementary bound on the dimension of the variety of secant planes to a curve in projective space. Theorem 10.1 allows us to sharpen the elementary bound:

Lemma 10.13 (strong secant plane lemma). *Let $C \subset \mathbb{P}^n$ be a smooth, irreducible and nondegenerate curve. If we denote by $\Sigma_d^r \subset C_d$ the locus of effective divisors D of degree d on C with $\dim \overline{D} \leq d - r - 1$, then for any $d \leq r$ and $r > 0$,*

$$\dim \Sigma_d^r \leq d - r - 1.$$

Proof. Consider the incidence correspondence

$$\Gamma := \{(D, H) \in \Sigma_d^r \times (\mathbb{P}^r)^* \mid \overline{D} \subset H\}.$$

The curve being nondegenerate, the projection map $\Gamma \to (\mathbb{P}^n)^*$ is finite. But the fibers of Γ over Σ_d^r have dimension at least $n - d + r$; if we had $\dim \Sigma_d^r \geq d - r$, it would follow that $\dim \Gamma \geq n$, and hence that the projection map $\Gamma \to (\mathbb{P}^r)^*$ is dominant, contradicting Theorem 10.1. □

Proof of the case of equality in Martens theorem. Now, if a curve C is non-hyperelliptic, we can apply the strong secant plane lemma (Lemma 10.13) to the canonical curve. Except for the trivial cases $d = r = 0$ and $d = 2g - 2, r = g-1$, we can apply the lemma to conclude that $\dim W_d^r(C) \leq d-2r-1$; it follows that if $\dim W_d^r(C) = d - 2r$ the curve C in question must be hyperelliptic. □

Using the case of equality in Martens' theorem, we can analyze equality in Clifford's theorem:

Corollary 10.14. *If C is a smooth curve of genus g and D is a divisor on C whose degree satisfies $\deg D = 2r(D) \leq 2g - 2$, then either $D = 0$ or $D = K_C$ or C is hyperelliptic.*

Proof. If C has a divisor D with $\deg D = 2r(D)$ then by the Riemann–Roch theorem, $\deg D = 2(\deg D - g + h^1(D))$, so $\deg D = 2g - 2h^1(D)$. If also $\deg D \leq 2g - 2$, then $h^1(D) \geq 1$ and thus $r(D) > \deg D - g$. Since $W_{\deg D}^{r(D)}$ contains D its dimension is ≥ 0, and we have a case of equality in Martens' theorem. □

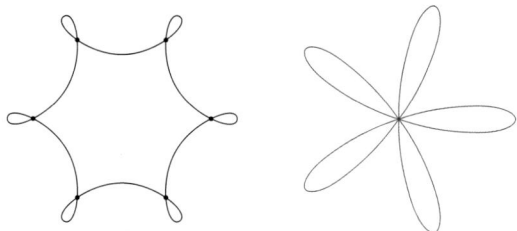

Figure 10.4. A general linear series of degree $g + 2$ maps a nonhyperelliptic curve of genus g to a plane curve with $\binom{g}{2}$ nodes but the image of a hyperelliptic curve has just one singular point, where g smooth branches intersect with distinct tangents.

The $g + 2$ theorem. Now that we have the strong form of Martens' theorem, we can prove the analogue of Theorem 5.13 for general linear series of degree $g + 2$. This was stated in Section 5.5, but we reproduce the statement here.

Theorem 10.15. *Let C be any smooth projective curve of genus g, and let D be a general divisor of degree $g+2$ on C. The complete linear series $|D|$ is basepoint free of dimension 2, and the associated map $\phi_D : C \to \mathbb{P}^2$ is a birational immersion onto its image C_0 and no two branches of C_0 share a tangent line. Moreover:*

(1) *If C is not hyperelliptic, then C_0 has only $\binom{g}{2}$ nodes and no other singularities.*

(2) *If C is hyperelliptic, then C_0 has only one singular point, which is an ordinary g-fold point.*

Figure 10.4 illustrates the two possibilities.

Proof. Since D is general and $\deg D > h^0(K)$ we see that D is nonspecial, whence $h^0(D) = (g + 2) - g + 1 = 3$. This yields the dimension claim.

If $|D|$ has a basepoint $p \in C$, then $3 = h^0(D) = h^0(D - p) = (g + 1) - g + 1 + h^0(K_C - D + p)$, so $K_C - D$ is effective of degree $2g - 2 - (g+1) = g - 3$; thus the family of such D depends on only $g - 3 + 1$ parameters, and $\mu(D)$ would lie in a proper subvariety of $\mathrm{Pic}_{g+2}(C)$, contradicting generality.

Next we show that there are only finitely many pairs $p, q \in C$ such that $\phi_D(p) = \phi_D(q)$; that is, ϕ_D is birational onto its image. Indeed, $\phi_D(p) = \phi_D(q)$ means that $h^0(D - p - q) = 2$. By the Riemann–Roch theorem, $D - p - q = K - E$ for some effective divisor E of degree $g - 2$; thus $\mathcal{O}_C(D)$ is in the image of the map

$$\nu : C_2 \times C_{g-2} \to \mathrm{Pic}_{g+2}(C)$$

sending $(p + q, E)$ to $K_C - E + p + q$. But ν is the composition of two surjective maps, $C_2 \times C_{g-2} \to C_g$ and $C_g \to \mathrm{Pic}_g \to \mathrm{Pic}_{g+2}$ by duality. Since the source and target of ν have the same dimension, the general fiber of ν is finite.

We turn to the types of singularities. First we eliminate tacnodes. To say that a pair of points $p, q \in C$ map to a point of C_0 such that the two branches have a common tangent means two things: that $h^0(D - p - q) \geq 2$; and that $h^0(D - 2p - 2q) \geq 1$. If this is the case, set $E = D - 2p - 2q$. The condition $h^0(D - 2p - 2q) \geq 1$ is equivalent to E being an effective divisor. Considering the canonical map $\phi_K : C \to \mathbb{P}^{g-1}$ the geometric Riemann–Roch theorem,

$$r(E) = \deg E - 1 - \dim \overline{\phi_K(E)},$$

implies that $\overline{\phi_K(E)}$ has dimension $g-3$. The further condition $h^0(D-p-q) \geq 2$ says that $\dim \overline{\phi_K(E + p + q)} = g - 2$, so the secant line $\overline{pq}$ meets the $(g - 3)$-plane $\overline{\phi_K(E)}$. Now, not every secant line to C can meet a given linear subspace $\Lambda \subset \mathbb{P}^{g-1}$ of dimension $g - 3$ — otherwise, the projection $\pi_\Lambda : C \to \mathbb{P}^1$ would be constant — so we see that $\mu(D)$ would have to lie on the image of a proper subvariety of $C_{g-2} \times C_2$ under the map $(E, p + q) \mapsto \mathcal{O}(E + 2p + 2q)$ from $C_{g-2} \times C_2$ to $\mathrm{Pic}_{g+2}(C)$. Since the dimension of this subvariety is $< g$ we see that if C' has a point at which two branches are tangent, then $\mathcal{O}_C(D)$ lies in a proper subvariety of $\mathrm{Pic}_{g+2}(C)$. Since the Abel–Jacobi map $C_{g+2} \to \mathrm{Pic}_{g+2}(C)$ is surjective, D would not be general.

Now let's suppose that C is nonhyperelliptic. To prove claim (1) of the theorem, we have to show that the image $C_0 = \phi_D(C)$ has only nodes as singularities, rather than cusps or triple points. That there are exactly $\binom{g}{2}$ nodes then follows from the adjunction formula.

(no cusps) To say that a point $p \in C$ maps to a cusp of C_0 (that is, the differential $d\phi_D$ is zero at p) amounts to saying that $h^0(D - 2p) \geq 2$; that is, $D - 2p$ is a g_g^1. But by the Riemann–Roch theorem, $W_g^1 = K_C - W_{g-2}$; so to say ϕ_D has a cusp means that

$$\mu(D) \in 2W_1 + K_C - W_{g-2},$$

and since the locus on the right has dimension at most $g - 1$, a general point of $J(C)$ will not lie in it. Note that this subsumes the fact that $|D|$ has no basepoints.

(no triple points) To say that C_0 has a triple point means that for some divisor $E = p + q + r$ of degree 3, $h^0(D - E) \geq 1$; thus we must have

$$\mu(D) \in W_3 + W_{g-1}^1.$$

To argue that this is not the case, we need to know that $\dim W_{g-1}^1 \leq g - 4$; streamlined if C is nonhyperelliptic.

This concludes the proof in the nonhyperelliptic case. Now suppose that C is hyperelliptic, and let $|E|$ be the g_2^1 on C. Since D is a divisor of degree $g + 2$, the divisor $D - E$ will have degree g, and so be effective; thus we can write $D \sim E + p_1 + \cdots + p_g$ for some g points p_i, which must be distinct since D was assumed general.

The fact that

$$h^0(D - p_1 - \cdots - p_g) = h^0(E) = 2 = h^0(D) - 1$$

implies that ϕ_D maps all the points p_i to the same point. The image curve C_0 thus has a point with at least g branches. Since these branches cannot share a tangent line, they are smooth and their tangents are distinct; that is, the point is an ordinary multiple point. By the adjunction formula $p_a(C) = \binom{g+1}{2}$ and since C has genus g the singularity must have δ invariant $\binom{g+1}{2} - g = \binom{g}{2}$. Such a singularity cannot have multiplicity $> g$, and is thus a g-fold point. Since the δ invariant of such a point is at least $\binom{g}{2}$ by the discussion following Theorem 2.39 (see also Proposition 15.9), the g-fold point is ordinary and there can be no other singularities. $\square$

10.4. Exercises

Exercise 10.1. Let $C \subset \mathbb{P}^r$ be a smooth curve over an algebraically closed field of arbitrary characteristic. We can re-embed C by a Veronese map, that is, consider $\widetilde{C} = \nu_m(C)$ where $\nu_m : \mathbb{P}^r \to \mathbb{P}^N$ is the m-th Veronese map. Prove:

Proposition 10.16 (compare Proposition 10.11). *If m is high enough then the projection of $\widetilde{C}$ from a general $\mathbb{P}^{N-3}$ is a birational map onto a nodal plane curve.*

$\blacklozenge$

Exercise 10.2 [Rathmann 1987]. Let k be an algebraically closed field of characteristic $p > 0$, and let $q = p^e$ for some $e \geq 1$. Let $C \subset \mathbb{P}^r$ be the closure of the image C_0 of the morphism

$$\mathbb{A}^1 \ni t \mapsto (t, t^q, t^{q^2}, \ldots, t^{q^r}) \in \mathbb{A}^r$$

where $\mathbb{A}^r \subset \mathbb{P}^r$ is the open set $x_0 = 1$.

(1) Show that C is a complete intersection, defined by the equations

$$x_0^{q-1} x_2 - x_1^q, x_0^{q-1} x_3 - x_2^q, \ldots, x_0^{q-1} x_r - x_{r-1}^q.$$

(2) Show that C is singular unless $q = r = 2$.

(3) Show every secant line to C_0 contains q points of C_0; more generally, if $a_1, \ldots, a_r$ are linearly independent points of C_0, show that the linear span of $a_1, \ldots, a_r$ contains $q^{(r-1)}$ points of C_0. Compare this with the configuration of points in affine r-space over a field of q elements.

Exercise 10.3. Here is another approach to the $g + 2$ theorem in the hyperelliptic case: Let C be a hyperelliptic curve of genus g and D a general divisor of degree $g + 1$ on C; let $|E|$ be the g_2^1 on C. Consider the map $\phi : C \to \mathbb{P}^1 \times \mathbb{P}^1$ given as the product of the maps $\phi_D : C \to \mathbb{P}^1$ and $\phi_E : C \to \mathbb{P}^1$ given by the pencils $|D|$ and $|E|$.

(1) Show that ϕ embeds the curve C as a curve of bidegree $(g + 1, 2)$ on $\mathbb{P}^1 \times \mathbb{P}^1$.

(2) Now embed $\mathbb{P}^1 \times \mathbb{P}^1$ into $\mathbb{P}^3$ as a quadric surface Q; pick a general point $p \in C \subset Q$ and project C from the point p. Show that the image curve C_0 is a plane curve of degree $g + 2$ with one ordinary g-fold point.   ◆

Exercise 10.4. Establish the analogue of Proposition 3.7 for hypersurfaces of any degree m, that is to say no irreducible, nondegenerate curve in $\mathbb{P}^d$ lies on more hypersurfaces of degree m than the rational normal curve. To do this, let $C \subset \mathbb{P}^d$ be any irreducible nondegenerate curve. Let Γ be a general hyperplane section of C, and use the exact sequences

$$0 \to \mathcal{I}_{C/\mathbb{P}^d}(l-1) \to \mathcal{I}_{C/\mathbb{P}^d}(l) \to \mathcal{I}_{\Gamma/\mathbb{P}^{d-1}}(l) \to 0.$$

with $2 \le l \le m$ to show that

$$h^0(\mathcal{I}_{C/\mathbb{P}^d}(m)) \le \binom{d+m}{m} - (md+1)$$

with equality only if C is a rational normal curve.

Exercise 10.5. Let $D \subset \mathbb{P}^r$ be a rational normal curve. If $\Gamma \subset D$ is any collection of d points on D (or for that matter any subscheme of D of degree d) then the Hilbert function of Γ is $h_\Gamma(m) = \min\{d, mr+1\}$.   ◆

Exercise 10.6. Let C be a reduced irreducible curve with general hyperplane section Γ, and write $d - 1 = M(r - 1) + \epsilon$ with $\epsilon < r - 1$ as in Castelnuovo's theorem. Prove that

(1) if $\epsilon > 0$, then $\mathcal{O}_C(M)$ is nonspecial, but $\mathcal{O}_C(M-1)$ is special; and

(2) if $\epsilon = 0$, then $\mathcal{O}_C(M-1)$ is nonspecial, but $\mathcal{O}_C(M-2)$ is special.   ◆

Exercise 10.7. If $C \subset \mathbb{P}^r$ is a Castelnuovo curve of degree $d \ge 2r$, show that $|D| = |\mathcal{O}_C(1)|$ is the unique g_d^r on C.   ◆

Exercise 10.8. We have seen that complete intersections $C = Q \cap S \subset \mathbb{P}^3$ of a quadric surface Q and a surface S of degree k achieve Castelnuovo's bound $g = \pi(2k, 3)$ on the genus of curves of degree $2k$ in $\mathbb{P}^3$. In fact, we will see in Chapter 17 that any curve $C \subset \mathbb{P}^3$ of degree $2k$ and genus $g = \pi(2k, 3) = (k-1)^2$ is of this form.

(1) Find the dimension of the subvariety $\Gamma \subset M_g$ consisting of Castelnuovo curves.

(2) Find the dimension of the subvariety $H \subset M_g$ of hyperelliptic curves, and compare this to the result of the first part.   ◆

Chapter 11

Monodromy of hyperplane sections

11.1. Uniform position and monodromy

We now return to the situation where the ground field k is algebraically closed of characteristic 0.

The central result of this chapter is the *uniform position lemma*, which deals with the monodromy group of the points of a general hyperplane section of a curve $C \subset \mathbb{P}^r$. To define the monodromy group and prove the uniform position lemma, we will use the classical topology. An equivalent algebraic definition is described in Cheerful Fact 11.1 below, but the strong form of uniform position can fail in positive characteristic, as shown by the examples in Exercise 10.2.

We may describe the monodromy group informally as follows: Suppose that $C \subset \mathbb{P}^r$ is an irreducible curve of degree d over $\mathbb{C}$, and $H_0 \subset \mathbb{P}^r$ a hyperplane transverse to C; say $C \cap H_0 = \{p_1, \ldots, p_d\}$. As we vary H_0 continuously along a real arc $\{H_t\}$, staying within the open subset $U \subset (\mathbb{P}^r)^*$ of hyperplanes transverse to C, we can "follow" each of the points $p_i(t)$ of intersection of C with the hyperplane H_t.

Now imagine that the hyperplanes H_t come back to the original H_0 at time $t = 1$; that is, we have a continuous family $\{H_t\}_{0 \le t \le 1} \subset U$ with $H_1 = H_0$. Each of the points p_i then traces out a continuous real arc $\{p_i(t) \in C \cap H_t\}_{0 \le t \le 1}$. Since $H_1 = H_0$, the end point $p_i(1)$ is one of the original points $p_j \in C \cap H_0$. In this way, we get a permutation of the set $C \cap H_0$; the group of all permutations arrived at in this way is called the *monodromy group* of the points $C \cap H_0$.

We will now give a precise definition of the monodromy group in a more general setting, and prove that the monodromy group of the points of a general hyperplane section of an irreducible curve is the full symmetric group; this is the uniform position lemma. The rest of the chapter has a series of applications.

The monodromy group of a generically finite morphism. Let $f : Y \to X$ be a dominant map between varieties of the same dimension over $\mathbb{C}$, and suppose that X is irreducible. There is then a Zariski open subset $U \subset X$ such that U and its preimage $V = f^{-1}(U)$ are smooth, and the restriction of f to V is a covering space in the classical topology. Let d be the number of sheets. This is the degree of the extension $K(Y)/K(X)$.

Homotopy theory associates a monodromy group to any finite topological covering map $f : V \to U$, defined as follows: Choose a basepoint $p_0 \in U$, and suppose $\Gamma := f^{-1}(p_0) = \{q_1, \ldots, q_d\}$. If γ is any loop in U with basepoint p_0, for any $i = 1, \ldots, d$ there is a unique lifting of γ to an arc $\tilde{\gamma}_i$ in V with initial point $\tilde{\gamma}_i(0) = q_i$ and end point $\tilde{\gamma}_i(1) = q_j$ for some $j \in \{1, 2, \ldots, d\}$. Since we could traverse the loop in the opposite direction, the index j determines i, and the map $i \mapsto j$ is a permutation of $\{1, 2, \ldots, d\}$. Since the set Γ is discreet, the permutation depends only on the class of γ in $\pi_1(U, p_0)$, so we have defined a homomorphism to the symmetric group:

$$\pi_1(U, p_0) \to \operatorname{Perm} \Gamma \cong S_d.$$

The image of this map, denoted in this chapter by M, is called the *monodromy group* of the map f. It depends on the labeling of the points of Γ, but a change in labeling only changes the group by conjugation with the corresponding permutation, so the monodromy group is well defined up to conjugation in S_d.

Cheerful Fact 11.1. In our setting the monodromy group is independent of the choice of open set U: if $U' \subset U$ is a Zariski open subset, the complement of $U \setminus U'$ has real codimension ≥ 2 so the map $\pi_1(U', p_0) \to \pi_1(U, p_0)$ is surjective. Thus the image of $\pi_1(U', p_0)$ in S_d is the same.

The theory of finite coverings of algebraic varieties is not only analogous to it *is* Galois theory: In the situation described above, if Y is irreducible, then the pullback map f^* expresses the function field $K(Y)$ as a finite algebraic extension of $K(X)$. The monodromy group of f is the Galois group of the Galois normalization of $K(Y)$ over $K(X)$ (see [Harris 1979]). Indeed, in early treatments of Galois theory, such as Camille Jordan's famous *Traité des substitutions* [1870], function fields played as large a role as number fields.

Since we assumed that X is irreducible, the space U is (path) connected, and it follows that the monodromy group is transitive if and only if the space V

is (path) connected. But V is a smooth variety, so this is the case if and only if V is irreducible. For example, the monodromy in the family of smooth quadric surfaces in $\mathbb{P}^3$ interchanges the two rulings:

Example 11.2. Consider the family

$$X := \{(p, Q) \mid p \in Q \text{ a point on a smooth quadric surface in } \mathbb{P}^3\}$$

of pointed smooth quadric surfaces in $\mathbb{P}^3$ and the double covering by

$$V := \{((p, Q), L) \mid (p, Q) \in X, \, L \text{ a line with } p \in L \subset Q\}.$$

The variety V is irreducible: if we project V to the (irreducible) variety of lines in $\mathbb{P}^3$, the fiber consists of a point on the line and quadric containing the line — that is, the product of $\mathbb{P}^1$ and the (dual of the) projective space on the space of quadratic forms in the ideal of the line. Thus the monodromy of the family is transitive, and exchanges the two lines through the point.

Uniform position. We will next compute the monodromy group of the universal hyperplane section of a curve, constructed as follows: Let $C \subset \mathbb{P}^r$ be an irreducible, nondegenerate curve of degree d, and let $X = (\mathbb{P}^r)^*$ be the space of hyperplanes in $\mathbb{P}^r$. We define the *universal hyperplane section* of C to be the projection $f : Y \to (\mathbb{P}^r)^*$ of the incidence variety

$$Y = \{(H, p) \in (\mathbb{P}^r)^* \times C \mid p \in H\}.$$

The fibers of f are the hyperplane sections of C, so f is a finite surjective map. If we let $U \subset (\mathbb{P}^r)^*$ be the open subset of hyperplanes meeting C transversely, then the restriction of f to the preimage V of U is a covering space whose fibers each consist of d distinct points. The preimage in Y of a point $p \in C$ is the set of hyperplanes containing p, a copy of $\mathbb{P}^{r-1}$, and thus Y is irreducible. Thus the monodromy group of f is transitive. But much more is true:

Theorem 11.3 (uniform position lemma). *The monodromy group of the universal hyperplane section of an irreducible curve $C \subset \mathbb{P}^r$ is the full symmetric group S_d.*

We postpone the proof to develop some necessary tools.

Theorem 11.3 fails over fields of finite characteristic, though there is no known counterexample for smooth curves; see Exercise 10.2 for singular examples, and [Rathmann 1987] and [Kadets 2021] for what is known.

Theorem 11.3 implies that two subsets of the same cardinality in the general hyperplane section of C are indistinguishable from the point of view of any discrete invariant that is semicontinuous in the Zariski topology. To make this precise, we introduce a definition:

Definition 11.4. Let $\phi : Y \to X$ be a finite morphism. By the *restricted fiber power* $\tilde{Y}^n/X$ we will mean the complement of all diagonals in the ordinary fiber power; that is,

$$\tilde{Y}^n/X := \{(x, y_1, \ldots, y_n) \in X \times Y^n \mid \phi(y_i) = x \text{ and } y_i \neq y_j \ \forall i \neq j\}$$

In down-to-earth terms, a point of $\tilde{Y}^n/X$ is a set of n distinct points in a fiber of ϕ together with the choice of a total order on these points.

Lemma 11.5. *Let $f : Y \to X$ be a generically finite cover of degree d, with monodromy group $M \subset S_d$. M is n-transitive if and only if the restricted fiber power $\tilde{Y}^n/X$ is irreducible.*

Proof. If we restrict ourselves to open subsets $U \subset X$ and $V = f^{-1}(U) \subset Y$ such that U and V are smooth and the restriction $f|_V : V \to U$ is a covering in the classical topology, then the restricted fiber powers are unions of connected components of the usual fiber powers Y^n/X. The condition that the monodromy is n-transitive is equivalent to the condition that the restricted fiber power $\tilde{Y}^n/X$ is connected; since the fiber powers are all smooth, this is equivalent to $\tilde{Y}^n/X$ being irreducible. $\qquad\square$

11.2. Flexes and bitangents are isolated

There are further general position results that we need for the proof of the uniform position lemma. We have separated them from the results on linear general position because these require characteristic 0; they are really local analytic statements.

Not every tangent line is tangent at a flex. Recall that a smooth point p on a curve $C \subset \mathbb{P}^r$ is called a *flex point* if the tangent line to C at p has contact of order 3 or more with C at p.

Lemma 11.6. *If $r > 1$ and $C \subset \mathbb{P}^r$ is a smooth, irreducible and nondegenerate curve, then not every point of C is a flex point.*

The proof applies to any linear series on a curve.

Proof. We begin by lifting the inclusion $C \hookrightarrow \mathbb{P}^r$ to an arc $v(t)$ in $\mathbb{C}^{r+1}$, so that the tangent line at the point $t = 0$ in projective space is represented by the span of $v(0)$ and $v'(0) \in \mathbb{C}^{r+1}$. To say that $v(t)$ is a flex point is to say that the vectors $v(t), v'(t)$ and $v''(t)$ are linearly dependent. If this holds for all t then

$$v(t) \wedge v'(t) \wedge v''(t) \equiv 0.$$

When we take the derivative of the wedge product $v(t) \wedge v'(t) \wedge v''(t)$ by applying the product rule we see that the first two terms are zero because they contain a repeated factor; it follows that

$$v(t) \wedge v'(t) \wedge v'''(t) \equiv 0,$$

so that $v'''(t)$ lies in the span of $v(t)$ and $v'(t)$ as well. Indeed, as we continue to take derivatives, we see in each case that all but one term is zero, and we deduce that $v^{(l)}(p)$ lies in the span of $v(t)$ and $v'(t)$ for all l. This being characteristic 0, it follows that C lies in a line, contrary to our hypothesis. $\qquad\square$

Not every tangent is bitangent. This statement seems even more obvious than Lemma 11.6 above, but, like that lemma, it is false in characteristic p — see Exercise 11.3.

Lemma 11.7. *Let $C \subset \mathbb{P}^r$ be a smooth, irreducible, nondegenerate curve, with $r > 1$. If $p \in C$ is a general point, then the tangent line $\mathbb{T}_p(C) \subset \mathbb{P}^r$ is not tangent to C at any other point.*

Proof. Again the result is local, this time with a pair of germs $D \xrightarrow{v} \mathbb{P}^r$ and $D \xrightarrow{w} \mathbb{P}^r$. If every tangent line to the two curves is a bitangent, we will show that they are both contained in a line.

Let C_1, C_2 be the images of v, w respectively, and let $\tilde{v}, \tilde{w}$ be lifts to $\mathbb{C}^{r+1}$. Let

$$\Sigma := \left\{(p, q) \in D \times D \mid p \neq q \text{ and } \mathbb{T}_{v(p)}(C_1) = \mathbb{T}_{w(q)}(C_2)\right\}$$

be the variety parametrizing bitangents to the two germs.

The statement that the tangent lines to C at the points $v(t)$ and $w(t)$ are equal then says that the vectors $\tilde{v}(t), \tilde{v}'(t), \tilde{w}(t)$ and $\tilde{w}'(t)$ all lie in a 2-dimensional subspace $\Lambda \subset \mathbb{C}^{r+1}$; in particular,

$$\tilde{v}(t) \wedge \tilde{v}'(t) \wedge \tilde{w}(t) \equiv 0 \quad \text{and likewise} \quad \tilde{v}(t) \wedge \tilde{w}(t) \wedge \tilde{w}'(t) \equiv 0.$$

We proceed now exactly as in the proof of Lemma 11.6: taking derivatives, we see that all derivatives of $\tilde{v}(t)$ and $\tilde{w}$ at $t = 0$ lie in Λ and hence C_1 and C_2 are both contained in the line in $\mathbb{P}^r$ corresponding to Λ. $\qquad\square$

In characteristic zero, [Kaji 1986, Theorem 3.1] shows that a general tangent line to a smooth projective curve C can intersect C in a point other than p, except of course in case $r = 2$, although this is possible in characteristic $p > 0$.

11.3. Proof of the uniform position lemma

Let $C \subset \mathbb{P}^r$ be an irreducible, nondegenerate curve of degree d and let

$$f : Y \subset (\mathbb{P}^r)^* \times C \to X = (\mathbb{P}^r)^*$$

be its universal hyperplane section; let $U \subset (\mathbb{P}^r)^*$ be the open subset of hyperplanes transverse to C and $V = f^{-1}(U)$; let $M \subset S_d$ be the monodromy group of V over U. To show that M is the full symmetric group, it suffices to show

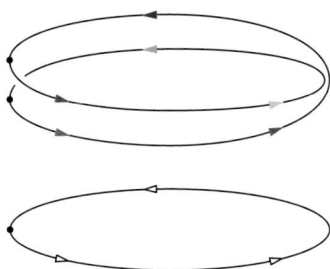

Figure 11.1. Monodromy action around the ramification point of a double cover.

that M contains all transpositions, and for this it is enough to show that M is doubly transitive and contains one transposition.

To prove that M is doubly transitive, we can give a concrete description of the restricted fiber power $\tilde{V}^2/U$: let

$$\Sigma := \{(H, p, q) \in (\mathbb{P}^r)^* \times C \times C \mid p, q \in H \text{ and } p \neq q\}.$$

Projection on the second and third factors expresses Σ as a $\mathbb{P}^{r-2}$-bundle over the complement $C \times C \setminus \Delta$ of the diagonal in $C \times C$. Thus Σ is irreducible, and it follows that the restricted fiber square $\tilde{V}^2/U$, which is a Zariski open subset of Σ, is as well. Note that this part of the argument does not rely on any assumption about the characteristic.

Next we give a criterion for a monodromy group to contain a transposition:

Lemma 11.8. *Let $f : Y \to X$ be a generically finite cover of degree d over an irreducible variety X, with monodromy group $M \subset S_d$. If, for some smooth point $p \in X$ the fiber $f^{-1}(p) \subset V$ consists of $d - 2$ reduced points $p_1, \ldots, p_{d-2}$ and one point q of multiplicity 2, where q is also a smooth point of Y, then M contains a transposition.*

See Figure 11.1 for the case $d = 2$.

Proof. The hypothesis implies that Y is smooth locally near the fiber over p. Let $U \subset X$ be a Zariski open subset of the smooth locus in X, as in the definition of the monodromy group, so that $V := f^{-1}(U)$ is also smooth and the restriction $f|_V : V \to U$ expresses V as a finite d-sheeted covering space of U, with U, V smooth and $p \in X$.

Let $B_i \subset Y$ be disjoint small connected closed neighborhoods of the points in $f^{-1}(p)$. Because $f|_V$ is finite, the image $A := \bigcap f(B_i)$ is a locally closed subset of the same dimension as X and thus A contains a neighborhood of p in the classical topology.

Let $p' \in A \cap U$. Two of the d points of $f^{-1}(p')$ lie in the component of B containing q; call these q' and q''. Since $B \cap V$ is the complement of a proper subvariety in B it is connected, and we can draw a real arc $\gamma : [0, 1] \to B \cap V$ joining q' to q''; by construction, the permutation of $f^{-1}(p')$ associated to the loop $f \circ \gamma$ exchanges q' and q'' and fixes each of the remaining $d - 2$ points of $f^{-1}(p')$. □

Completion of the proof of Lemma 11.3. We must show that a reduced irreducible curve C of degree d (in characteristic 0) has a hyperplane section consisting of $d - 2$ smooth points and one double point; that is, a hyperplane simply tangent to the curve at one point and transverse everywhere else. By the results of Section 11.2 there is a tangent line L to C at a smooth point that is not flex tangent, and is not tangent to C at any other point. A general hyperplane H containing L meets C doubly in the point of tangency. At any other point p of $L \cap C$, if any, the intersection $C \cap H$ is transverse unless H contains the tangent line at p. Containing such a line is a proper codimension 1 condition on the hyperplanes containing L, and since there can only be finitely many such points, a general H will be transverse at all such p. On the other hand, at points not on L the intersection $H \cap C$ is transverse by Bertini's theorem. This completes the proof. □

Uniform position for higher-dimensional varieties. There is a generalization of Theorem 11.3 for irreducible varieties $X \subset \mathbb{P}^r$ of any dimension k. To set this up, let $\mathbb{G}(r - k, r)$ be the Grassmannian parametrizing $(r - k)$-planes $\Lambda \subset \mathbb{P}^r$. We introduce the *universal $(r - k)$-plane section of X*:

$$Y = \{(\Lambda, p) \in \mathbb{G}(r - k, r) \times X \mid p \in \Lambda\}.$$

Via projection on the first factor, Y is expressed as a generically finite cover of $\mathbb{G}(r-k, r)$, and we can ask for its monodromy. The answer is the same as for curves:

Theorem 11.9. *The monodromy group of the universal $(r - k)$-plane section of an of the universal $(r - k)$-plane section of an irreducible k-dimensional variety $X \subset \mathbb{P}^r$ is the full symmetric group S_d.*

Proof. This follows from Theorem 11.3. To see this, fix a general $(r - k + 1)$-plane $\Gamma \subset \mathbb{P}^r$; since a general hyperplane section of an irreducible variety $X \subset \mathbb{P}^r$ of dimension $k \geq 2$ is again irreducible, we see that $C := \Gamma \cap X$ is an irreducible curve. The restriction of the universal $(r - k)$-plane section of X to the sub-Grassmannian $\mathbb{G}(r-k, \Gamma) \cong (\mathbb{P}^r)^*$ of $(r-k)$-planes contained in Γ is just the universal hyperplane section of the curve C, which we know has monodromy S_d; since the monodromy group of a cover can only get smaller under restriction to a subvariety of the target, the result follows. □

We will see an application of this result in Proposition 11.14.

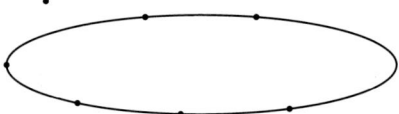

Figure 11.2. Seven points in the plane, of which six lie on a conic. The seven are in linearly general position but not in numerical uniform position.

11.4. Applications of uniform position

Irreducibility of fiber powers. If we apply the uniform position lemma to the universal hyperplane section of a curve $C \subset \mathbb{P}^r$ we get an irreducibility result:

Corollary 11.10. *If $C \subset \mathbb{P}^r$ is a smooth curve of degree d, then for $k \leq d$, the restricted fibered powers $\tilde{Y}^k/X$ of the universal hyperplane section of C are irreducible.*

Numerical uniform position. The property of uniform position, as exposed above, is really a property of a family — the irreducibility of the restricted fiber powers — and not of a given set of points. There is a useful weaker property that can be applied to a given set of points:

Definition 11.11. A set of points $\Gamma \subset \mathbb{P}^n$ is in *numerical uniform position* if any two subsets of Γ with the same cardinality impose the same number of conditions on forms of every degree d; that is, any two subsets of the same cardinality have the same Hilbert function.

This is strictly stronger than linearly general position, which involves only forms of degree $d = 1$. The difference is illustrated in Figure 11.2.

Corollary 11.12 (numerical uniform position lemma). *The general hyperplane section of an irreducible curve $C \subset \mathbb{P}^r$ is in numerical uniform position.*

Proof. Let $U = (\mathbb{P}^r)^* \setminus C^*$ be the open subset of hyperplanes transverse to C, and let $Y \to U$ be the universal hyperplane section. Corollary 11.10 says that the restricted fiber powers $\tilde{V}^n/U$ are irreducible.

Now, for each m the number of conditions that Γ imposes on forms of degree m is lower semicontinuous, so it achieves its maximum on a Zariski open subset of $\tilde{V}^n/U$. Since $\tilde{V}^n/U$ is irreducible, the complement Z of this open set has dimension strictly less than $\dim \tilde{V}^n/U = \dim U$. Thus a general hyperplane $H \in (\mathbb{P}^r)^*$ lies outside the image of Z, meaning that the number of conditions imposed by all the k-element subsets $\Gamma \subset C \cap H$ have this maximal value. $\square$

This result may be seen as an important strengthening of Theorem 10.1, since if C is a reduced, irreducible nondegenerate curve in $\mathbb{P}^n$ then a general subset of n points of C is linearly independent and spans a hyperplane; Corollary 11.12 says that this holds for every subset of n points of every general hyperplane section, which reproves Theorem 10.1, though only in characteristic 0.

Sums of linear series. Another consequence of the uniform position lemma is a result about sums of linear series. Recall that if D is a divisor on a curve C we write $r(D) = \dim |D| = h^0(\mathcal{O}_C(D)) - 1$.

Corollary 11.13. *If D, E are effective divisors on a curve C then*

$$r(D + E) \geq r(D) + r(E).$$

If the genus of C is > 0 and $|D + E|$ is birationally very ample, then the inequality is strict.

On $\mathbb{P}^1$, by contrast, any effective divisor D has $r(D) = \deg D$, so the inequality above is always an equality for $C = \mathbb{P}^1$.

Proof. The inequality follows in general because the sums of divisors in $|D|$ and divisors in $|E|$ already move in a family of dimension $r(D) + r(E)$; the key point is the strict inequality in case $D + E$ is birationally very ample.

If $D+E$ is birationally very ample and $r(D+E) = r(D)+r(E)$ then restricting to an open set we may identify C with its image under the complete linear series $|D + E|$, and we see that a general hyperplane section $H \cap C$ contains a divisor equivalent to D.

Let Y be the $\deg D+\deg E$ restricted fiber power of the universal hyperplane. A point $y \in Y$ is a hyperplane section plus an ordering of its points. Let $\phi : Y \to \mathrm{Pic}_d(C)$ be the Abel–Jacobi map taking y to the class of the divisor that is the sum of first d points in this order. The preimage Y' of the point of $\mathrm{Pic}_d(C)$ corresponding to the class of D is a closed subset, and since every divisor in the class of a hyperplane section contains a divisor linearly equivalent to D, the subvariety Y' dominates $(\mathbb{P}^n)^*$, and thus has the same dimension as Y. Consequently $Y' = Y$, and the sum of the first d points in any ordering of the general hyperplane section — that is, the sum of any d of the points — is equivalent to D.

Thus if $p \in D$ and $q \notin D$, then $D - p + q \equiv D$, whence $q \equiv p$. Thus $r(p) \geq 1$, so $C \cong \mathbb{P}^1$. $\qquad\square$

Nodes of plane curves. In Section 8.6, we introduced the *Severi variety* $V_{d,g}$; this is the locally closed subset of the projective space $\mathbb{P}^N$ of all plane curves of degree d parametrizing irreducible plane curves of degree d having $\binom{d-1}{2} - g$ nodes and no other singularities. We proved there that $V_{d,g}$ was smooth, and by Cheerful Fact 8.13 it is irreducible for all d and g.

To compute the monodromy we introduce the incidence correspondence

$$\Phi := \{(C, p) \in V_{d,g} \times \mathbb{P}^2 \mid p \in C_{\text{sing}}\}.$$

This is a covering space of $V_{d,g}$ with $\binom{d-1}{2} - g$ sheets, and we compute its monodromy. We start with the extremal case $g = 0$:

Proposition 11.14. *The monodromy group of Φ over $V_{d,0}$ is the full symmetric group S_δ, with $\delta = \binom{d-1}{2}$.*

Proof. Every nodal curve $C \subset \mathbb{P}^2$ is the projection of a rational normal curve $\tilde{C} \subset \mathbb{P}^d$ from a $(d-3)$-plane $\Lambda \subset \mathbb{P}^d$. Moreover, if Λ is general, the nodes of the projection correspond to the points of intersection of Λ with the secant variety $X \subset \mathbb{P}^d$ of the rational normal curve $\tilde{C}$. Applying Theorem 11.9, the result follows.   □

This result has an immediate consequence, which played a major role in the proof of Cheerful Fact 8.13: given the description in Section 8.6 of $V_{d,g}$ in a neighborhood of a point in $V_{d,0}$, it follows that there is a unique irreducible component of $V_{d,g}$ containing $V_{d,0}$ in its closure. Thus, to prove the irreducibility of $V_{d,g}$, it is sufficient to show that every component of $V_{d,g}$ contains $V_{d,0}$ in its closure.

Going in the other direction, note also that if we assume Cheerful Fact 8.13, then we can deduce the analogous result for all g:

Proposition 11.15. *The monodromy group of Φ over $V_{d,g}$ is the full symmetric group S_δ, with $\delta = \binom{d-1}{2} - g$.*

11.5. Exercises

As a consequence of Theorem 11.9, we can deduce the Bertini irreducibility theorem:

Exercise 11.1. Let $X \subset \mathbb{P}^r$ be an irreducible variety of dimension $k \geq 2$. Show that a general hyperplane section of X is irreducible.   ◆

Exercise 11.2. In Example 11.2, we gave a global argument to say that in the family of smooth quadric surfaces in $\mathbb{P}^3$ the monodromy exchanges the two rulings of a quadric by lines. Prove this with a local calculation, analyzing the family

$$Q_t := V(X^2 + Y^2 + Z^2 + tW^2)$$

in a neighborhood of $t = 0$.   ◆

Exercise 11.3 (tangential degeneracy [Kaji 1986, Example 4.1]). Show that the linear series in Exercise 3.1 defines an isomorphism with a smooth curve C of type $(1, d-1)$ on a nonsingular quadric in $\mathbb{P}^2_k$ no matter what field k is chosen. Now suppose that k has characteristic $p > 0$ and $d - 1 = p^\ell n$ with

$\ell \geq 2$ and n relatively prime to p. Show that the tangent lines to C are the rulings of the quadric in the family that meet C with multiplicity $d - 1$, and that the intersection of each tangent line meets C in n distinct points, each with multiplicity p^ℓ. Thus Lemma 11.7 fails in positive characteristic.

Exercise 11.4. Let $C \subset \mathbb{P}^r$ be a union of irreducible curves C_i of degrees d_i. Prove that the monodromy group of the points of a general hyperplane section of C is the product $\prod S_{d_i}$. ◆

Exercise 11.5. Let d and e be positive integers, $\mathbb{P}^M$ the space of plane curves of degree d and $\mathbb{P}^N$ the space of plane curves of degree e, and

$$\Phi := \{(D, E, p) \in \mathbb{P}^M \times \mathbb{P}^N \times \mathbb{P}^2 \mid p \in D \cap E\}.$$

Prove that the monodromy group of the projection $\pi : \Phi \to \mathbb{P}^M \times \mathbb{P}^N$ is the symmetric group S_{de} on de letters

(1) by applying Theorem 11.3; and

(2) from scratch, using the method used in the proof of Theorem 11.3. ◆

Exercise 11.6. If $E \subset \mathbb{P}^2$ is a smooth plane cubic curve, then a point $p \in E$ is called a *flex* if $\mathcal{O}_E(3p) \cong \mathcal{O}_E(1)$ (see Chapter 13 for a generalization). By Corollary 5.9 in case $g = 1$, there are nine of them. Now let $\mathbb{P}^9$ be the space of plane cubics, and let

$$\Phi := \{(E, p) \in \mathbb{P}^9 \times \mathbb{P}^2 \mid p \text{ is a flex of } E\}.$$

Show that the monodromy group of $\Phi \to \mathbb{P}^9$ is a proper subgroup of S_9. ◆

Exercise 11.7. Let C be a smooth projective curve. Show that if $\pi : C \to \mathbb{P}^1$ is a simply branched cover of degree n, then the monodromy of the map π is the full symmetric group S_n. ◆

Chapter 12

Brill–Noether theory and applications to genus 6

12.1. What linear series exist?

Let's start with a naive question: when does there exist a curve C of genus g and a g_d^r on C — equivalently, a line bundle $\mathcal{L}$ of degree d on C with $h^0(\mathcal{L}) \geq r + 1$? The Riemann–Roch and Clifford theorems together provide a complete answer to this question:

Theorem 12.1. *There exists a curve C of genus g and a line bundle $\mathcal{L}$ of degree d on C with $h^0(\mathcal{L}) \geq r + 1$ if and only if*

$$r \leq \begin{cases} d - g & \text{if } d \geq 2g - 1; \text{ and} \\ d/2 & \text{if } 0 \leq d \leq 2g - 2. \end{cases}$$

For the perhaps more interesting question of when there exists a curve of genus g with a birationally very ample g_d^r, Castelnuovo's theorem gives a quadratic bound, roughly $d \geq \sqrt{g(2r-2)}$.

In both these situations, the curves that achieve the bounds are quite special. Perhaps the most interesting question is, for which r, d do *all* curves of genus g have a g_d^r, and what is the behavior of these series on a general curve? Brill–Noether theory provides some answers to both these questions.

12.2. Brill–Noether theory

The following result was stated by Brill and Noether in 1874, and finally proven in a series of works by Kempf [1971], Kleiman and Laksov [1972; 1974], and Kleiman [1976], culminating in a paper by Griffiths and the second author [Griffiths and Harris 1980].

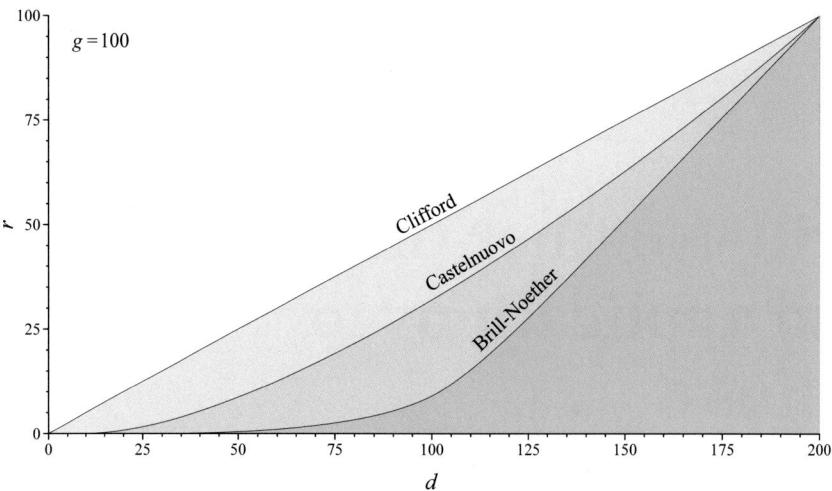

Figure 12.1. For smooth curves of genus 100, these are bounds on (d, r) for all linear series (Clifford), birationally very ample series (Castelnuovo), and all linear series on general curves (Brill–Noether).

Theorem 12.2 (basic Brill–Noether). *If $r \geq 0$ and*

$$\rho(g, r, d) := g - (r+1)(g-d+r) \geq 0,$$

then every smooth projective curve of genus g possesses a g^r_d. Conversely, if $\rho < 0$ then a general curve C of genus g does not possess a g^r_d.

It is interesting to compare the values of d, r that are possible on special and general curves; see Figure 12.1.

Gathering the inequalities, and putting them all in terms of lower bounds on d given g, r, we get

$d \geq \min\{r + g, 2r\}$ (by the Riemann–Roch and Clifford theorems),

$d \geq \sqrt{(2r-2)g}$ (by an approximation to the Castelnuovo theorem),

$d \geq r + g - \dfrac{g}{r+1}$ (for a general curve).

In the following sections, we'll explain the heuristic argument that led Brill and Noether to the statement of Theorem 12.2 and discuss some refinements. In Chapter 13 we'll give a proof based on the study of inflections and on families of Jacobians.

The case $r = 1$ is already interesting:

Corollary 12.3. *If C is any curve of genus g, then C admits a map to $\mathbb{P}^1$ of degree d for some $d \leq \lceil \frac{g+2}{2} \rceil$.*

Thus any curve of genus 2 is hyperelliptic, any curve of genus 3 or 4 is either hyperelliptic or trigonal (admits a 3-1 map to $\mathbb{P}^1$), and so on. We have already verified this assertion in genus $g \leq 5$ by analyzing the geometry of the canonical map; for higher genera, though, this is not feasible.

Note also that this is exactly the converse to Corollary 8.11 of Chapter 7.

12.2.1. A Brill–Noether inequality. The proof of the Brill–Noether theorem starts with a dimension estimate that was first carried out in [Brill and Nöther 1874]. The estimate provides an inequality on the dimension of the variety W_d^r, and the assertion of the theorem is that this is sharp for a general curve.

Let C be a smooth projective curve of genus g, and $D = p_1 + \cdots + p_d$ a divisor on C. Assume for simplicity that the points p_i are distinct; the same argument can be carried out in general, but requires more complicated notation.

When does the divisor D move in an r-dimensional linear series? By the Riemann–Roch theorem $h^0(D) \geq r + 1$ if and only if the vector space $H^0(K - D)$ of 1-forms vanishing on D has dimension at least $g - d + r$ — that is, if and only if the evaluation map

$$H^0(K) \to H^0(K|_D) = \bigoplus k_{p_i}$$

has rank at most $d - r$.

We can represent this map by a $g \times d$ matrix. Choose a basis $\omega_1, \ldots, \omega_g$ for the space $H^0(K)$ of 1-forms on C; choose an analytic open neighborhood U_j of each point $p_j \in D$ and choose a local coordinate z_j in U_j around each point p_j, and write

$$\omega_i = f_{i,j}(z_j) \, dz_j$$

in U_j. We have $r(D) \geq r$ if and only if the matrix-valued function

$$A(z_1, \ldots, z_d) = \begin{pmatrix} f_{1,1}(z_1) & f_{2,1}(z_1) & \cdots & f_{g,1}(z_1) \\ f_{1,2}(z_2) & f_{2,2}(z_2) & \cdots & f_{g,2}(z_2) \\ \vdots & \vdots & & \vdots \\ f_{1,d}(z_d) & f_{2,d}(z_d) & \cdots & f_{g,d}(z_d) \end{pmatrix}$$

has rank $d - r$ or less at $(z_1, \ldots, z_d) = (0, \ldots, 0)$.

In the space $M_{d,g}$ of $d \times g$ matrices, the subset of matrices of rank $d - r$ or less has codimension $r(g - d + r)$ [Eisenbud 1995, Exercise 10.9]. It follows that if an effective divisor D of degree d with $h^0(D) \geq r + 1$ exists, then in a neighborhood of the point $D \in C_d$ the locus C_d^r of such divisors must have dimension at least $d - r(g - d + r)$, with equality if the map A is dimensionally transverse to the locus in $M_{d,g}$ of matrices of rank at most $d - r$. Since a general fiber of the map $\mu : C_d^r \to W_d^r(C)$ has dimension r, it follows that

$$\dim W_d^r(C) \geq d - r(g - d + r) - r = g - (r + 1)(g - d + r)$$

and this is exactly the Brill–Noether (in)equality. We will give a proof of the Brill–Noether theorem in Chapter 14.

12.2.2. Refinements of the Brill–Noether theorem. Theorem 12.2 suggests a slew of questions, both about the geometry of the schemes $W_d^r(C)$ parametrizing linear series on a general curve C (are they irreducible? what are their singular loci?), and about the geometry of the linear series themselves (do they give embeddings? what's the Hilbert function of the image?). This is an active area of research. Here is some of what is currently known, starting with results about the geometry of $W_d^r(C)$:

Theorem 12.4. *Let C be a general curve of genus g, and set*

$$\rho = g - (r+1)(g-d+r).$$

Assume that $d \leq g + r$. Then:

(1) $\dim(W_d^r(C)) = \rho$ [Griffiths and Harris 1980].

(2) *The singular locus of $W_d^r(C)$ is exactly $W_d^{r+1}(C)$* [Gieseker 1982; Lazarsfeld 1986].

(3) *If $\rho > 0$ then $W_d^r(C)$ is irreducible* [Fulton and Lazarsfeld 1981].

(4) *If $\rho = 0$ then $W_d^r(C)$ consists of a finite set of points of cardinality*

$$\#W_d^r(C) = g! \prod_{\alpha=0}^{r} \frac{\alpha!}{(g-d+r+\alpha)!}$$

and the monodromy of the generically finite covering of M_g by the universal family $\mathcal{W}_d^r$ of W_d^rs is transitive [Eisenbud and Harris 1987b].

(5) *If $\mathcal{L}$ is an invertible sheaf on C, then the multiplication map*

$$m : H^0(L) \otimes H^0(\omega_C \otimes L^{-1}) \to H^0(\omega_C)$$

is injective, and the Zariski tangent space to the scheme $W_d^r(C)$ at the point L, as a subspace of the tangent space $T_L \operatorname{Pic}_d(C) = H^0(\omega_C)^*$, is the annihilator of the image of m or, equivalently, the kernel of the dual of m [Gieseker 1982].

Corollary 12.5. *If C is a general curve and $\mathcal{L}$ is a general point of $W_d^r(C)$ with $r \geq 2$, then $\mathcal{L}^m$ is nonspecial for all $m \geq 2$.*

Proof. If $\mathcal{L}^m$ were special — that is, if $\omega_C \otimes \mathcal{L}^{-m} = E$ were effective — then we would have an inclusion $H^0(\mathcal{L}) = H^0(\omega_C \otimes \mathcal{L}^{-m+1}(-E)) \hookrightarrow H^0(\omega_C \otimes \mathcal{L}^{-m+1})$. By part 5 of Theorem 12.4, the map

$$m : H^0(\mathcal{L}^{m-1}) \otimes H^0(\omega_C \otimes \mathcal{L}^{-m+1}) \to H^0(\omega_C)$$

is injective, so the map

$$H^0(\mathcal{L}^{m-1}) \otimes H^0(\mathcal{L}) \subset H^0(\mathcal{L}^{m-1}) \otimes H^0(\omega_C \otimes \mathcal{L}^{-m+1})$$

obtained by restriction would likewise be injective. However if $\sigma, \tau \in H^0(\mathcal{L})$ are two linearly independent sections, then $\sigma^{m-1} \otimes \tau - \sigma^{m-2}\tau \otimes \sigma$ lies in the kernel, contradicting the specialness of $\mathcal{L}^m$. $\qquad\square$

Remark 12.6. As a special case of part 4 of the theorem we see that the number of g_d^1s in the case $\rho = 0$ (that is to say, $g = 2d - 2$) is the Catalan number $C_{d-1} := \binom{2d}{d}/d$.

We have already seen this in the first two cases: in genus 2, it says the canonical series $|K|$ is the unique g_2^1 on a curve of genus 2, and in the case of genus 4 we have already seen that there are exactly two g_3^1s on a general curve of genus 4. In genus 6, it says that a general curve of genus 6 has 5 g_4^1s; we'll describe these in Section 12.3 below.

Remark 12.7. Parts (5) and (1) of the theorem imply part (2). [Arbarello et al. 1985, Section IV.4] shows that at a point $\mathcal{L} \in W_d^r(C) \setminus W_d^{r+1}(C)$, the tangent space to W_d^r at the point $\mathcal{L}$ is the annihilator in $(H^0(\omega_C))^*$ of the image of μ; given that μ is injective, we can compare dimensions and deduce that W_d^r is smooth at $\mathcal{L}$.

Remark 12.8. For any curve C, there exists a scheme $G_d^r(C)$ parametrizing linear series of degree d and dimension r; that is, in set-theoretic terms,

$$G_d^r(C) = \{(\mathcal{L}, V) \mid \mathcal{L} \in Pic_d(C), \text{ and } V \subset H^0(\mathcal{L}) \text{ with } \dim V = r + 1\}.$$

$G_d^r(C)$ maps to $W_d^r(C)$; the map is an isomorphism over the open subset $W_d^r(C) \setminus W_d^{r+1}(C)$ and has positive-dimensional fibers over $W_d^{r+1}(C)$. It was conjectured by Petri and proven in [Gieseker 1982] that for a general curve the scheme $G_d^r(C)$ is smooth for any d and r.

Recall that in theorems 5.9, 5.14, 5.13 we proved that general invertible sheaves of degrees $g+1$, $g+2$ and $g+3$ on any curve give the nicest possible maps to (respectively) $\mathbb{P}^1$, $\mathbb{P}^2$ and $\mathbb{P}^3$. These linear series, being general of degree $\geq g$, are nonspecial and have respectively 2, 3, or 4-dimensional spaces of sections. The following result shows that something similar is true on a general curve for general linear series with 2,3, or 4-dimensional spaces of sections, though they may have degrees much less than $g + 1, g + 2, g + 3$:

Theorem 12.9 [Eisenbud and Harris 1983, Proposition 5.4]. *Let C be a general curve of genus g, and suppose that $|D|$ is a general g_d^r on C.*

(1) *If $r \geq 3$ then D is very ample; that is, the map $\phi_D : C \to \mathbb{P}^r$ embeds C in $\mathbb{P}^r$;*

(2) *If $r = 2$ the map $\phi_D : C \to \mathbb{P}^2$ gives a birational embedding of C as a nodal plane curve; and*

(3) *If $r = 1$, the map $\phi_D : C \to \mathbb{P}^1$ expresses C as a simply branched cover of $\mathbb{P}^1$.*

In case $\rho = 0$ — so that there are a finite number of g_d^rs on a general curve C — these statements hold for *all* the g_d^rs on C.

In the course of investigating embeddings of a curve $C \subset \mathbb{P}^n$ we have again and again asked about the ranks of the maps $H^0(\mathcal{O}_{\mathbb{P}^n}(d)) \to H^0(\mathcal{O}_C(d))$. In

the case of a general curve, the following theorem of E. Larson gives a comprehensive answer, which is the same as giving the Hilbert function of the image
curve:

Theorem 12.10 (maximal rank theorem [Larson 2017]). *If C is a general curve
of genus g and $\mathcal{L} \in W_d^r(C)$ is a general point, then for each $m > 0$ the multiplication map*

$$\rho_m : \mathrm{Sym}^m H^0(\mathcal{L}) \to H^0(\mathcal{L}^m)$$

*has maximal rank; that is, it is injective if $\binom{m+r}{r} \leq h^0(\mathcal{L}^m)$ and surjective if
$\binom{m+r}{r} \geq h^0(\mathcal{L}^m)$.*

If $\mathcal{L} \in W_d^r(C)$ is a general point, then Corollary 12.5 shows that $h^0(\mathcal{L}^m) =
md - g + 1$ for all $m \geq 2$, and this allows us to compute the Hilbert function of
a general embedding as a curve of degree d as

$$h_C(m) = \min\left(\binom{m+r}{r}, \, md - g + 1\right).$$

A key step in Larson's proof is an interpolation theorem [Atanasov et al.
2019], later extended as follows:

Theorem 12.11 [Larson and Vogt 2023]. *Let d, g and r be nonnegative integers
with $\rho(d, g, r) \geq 0$. There is a general curve of degree d and genus g through n
general points in $\mathbb{P}^r$ if and only if*

$$(r - 1)n \leq (r + 1)d - (r - 3)(g - 1)$$

except in the four cases $(d, g, r) = (5, 2, 3), (6, 4, 3), (7, 2, 5)$ and $(10, 6, 5)$.

There is a possible extension of the maximal rank theorem. If $C \subset \mathbb{P}^r$ is a
general curve embedded by a general linear series, the maximal rank theorem
tells us the dimension of the m-th graded piece of the ideal of C, for any m: this
is just the dimension of the kernel of ρ_m. But it doesn't tell us the degrees of
generators of the homogeneous ideal of C. For example, if m_0 is the smallest m
for which $I(C)_m \neq 0$, or numerically the smallest m such that $\binom{m+r}{r} > md -
g + 1$, we can ask: is the homogeneous ideal $I(C)$ generated by $I(C)_{m_0}$? This
can't always be the case, since there are examples where the smallest nonzero
graded piece of $I(C)$ has dimension 1. But one might conjecture that $I(C)$ is
always be generated by its graded pieces of degrees m and $m + 1$; this is an open
problem.

To answer this, given that we know the dimensions of $I(C)_m$ for every m,
we would need to know the ranks of the multiplication maps

$$\sigma_m : I(C)_m \otimes H^0(\mathcal{O}_{\mathbb{P}^r}(1)) \to I(C)_{m+1}$$

for each m. In particular, we may conjecture that *the maps σ have maximal
rank*; if this were true we could deduce the degrees of a minimal set of generators for the homogeneous ideal $I(C)$.

Another recent strand of work on Brill–Noether theory was developed in the thesis [Larson 2021] and in [Larson et al. 2008], providing analogues of many of the parts of the classical Brill–Noether theorem for general curves of given gonality in the cases $\rho \geq 0$.

There are many remaining questions! One is the question of *secant planes*: a naive dimension count would suggest that an irreducible, nondegenerate curve $C \subset \mathbb{P}^r$ should have an s-secant t-plane if and only if $s(r - t - 1) \leq (t + 1)(r - t)$ (for example a curve $C \subset \mathbb{P}^3$ has 4-secant lines, but no 5-secant lines; see Exercise 12.1). Is this true for a general curve embedded in $\mathbb{P}^r$ by a general linear series?

12.3. Linear series on curves of genus 6

We have seen in our analysis of curves of genus up to 5 that curves of the same genus can look quite different from the point of view of the linear series they possess: the existence of g^r_ds, the geometry of the schemes $W^r_d(C)$ parametrizing them, and the geometry of the associated maps to projective space, can look quite different on different curves.

The variety of possible behaviors has increased modestly with the genus. Genus 6 is a tipping point: we could still enumerate all the possible behaviors of the schemes $W^r_d(C)$ — as distinguished by the number of components, dimension and singularities of the various schemes $W^r_d(C)$, and the geometry of the associated maps — but it's quite a long list, and we will actually study just a few cases. For genus 7 and higher a full analysis has probably never been carried out.

In lower genus we tacitly verified the statements of the Brill–Noether theorem from our descriptions of the canonical models. In genus 6, by contrast, we cannot deduce the Brill–Noether theorem from studying the geometry of the canonical curve — though we can easily see that a canonical curve $C \subset \mathbb{P}^5$ of genus 6 lies on a 6-dimensional vector space of quadrics, that doesn't tell us much about its geometry.

Instead we will appeal directly to the Brill–Noether theorems. Here is a summary of what we will use:

Theorem 12.12. *Every smooth curve C of genus 6 has at least one g^2_6. If C is general, then $W^2_6(C)$ and $W^1_4(C)$ each consist of 5 reduced points, while $W^2_5 = W^1_3 = \emptyset$. Less formally, C has precisely 5 g^2_6s and 5 g^1_4s, but no g^2_5s and no g^1_3s. The image of the map associated to each g^2_6 is a nodal plane curve and its nodes are in linearly general position, that is, no three are collinear.*

All these assertions except for the linear general position of the nodes follow immediately from Theorem 12.2 and Theorem 12.4; we will deduce the linear general position of the nodes from the relationship of the different linear series on a general curve that are given by these theorems.

12.3.1. General curves of genus 6. *We suppose for the rest of this section that C is a general smooth curve of genus 6.*

By Theorem 12.12 we can map C birationally to a plane sextic C_0 with only nodes as singularities. Since the arithmetic genus of a plane sextic is $\binom{6-1}{2} = 10$, the curve C_0 must have exactly 4 nodes.

Once we have exhibited one birational map of C to a plane sextic with 4 nodes, we can describe all five g_6^2s and all five g_4^1s in terms of this plane model. For example, composing a g_6^1 corresponding to $f : C \to \mathbb{P}^1$ with the projections from the 4 nodes gives four g_4^1s. To see the fifth g_4^1 we introduce some terminology:

Suppose $f : X \to S$ is a regular map from a smooth curve X to a surface S. To a line bundle $\mathcal{L}$ on S and a vector space $V \subset H^0(\mathcal{L})$ of sections, we can associate a linear series on X by taking the pullback linear series $f^*V \subset H^0(f^*\mathcal{L})$ on X and subtracting the basepoints; this is called the *linear series cut out on X by V*.

To see the g_4^1s on C in this way, suppose again that C is a general curve of genus 6 as above and $f : C \to \mathbb{P}^2$ is a birational map onto a sextic curve C_0 with four nodes; let $p \in C_0$ be one of the nodes and consider the linear series $(\mathcal{O}_{\mathbb{P}^2}(1), V)$ of lines in $\mathbb{P}^2$ through p. The pullback $f^*\mathcal{O}_{\mathbb{P}^2}(1)$ of course has degree 6, but the pullback linear series f^*V has two basepoints, at the points $q, r \in C$ lying over p. The linear series cut on C by V is thus a g_4^1; see Figure 12.2, left.

To produce the fifth g_4^1, consider the linear series cut on C by conic plane curves passing through all four nodes of C_0 (Figure 12.2, right). There is a pencil of such conics, and the pullback $f^*\mathcal{O}_{\mathbb{P}^2}(2)$ has degree 12. If the nodes are linearly independent then the pullback series has eight basepoints; thus we

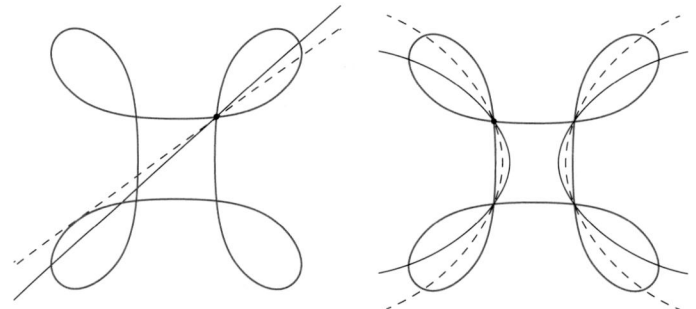

Figure 12.2. Left: A g_4^1 as (left) the projection from a node of a plane sextic and (right) a pencil of conics through the four nodes of a plane sextic.

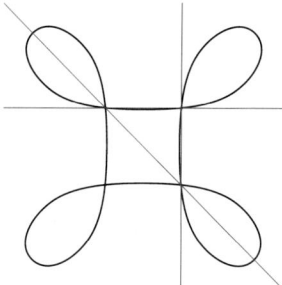

Figure 12.3. A sextic with 4 nodes and the fundamental triangle of the quadratic transformation giving a different g_6^2.

arrive at another g_4^1 on C. Not all the nodes can be contained in a line, since then, by Bézout's theorem, the line would be a component of C_0. Thus if the nodes are linearly dependent, then exactly 3 lie on a line so the linear series cut by the conics containing the nodes coincides with the projection from the fourth node. This would represent a nonreduced point of the scheme $W_4^1(C)$, the subject of Exercise 12.9 below. Thus the nodes are independent.

For another example, consider the linear series cut on C by cubics passing through all four nodes. This has degree $3 \cdot 6 - 8 = 10$ and dimension 6. It follows that this is the complete canonical series on C. (In Chapter 15 we will see directly that this is the case.)

Given the degree six map $f : C \to C_0 \subset \mathbb{P}^2$ corresponding to one g_6^2 we can use the fact that the five g_6^2s on C are residual to the five g_4^1s in the canonical series to construct the other four g_6^2s: they are cut out on C by the linear series of plane conics passing through three of the four nodes of C_0 Equivalently, their images are the curves obtained from C_0 by the quadratic transformation of $\mathbb{P}^2$ centered at 3 of the 4 nodes, which blows up these 3 nodes and blows down the three lines joining them (Figure 12.3).

In previous chapters we have seen that in genus ≤ 5 a general canonical curve is a complete intersection, but this fails for a canonical curve C of genus 6. There is a 21-dimensional vector space of quadratic forms on $\mathbb{P}^5$, and $h^0(\mathcal{O}_C(2)) = 2(2g-2)-g+1 = 15$, so C lies on at least 6 quadrics, and we will show that its ideal sheaf is generated by exactly 6 quadrics. Since $6 > \operatorname{codim} C$, the canonical curve of genus 6 is not a complete intersection. However, such curves lie on a quintic del Pezzo surface, which may be described as follows.

12.3.2. Del Pezzo surfaces. We met the del Pezzo surface of degree 5 in Section 4.6. We briefly sketch, without proofs, a little of the rich classical theory of del Pezzo surfaces in general. The basics are well treated in [Beauville 1996, pp. 45–50]; for more, see the beautiful book [Manin 1986], which also goes into

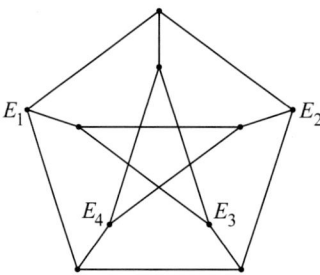

Figure 12.4. Dual graph of the configuration of 10 lines on a quintic del Pezzo surface, the plane blown up at 4 points showing 4 pairwise disjoint exceptional divisors.

some of the arithmetic theory. We will use only the case of the del Pezzo surface of degree 5, which lies in $\mathbb{P}^5$.

By definition, a *del Pezzo* surface is a smooth surface embedded in $\mathbb{P}^n$ by its complete anticanonical series $-K_S$. These exist only for $3 \leq n \leq 9$. The best-known example is a smooth cubic surface in $\mathbb{P}^3$. That it is a del Pezzo surface follows from the adjunction formula.

A del Pezzo surface in $\mathbb{P}^n$ has degree n, and is isomorphic to the blowup of $\mathbb{P}^2$ at $9 - n$ points of which no 3 lie on a line and no 6 lie on a conic, embedded by the linear series on $\mathbb{P}^2$ consisting of the cubics passing through the $9 - n$ points — except when $n = 8$, in which case the linear series of curves of type $(2, 2)$ on $\mathbb{P}^1 \times \mathbb{P}^1$ provides another example.

Comparing the linear series of cubic forms containing $p_1, \ldots, p_4$ with the linear series of sextic forms vanishing to order 2 at $p_1, \ldots, p_4$, we see that a quintic del Pezzo surface $S \subset \mathbb{P}^5$ lies on at least 5 quadrics. In fact, its homogeneous ideal is generated by exactly 5 quadrics.

A quintic del Pezzo surface $S \subset \mathbb{P}^5$ contains exactly 10 lines, which (in terms of the description of S as the blow up of $\mathbb{P}^2$ at four points $p_1, \ldots, p_4 \in \mathbb{P}^2$) are the 4 exceptional divisors and the 6 proper transforms of the lines joining the p_i pairwise. The configuration of these lines is shown in Figure 12.4. It is the intersection of a $\mathbb{P}^5$ with the Grassmannian $G(2, 5) \subset \mathbb{P}^9$, and correspondingly the five quadrics containing S can be realized as the Pfaffians of a 5×5 skew-symmetric matrix of linear forms on $\mathbb{P}^5$.

There is also a notion of a *weak del Pezzo* surface; this is a smooth surface whose anticanonical sheaf is nef but not necessarily ample. We get such a surface if we blow up $\mathbb{P}^2$ at a configuration of points of which three are collinear; in this circumstance the anticanonical sheaf on the blowup S has degree 0 on the proper transform of the line containing the three points, and this proper transform is correspondingly collapsed to a rational double point of the image

$\phi_{-K}(S)$. In general, the description of del Pezzo surfaces as blowups of the plane extends to the case of weak del Pezzos.

12.3.3. The canonical image of a general curve of genus 6. Using the Brill–Noether theorem, we have seen that a general curve C of genus 6 is the normalization of a plane sextic C_0 with four nodes, and that the canonical series on C is cut out by cubics in the plane passing through the four nodes. Thus the canonical model lies on the surface $S \subset \mathbb{P}^5$ that is the image of the plane under the (rational) map given by cubics through these four points, which we now recognize as a quintic del Pezzo surface.

Theorem 12.13. *A general canonical curve C of genus 6 is the intersection of a quintic del Pezzo surface and a quadric.*

Proof. We have seen that $C \subset \mathbb{P}^5$ lies on a quintic del Pezzo surface $S \subset \mathbb{P}^5$. The surface is cut out by 5 quadrics, and we know that C lies on 6 independent quadrics, so C is contained in the complete intersection of S with a quadric. Since this scheme has degree 10, which is the degree of C, they are equal. $\square$

For corresponding theorems for general curves with $g = 7, 8, 9$ see [Mukai 1995a], [Mukai 2010], and [Mukai 1995b].

12.4. Classification of curves of genus 6

From Theorem 12.12 we know that every curve C of genus 6 has a g_6^2. For a general curve the corresponding morphism ϕ_D maps C birationally onto a plane sextic with 4 nodes and exactly 5 g_4^1s. In this section we will analyze some of the other curves of genus 6, starting with the question "what could go wrong?" with the birational map to $\mathbb{P}^2$.

Since we already have a complete picture in the hyperelliptic case, we'll assume that C is nonhyperelliptic. Clifford's theorem then rules out the existence of a g_6^3 so the g_6^2 is a complete linear series $|D|$ for some divisor D of degree 6.

Further analysis can be divided as follows:

Case 1: C is not trigonal.

(a) $|D|$ has a basepoint; the canonical image of C lies on the Veronese surface.

(b) ϕ_D maps C two-to-one onto a cubic $E \subset \mathbb{P}^2$ of genus 1; the canonical image is the complete intersection of a quadric and the cone over the elliptic quintic $E \subset \mathbb{P}^4$.

(c) ϕ_D is birational onto a plane sextic with no triple point.

Case 2: C is trigonal. The canonical image lies on a rational normal scroll $S(a, 4-a)$; see Chapter 17. In this case $|D|$ is basepoint free and ϕ_D either maps C three to one onto a conic or birationally onto a sextic with a triple point.

In the rest of this section we will examine cases 1(a) and 1(b). Case 2 can be further divided by the value of a. Case 1(c) may be divided into many parts according to the various configurations of double points of the plane sextic, which may be nodes, cusps, tacnodes or higher-order double points, corresponding to various possible schemes W_4^1.

The analysis of the other cases is lengthy, but largely accessible with the tools we've introduced.

$|D|$ has a basepoint. Clifford's theorem shows that a nonhyperelliptic curve of genus 6 cannot have a g_4^2, so $|D|$ has exactly one basepoint, and when we subtract the basepoint we get a basepoint free g_5^2.

If $C \to C_0 \subset \mathbb{P}^r$ is the map given by a basepoint free g_d^r, the degree d of the linear series is the degree of the image curve C_0 times the degree of the map $C \to C_0$. Since 5 is prime, the associated map $\phi_D : C \to \mathbb{P}^2$ is birational onto a quintic curve. Moreover, since plane quintic curves have arithmetic genus 6, the image $\phi_D(C)$ is smooth; thus C is isomorphic to a smooth plane quintic.

This allows us to describe the other special linear series on C. By the adjunction formula, the canonical series $|K_C|$ is cut on C by conics in the plane.

The plane $\mathbb{P}^2$ is embedded by the complete linear series of quadrics as the Veronese surface in $\mathbb{P}^5$, of C is cut out by quadrics, the canonical model of C lies on this surface, and the canonical ideal is generated by the 6-dimensional family of quadrics containing the Veronese surface — the 2×2 minors of the generic symmetric 3×3 matrix corresponding to the multiplication map

$$H^0(\mathcal{O}_{\mathbb{P}^2}(1)) \otimes H^0(\mathcal{O}_{\mathbb{P}^2}(1)) \to H^0(\mathcal{O}_{\mathbb{P}^2}(2))$$

as explained in Proposition 17.6– together with a 3-dimensional family of cubics, the image of the 3 dimensional family of forms of degree 6 that are multiples of the quintic form defining C in $\mathbb{P}^2$.

As we showed in Chapter 4, the g_4^1s on C are exactly the projections from points of C, so $W_4^1(C) \cong C$, and there are no g_3^1s.

C is not trigonal and the image of ϕ_D is two-to-one onto a plane curve of degree 3. The cubic curve E is smooth since otherwise it would have geometric genus 0 and C would be hyperelliptic. Thus C is a double cover of a smooth curve of genus 1; we say that C is *bielliptic*.

In this case the canonical divisor class K_C is the pullback of an invertible sheaf $\mathcal{O}_E(F)$ for some divisor class of degree 5 on E. But it is not the case that the canonical series $|K_C|$ is the pullback of the linear series $|\mathcal{O}_E(F)|$: by the Riemann–Roch theorem, the latter has dimension 4, rather than 5. Indeed, if we recall that the target of the canonical map $\phi_K : C \to \mathbb{P}^5$ is the projective space $\mathbb{P}H^0(K_C)$, there is be a point $X \in \mathbb{P}^5$ corresponding to the hyperplane $\pi^* H^0(F) \hookrightarrow H^0(K_C)$, and projection of the canonical curve from this point maps C 2-to-1 onto the $\overline{\text{image}}$ $\phi_F(E) \subset \mathbb{P}^4$. In other words, the canonical model of C lies on a cone $S = \overline{X,E}$ over an elliptic normal quintic curve $E \subset \mathbb{P}^4$.

As we saw in Chapter 4, the quintic curve $\phi_F(E) \subset \mathbb{P}^4$ lies on 5 quadrics, as does the cone $S \subset \mathbb{P}^5$ over it. Thus there is a quadric $Q \subset \mathbb{P}^5$ containing C but not containing S. Bézout's theorem shows that in this case, the canonical model of a bielliptic curve of genus 6 is the intersection of the cone over an elliptic quintic curve with a quadric.

12.5. Exercises

Exercise 12.1. We saw in Chapter 5 that if C is any curve of genus g and D a general divisor of degree $g + 3$ on C, then $\phi_D : C \hookrightarrow \mathbb{P}^3$ is an embedding. Using the Brill–Noether theorem, show that if C is general then the image curve in $\mathbb{P}^3$ has no 5-secant lines.

Exercise 12.2. Let C be a bielliptic curve of genus 6; that is, a double cover of a smooth projective curve E of genus 1.

(1) Show that C cannot be hyperelliptic (going forward, we will identify C with its canonical image in $\mathbb{P}^5$).

(2) Let F an invertible sheaf of degree 5 on E and $\phi_F(E) \subset \mathbb{P}^4$ the corresponding elliptic normal quintic curve. Show that $\phi_F(E)$ lies on 5 quadrics, as does the cone $S \subset \mathbb{P}^5$ over it.   ◆

(3) Deduce that there is a quadric $Q \subset \mathbb{P}^5$ containing C but not containing S. Now invoke Bézout's theorem to deduce that in this case, the canonical model of a bielliptic curve of genus 6 is the intersection of the cone over an elliptic quintic curve with a quadric.

Exercise 12.3. Use the preceding exercise to show that if C is a bielliptic curve of genus 6, a 2-sheeted cover of an elliptic curve E, then every g^1_4 on C is the pullback of a g^1_2 on E, and likewise every g^2_6 on C is the pullback of a g^2_3 on E. Deduce that in this case, $W^1_4(C)$ and $W^2_6(C)$ are each isomorphic to E.   ◆

Exercise 12.4. Let C be a trigonal curve of genus 6 with g^1_3 $|E|$, and $p \in C$ a general point. Show that the linear series $|K_C - E - p|$ is a g^2_6, and that the corresponding map $C \to \mathbb{P}^2$ maps C birationally onto a plane sextic curve with a triple point.

Exercise 12.5. Let C be the normalization of a plane sextic C_0 with four nodes, three of which are collinear. Show that by choosing a different g^2_6 on C, we can express it as the normalization of a plane sextic with two nodes and a tacnode.

◆

Exercise 12.6. Show that if C is the normalization of a plane sextic C_0 with only double points, then $W^1_4(C) \cong W^2_6(C)$ is zero-dimensional (so in particular this case does not overlap with any of the previous cases)

Exercise 12.7. Find an example of a curve C of genus 6 such that $W_4^1(C)$ consists of one point of degree 5. Bonus points for showing that in this case the scheme $W_4^1(C) \cong \operatorname{Spec} \mathbb{C}[\epsilon]/(\epsilon^5)$. ◆

Exercise 12.8. Show that if C is a smooth plane quintic, then the g_6^2s on C all have a basepoint; that is, they are all of the form $|K_C| + p$ for $p \in C$.

Furthermore, the canonical model of C will lie on a quadratic Veronese surface S; and the six quadrics containing the canonical curve C are the six quadrics containing S. In particular, the intersection of the quadrics containing C will be S, so the ideal of C requires generators of degree > 2.

Exercise 12.9. Show that if the nodes of the curve C_0 are in linear general position — that is, no three are collinear — then indeed the map

$$\mu : H^0(D) \otimes H^0(K - D) \to H^0(K)$$

is an isomorphism for each of the five g_4^1s on C.

Chapter 13

Inflection points

Generalizing the ramification points of a map from a smooth curve C to $\mathbb{P}^1$, there are finitely many "special" points determined by any linear series on any curve, the *inflection points*. We love this topic for its echoes of classical algebraic geometry and it will provide the tools to give a proof of the Brill–Noether theorem in the following chapter. In characteristic 0 every linear series has finitely many inflection points, and the number of these, properly counted, depends only on the genus of the curve and the degree of the linear series.

13.1. Inflection points, Plücker formulas and Weierstrass points

Definitions. To characterize the inflection points of a linear series $\mathcal{D} = (\mathcal{L}, V)$ on a curve C, we will use the following result:

Proposition 13.1. *Let V be a finite-dimensional vector space of global sections of an invertible sheaf $\mathcal{L}$ on a smooth curve C, and let $p \in C$ be a point. There exists a basis $\sigma_0, \ldots, \sigma_r$ of V consisting of sections vanishing to different orders at p. Thus the set*

$$\{\operatorname{ord}_p(\sigma) \mid \sigma \neq 0 \in V\}$$

has cardinality $\dim V$.

Proof. Let $\tau_0, \ldots, \tau_r$ be any basis of V. If τ_i and τ_j vanish to the same order at p, then some nonzero linear combination $\tau_i' := a\tau_i + b\tau_j$ vanishes to strictly higher order. Since the coefficients a and b are both necessarily nonzero we may modify the basis, replacing τ_i with τ_i', strictly increasing the sum of the orders. The order of vanishing of each σ_i is bounded above by $\deg \mathcal{L}$, so the sum of the orders is bounded by $(r+1) \deg \mathcal{L}$. Thus the process must terminate, and when it does, the orders must be distinct. $\qquad \square$

According to Proposition 13.1, if $\mathcal{D} = (\mathcal{L}, V)$ is any g^r_d on C, we may write

$$\{\text{ord}_p(\sigma) \mid \sigma \neq 0 \in V\} = \{a_0, \ldots, a_r\} \text{ with } 0 \leq a_0 < a_1 < \cdots < a_r.$$

The sequence $a_i = a_i(\mathcal{D}, p)$ is called the *vanishing sequence* of $\mathcal{D}$ at p. Since $a_i \geq i$, the numbers $\alpha_i = \alpha_i(\mathcal{D}, p) := a_i - i$ are often more interesting, and the sequence $0 \leq \alpha_0 \leq \alpha_1 \leq \cdots \leq \alpha_r$ is called the *ramification sequence* of $\mathcal{D}$ at p.

We say that p is an *inflection point* of the linear series $\mathcal{D}$ if $(\alpha_0, \ldots, \alpha_r) \neq (0, \ldots, 0)$ — equivalently, if $\alpha_r > 0$ — and we define the *weight* of p to be

$$w(\mathcal{D}, p) = \sum_{i=0}^{r} \alpha_i(\mathcal{D}, p)$$

and the on C to be

$$\sum_{p \in C} w(\mathcal{D}, p)p.$$

If $\mathcal{D}$ is very ample, so that it may be viewed as the linear series cut on C by hyperplanes for some embedding $C \subset \mathbb{P}^r$, then p is an inflection point if $a_r > r$; that is, if there is a hyperplane $H \subset \mathbb{P}^r$ having contact of order $r + 1$ or more with C at p.

The first two terms in the ramification sequence are particularly important: $\alpha_0(\mathcal{D}, p)$ is nonzero if and only if p is a basepoint of $\mathcal{D}$; and if $\alpha_0(\mathcal{D}, p) = 0$, then $\alpha_1(\mathcal{D}, p) = 0$ if and only if, in addition, the map $\phi_\mathcal{D}$ is an immersion (that is, has nonzero derivative) at p.

The Plücker formula. In characteristic 0, a linear series on a smooth projective curve can have only finitely many inflection points (this can be proven directly just as in Lemma 11.6), and indeed the sum of the weights of all the inflection points depends only on the genus of the curve and the degree and dimension of the linear series. This is the Plücker formula:

Theorem 13.2. *If C be a smooth curve of genus g and $\mathcal{D}$ is a linear series of degree d and dimension r, then*

$$\sum_{p \in C} w(\mathcal{D}, p) = (r + 1)d + r(r + 1)(g - 1).$$

For a proof, see for example [Eisenbud and Harris 2016, Theorem 7.13].

Theorem 13.2 also holds in positive characteristic under the positive characteristic hypothesis that the number of inflection points is finite (equivalently, not every point is an inflection point). This may seem like an unnecessary hypothesis — it's hard to imagine a plane curve in which every point is a flex! — but in positive characteristic there are such curves; see Exercise 13.1.

As an immediate consequence of the Plücker formula, we have:

Corollary 13.3. *If $C \subset \mathbb{P}^r$ is a smooth nondegenerate curve with no inflection points, then C is the rational normal curve of degree r.*

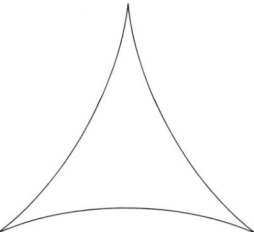

Figure 13.1. The deltoid is a quartic plane curve with three cusps and S_3 symmetry. Its affine equation is $(x^2+y^2)(x^2+y^2+18)+8y(y^2-3x^2)-27 = 0$.

Proof. Suppose $C \subset \mathbb{P}^r$ is a curve of degree d and genus g. If C has no inflection points then, by the Plücker formula, we must have

$$(r + 1)d + r(r + 1)(g - 1) = 0.$$

This immediately implies that $g = 0$, so that we must have $(r + 1)(d - r) = 0$ and hence $d = r$; thus C is a rational normal curve. $\qquad\square$

Corollary 13.4. *A nondegenerate smooth curve $C \subset \mathbb{P}^r$ is the rational normal curve of degree r if and only if any of the following equivalent properties holds:*

(1) *C is projectively homogeneous.*

(2) *C has no inflection points.*

(3) *Every subscheme of length $r + 1$ contained in C spans $\mathbb{P}^r$.*

Proof. Corollary 13.3 implies that the rational normal curves are the only ones without any inflection points.

By Proposition 3.10, the rational normal curve is projectively homogeneous. On the other hand, since not every point on C is an inflection point, a curve with an inflection point cannot be projectively homogeneous.

If $p \in C$ is an inflection point, then $(r + 1)p$ is a subscheme that lies in a hyperplane whereas in Proposition 1.9 we showed that if $C \subset \mathbb{P}^r$ is a rational normal curve, and $\Gamma \subset C$ any proper subscheme of C of degree $r + 1$, then Γ spans $\mathbb{P}^r$. $\qquad\square$

The Plücker formula leaves many questions about the possible configurations of flex points unanswered. For example, how many cusps can there be on a plane curve of geometric genus g and degree d? See Figure 13.1 for the maximum possible on a curve of degree 4. The answer is known only up to degree 8: Zariski [1931] proved that a plane curve of degree 8 can have 15 but not 16 cusps. See [Calabri et al. 2014] and [Kulikov and Shustin 2015] for recent contributions and [Kharlamov and Sottile 2003] for a study of what is possible over the real numbers.

One thing we do know is the behavior of the inflection points for a general linear series on a general curve; we will state this here and prove it as a corollary to the proof of the Brill–Noether theorem in Chapter 14:

Theorem 13.5. *If C is a general curve of genus g, $\mathcal{L} \in W_d^r(C) \subset \mathrm{Pic}_d(C)$ a general line bundle of degree d with $h^0(\mathcal{L}) = r + 1$ and $V = H^0(\mathcal{L})$, then every inflection point of the linear series $\mathcal{D} = (\mathcal{L}, V)$ has weight 1 and hence ramification sequence $(0, \ldots, 0, 1)$.*

Flexes of plane curves. Specializing Theorem 13.2 to a smooth curve of degree d in the plane, and using the formula $g = \binom{d-1}{2}$, we see that the total number of flexes is $3(d - 2)d$.

It turns out that the inflectionary divisor $\sum w(C, p)p$ is the intersection of C with a curve of degree $3(d - 2)$, called the *Hessian*: If C is defined by a form $F(x_0, x_1, x_2)$ of degree d, then the Hessian of C is the curve defined by the determinant of the *Hessian matrix* of partial derivatives

$$\mathrm{Hess}(C) := \begin{pmatrix} \partial^2 F/\partial x_0 \partial x_0 & \partial^2 F/\partial x_0 \partial x_1 & \partial^2 F/\partial x_0 \partial x_2 \\ \partial^2 F/\partial x_1 \partial x_0 & \partial^2 F/\partial x_1 \partial x_1 & \partial^2 F/\partial x_1 \partial x_2 \\ \partial^2 F/\partial x_2 \partial x_0 & \partial^2 F/\partial x_2 \partial x_1 & \partial^2 F/\partial x_2 \partial x_2 \end{pmatrix}$$

Theorem 13.6. *If C is a smooth plane curve then the flex divisor of C is the intersection of C with the Hessian curve defined by $\det \mathrm{Hess}(C)$.*

The proof is an exercise in Euler's formula and matrix manipulation; see Exercise 4.3.

Weierstrass points. As with any extrinsic invariant of a curve in projective space, we can derive an intrinsic invariant of an abstract curve by applying the invariant to the canonical linear series. We define a *Weierstrass point* of a curve C to be an inflection point of the canonical linear series $|K_C|$.

Thus p is a Weierstrass point of C if there exists a differential form on C vanishing to order g or more at p. The *weight* w_p of a Weierstrass point $p \in C$ is defined to be the weight $w(|K_C|, p)$ of p as an inflection point of the canonical series.

The Plücker formula tells us the total weight of the Weierstrass points on a given curve C:

Corollary 13.7. *The sum of the weights of the Weierstrass points on a curve C of genus g is*

$$\sum_{p \in C} w_p = g^3 - g. \qquad \square$$

Theorem 13.5 implies that on a general curve C of genus g, every Weierstrass point has weight 1; thus there are $g^3 - g$ distinct Weierstrass points on C.

For example, suppose C is a curve of genus 2. The canonical series on C gives a map $\phi_K : C \to \mathbb{P}^1$ of degree 2; the Weierstrass points of C are the 6 ramification points of this map.

In genus 3, if C is hyperelliptic then the Weierstrass points are exactly the 8 ramification points of the 2-sheeted cover $C \to \mathbb{P}^1$, and each has weight 3. If C is nonhyperelliptic, then it is a plane quartic curve. A general such curve has 24 ordinary flexes, which are Weierstrass points of weight 1; special quartics may have some number α of *hyperflexes* — points where the tangent line has contact of order 4 with the curve — which are Weierstrass points of weight 2; in this case C has α Weierstrass points of weight 2 and $24 - 2\alpha$ Weierstrass points of weight 1. (It has been shown that α cannot be 11, but all other values between 0 and 12 occur; see [Vermeulen 1983].)

Another characterization of Weierstrass points. The Riemann–Roch formula tells us that

$$h^0(\mathcal{O}_C(gp)) = g - g + 1 + h^0(K_C(-gp)),$$

so the condition $h^0(K_C(-gp)) \neq 0$ that p be a Weierstrass point is equivalent to the condition $h^0(\mathcal{O}_C(gp)) > 1$. In other words:

Proposition 13.8. *A point $p \in C$ is a Weierstrass point if and only if there exists a nonconstant rational function on C, regular on $C \setminus \{p\}$ and having a pole of order at most g at p.*

The Weierstrass semigroup. Proposition 13.8 suggests that we look at the set of all possible orders of pole at p of rational functions regular on $C \setminus \{p\}$; that is,

$$W(C, p) := \{-\operatorname{ord}_p(f) \mid f \in K(C) \text{ with } f \text{ regular on } C \setminus \{p\}\}.$$

This is clearly a subsemigroup of the natural numbers $\mathbb{N}$; it is called the *Weierstrass semigroup* of the point p.

Another way to characterize the condition that there exists a rational function on C, regular on $C \setminus \{p\}$, with a pole of order exactly k at p is to say that

$$h^0(\mathcal{O}_C(kp)) = h^0(\mathcal{O}_C((k-1)p)) + 1.$$

Applying the Riemann–Roch theorem to both sides of this equation, we see that it is equivalent to the condition

$$h^0(K_C(-kp)) = h^0(K_C((-k+1)p)).$$

In English: there exists a rational function on C, regular on $C \setminus \{p\}$, with a pole of order exactly k at p, if and only if there does *not* exist a regular differential on C with a zero of order exactly $k-1$ at p. In other words, the complement $\mathbb{N} \setminus W(C, p)$ is exactly the vanishing sequence of the canonical series at p, shifted by 1; in particular, it has cardinality exactly g. This is called the *Weierstrass gap sequence* of the point p.

There is still much we don't know about Weierstrass points in general. Most notably, we don't know what semigroups of finite index in $\mathbb{N}$ occur as Weierstrass semigroups; an example of Buchweitz shows that not all semigroups occur, but there are also positive results, such as the statement in [Eisenbud and Harris 1987a] that every semigroup of weight $w \leq g/2$ occurs, and its refinement and strengthening in [Pflueger 2018]).

13.2. Finiteness of the automorphism group

Because the Weierstrass points of a smooth projective curve C are defined intrinsically, any automorphism of C must carry Weierstrass points to Weierstrass points. We can use this observation to conclude:

Theorem 13.9. *If C is a smooth curve of genus ≥ 2 then the automorphism group* Aut C *is finite.*

We will actually prove a strong form of the assertion: the subgroup of Aut C of automorphisms that fix each of the finitely many Weierstrass points is either trivial or, in the case of hyperelliptic curves, just the $\mathbb{Z}/2$ generated by the hyperelliptic involution.

We will study the fixed points of automorphisms by studying the intersection of the graph of an automorphism with the diagonal divisor in $C \times C$ (an alternative proof uses the Lefschetz fixed point formula). For this we use the *Néron–Severi group* $N(S)$ of a surface S. This consists of divisors modulo *numerical equivalence* — that is, in $N(S)$ we identify divisors H, H' if for all divisors D on S we have $H \cdot D = H' \cdot D$. The group $N(S)$ is a finitely generated free abelian group for any surface — in characteristic 0 it is a subgroup of the quotient of the second integral homology group by its torsion elements. The rank of $N(S)$ is called the *Picard number* of S.

Theorem 13.10 (Hodge index theorem). *If $H \subset S$ is an ample divisor on a smooth projective surface, and $D \neq 0 \in N(S)$ is a divisor class with $D \cdot H = 0$, then $D^2 < 0$; that is, the intersection pairing is negative definite on the orthogonal complement of an ample divisor.*

Proof. The result follows easily from the Riemann–Roch theorem for surfaces; see [Hartshorne 1977, Theorem V.1.9]. □

There is also a stronger form of the Hodge index theorem, in which the condition that D is ample is weakened to the requirement that $D^2 > 0$. The proof of this stronger form relies on Hodge theory; see for example [Voisin 2003] or [Griffiths and Harris 1978].

Together, the next two lemmas imply the stronger form of Theorem 13.9:

Lemma 13.11. *Let C be a smooth projective curve of genus $g \geq 2$, and $f : C \to C$ an automorphism of C. If f has $2g + 3$ or more distinct fixed points, then f is the identity.*

Proof. Let $S = C \times C$, and let Δ and $\Gamma \subset S$ be the diagonal and the graph of f respectively, and let Φ_1 and $\Phi_2 \subset S$ be fibers of the two projection maps. Let $\delta, \gamma, \varphi_1$ and φ_2 be the classes of these curves in $N(S)$. The number of fixed points of f (counted with multiplicities) is the intersection number $b = \delta \cdot \gamma$.

We know all the other pairwise intersection numbers of these classes. To begin with, the ones involving φ_1 or φ_2 are obvious. From the sequence

$$0 \to \mathcal{T}_\Delta \to \mathcal{T}_{C \times C}|_\Delta \to \mathcal{N}_{\Delta/C \times C} \to 0$$

we see that the normal bundle of the diagonal is isomorphic to the tangent bundle of C, so

$$\delta^2 = 2 - 2g.$$

Since the automorphism $id_C \times f : C \times C \to C \times C$ carries Δ to Γ, we see that $\gamma^2 = 2 - 2g$ as well.

We can now apply the index theorem for surfaces to deduce our inequality. To keep things relatively simple, let's introduce two new classes: set

$$\delta' = \delta - \varphi_1 - \varphi_2 \quad \text{and} \quad \gamma' = \gamma - \varphi_1 - \varphi_2,$$

so that δ' and γ' are orthogonal to the class $\varphi_1 + \varphi_2$. Since $\varphi_1 + \varphi_2$ has positive self-intersection, the index theorem tells us that the intersection pairing must be negative definite on the span $\langle \delta', \gamma' \rangle \subset N(S)$. In particular, the determinant of the intersection matrix

	δ'	γ'
δ'	$-2g$	$b - 2$
γ'	$b - 2$	$-2g$

(where again $b = \gamma \cdot \delta$) is nonnegative, or equivalently, $b \leq 2g + 2$. $\square$

Having established an upper bound on the number of fixed points an automorphism f of C (other than the identity) may have, it remains to find a lower bound on the number of distinct Weierstrass points; this is the content of the next lemma.

Lemma 13.12. *If C is a smooth projective curve of genus $g \geq 2$, then C has at least $2g + 2$ distinct Weierstrass points; and if it has exactly $2g + 2$ Weierstrass points it is hyperelliptic.*

Proof. Let $p \in C$ be any point, and $w_1 = w_1(p), \ldots, w_g = w_g(p)$ the ramification sequence of the canonical series $|K_C|$ at p. By definition,

$$h^0(K_C(-(w_i + i)p)) = g - i.$$

Applying Clifford's theorem we have

$$g - i \leq \frac{2g - 2 - w_i - i}{2} + 1;$$

solving, we see that $w_i \leq i$ and hence

$$w_p \leq \binom{g}{2},$$

where w_p is the total weight of p as a Weierstrass point. Since the total weight of the Weierstrass points on C is $g^3 - g$ by the Plücker formula (Theorem 13.2), the number of distinct Weierstrass points is at least

$$\frac{g^3 - g}{\binom{g}{2}} = 2g + 2.$$

Finally, by the strong form of Clifford's theorem (2.34), equality implies that the curve is hyperelliptic. □

We can deduce Theorem 13.9 from Lemmas 13.11 and 13.12 as follows. If C is nonhyperelliptic, then Lemma 13.12 says it must have at least $2g + 3$ Weierstrass points, and then Lemma 13.11 says that any automorphism fixing all the Weierstrass points must be the identity. If, on the other hand, C is hyperelliptic, with degree 2 map $\pi : C \to \mathbb{P}^1$, then since the g_2^1 on C is unique, any automorphism of C must carry fibers of π to fibers of π; that is, it must commute with an automorphism of $\mathbb{P}^1$. In other words, we have an exact sequence

$$0 \to \mathbb{Z}/2 \to \operatorname{Aut} C \to G \to 0$$

where G is a subgroup of the group of automorphisms of $\mathbb{P}^1$ preserving the set of branch points of π. Since there are at least 6 such branch points, this group is finite.

The argument here gives the explicit bound, $(g^3 - g)!$, for the size of $\operatorname{Aut} C$. This is far from sharp: Exercise 13.4 outlines a proof of the inequality $|\operatorname{Aut} C| \leq 84(g - 1)$, which is sharp for infinitely many g.

13.3. Curves with automorphisms are special

In genus $g > 2$, curves with nontrivial automorphisms are rare. The general curve has none, and more is true:

Lemma 13.13. *In the moduli space M_g of smooth curves of genus $g \geq 3$, the locus of curves with nontrivial automorphisms has codimension $g - 2$; and the only component of this codimension is the locus of hyperelliptic curves.*

Proof. Let C be a smooth curve of genus g having a nontrivial automorphism $\phi : C \to C$. Taking a power, we may assume ϕ has prime order p. Let $\langle \phi \rangle = \{1, \phi, \phi^2, \ldots, \phi^{p-1}\}$ be the group of automorphisms of C generated by ϕ, and let $B = C/\langle \phi \rangle$ be the quotient of C by this group, so that the quotient map $\pi : C \to B$ expresses C as a p-sheeted branched cover of B, having b branch points $q_1, \ldots, q_b$ of multiplicity $p - 1$ and otherwise unramified. As we have seen in Chapter 6, if we know B and the set of branch points on B then we know C up to a finite set of choices.

The key fact is that, in a neighborhood of $[C] \in M_g$, the locus of curves with an automorphism of order p has dimension at most $3h - 3 + b$. Our claim is thus that

$$3g - 3 - (3h - 3 + b) \geq g - 2$$

or equivalently

$$2g - 1 - (3h - 3) - b \geq 0.$$

By applying the Riemann–Hurwitz formula to the map $C \to B$, we get

$$2g - 2 = p(2h - 2) + b(p - 1).$$

Plugging this in, our claim is equivalent to the assertion that

$$p(2h - 2) + b(p - 1) + 1 - (3h - 3) - b \geq 0$$

that is,

$$(2p - 3)h - 2p + b(p - 2) + 4 \geq 0.$$

Now, the expression on the left could be negative only if $h = 0$, in which case the expression reduces to

$$(b - 2)(p - 2) \geq 0;$$

and since $h = 0$ the number b of branch points must be at least 2, so we're done. $\qquad\square$

13.4. Inflections of linear series on $\mathbb{P}^1$

We are in a position to describe all possible inflectionary behavior of linear series on $\mathbb{P}^1$. This will provide the essential ingredient in our proof of the Brill–Noether theorem in the next chapter.

Giving a g^r_d on $\mathbb{P}^1$ amounts to choosing an $(r + 1)$-dimensional subspace W of $H^0(\mathcal{O}_{\mathbb{P}^1}(d))$ and thus the family of g^r_ds is parametrized by the Grassmannian $G(r + 1, H^0(\mathcal{O}_{\mathbb{P}^1}(d))) = G(r + 1, d + 1)$.

Theorem 13.14. *The set of g^r_ds on $\mathbb{P}^1$ with ramification sequence that is termwise at least as big as* $\alpha = (\alpha_0, \ldots, \alpha_r)$ *at a given point* $p \in \mathbb{P}^1$ *is an irreducible subvariety of* $G(r + 1, d + 1)$ *of codimension* $|\alpha| := \sum \alpha_i$. *The intersection of these subvarieties for any set of distinct points and any ramification sequences at* *those points is either empty or dimensionally transverse, depending only on the collection of ramification sequences.*

We will restate and prove this theorem, giving the combinatorial condition on the ramification indices, as Theorem 13.22 below. To simplify the notation, we will write $\ell := r + 1$ and $e := d + 1$.

Let $V = H^0(\mathcal{O}_{\mathbb{P}^1}(d))$ and, for $p \in \mathbb{P}^1$, write $V_i(p)$ for the space of forms of degree d that vanish to order $\geq e - i$ at p, and defined the *vanishing flag* $\mathcal{V}(p)$ at p to be the chain of subspaces

$$0 \subset V_1(p) \subsetneq V_2(p) \subsetneq \cdots \subsetneq V_e(p) = V.$$

Note that $\dim V_i(p) = i$. A subspace W of dimension $\ell = r + 1$ has vanishing sequence termwise not less than $\boldsymbol{a} = a_0, \ldots a_r$ and thus ramification sequence termwise not less than

$$\alpha = (\alpha_0 = a_0 - 0, a_1 - 1 \ldots, \alpha_r = a_r - r)$$

at p if and only if $\dim W \cap V_{e - a_{\ell - i}}(p) \geq i$ for $i = 0, \ldots, \ell$. Such a condition is called a *Schubert condition* on W, and we pause to describe the *Schubert varieties* in the Grassmannian $G(\ell, e)$ that consist of subspaces satisfying such a condition. See [Eisenbud and Harris 2016, Chapters 3 and 4] for an exposition. To accord with the notation there, which indexes Schubert cycles by certain decreasing sequences, we define $\beta_i = \alpha_{\ell - i}$ so that

$$d - r = e - \ell \geq \beta_1 \geq \cdots \geq \beta_\ell \geq 0,$$

and we write β for the sequence of β_i. Putting this together, we have:

Proposition 13.15. *The condition for a subspace $W \subset H^0(\mathcal{O}_{\mathbb{P}^1}(d))$ of dimension ℓ to define a g_d^r with ramification sequence at least $\alpha = (\alpha_0, \ldots, \alpha_r)$ at p is*

$$\dim W \cap V_{e - \ell + i - \beta_i}(p) \geq i \text{ for } 0 \leq i \leq \ell.$$

where $e := d + 1$ and $\beta_i = \ell - \alpha_i$.

Schubert cycles.

Definition 13.16. A *complete flag* $\mathcal{V}$ in an e-dimensional vector space V is a nested sequence of vector spaces

$$0 \subset V_1 \subset V_2 \subset \cdots \subset V_e = V.$$

with $\dim V_i = i$.

Given a complete flag in V and any ℓ-dimensional subspace $W \subset V$, we can derive the nested sequence of e subspaces of W:

$$0 \subset W \cap V_1 \subset W \cap V_2 \subset \cdots \subset W \cap V_e = W.$$

Each term in this sequence is either equal to the previous one, or of dimension 1 greater; the former occurs $e - \ell$ times, and the latter ℓ times. For a general

ℓ-plane W, the jumps occur at the end; that is, we have $W \cap V_{e-\ell} = 0$, and thereafter the dimension of the intersection goes up by 1 each time. This makes it natural to describe the special position of a given ℓ-plane by how early the i-th jump occurs:

Definition 13.17. A *Schubert index* for $G(\ell, e)$ is a sequence $\beta = (\beta_1, \ldots, \beta_\ell)$ of integers with $e - \ell \geq \beta_1 \geq \beta_2 \geq \cdots \geq \beta_\ell \geq 0$. The *Schubert cycle* $\Sigma_\beta(\mathcal{V}) \subset G$ associated to a complete flat $\mathcal{V}$ in $\mathbb{C}^e$ and Schubert index β is

$$\Sigma_\beta(\mathcal{V}) := \left\{ W \in G \mid \dim(W \cap V_{e-\ell+i-\beta_i}) \geq i \ \forall i \right\}.$$

Figure 13.2 illustrates some of the possibilities for lines in $\mathbb{P}^3$.

In particular, if $\mathcal{V}(p)$ is the vanishing flag for forms of degree d at a point $p \in \mathbb{P}^1$, then a g_d^r has ramification sequence (termwise) $\geq \alpha$ at p if and only if W belongs to the Schubert variety $\Sigma_\beta(\mathcal{V}(p))$, where $\beta_i = \alpha_{\ell-i}$ as above.

For example, the Schubert cycle $\Sigma_{0,\ldots,0}$ is the whole Grassmannian, and $\Sigma_{1,0,\ldots,0}(\mathcal{V})$ is the set of ℓ-planes that meet $V_{e-\ell}$ nontrivially, which is a hyperplane section of the Grassmannian in its Plücker embedding. More generally, the *special Schubert cycle* $\Sigma_\gamma(\mathcal{V}) := \Sigma_{\gamma,0,\ldots,0}(\mathcal{V})$ is the set of ℓ-planes meeting $V_{\gamma-\ell-\beta+1}$ nontrivially. Since this condition really involves only the single space $U = V_{e-\ell-\gamma+1}$, we sometimes write it as $\Sigma_\gamma(U)$.

For any Schubert index β, the codimension of $\Sigma_\beta(\mathcal{V})$ in $G(\ell, e)$ is $|\beta| := \sum \beta_i$ (Exercise 13.5).

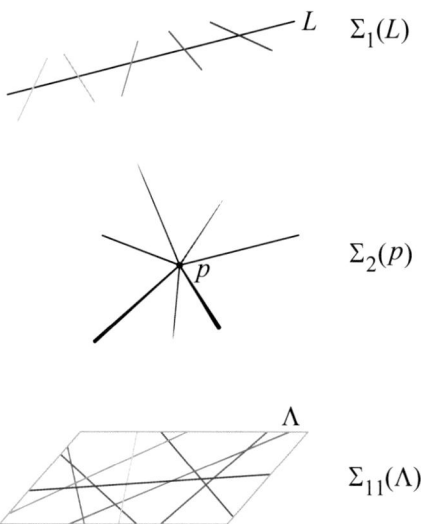

Figure 13.2. Schubert cycles in $G(2, 4)$: $\Sigma_1(L)$ consists of lines meeting $L \subset \mathbb{P}^3$ (top); $\Sigma_2(p)$ consists of lines passing through $p \in \mathbb{P}^3$ (middle); $\Sigma_{1,1}(\Lambda)$ consists of lines in the plane $\Lambda \subset \mathbb{P}^3$ (bottom).

Special Schubert cycles and Pieri's formula.

Cheerful Fact 13.18. The variety of complete flags in $\mathbb{C}^e$ is rational, and it follows that the class of $\Sigma_\beta(\mathcal{V}) \subset G$ in the Chow ring $A^*(G(\ell, e))$ of the Grassmannian is independent of the flag $\mathcal{V}$. It is typically denoted σ_β. Moreover, the classes σ_β form a basis for the Chow ring as a free abelian group.

Thus the product $\sigma_\beta \cdot \sigma_{\beta'}$ is a linear combination of Schubert classes, given combinatorially by the *Littlewood–Richardson rule* — see for example [Vakil 2006]. *Pieri's formula* is the special case of the Littlewood–Richardson rule that expresses the product in the Chow ring $A^*(G(\ell, e))$ of a special Schubert class with an arbitrary Schubert class. In Chapter 14 we will use it to describe the possible ramification behavior of rational curves with cusps.

Proposition 13.19 (Pieri's formula). *If σ_γ is a special Schubert class and σ_β is an arbitrary Schubert class, then*

$$\sigma_\gamma \cdot \sigma_\beta = \sum \sigma_\delta$$

where the sum ranges over all Schubert indices $\delta = (\delta_1, \ldots \delta_\ell)$ with

$$\sum \delta_i = \gamma + \sum \beta_i \quad and \quad \beta_i \le \delta_i \le \beta_{i-1} \quad for\ all\ i.$$

For a proof, see for example [Eisenbud and Harris 2016, Section 4.2.4].

Figure 13.3 illustrates the degeneration of $\Sigma_1 \cdot \Sigma_1$ to $\Sigma_2 \cup \Sigma_{1,1}$.

To understand Proposition 13.19, represent the Schubert class σ_β by stacks of coins, with β_1 coins in the first stack, β_2 coins in the second stack, and so on. We now want to add a total of γ coins to the stacks; we can add any number of them to any stack (including a stack that was previously empty), with the one condition that the new height of each stack can't be larger than the previous

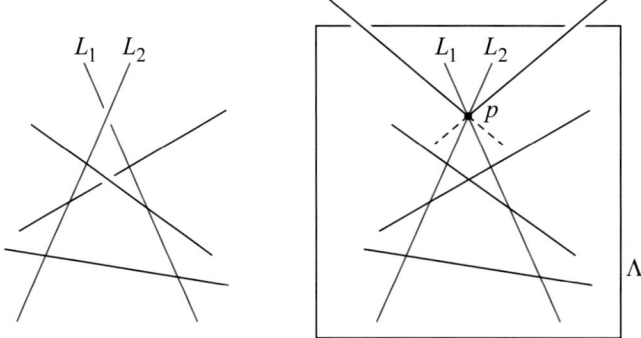

Figure 13.3. When the lines L_1 and L_2 move to meet each other at P they become coplanar in Λ, and the intersection $\Sigma_1(L_1) \cap \Sigma(L_2)$ degenerates to the union $\Sigma_2(P) \cup \Sigma_{1,1}(\Lambda)$.

height of the stack to its left. This interpretation makes the following corollary clear:

Corollary 13.20. *If σ_γ is any special Schubert class, σ_β is any Schubert class, and $m \geq 0$ is an integer with $m\gamma + \sum \beta_i \leq \dim G(\ell,e) = \ell(e-\ell)$, then*

$$(\sigma_\gamma)^l \cdot \sigma_\beta \neq 0 \in A^*(G(\ell,e),\mathbb{Z}).$$

The next result is expressed in terms of the *osculating flag* $\mathcal{V} = 0 \subset V_1 \subset \cdots \subset V_d = \mathbb{P}^d$ to the rational normal curve $C \subset \mathbb{P}^d$ at a point $p \in C$, which we will now define.

Let $\widetilde{\mathcal{V}} = (\widetilde{V}_1, \ldots, \widetilde{V}_{d+1})$ be the ramification flag defined previously (leaving out $\widetilde{V}_0$, which corresponds to the empty projective space), and let V_i be the projectivization of V_{i+1}. That is, V_i is the projective subspace of dimension i that meets C at p with maximal order of contact.

Proposition 13.21. *Let $C \subset \mathbb{P}^d$ be a rational normal curve, and $(V_W, \mathcal{O}_C(1))$ be the g_d^r on $\mathbb{P}^1$ cut by hyperplanes in $\mathbb{P}^d$ containing a plane $W \subset \mathbb{P}^d$ of dimension $d - r - 1$, and write $\mathcal{V} = 0 \subset V_1 \subset \cdots \subset V_d = \mathbb{P}^d$ for the osculating flag of C at p. The ramification sequence $\alpha(V_W, p)$ is determined by the formula*

$$W \in \Sigma_a(\mathcal{V}) \iff \alpha_i(V_W, p) \geq a^*_{r+1-i} = \#\{j \mid a_j \geq r + 1 - i\}.$$

In other words, the ramification sequence $\alpha(V_W, p)$ of the linear series V_W at p is exactly the reverse of the transpose of the Schubert index of the smallest Schubert cycle containing $\widetilde{W}$, the vector space associated to W, in $G(d-r, d+1)$.

Proof. The condition $\widetilde{W} \in \Sigma_a(\widetilde{\mathcal{V}})$ means that

$$\dim \widetilde{W} \cap \widetilde{V}_{r+1+i-a_i} \geq i$$

for each i. It follows that the codimension of the space of hyperplanes containing $\widetilde{W}$ has codimension $\leq r - a_i + 1$. since the projective space corresponding to $\widetilde{V}_{r+i-a_i}$ is the linear span of the divisor $(r + i - a_i + 1)p \subset C$, a codimension $\leq r - a_i + 1$ space of sections of $\widetilde{V}_{\widetilde{W}}$ vanishes to order $\geq r - a_i + 1 + i$ at p.

If we write the distinct orders of vanishing of the sections in $\widetilde{V}_{\widetilde{W}}$ as $b_0 < b_1 < \cdots < b_r$, we see that $\widetilde{W} \in \Sigma_a$ if and only if a_i of the b_j are $\geq r + i + a_i + 1$, that is, $b_{r-a_i+1} \geq r - a_i + 1 + i$ or equivalently $\alpha_{r-a_i+1} \geq i$ for each i. Since $\alpha_i \leq \alpha_{i+1} \leq \alpha_r$, this is equivalent to saying that the number of $\{j \mid \alpha_j \geq i\}$ is at least a_i, or, for the reverse sequence $\alpha' = \alpha_r \geq \alpha_{r-1} \geq \cdots \geq 0$, that the a_i-th term is $\geq i$. On the other hand $a'_{a_i} = \#\{j \mid a_j \geq a_i\} = i$ since the sequence a is weakly decreasing, so α' is termwise $\geq a'$. $\square$

Conclusion. Using these ideas we can rephrase and prove Theorem 13.14 in terms of Schubert cycles, adding the precise condition for the existence of a g_d^r with prescribed ramification sequences at an arbitrary collection of distinct points in $\mathbb{P}^1$:

Theorem 13.22. *Let $p_1, \ldots, p_\delta \in C$ be distinct points on a rational normal curve $C \subset \mathbb{P}^d$, and $V^1, \ldots, V^\delta$ the corresponding vanishing flags. If $\beta^1, \ldots, \beta^\delta$ are δ Schubert indices for $G(d-r, d+1)$, the Schubert cycles $\Sigma_{\beta^1}(V^1), \ldots, \Sigma_{\beta^\delta}(V^\delta) \subset G(d-r, d+1)$ intersect properly; that is, the intersection is either empty or has codimension exactly $\sum_{j=1}^{\delta} |\beta^j|$, the sum of the codimensions of the cycles Σ_{β^j}. Moreover, the intersection is nonempty if and only if the intersection product of the classes $[\Sigma_{\beta^j}]$ is nonzero in $A^*(G(d-r, d+1))$.*

Proof. If the intersection is empty then the Chow class is 0, so it suffices to show that the intersection is proper, and we may assume that it is nonempty. Because the Grassmannian is smooth, the codimension of the intersection of any subvarieties is at most the sum of their codimensions, so it is enough to show that the codimension of the intersection cannot be too small.

The Schubert cycle Σ_1 is a hyperplane section of $G(d-r, r+1)$, so that if $\Phi \subset G$ is any subvariety of dimension m, its intersection with m Schubert cycles Σ_1 is nonempty. Thus, if the intersection

$$X := \bigcap_{i=1}^{\delta} \Sigma_{\beta^i}(V^i)$$

had dimension strictly bigger than the expected

$$\rho := (r+1)(d-r) - \sum_{i=1}^{\delta} |\beta^i|,$$

we could choose $\rho + 1$ additional points $q_1, \ldots, q_{\rho+1}$ on C, with vanishing flags $V^1, \ldots, V^{\rho+1}$ and the intersection of X with the Schubert cycles $\Sigma_1(V^i)$ would still be nonempty.

It thus suffices to show that the intersection is empty if

$$\sum_{i=1}^{\delta} |\beta^i| > (r+1)(d-r) = \dim G(d-r, d+1).$$

If on the contrary

$$W \in \bigcap_{i=1}^{\delta} \Sigma_{\beta^i}(V^i),$$

then, by Proposition 13.21, the linear series $(W, \mathcal{O}_{\mathbb{P}^1}(d))$ would have ramification of weight $|\beta_i|$ at p_i, and the sum of the weights would be strictly greater than $(r+1)(d-r)$. This contradicts the Plücker formula, Theorem 13.2. $\quad\square$

Using Proposition 13.21 this result becomes a characterization of the sets of possible ramification sequences for linear series at specified points on $\mathbb{P}^1$.

Explicitly, suppose we are given a collection of distinct points $p_1, \ldots, p_\delta \in \mathbb{P}^1$, and for each point p_i a ramification sequence

$$\alpha^i = (\alpha_0^i, \alpha_1^i, \ldots, \alpha_r^i) \quad \text{with} \quad 0 \leq \alpha_0^i \leq \alpha_1^i \leq \cdots \leq \alpha_r^i \leq d - r.$$

Let β^i be this sequence in reverse (and relabeled); that is

$$\beta^i = (\alpha_r^i, \alpha_{r-1}^i, \ldots, \alpha_1^i, \alpha_0^i)$$

and finally let b^i be the transpose of the Schubert index β^i.

Corollary 13.23. *With α^i and b^i related as above, there exists a g_d^r on $\mathbb{P}^1$ with ramification sequence at p_i equal to $(\alpha_0^i, \alpha_1^i, \ldots, \alpha_r^i)$ if and only if*

$$\prod_{i=0}^{\delta} \sigma_{b^i} \neq 0 \quad in \quad A^*(G(d - r, r + 1)).$$

Proof. If the product is nonzero, Proposition 13.21 and Theorem 13.22 immediately show the existence of a g_d^r on $\mathbb{P}^1$ with ramification sequence greater than or equal to α^i at p_i. But by Theorem 13.22, the ones with ramification strictly greater than the α^i form a family of strictly smaller dimension. Thus a general g_d^r with ramification sequence greater than or equal to α^i at p_i has ramification sequence exactly equal to α^i at p_i. □

We can deduce a result about general secant loci strengthening one that was originally proven in [Griffiths and Harris 1980]: Given a curve $C \subset \mathbb{P}^d$, we say that a *secant flag* in $\mathbb{P}^d$ is a flag

$$0 \subset V_1 \subset V_2 \subset \cdots \subset V_{d-1} \subset V_d = \mathbb{P}^d$$

where each V_i is spanned by its (scheme-theoretic) intersection with C. In other words, there is a sequence of points $p_1, p_2, \ldots, p_{d+1} \in C$ such that

$$V_i = \overline{p_1 + p_2 + \cdots + p_{i+1}}$$

(An osculating flag is just the special case where all p_i are equal.) Since any secant flag can specialize to an osculating flag, we can deduce:

Corollary 13.24. *Schubert cycles defined relative to general secant flags to a rational normal curve intersect properly.*

Unlike the case for osculating flags, the hypothesis of generality is necessary for secant flags; see Exercise 13.9.

13.5. Exercises

Exercise 13.1. Let k be a field of characteristic p. Show that every point of the affine curve $y = x^{p+1} + 1$ over k is a flex point.

Exercise 13.2. Show that if C is a smooth projective curve of genus $g \geq 2$ and $f : C \to C$ an automorphism fixing $2g + 2$ points, then C must be hyperelliptic and f the hyperelliptic involution.   ◆

Exercise 13.3. Suppose that $C \subset D \times E$ is a curve contained in the product of smooth curves D, E of genera g and h respectively. Prove that if the projection maps $C \to D$ and $C \to E$ have degrees d and e, then $p_a(C) \leq (d-1)(e-1) + dg + eh$.   ◆

Exercise 13.4. Show that if C is a smooth projective curve of genus $g \geq 2$ then $|\text{Aut}\, C| \leq 84(g-1)$.   ◆

Exercise 13.5. Show that the codimension of $\Sigma_\beta(V)$ in the Grassmannian G is equal to $\sum \beta_i$.   ◆

Exercise 13.6. Let β^* be the transpose Schubert index to β.

(1) Show that $|\beta^*| = |\beta|$.
(2) Show that $(\beta^*)^* = \beta$.
(3) Show that the isomorphism $\phi : G(k+1, V) \xrightarrow{\cong} G(n-k, V^*)$ carries the Schubert cycle $\Sigma_\beta(V)$ to the Schubert cycle $\Sigma_{\beta^*}(V^*)$.

(Of course, the first two parts follow from the third; think of the first two as warmups.)

Exercise 13.7 (ramification and osculating planes). Let $p \in C \subset \mathbb{P}^d$ be a point on a rational normal curve of degree d, and let $\Lambda \subset \mathbb{P}^d$ be a $d-r-1$ plane, and let $\mathcal{U} := (\mathcal{O}_{\mathbb{P}^1}(d), V)$ be the g_d^r, possibly with basepoints, cut out by hyperplanes containing Λ. Consider the complete flag

$$U(p) := \{U_1, \ldots, U_d\}$$

of *osculating spaces* at p, where U_t is the linear span of the divisor tp considered as a subscheme of length t in C. Show that the ramification sequence of $\mathcal{U}$ at p is determined by the formula

$$W \in \Sigma_a(U(p)) \iff \alpha_i(\mathcal{U}, p) \geq a^*_{r+1-i} = \#\{j \mid a_j \geq r+1-i\}.$$

In other words, the ramification sequence $\alpha(\mathcal{U}, p)$ of the linear series $\mathcal{U}$ at p is the reverse of the transpose of the Schubert index of the smallest Schubert cycle containing L, the vector space associated to Λ, in $G(d-r, d+1)$. Conclude from Theorem 13.22 that any collection of Schubert cycles associated to osculating flags of C at distinct points have intersection that is dimensionally transverse or empty.

Exercise 13.8. Let C be the rational normal curve in $\mathbb{P}^r$. And let $(p_i, q_i) \in C^2$ be t pairs of points, all distinct. For each i, let S_i be the Schubert variety of k-planes in $\mathbb{P}^r$ that meet the secant line $\overline{p_i q_i}$. Show that if the p_i, q_i are sufficiently general, then the intersection of the S_i is dimensionally transverse.   ◆

Exercise 13.9. Show by example that Schubert cycles defined relative to arbitrary secant flags to a rational normal curve may fail to intersect properly. ◆

Exercise 13.10. We can define the osculating flag to any nondegenerate curve $C \subset \mathbb{P}^d$ at any smooth point p. If p is an inflection point then $\overline{mp}$ may not be an $(m-1)$-plane, but we can look at the nested sequence of subspaces

$$\{p\} \subset \overline{2p} \subset \overline{3p} \subset \cdots$$

and pick out exactly the terms whose dimension is strictly greater than the preceding term; this gives a complete flag

$$\{p\} \subset \overline{b_2 p} \subset \overline{b_3 p} \subset \cdots \subset \overline{b_{d+1} p} = \mathbb{P}^d.$$

Show that the sum $\sum b_i - i$ is equal to the weight of the point p as an inflection point of the hyperplane series.

Exercise 13.11. The Plücker formula shows that an elliptic normal curve $E \subset \mathbb{P}^{n-1}$ has n^2 inflection points. Show that they are all simple, and that if you choose one as the origin for the group law on E, then they are the n-torsion points of $E \cong \mathrm{Jac}(E)$. ◆

Exercise 13.12 (Buchweitz). It was long believed that every semigroup $H \subset \mathbb{N}$ of finite index $g = \#(\mathbb{N} \setminus H)$ could occur as the Weierstrass semigroup of a Weierstrass point on a curve of genus g, but this is not the case. The following counterexample for a curve of genus 16 was found by Buchweitz (unpublished): Show that on a curve of genus 16 there can be no Weierstrass point p with ramification sequence $(0^{12}, 6, 7, 9, 9)$ by showing that there would be too many vanishing orders at p of quadratic differentials.

Chapter 14

Proof of the Brill–Noether theorem

In this chapter we give a proof of the Brill–Noether theorem based on the analysis of inflections of linear series on $\mathbb{P}^1$ of Theorem 13.22. We will focus on proving what we call the "basic" Brill–Noether theorem (Theorem 12.2), which we reproduce for convenience:

Theorem 14.1 (basic Brill–Noether). *If*

$$\rho(g, r, d) := g - (r + 1)(g - d + r) \geq 0,$$

then every smooth projective curve of genus g possesses a g_d^r; and for a general curve C, $\dim W_d^r(C) = \rho$. Conversely, if $\rho < 0$ then a general curve C of genus g does not possess a g_d^r.

In fact, a closer examination of our proof will yield some of the assertions of the stronger Theorems 12.4 and 12.9, as well as additional results on the existence of linear series with specified inflectionary behavior; we will discuss these at the end of the chapter.

14.1. Castelnuovo's approach

The original argument of Brill and Noether reduced the proof to the assertion that a matrix depending on an invertible sheaf on a general curve behaves in a certain sense like a matrix of indeterminates. This was finally proven in [Griffiths and Harris 1980] using an idea that goes back to Castelnuovo [1889].

Interestingly, Castelnuovo's goal was not to prove the Brill–Noether theorem, which was considered established at the time (or at least not in need of further demonstration). Given that a general curve of genus $2d - 2$ has finitely many g_d^1s, Castelnuovo asked: how many? We have answered this question in the cases $g = 2, 4$ and 6 in earlier chapters, and these cases were certainly known to Castelnuovo, but the cases of higher g were unknown. His idea was to specialize to a general curve C_0 with g nodes whose normalization is $\mathbb{P}^1$, and count the number of g_d^1s on that curve. Appealing to the (then) vague "principle of conservation of number", Castelnuovo felt that the result probably reflected the number an a general smooth curve as well.[1]

By way of notation, we'll write $r_1, \ldots, r_g$ for the nodes of C_0, and let $p_i, q_i \in \mathbb{P}^1$ be the two points lying over r_i. Castelnuovo counted the g_d^1s on C_0 by observing that any pencil on C_0 can be pulled back to a pencil on the normalization $\mathbb{P}^1$; if we embed $\mathbb{P}^1$ in $\mathbb{P}^d$ as a rational normal curve of degree d, such a pencil is cut out by the hyperplanes containing a $(d-2)$-plane $\Lambda \subset \mathbb{P}^d$. To say that such a pencil is a pullback from C_0 means that every divisor of the pencil that contains p_i contains q_i and vice versa. This is equivalent to the condition that $\Lambda \cap \overline{p_i q_i} \neq \emptyset$.

In the Grassmannian $G(d-1, d+1)$ of $(d-2)$-planes in $\mathbb{P}^d$, the locus of those that meet the line $L_i = \overline{p_i q_i}$ is what we called in the last chapter the Schubert cycle $\Sigma_1(L_i)$. In these terms the set of g_d^1s on C_0 is the intersection

$$W_d^1(C_0) = \bigcap_{i=1}^{2d-2} \Sigma_1(L_i).$$

Castelnuovo proposed that if the points $p_i, q_i \in \mathbb{P}^1$ were chosen generally, then the Schubert cycles L_i would meet transversely, and thus that the cardinality of this intersection is the degree of the power σ_1^{2d-2} in the Chow ring $A(G(d-1, d+1))$. Castelnuovo evaluated this power, and came to the conclusion that a general curve C of genus $g = 2d-2$ has

$$\#W_{d+1}^1(C) = \frac{(2d-2)!}{(d-1)! \, d!}$$

pencils of degree d. Indeed, we see in Exercise 14.3 that if the points p_i, q_i are general, then the Schubert cycles $\Sigma_1(L_i)$ at least intersect properly, so the given number is the number of g_d^1s counted with appropriate multiplicities.

Kleiman and Laksov [1972; 1974] and Kempf [1971] used a different idea, which we'll describe below, to prove the "existence" part of Theorem 14.1 — that is, that $W_d^r(C)$ is nonempty when $\rho \geq 0$. In [Kleiman 1976], the general "nonexistence" part is reduced to the proof of Exercise 14.3 and completed for $r = 1$. The general statement was finally proven in [Griffiths and Harris 1980].

---

[1] Castelnuovo presented his computation as heuristic, not claiming it was a full proof, and the reviewer of his paper in the *Zentralblatt für Mathematik* wrote very politely: "Das Resultat, welches er bekommen hat, gibt mit grosser Wahrscheinlichkeit den wahren Wert von N ...; daher sind wir mit dem Verf. einverstanden, wenn er seinen Versuch nicht für wertlos hält" (The result that he obtained gives the true value of [the number of g_d^1s] with high probability, and thus we agree with the author that his work is not worthless...)

We will also adopt this general approach, and we will prove the existence and nonexistence parts together for all g, d, r. However, the strength of Theorem 13.22 compared to Exercise 14.3 suggests specializing to a g-cuspidal curve C_0 rather than a g-nodal one, and we will use this refinement.

Upper bound on the codimension of $W_d^r(C)$. Let C be a reduced irreducible projective curve of arithmetic genus g. In Section 12.2.1 we showed how we might arrive at the "expected" dimension of the locus $W_d^r(C)$ by estimating the dimension of the subvariety $C_d^r \subset C_d$ of divisors moving in an r-dimensional linear series; here we'll give a similar argument using the Picard variety $\mathrm{Pic}(C)$. Again the heart of the proof is the estimation of the codimension of an ideal of minors of a certain matrix of functions. Since the proof uses degeneration to a cuspidal curve, we will work with flat families of curves that may include singular fibers. For simplicity we will take the base of the family to be a complex disk Δ centered at the origin in $\mathbb{C}$; the more algebraically minded reader can substitute Spec of a discrete valuation ring.

Theorem 14.2. *Let $\mathcal{C}/\Delta$ be a family of reduced irreducible projective curves of arithmetic genus g. If the fiber C_0 of the family has $\dim W_d^r(C_0) = \rho(g, r, d) \geq 0$ locally at a particular invertible sheaf $\mathcal{L}_0$, then $\dim W_d^r(C_b) = \rho$ locally for all b in a neighborhood of $0 \in \Delta$ and invertible sheaves in a neighborhood of $\mathcal{L}_0$ in the relative Picard variety $\mathrm{Pic}_d(\mathcal{C}/\Delta)$. In particular, $W_d^r(C_b)$ is nonempty for all b in a neighborhood of $0 \in \Delta$.*

The idea of the following proof is a modern form of the discussion in the original paper of Brill and Noether.

Proof. Possibly after pulling back the family $\mathcal{C}/\Delta$ along a ramified covering $\Delta \to \Delta$ and restricting to a smaller disk around 0, we may choose m sections $p_i(b)$ with distinct values in the smooth locus of each fiber, and m as large as we like. We define a family of divisors $D_b = \sum_i p_i(b)$. By the semicontinuity of fiber dimension, we may choose m so large that $h^1(\mathcal{O}_{C_b}(D_b)) = 0$ and that for every invertible sheaf $\mathcal{L}_b$ of degree d on C_b we have $h^1(\mathcal{L}_b(D_b)) = 0$. By Theorem 2.3 we have

$$h^0(\mathcal{L}_b(D_b)) = d + m - g + 1.$$

The pushforward of the Poincaré sheaf $\mathcal{P}$ on $\mathrm{Pic}_{d+m}(\mathcal{C}/\Delta) \times \mathcal{C}$ to $\mathrm{Pic}_{d+m}(\mathcal{C}/\Delta)$ is thus a vector bundle $\mathcal{E}$ of rank $d + m - g + 1$ on Δ whose fiber over a point representing $\mathcal{L}_b$ is the vector space $H^0(\mathcal{L}(D_b))$. Similarly, the pushforward $\pi_{1_*}(\mathcal{P}|_{D_b})$ is a trivial vector bundle $\mathcal{F}$ whose fiber at every point is the m-dimensional vector space $\bigoplus_{i=1}^m \mathcal{L}(E)|_{p_i}$. Regarding D as a family of finite schemes inside $\mathcal{C}$, the restriction map

$$\mathcal{P} \to \mathcal{P}_D$$

pushes forward to give a map of vector bundles $\phi : \mathcal{E} \to \mathcal{F}$ on $\mathcal{C}$ which, on a fiber over b, is the evaluation of sections $\sigma \in H^0(\mathcal{L}_b(D_b))$ at the points p_i.

If $\mathcal{L}_b$ is an invertible sheaf of degree d on C_b then the space $H^0(\mathcal{L}_b)$ is the kernel of the map ϕ at the point corresponding to $\mathcal{L}_b(D_b) \in \mathrm{Pic}_{d+m}(\mathcal{C}/\Delta)$. Locally on $\mathcal{C}$ the map ϕ can be defined by a $(d+m) \times (d+m-g+1)$ matrix of regular functions. The locus $W_d^r(C_b)$ is thus the translate by $\otimes \mathcal{O}_\mathcal{C}(D)$ of the locus where ϕ has rank $d+m-g-r$ or less, and by [Eisenbud 1995, Exercise 10.9] this locus is either empty or its components have codimension $\leq (r+1)(g-d+r)$ in $\mathrm{Pic}_{d+m}(C)$, which has dimension g. Consequently, every component of $W_d^r(C)$ has dimension at least ρ. $\square$

It was in fact this set-up that was used by Kleiman-Laksov and Kempf to prove the existence half of Brill–Noether: they determined the characteristic classes of the bundles involved and applied the *Thom–Porteous formula*, a formula for the class of the degeneracy locus of a map of vector bundles. An account of this proof is given in Appendix D of [Eisenbud and Harris 2016].

We will see that if C_0 is any rational curve with g cusps, and $\rho(g,r,d) \geq 0$, then $\dim W_d^r(C_0) = \rho$ locally at every invertible sheaf on C_0.

14.2. Specializing to a g-cuspidal curve

Our first goal is to find a family $\{C_t\}$ of curves of arithmetic genus g, with C_t smooth for $t \neq 0$ and C_0 a rational curve with g cusps. To do this we show how to construct a rational curve C_0 with g cusps and deform it to a smooth curve.

Constructing curves with cusps.

Proposition 14.3. *Let C be any curve and $p \in C$ a smooth point. There exists a curve C_0 and a bijective morphism $f : C \to C_0$ such that f maps $C \setminus \{p\}$ isomorphically to $C_0 \setminus \{r\}$ and the image $r = f(p) \in C_0$ is a cusp of C_0.*

We will say that we *crimp* C at p to obtain C_0. For the corresponding result with nodes instead of cusps, see Exercise 14.2.

Proof. We can construct C_0 explicitly as a topological space homeomorphic to C, with structure sheaf $\mathcal{O}_{C_0}$ that is the subsheaf of $\mathcal{O}_C$ consisting of functions on C whose derivative at p is 0. $\square$

If we start with $\mathbb{P}^1$, pick any g points $p_1, \ldots, p_g \in \mathbb{P}^1$ and crimp at each p_i, we arrive at a g-cuspidal curve C_0.

Smoothing a cuspidal curve.
Given a curve C_0 with a finite number of cusps and no other singularities, we can find a proper flat family $\mathcal{C} \to \Delta$ with special fiber C_0 and all other fibers smooth; that is, we can smooth C_0.

To begin with, we can do this locally in the complex analytic setting: if $p \in C_0$ is a cusp, we can find an analytic neighborhood of p in which C_0 is given by the equation $y^2 = x^3$; we can smooth this by taking the family

$$y^2 = x(x - t)(x - 2t)$$

for $t \in \Delta$. (See Figure 8.2.)

The next step is to argue that we can glue together these local smoothings to obtain a proper family $\mathcal{C} \to \Delta$, and this is where we need to invoke a result from deformation theory of projective schemes that are locally complete intersections:

Lemma 14.4 [Fantechi and Göttsche 2005, Proposition 6.5.2]. *Let* $p_1, \ldots, p_g \in \mathbb{P}^1$ *be distinct points, and let* C_0 *be the curve obtained by crimping* $\mathbb{P}^1$ *at each* p_i. *There exists a family of curves* $\pi : \mathcal{C} \to \Delta$, *where*

(1) Δ *is a disk centered at the origin in* $\mathbb{C}$.

(2) *for all* $b \neq 0 \in \Delta$, *the fiber* $C_b = \pi^{-1}(b)$ *is a smooth, projective curve of genus* g; *and*

(3) *the fiber over 0 is the curve* C_0.

Sketch of proof. The cuspidal curve C_0 can be embedded in projective space $\mathbb{P}^n$ and there it is locally a complete intersection. Thus, as with the case of one cusp, above, the local obstructions to smoothing vanish. The global obstruction is the first cohomology of a sheaf supported at the cusps, and therefore vanishes as well. □

14.3. The family of Picard varieties

By Lemma 14.4 there is a flat family $\mathcal{C} \to \Delta$ of curves, specializing from a smooth curve of genus g to a g-cuspidal curve. The next step is to relate linear series on the general fiber of our family to their limits on C_0.

The Picard variety of a cuspidal curve. We describe the invertible sheaves on a cuspidal curve C_0 in terms of the invertible sheaves on its normalization:

Proposition 14.5. *Let* $p \in C_0$ *be a cusp that is the result of crimping a reduced irreducible projective curve* C *at a smooth point* $q \in C$, *and let* $\nu : C \to C_0$ *be the natural morphism. There is an exact sequence of groups*

$$0 \to (\mathbb{C}, +) \to \mathrm{Pic}_0(C_0) \xrightarrow{\nu^*} \mathrm{Pic}_0(C) \to 0.$$

Proof. The preimage $\nu^{-1}(p)$ is the nonreduced scheme $2q \subset C$. If $\mathcal{L}$ is an invertible sheaf on C_0 then $\nu^*(\mathcal{L})$ is invertible on C, and thus locally trivial near q and, in particular, trivial when restricted to the subscheme $2q$. Conversely, if $\mathcal{L}'$ is an invertible sheaf on C and we choose a trivialization of $a : \mathcal{L}'|_U \cong$

$\mathcal{O}_U$ on a neighborhood $U \subset C$ of q, then there is an invertible sheaf $\mathcal{L}$ on C_0 together with a trivialization $a : \mathcal{L}|_{\nu(U)} \cong \mathcal{O}_{\nu(U)}$ on $\nu(U)$, unique up to an automorphism of $\mathcal{L}'$ (that is, up to multiplication by a scalar), that pulls back to $\mathcal{L}'$ with the given trivialization a. Thus the data of an invertible sheaf $\mathcal{L}$ on C_0 is equivalent to the data of an invertible sheaf $\mathcal{L}'$ on C, together with a trivialization of $\tilde{\mathcal{L}}$ on the preimage $\nu^{-1}(p) = 2q$ up to multiplication by a nonzero scalar.

Of course every invertible sheaf on the 0-dimensional scheme $2q$ is trivial, and a change of trivialization thus corresponds to an automorphism of the structure sheaf $\mathcal{O}_{2q} \cong \mathbb{C}[\epsilon]/(\epsilon^2)$ — not as an algebra, but as a rank 1 free module over $\mathbb{C}[\epsilon]/(\epsilon^2)$. Such a map is determined by its effect on the generator 1, and can take this element to any other generator $a + b\epsilon$ with $a \neq 0$ and $b \in \mathbb{C}$ arbitrary. After multiplying $\mathcal{L}'$ by a^{-1} we may suppose that $a = 1$ and we may take the automorphism to induce the identity on $\mathbb{C}$. In this case, the family of trivializations, modulo multiplication by a nonzero scalar, is determined by the element $b \in \mathbb{C}$. Composing the isomorphisms $1 \mapsto 1 + b\epsilon$ and $1 \mapsto 1 + b'\epsilon$ gives $1 \mapsto 1 + (b + b')\epsilon$, so the kernel of the map $\nu^* : \mathrm{Pic}_d(C_0) \to \mathrm{Pic}_d(C)$ is the additive group $(\mathbb{C}, +)$. □

Corollary 14.6. *If C_0 is a g-cuspidal rational curve, then $\mathrm{Pic}_d(C_0) \cong \mathbb{C}^g$. More precisely, $\mathrm{Pic}_d(C_0)$ is a principal homogeneous space for the additive group $\mathbb{C}^g$, in the sense that this group acts faithfully and transitively.*

Proof. We can construct one invertible sheaf on C_0 as the inverse of the ideal sheaf of a divisor of d smooth points, and any other will differ from this one by the choice of an element of $\mathrm{Pic}_0(C_0)$, corresponding to a choice of g trivializations, as in the proof of the proposition. □

Thus we see that $\mathrm{Pic}_d(C_0)$ has dimension g just as does the Picard variety of a smooth curve of genus g. An important difference is that $\mathrm{Pic}_d(C_0)$ is not compact.

The relative Picard variety. Returning to the family $\pi : \mathcal{C} \to \Delta$ mentioned at the start of Section 14.3, the Picard varieties $\mathrm{Pic}_d(C_t)$ form a family $\mathrm{Pic}_d(\mathcal{C}/\Delta)$, and the varieties $W_d^r(C_t)$ form a subfamily $\mathcal{W}_d^r(\mathcal{C}/\Delta)$. In the argument of Theorem 14.2 bounding the dimensions of the W_d^r we may replace the points p_i by sections of the family, and thus the codimension of $\mathcal{W}_d^r(\mathcal{C}/\Delta) \subset \mathrm{Pic}_d(\mathcal{C}/\Delta)$ is at most $(r + 1)(g - d + r)$ locally at each point. See Cheerful Fact 5.6.

We will soon show that $W_d^r(C_0)$, the fiber of $\mathcal{W}_d^r(\mathcal{C}/\Delta)$ over 0, is nonempty of dimension $\rho(g, r, d)$ when $\rho(g, r, d) \geq 0$ and otherwise empty. If the family of Picard varieties over Δ were proper, so that limits of invertible sheaves were invertible as in Figure 14.1, this would prove Theorem 14.1; but it is not proper, and thus it is a priori possible that $W_d^r(C_0) = \emptyset$ but $W_d^r(C_t) \neq \emptyset$ for $t \neq 0$. This would mean that families of invertible sheaves $\mathcal{L}_t$ on C_t simply don't have limits in $\text{Pic}_d(C_0)$. Indeed, the limit of a family of invertible sheaves need not be an invertible sheaf. Nevertheless we can describe these limits quite precisely.

Limits of invertible sheaves. Suppose that $\pi : \mathcal{C} \to \Delta$ is a family of smooth genus g curves specializing to a rational curve C_0 with g cusps as in Lemma 14.4. Let

$$\pi^\circ : \mathcal{C}^\circ := \mathcal{C} \setminus C_0 \to \Delta^\circ := \Delta \setminus 0,$$

and suppose that there is a invertible sheaf $\mathcal{L}^\circ$ on $\mathcal{C}^\circ$ such that $h^0(\mathcal{L}^\circ|_{C_b}) \geq r+1$ for each $b \neq 0 \in \Delta$ so that $\mathcal{L}^\circ|_{C_b} \in W_d^r(C_b)$. We would like to describe the "limit" of $\mathcal{L}$ as $b \to 0$.

Lemma 14.7. *In the situation above, there exists a torsion-free sheaf $\mathcal{L}$ of rank 1 on $\mathcal{C}$, flat over Δ and locally isomorphic to an ideal sheaf of $\mathcal{C}$, such that $\mathcal{L}|_{\mathcal{C}^\circ} \cong \mathcal{L}^\circ$.*

In fact, any torsion-free sheaf of rank 1 on C_0 is locally isomorphic to an ideal sheaf; this follows from the fact that C_0 is generically Gorenstein. However, the argument we will give shows directly that the extension is isomorphic to an ideal sheaf tensored with an invertible sheaf, and the fact that it is locally isomorphic to an ideal sheaf follows at once.

Proof. Choose an auxiliary invertible sheaf $\mathcal{M}$ on $\mathcal{C}$ with relative degree $e > d + 2g$ and let $\mathcal{M}^\circ$ be the restriction of $\mathcal{M}$ to $\mathcal{C}^\circ$. Consider the invertible sheaf

$$\mathcal{N}^\circ = (\mathcal{L}^\circ)^* \otimes \mathcal{M}^\circ.$$

The bundle $\mathcal{N}^\circ$ has lots of sections: the direct image, as a sheaf on B, is locally free of rank $e - g + 1 > 0$, and after restricting to an open neighborhood of $0 \in B$ we can assume it's generated by them.

Choose a section σ of $\mathcal{N}^\circ$; let $D^\circ \subset \mathcal{C}^\circ$ be its divisor of zeros, and let $D \subset \mathcal{C}$ be the closure of D° in $\mathcal{C}$. Because it is the closure, the scheme D has no embedded 0-dimensional components, and thus $\mathcal{O}_{\mathcal{C}}/\mathcal{I}_{D/\mathcal{C}}$ has no $\mathcal{O}_\Delta$-torsion.

Since Δ is a smooth curve, this implies that $\mathcal{O}_{\mathcal{C}}/\mathcal{I}_{D/\mathcal{C}}$ is flat over Δ, and thus $\mathcal{I}_{D/\mathcal{C}}|_{C_0}$ is an ideal sheaf, the kernel of $\mathcal{O}_{\mathcal{C}} \to \mathcal{O}_{\mathcal{C}}/\mathcal{I}_{D/\mathcal{C}}$. Thus

$$\mathcal{L} := \mathcal{I}_{D/\mathcal{C}} \otimes \mathcal{M}$$

has the desired properties. □

Fortunately the ideal sheaves on cuspidal curves have a simple local structure. The reason lies in the relation of the local ring R_0 of a cusp to its integral closure.

Definition 14.8. The *conductor* of an integral domain R_0 is the annihilator of the R_0-module R/R_0, where R is the integral closure of R_0.

Figure 14.1. Cartier divisor on a smooth total space of a family of curves that degenerates to a cuspidal curve; in this case the limit of the associated family of invertible sheaves is invertible.

It follows at once from the definition that the conductor of R_0 is also an ideal of R, and that it is the largest ideal of R_0 that is also an ideal of R. The following result is a restatement of Example 2 after Proposition 2.26:

Proposition 14.9. *If R_0 is the local ring of an ordinary cusp singularity of a curve, then $R/R_0 \cong k$, the residue field of R, and thus the conductor of R_0 is the maximal ideal $\mathfrak{m}_0$ of R_0.* □

Proof. These properties can be verified after completing at the maximal ideal of R_0. To say that R_0 has an ordinary cusp singularity means that the completion of $R_0 \subset R$ is $k[\![x^2, x^3]\!] \subset k[\![x]\!]$, and the quotient is $kx \cong k$. □

Theorem 14.10. *Let p be an ordinary cusp of a curve C_0 with normalization $\pi : C \to C_0$. Let $R_0 = \mathcal{O}_{C_0,p}$ be the local ring of the cusp, with maximal ideal $\mathfrak{m}_0 = \mathfrak{J}_p R_0$, and let $R_0 \xrightarrow{\pi^*} R$ be its normalization. If $I_0 \neq 0$ is an ideal of R_0 then either $I_0 \cong R_0$ or $I_0 \cong \mathfrak{m}_0$.*

In the latter case, $I_0 \cong Ra$, and if R_0/I_0 has length v then $R/RI_0 = R/I_0$ has length $v + 1$. Moreover, there is a split exact sequence

$$0 \to R/\mathfrak{m}_0 R \to R \otimes I_0 \to RI_0 \to 0$$

with $RI_0 \cong R$.

Interpreting Theorem 14.10 in the context of an ideal sheaf on a cuspidal curve we get:

Corollary 14.11. *Let $p \in C_0$ be an ordinary cusp in a reduced irreducible curve and let $\pi : C \to C_0$ be the partial normalization of C_0 at p, so that $p_a(C) = p_a(C_0) - 1$. Let $q \in C$ be the point lying over $p \in C_0$.*

If $\mathcal{F}_0$ is locally isomorphic to a nonzero ideal sheaf on C_0 and $\mathcal{F}_0$ is not locally free at p, then there is a unique locally free sheaf $\mathcal{F}$ on C and a short exact sequence

$$0 \to \mathcal{O}_{\pi^{-1}(p)} \to \pi^*\mathcal{F}_0 \to \mathcal{F} \to 0.$$

Thus $\chi(\mathcal{F}) = \chi(\mathcal{F}_0) - 2$, so

$$\deg \mathcal{F} = \chi(\mathcal{F}) - \chi(\widetilde{\mathcal{O}}_C) = \deg \mathcal{F}_0 - 1.$$

Moreover, the map $H^0(\mathcal{F}_0) \xrightarrow{\pi^*} H^0(\mathcal{F}(-q))$ is a monomorphism. □

Proof of Theorem 14.10. The endomorphism ring of I_0 is commutative, it contains R_0, and (since it stabilizes the finitely generated module) it is integral over R_0. Thus

$$R_0 \subset \operatorname{End} I_0 \subset R.$$

Since $R/R_0 \cong k$, the ring $\operatorname{End} I_0$ is equal to either R_0 or R.

First, suppose $\operatorname{End} I_0 = R$, which is a discrete valuation ring. Every ideal of R is principal, so we may write $I_0 = Ra$. Since $\mathfrak{m}_0$ is also an ideal of R, it is isomorphic to R as an R-module, and since $R_0 \subset R$, $I \cong \mathfrak{m}_0$ as R_0-modules. After completing we have $\widehat{R}_0 \cong \mathbb{C}[[t^2, t^3, \ldots]]$ and it is evident that if a has valuation d in R then the length of $R_0/(aR)$ is $d - 1$, proving the claims in this case.

On the other hand, suppose $\operatorname{End} I = R_0$ and consider the inclusions

$$\mathfrak{m}_0 I \subset I \subset RI.$$

The left- and right-hand modules both have endomorphism ring R, so both containments must be strict. Since $R/\mathfrak{m}_0$ has length 2, we see that $I/\mathfrak{m}_0 I$ is principal. By Nakayama's lemma $I \cong R_0$. □

Returning once again to the family $\mathcal{C} \to \Delta$ at the beginning of Section 14.3, assume that for general t the scheme $W_d^r(C_t) \neq \emptyset$. After replacing Δ with a ramified covering we may assume that $\mathcal{W}_d^r(\mathcal{C}/\Delta) \to \Delta$ has a section, that is, an invertible sheaf $\mathcal{L}$ on $\mathcal{C}$ such that the restriction of $\mathcal{L}$ to each fiber C_t with $t \neq 0$ has at least $r + 1$ sections, so the g-cuspidal curve C_0 that is the limit of this family has a torsion-free sheaf $\mathcal{L}_0$ of degree d. By the semicontinuity of cohomology, $\mathcal{L}_0$ also has at least $r + 1$ sections.

At each cusp r_i of C_0, $\mathcal{L}_0$ is either locally free or isomorphic to the maximal ideal of the cuspidal curve. Thus, if we denote by $p_1, \ldots, p_k$ the cusps of C_0 at which $\mathcal{L}_0$ is isomorphic to the maximal ideal, and let C_0' be the partial normalization of C_0 at those cusps, then C_0' is a curve of arithmetic genus $g - k$ having $g - k$ cusps, and the pullback of $\mathcal{L}_0$ to C_0' has at least r sections with basepoints at the points lying over $p_1, \ldots, p_k$; this is a g_{d-k}^r on the $(g-k)$-cuspidal curve D_0.

14.4. Putting it all together

Nonexistence. We can now prove that if $\rho(g,r,d) < 0$ then $W_d^r(C) = \emptyset$ for curves in an open dense subset of M_g — that is, a general curve of genus g does not possess a g_d^r. Here we use Theorem 13.22 from the last chapter, which implies exactly this statement for invertible sheaves on an arbitrary g-cuspidal curve C_0. If it were the case that a general curve had a g_d^r with $\rho(g,r,d) < 0$, then by the results of Section 14.3, there would be an invertible sheaf of degree d with $\geq r+1$ sections on a $(g-k)$-cuspidal curve C_0' for some $k \geq 0$. But since

$$\rho(g-k,r,d-k) = \rho(g,r,d) - k < 0,$$

this is impossible by Theorem 13.22.

Existence. Once more we let $\mathcal{C} \to \Delta$ be a family of smooth curves specializing to a g-cuspidal curve C_0. We can combine Corollary 13.20 with Theorem 13.22 to show that the variety $W_d^r(C_0)$ is nonempty of dimension $\rho(g,r,d)$. Since the codimension of $\mathcal{W}_d^r(\mathcal{C}/\Delta) \subset \mathrm{Pic}_d(\mathcal{C}/\Delta)$ is at most $(r+1)(g-d+r)$ locally everywhere, we may conclude, by the semi-continuity of fiber dimension, that likewise for general t the variety $W_d^r(C_t)$ is nonempty of dimension exactly $\rho(g,r,d)$.

14.5. Brill–Noether with inflection

The approach we've taken here to the proof of Brill–Noether is well-suited to analyzing the inflectionary behavior of linear series on a general curve; indeed, a small modification of the argument above allows us to prove a stronger form of the Brill–Noether statement, concerning the existence of g_d^rs on a general curve C that are required to have inflection points of given weights.

Definition 14.12. Let C be a smooth curve of genus g and $p_1,\dots,p_n \in C$ distinct points of C. If $\mathcal{D} = (L,V)$ is a linear series on C of degree d and dimension r, we define the *adjusted Brill–Noether number* of $\mathcal{D}$ relative to the points p_k to be

$$\rho(\mathcal{D}; p_1,\dots,p_k) := g - (r+1)(g-d+r) - \sum_{k=1}^{n} w(\mathcal{D},p_k).$$

In these terms, we can prove a generalization of the "nonexistence" half of Brill–Noether:

Theorem 14.13. *Let $(C; p_1,\dots,p_n)$ be a general n-pointed curve of genus g (that is, let C be a general curve and $p_1,\dots,p_n \in C$ general points). If $\mathcal{D}$ is any linear series on C, then*

$$\rho(\mathcal{D}; p_1,\dots,p_n) \geq 0.$$

Proof. To start, let $\mathcal{C} \to \Delta$ be a family of curves as in the proof of Lemma 14.4. As in the argument at the start of Section 14.4, possibly after pulling back the family along a ramified map $\Delta \to \Delta$, we can choose sections $\sigma_1, \ldots, \sigma_n : \Delta \to \mathcal{C}$ of $\mathcal{C} \to \Delta$ such that the $\sigma_k(0)$ are distinct smooth points of C_0. If the general curve C_b in our family admits a g_d^r $\mathcal{D}$ with

$$\rho(\mathcal{D}; \sigma_1(b), \ldots, \sigma_n(b)) < 0$$

then, possibly after pulling the family back along a ramified map $\Delta \to \Delta$, we can choose a family $\{\mathcal{D}_b\}$ of such linear series on the fibers C_b for $b \neq 0$ and, taking limits, we arrive at a g_d^r $\mathcal{D}_0$ on $\mathbb{P}^1$ with

$$w(\mathcal{D}_0, q_i) \geq r$$

for each of the g points $q_i \in \mathbb{P}^1$ lying over the cusps of C_0, and in addition

$$w(\mathcal{D}_0, r_k) \geq w(\mathcal{D}_b, \sigma_k(b))$$

where $r_k \in \mathbb{P}^1$ is the point in $\mathbb{P}^1$ lying over $\sigma_k(0) \in C_0$. Adding up, we have

$$\sum_{i=1}^{g} w(\mathcal{D}_0, q_i) + \sum_{k=1}^{n} w(\mathcal{D}_0, r_i) \geq rg + \sum_{k=1}^{n} w(\mathcal{D}_b, \sigma_k(b))$$

$$> rg + g - (r+1)(g - d + r) = (r+1)(d - r),$$

since we assumed that

$$\rho(\mathcal{D}_b; \sigma_1(b), \ldots, \sigma_n(b)) = g - (r+1)(g - d + r) - \sum_{k=1}^{n} w(\mathcal{D}_b, \sigma_k(b)) < 0.$$

But as before the Plücker formula for $\mathbb{P}^1$ tells us that

$$\sum_{p \in \mathbb{P}^1} w(\mathcal{D}_0, p) = (r+1)(d - r),$$

a contradiction. $\qquad\square$

This extension of the "nonexistence" part of the Brill–Noether theorem raises the question of a converse: if $(C; p_1, \ldots, p_n)$ is a general n-pointed curve of genus g, and we specify ramification sequences $\alpha^1, \ldots, \alpha^n$, can we say that there exists a g_d^r $\mathcal{D}$ on C with $\alpha_i(\mathcal{D}, p_k) \geq \alpha_i^k$ for $k = 1, \ldots, n$ and $i = 0, \ldots, r$? If the product of the corresponding Schubert classes in $G(d - r, d + 1)$ is nonzero, then we can indeed deduce the existence of such a linear series; and if the product is 0, we can deduce that no such linear series exists.

Here is one way to state what we know without getting lost in a thicket of Schubert calculus:

Theorem 14.14. *Let C be a smooth curve of genus g and $p_1, \ldots, p_n \in C$ distinct points; for $k = 1, \ldots, n$ let $\alpha^k = (\alpha_0^k, \ldots \alpha_r^k)$ be a nondecreasing sequence of nonnegative integers, and let*

$$G_d^r(p_1, \ldots, p_n; \alpha^1, \ldots, \alpha^n) = \{\mathcal{D} \in G_d^r(C) \mid \alpha_i(\mathcal{D}, p_k) \geq \alpha_i^k\}.$$

If $(C, p_1, \ldots, p_n)$ is a general n-pointed curve, then either

$$G_d^r(p_1, \ldots, p_n; \alpha^1, \ldots, \alpha^n)$$

is empty, or it has dimension

$$\rho(g, r, d) - \sum_{k+1}^{n} \sum_{i=0}^{r} \alpha_i^k.$$

Finally, from the codimensions of the Schubert varieties we can deduce the ramification behavior of general linear series:

Theorem 14.15. *If $\mathcal{D}$ is a general g_d^r on a general curve, then $\mathcal{D}$ has only simple ramification; that is,*

$$w(\mathcal{D}, p) \leq 1 \quad \text{for all } p \in C.$$

Applying this in case $d = 2g - 2$ and $r = g - 1$, we arrive at the statement made after Corollary 13.7: that a general curve C of genus g has only Weierstrass points of weight 1.

14.6. Exercises

Exercise 14.1. By Theorem 14.14, there is a g_4^1 on a general curve C of genus 3 ramified at any 2 general points $p, q \in C$. Construct it. ◆

The next three exercises give the results necessary to prove Theorem 14.1 using degeneration to nodal curves rather than to cuspidal curves.

Exercise 14.2 (construction of nodal curves). Let C be a smooth curve, and let $\{(p_i, q_i) \mid i = 1 \ldots h\}$ be $2h$ distinct points of C.

Show that if C is embedded in $\mathbb{P}^r$ by a complete linear series of sufficiently high degree, then the projection of C from a general $(r-4)$-plane Λ that meets each secant $\overline{p_i q_i}$ maps C to a curve with h ordinary nodes in $\mathbb{P}^3$. ◆

Exercise 14.3. Suppose that $C \subset \mathbb{P}^d$ is the rational normal curve of degree d, and $\{L_i = \overline{p_i q_i}\}$ is a collection of secants of C. Show that if the points $p_i, q_i \in C$ are chosen sufficiently generally, then any collection of Schubert cycles $\Sigma_{a_i}(L_i)$ meet dimensionally transversely. ◆

Exercise 14.4. Imitate the proof of Proposition 14.5 to show that if R_0 is the local ring of a node — that is, the completion $\widehat{R}$ is isomorphic to $\mathbb{C}[x, y]/(xy)$ — then every nonzero ideal of R is isomorphic either to R or to the maximal ideal of R. Deduce that if $C \to C_0$ is the partial normalization of a single node on a reduced irreducible projective curve, then there is an exact sequence $0 \to \mathbb{C}^* \to \mathrm{Pic}_0(C_0) \to \mathrm{Pic}_0(C) \to 0$. ◆

Chapter 15

Using a singular plane model

In the first part of Chapter 4 we showed how to use an embedding of a smooth curve C in $\mathbb{P}^2$ to understand differentials and linear series on C. In that form, the technique has limited applicability, since most smooth curves cannot be embedded in $\mathbb{P}^2$. However, by Proposition 10.11, any smooth curve C can be projected birationally to a curve $C_0 \subset \mathbb{P}^2$ with only nodes. We open this chapter by showing how to use such a nodal model C_0 to describe differentials and linear series on C, a theory well-understood by Brill and Noether; and then explain what is necessary to adapt the technique to birational images of C with arbitrary singularities.

15.1. Nodal plane curves

The methods of Chapter 4 can be applied, with one change, when $C_0 \subset \mathbb{P}^2$ is a nodal curve.

Let $\nu : C \to C_0 \subset \mathbb{P}^2$ be the normalization morphism from a smooth curve, and let $B \subset C_0$ be a subscheme. It will be convenient to speak of *the linear series cut out on C* by curves of degree m containing B: though B is only a subscheme, $\nu^{-1}(B)$ may be considered as a Cartier divisor because C is smooth, and we define the linear series cut out on C by curves of degree m containing B to be the linear series $\mathcal{V}$ of divisors in C that are residual to $\nu^{-1}(B)$ in the pullbacks of the intersections of C_0 with curves of degree m. Formally, since $B' := \nu^{-1}(B)$ is a Cartier divisor on C, we can say that if L is the divisor of a line in $\mathbb{P}^2$ and L' its pullback to C then $\mathcal{V}$ is a space of divisors corresponding to sections of $\mathcal{O}_C(mL' - B')$; more precisely, it is the space of divisors corresponding to sections of $\mathcal{O}_C(mL' - B')$ in the image of the restriction/pullback map

$$H^0(\mathcal{J}_{B/\mathbb{P}^2}(m)) \to H^0(\mathcal{O}_C(mL' - B')).$$

15.1.1. Differentials on a nodal plane curve.

Let $C_0 \subset \mathbb{P}^2$ be a curve of degree d with δ nodes and no other singularities. By the adjunction formula (Proposition 2.8), Proposition 2.25, and the first example that follows it, the genus g of the normalization C of C_0 is the arithmetic genus $p_a(C_0) = \binom{d-1}{2}$ of C_0 minus δ, that is,

$$g = \binom{d-1}{2} - \delta.$$

We will make this explicit by exhibiting a vector space of g regular differential forms on C.

Choose homogeneous coordinates $[X, Y, Z]$ on $\mathbb{P}^2$ so that C_0 intersects the line $L = V(Z)$ in a divisor D consisting only of smooth points of C_0 other than $[0, 1, 0]$, and so that at each node of C_0 (necessarily contained in the affine plane $U = \mathbb{P}^2 \setminus L$) the tangents to C_0 have finite slope. Let the nodes of C_0 be $q_1, \ldots, q_\delta$, with $r_i, s_i \in C$ lying over q_i; we'll denote by Δ the divisor $\sum r_i + \sum s_i$ on C.

Let $F(X, Y, Z)$ be the homogeneous polynomial of degree d defining the curve C_0, and let $f(x, y) = F(x, y, 1)$ be the defining equation of the affine part $C_0^\circ := C_0 \cap U$ of C_0. Let $\nu : C \to C_0$ be the normalization map. We start by considering the rational differential $\nu^*(dx)$ on $C^\circ := \nu^{-1}(C_0^\circ)$.

In the smooth case where $C_0 = C$ we saw that this differential was regular and nonzero on C°; this followed from the fact that f_x and f_y had no common zeroes on C_0. But now f_x and f_y have common zeroes: they both vanish to order 1 at the points q_i and thus $\nu^*(f_x)$ and $\nu^*(f_y)$ have simple zeroes at the points r_i and s_i.

As before, the differential $\nu^*(dx)$ has double poles along the divisor D on C_0 lying over the point at infinity in $\mathbb{P}^1$ and we see that for a polynomial $e(x, y)$ of degree $\leq d - 3$, the differential

$$\nu^* \left(\frac{e(x, y)\, dx}{f_y} \right)$$

is regular except for simple poles at the points r_i and s_i.

We can get rid of these poles by requiring that e vanishes at the points q_i. We say in this case that e (and the curve defined by e) . *satisfies the conditions of adjunction.*

Theorem 15.1. *If C_0 is a nodal plane curve of degree d with normalization $\nu : C \to C_0$ then the regular differentials on C, in terms of the notation above, are precisely those of the form*

$$\nu^* \left(\frac{e(x, y)\, dx}{f_y} \right),$$

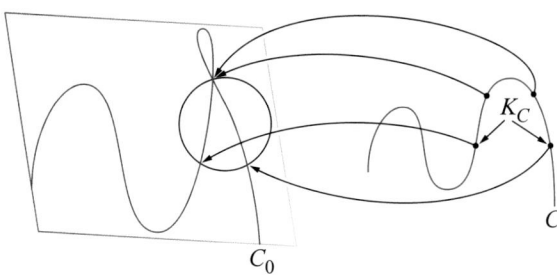

Figure 15.1. A curve C of geometric genus 2 represented as the normalization of a plane curve C_0 of degree 4 with a node, and a canonical divisor, represented by a conic containing the node.

where $e(x, y)$ ranges over the polynomials of degree $\leq d - 3$ vanishing at the nodes of C_0.

Thus if $\mathfrak{F}(C_0) \subset C_0$ denotes the union of the reduced points at the nodes of C, then $|\omega_C|$ is the linear series cut out on C by forms of degree $d - 3$ containing $\mathfrak{F}(C_0)$.

See Figure 15.1 for a picture in a case where C_0 has a single node.

Proof. The dimension of the space of polynomials $e(x, y)$ of degree at most $d - 3$ is $\binom{d-1}{2}$, and vanishing at δ nodes imposes at most δ linear conditions on e. The linear map sending $e \mapsto \nu^*(e\,dx/f_y)$ is injective, and the target has dimension $\binom{d-1}{2} - \delta$, so this must be an isomorphism. $\qquad\square$

We will give a more conceptual proof of this theorem in Section 15.2.

In particular, Theorem 15.1 shows that the linear series cut out on C by forms of degree $d - 3$ containing $\mathfrak{F}(C_0)$ is complete. (We will soon see that the linear series cut out on C by forms of degree m containing $\mathfrak{F}(C_0)$ is complete for every m.)

This gives another proof of Lemma 8.16.

Corollary 15.2. *If C is a nodal plane curve of degree d, then the nodes of C_0 impose independent conditions on forms of degree $d - 3$.*

Proof. Otherwise the space of differential forms on the normalization of C_0 would be too large. $\qquad\square$

One can use Theorem 15.1 to re-embed a plane curve of geometric genus g as a canonical curve in $\mathbb{P}^{g-1}$:

Corollary 15.3. *The canonical ideal of the normalization of a nodal plane curve of degree d is the ideal of polynomial relations among the forms of degree $d - 3$ that vanish at the nodes of the curve.* $\qquad\square$

15.1.2. Linear series on a nodal plane curve. Since C_0 is singular, not every effective divisor on C is the preimage of an effective Cartier divisor on C_0. As an example, one may take a single point lying over a node as in Figure 15.2.

However, we can still represent every divisor on C as the preimage of a divisor on C_0 up to linear equivalence, and the same goes for any reduced curve:

Lemma 15.4. *Let $\nu : C \to C_0$ be the normalization of any reduced projective curve. If D is any divisor on C, then D is linearly equivalent to the pullback of a divisor supported on the smooth locus of C_0. More precisely, every effective divisor on C containing Δ can be written as the pullback of a Cartier divisor on C_0.*

We will see a more general version in Theorem 15.12.

Proof. It suffices to prove the result locally on C_0, where it is geometrically obvious: if the node $p \in C_0$ has preimages q, r corresponding to branches Q and R of C_0 at p, then a divisor $aq + br$ with both a, b strictly positive, is locally the pullback of the intersection of C_0 with a curve C' meeting Q with multiplicity a at p and meeting R with multiplicity b at p. For example, assuming that $a \leq b$, we could take C' to be the union of $a - 1$ general lines through p and a smooth plane curve meeting R with multiplicity $b - a + 1$ at p.   □

Returning to the case of a nodal plane curve C_0 and its normalization $\nu : C \to C_0$, suppose that D is a divisor on C that is the pullback of a difference of Cartier divisors $D_+ - D_-$ on C_0. We will compute the complete linear series $|D|$. Let $\mathfrak{F}(C_0)$ be the set of nodes in C and let Δ be the preimage of $\mathfrak{F}(C_0)$ in C.

Theorem 15.5. *Let $D = D_+ - D_-$ be a divisor on C, and let G be a form on $\mathbb{P}^2$ that vanishes on $D_+ + \mathfrak{F}(C_0)$ but not identically on C_0.*

If G has degree m and $A = (\nu^*G) - D_+ - \Delta$, then every effective divisor on C linearly equivalent to D (if any) has the form $(\nu^*H) - D_- - A - \Delta$ for some H of degree m that vanishes on $\mathfrak{F}(C_0) + A$ but not identically on C_0.

The reason for including $\mathfrak{F}(C_0)$ is that an arbitrary linear series on C cannot be represented as the pullback of a linear series on C_0 without this. We will see a generalization in Theorem 15.13 and the surrounding discussion.

Corollary 15.6. *With notation as in Theorem* 15.5, *the ideal of the image of C with respect to the linear series $|D|$ is the ideal of polynomial relations among the forms of degree m in $\mathfrak{f}_{C/C_0}$ that vanish on $D_- + A$.*   □

Proof of Theorem 15.5. Choose an integer m big enough that there is a form G vanishing on D_+ and $\mathfrak{F}(C_0)$ so that $\nu^*(G)$ vanishes on $D_+ + \Delta$, but not everywhere on C. (If D_+ contains some positive multiple of a point p of Δ this means that G defines a curve sufficiently tangent to the corresponding branch of C_0.) Then, as before, we can write the zero locus of G pulled back to C as

$$(\nu^*G) = D_+ + \Delta + A.$$

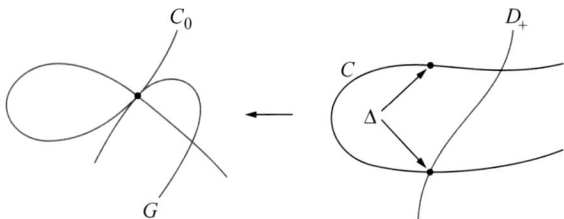

Figure 15.2. If G is tangent to a branch of C_0 at the node, then D_+ contains a point of Δ on C.

Next, we look for forms H of the same degree m, vanishing at $A + D_-$ and on $\mathfrak{F}(C_0)$ but not on all of C_0. If there are no such polynomials H then, as we shall show, there are no effective divisors equivalent to D. Supposing that there is such a form H, let D' be the divisor

$$D' = (v^*H) - (D_+ + \Delta),$$

that is, D' is residual to $(D_+ + \Delta)$ in (v^*H).

Since $v^*(G/H)$ is a rational function on C we have

$$D_- + \Delta + A + D' = (v^*H) \sim (v^*G) = D_+ + \Delta + A,$$

and thus D' is an effective divisor linearly equivalent to $D = D_+ - D_-$ on C.

To complete the argument we must show that we get *all* divisors D' in this way. In this case the curve C can be desingularized by blowing up the plane once at each node, and we can give a proof based on the resulting surface S. The same technique would work for any curve with only ordinary multiple points, in which case the total transform of C_0 on S has normal crossings. We will give a different proof, extending this theorem to curves with arbitrary singularities, in Section 15.2.

Proposition 15.7. *If C_0 is a reduced irreducible plane curve all of whose singularities are ordinary nodes, then for each integer m, the linear series cut out on the normalization C of C_0 by forms of degree m containing the nodes is complete.*

Proof. To prove Proposition 15.7, we work on the blowup $\pi : S \to \mathbb{P}^2$ of $\mathbb{P}^2$ at the nodes q_i of C_0. The proper transform of $C_0 \subset \mathbb{P}^2$ in S is the normalization of C_0, which we will again call C.

Let L be the class on S of the pullback of a line in $\mathbb{P}^2$ and let E be the sum of the exceptional divisors, the preimage of $\mathfrak{F}(C_0)$. We write $h = L \cap C$ and $e = E \cap C = \sum(p_i + q_i)$ for the corresponding divisors on C. Because C has double points at each q_i we have $C \sim dL - 2E$ and by Theorem 2.39 we have $K_S \sim -3L + E$.

The proper transform of a degree m curve $A \subset \mathbb{P}^2$ passing simply through the points q_i is $\pi^*A - E$; this gives an isomorphism

$$H^0(\mathcal{I}_{\{q_1,\ldots,q_\delta\}/\mathbb{P}^2}(m)) \cong H^0(\mathcal{O}_S(mL - E)).$$

In these terms we can describe the linear series cut on C by plane curves of degree m passing through the nodes of C_0 as the image of the map

$$H^0(\mathcal{O}_S(mL - E)) \to H^0(\mathcal{O}_C(mL - E)),$$

and we must show that this map is surjective.

From the long exact cohomology sequence associated to the exact sequence of sheaves

$$0 \to \mathcal{O}_S((m - d)L + E) \to \mathcal{O}_S(mL - E) \to \mathcal{O}_C(mL - E) \to 0,$$

we see that it will suffice to prove that $H^1(\mathcal{O}_S((m - d)L + E)) = 0$.

By Serre duality on S,

$$H^1(\mathcal{O}_S((m - d)L + E)) \cong H^1(\mathcal{O}_S((d - m - 3)L))^*.$$

The line bundle $\mathcal{O}_S((d-m-3)L)$ is the pullback to S of the bundle $\mathcal{O}_{\mathbb{P}^2}(d-m-3)$, which has vanishing H^1. Lemma 15.8 completes the proof. □

Lemma 15.8. *Let X be a smooth projective surface, and $\pi : S \to X$ the blowup of a finite set of reduced points. If $\mathcal{L}$ is any line bundle on X, then*

$$H^1(S, \pi^*\mathcal{L}) = H^1(X, \mathcal{L}).$$

Proof. Because $\mathbb{P}^2$ is normal, and $\pi_*(\mathcal{O}_S)$ is a finite birational algebra over $\mathcal{O}_{\mathbb{P}^2}$, we have $\pi_*(\mathcal{O}_S) = \mathcal{O}_{\mathbb{P}^2}$. Since any invertible sheaf $\mathcal{L}$ on $\mathbb{P}^2$ is locally isomorphic to $\mathcal{O}_{\mathbb{P}^2}$, is also an isomorphism.

The Leray spectral sequence (Theorem 2.26) gives an exact sequence

$$0 \to H^1(\pi_*(\mathcal{L})) \to H^1(\mathcal{L}) \to H^0(R^1(\pi_*(\mathcal{L})) \to 0$$

The restriction of $\pi^*(\mathcal{L})$ to any fiber of π is trivial and has vanishing H^1, so $H^0(R^1(\pi_*(\mathcal{L}))) = 0$, and $\pi_*\pi^*\mathcal{O}_{\mathbb{P}^2}(1)$ is an invertible sheaf. The natural map $\pi_*\pi^*(\mathcal{L}) \to \mathcal{L}$ is an isomorphism away from the codimension 2 set of points blown up. Thus these two sheaves are isomorphic, and

$$H^1(S, \pi^*\mathcal{L}) = H^1(\pi_*\pi^*\mathcal{L}) = H^1(\mathcal{L}),$$

completing the proof. □

This concludes the proof of Theorem 15.5 □

Proposition 15.9. *Let C be a curve on a smooth surface S, and let $\nu : S' \to S$ be the blowup of S at p. If C' is the strict transform of C, then*

$$p_a(C') = p_a(C) - \binom{m}{2},$$

where m is the multiplicity of $p \in C$.

Proof. This follows from comparing the adjunction formulas on S and S'. To start, we have

$$p_a(C) = \frac{C^2 + K_S \cdot C}{2} + 1.$$

On S' let E be the divisor class of the exceptional divisor. As we've seen,

$$K_{S'} = \nu^* K_S + E,$$

while the class of C' is given by

$$C' \sim \nu^* C - mE.$$

It follows that

$$(C')^2 = C^2 + m^2 E^2 = C^2 - m^2 \quad \text{and} \quad K_{S'} \cdot C' = K_S \cdot C + m.$$

Thus, applying adjunction on S', we find the desired equality:

$$p_a(C') = \frac{C'^2 + K_{S'} \cdot C'}{2} + 1 = \frac{C^2 + K_S \cdot C - m(m-1)}{2} + 1 = p_a(C) - \binom{m}{2}. \quad \square$$

Cheerful Fact 15.10. Any plane curve can be desingularized by iteratively blowing up of singular points of C, then of the strict transform, and so on. See for example [Fulton 1969] or [Brieskorn and Knörrer 1986]. This is in fact the same sequence of transforms as the one given in Exercise 2.15. The points on the various blowups that map to the original singular point are called *infinitely near points*.

This gives a nice formula for the δ invariant of any singularity:

Corollary 15.11. *The δ invariant of any singularity of a plane curve C_0 at a point p can be computed as the sum of the numbers $\binom{m_q}{2}$ over all infinitely near singular points q, where m_q denotes the multiplicity of the pullback of C_0 at q.* $\square$

15.2. Arbitrary plane curves

Throughout this section $C_0 \subset \mathbb{P}^2$ denotes a reduced and irreducible plane curve with arbitrary singularities. Let $\nu : C \to C_0$ be its normalization, and write L' for the pullback to C of the class L of a line in $\mathbb{P}^2$.

It is possible to carry out an analysis of linear series on the normalization of an arbitrary plane curve in a manner analogous to what we did in the preceding section for nodal curves, replacing the set of nodes by the *adjoint scheme* $\mathfrak{F}(C_0) \subset C_0$, which in the case of a plane curve is the scheme defined by the *conductor ideal* $\mathfrak{f}_{C/C_0}$, the annihilator in $\mathcal{O}_{C_0}$ of $\nu_*(\mathcal{O}_C)/\mathcal{O}_{C_0}$ (see Theorem 16.17).

The conductor ideal and linear series on the normalization. Let $\Delta \subset C$ be the divisor defined by the pullback of $\mathfrak{f}_{C/C_0}$ to C. The following result gives a simple way of expressing any divisor on C in terms of Δ and a Cartier divisor on C_0:

Theorem 15.12. *If D is an effective divisor on C that contains Δ, then D is the pullback of a Cartier divisor on C_0.*

Proof. The result is local, so it suffices to treat the affine case of an affine curve C_0 with coordinate ring $\mathcal{O}'$ contained in its integral closure $\mathcal{O}$, the coordinate ring of its normalization C. The ideal of Δ in $\mathcal{O}$ is the conductor $\mathfrak{f}_{\mathcal{O}/\mathcal{O}'}$ (which is contained in $\mathcal{O}'$, but stable under multiplication by elements of $\mathcal{O}$, and thus also an ideal of $\mathcal{O}$). Thus $\mathcal{I}_D$ may be regarded as an ideal — though not necessarily a principal ideal — of $\mathcal{O}'$. Since the ground field $\mathbb{C}$ is infinite, $\mathcal{I}_D \subset \mathcal{O}'$ is the integral closure of a principal ideal (x) of $\mathcal{O}'$. (See [Huneke and Swanson 2006, Chapter 8], for example.) This means that for sufficiently large n we have $x\mathcal{I}_D^n = \mathcal{I}_D^{n+1}$.

Let D_0 be the Cartier divisor on C_0 corresponding to x. Pulling everything back to C we have $\nu^*(D_0) + nD = (n+1)D$, whence $\nu^*(D_0) = D$. $\qquad\square$

In view Theorem 15.12 we can specify an arbitrary linear series on C as the difference $\mathcal{E} - \Delta$, where $\mathcal{E}$ is a linear series with Δ in its base locus, by specifying a linear series on C_0 containing $\mathfrak{F}(C_0) \subset C_0$. The following theorem makes this algorithmic.

Theorem 15.13. *Let $D = D_+ - D_-$ be a divisor on C, and let G be a form on $\mathbb{P}^2$, vanishing on $\mathfrak{F}(C_0)$, whose pullback to C vanishes on D_+ but not on all of C.*

If G has degree m and $(\nu^*G) = D_+ + A + \Delta$, then every effective divisor on C linearly equivalent to D (if any) has the form $(\nu^*H) - D_- - A - \Delta$ for some H of degree m vanishing on $\mathfrak{F}(C_0) + A$. Thus, the global sections of $H^0(\mathcal{O}_C(D))$ can be written as divisors cut by forms of degree m in $\mathfrak{f}_{C/C_0}$ with basepoints at $D_- + A + \Delta$.

It follows that one can compute the image of C under the map given by $|D|$ directly from C_0 and the conductor:

Corollary 15.14. *With notation as in Theorem 15.13, the ideal of the image of C with respect to the linear series $|D|$ is the ideal of polynomial relations among the forms of degree m in $\mathfrak{f}_{C/C_0}$ that vanish on $D_- + A$.* $\qquad\square$

Proof of Theorem 15.13. If $H \in \mathfrak{f}(C_0)$ vanishes on A but not on all of C_0, then as before

$$D_- + \Delta + A + D' = (\nu^*H) \sim (\nu^*G) = D_+ + \Delta + A,$$

So $D' = (\nu^*H) - D_- - A - \Delta$ is linearly equivalent to D. The proof that every divisor D' linearly equivalent to D has this form is the content of Theorem 15.15

below, which was known classically as the *completeness of the adjoint series.*

□

Theorem 15.15. *For every integer $m \geq 0$ the series cut out on C by forms of degree m on $\mathbb{P}^2$ containing $\mathfrak{F}(C_0)$ is complete.*

Proof. If $R_0 \subset R$ is an inclusion of commutative rings, then the ideal

$$\mathfrak{f}_{R/R_0} := \operatorname{ann}_{R_0}(R/R_0) \subset R_0$$

is called the *conductor* of $R_0 \subset R$. It is by definition an ideal of R_0, but is also an ideal of R; this follows because if $f \in \mathfrak{f}_{R/R_0}$ and $r \in R$, then $fR \subset R_0$ so $(rf)R = frR \subset fR \subset R_0$.

If R_0 is a domain and R is a subring of the quotient field $Q(R)$ of R, then $\mathfrak{f}_{R/R_0} \cong \operatorname{Hom}_{R_0}(R, R_0)$. To see this, note that R_0 and R become equal after tensoring with $Q(R_0)$ and thus $\operatorname{Hom}_{R_0}(R, R_0) \subset \operatorname{Hom}_Q(Q, Q) = Q$ may be identified with the set of elements $\{\alpha \in Q \mid \alpha R \subset R_0\}$. If α is in this set, then $\alpha \cdot 1 = \alpha \in R_0$, as required.

Returning to the case of the curve C_0, it follows that the global sections of $(\nu^*(\mathcal{O}_{C_0}(m))(-\Delta)$ on C are, on each affine open set U, represented by the elements of $\mathcal{O}_{C_0}(U)$ that are restrictions to U of forms of degree m contained in the ideal $\widetilde{\mathfrak{f}_{C/C_0}}$. Thus the global sections of the sheaf $\widetilde{\mathfrak{f}_{C/C_0}}(m)$ cut out a complete linear series on C.

Write $S = \mathbb{C}[x_0, x_1, x_2]$ for the homogeneous coordinate ring of $\mathbb{P}^2$. It remains to prove that the homogeneous ideal $\mathfrak{f}_{C/C_0}$ maps surjectively to $H^0_*(\widetilde{\mathfrak{f}_{C/C_0}})$, and this amounts to the statement that the depth of $\mathfrak{f}_{C/C_0}$ as an S-module is at least 2. Set $R_0 = H^0_*(\mathcal{O}_{C_0})$ and $R = H^0_*(\nu_*(\mathcal{O}_C))$. We see from the general considerations above that

$$\mathfrak{f}_{R/R_0} = \operatorname{Hom}_{R_0}(R, R_0).$$

Any nonzerodivisor on a module M is a nonzerodivisor on $\operatorname{Hom}(P, M)$ for any module P since $(a\phi)(p) = a(\phi(p))$ by definition. Since $R_0 = S/(F)$, it is a module of depth 2, and we may choose a regular sequence a, b of elements in R_0. From the short exact sequence

$$0 \to R_0 \xrightarrow{a} R_0 \to R_0/(a) \to 0$$

we get a left exact sequence

$$0 \to \operatorname{Hom}_{R_0}(R, R_0) \xrightarrow{a} \operatorname{Hom}_{R_0}(R, R_0) \to \operatorname{Hom}_{R_0}(R, R_0/(a)).$$

Thus

$$\operatorname{Hom}_{R_0}(R, R_0)/(a \operatorname{Hom}_{R_0}(R, R_0)) \subset \operatorname{Hom}_{R_0}(R, R_0/(a))$$

and since b is a nonzerodivisor on $\operatorname{Hom}_{R_0}(R, R_0/(a))$, it is a nonzerodivisor on $\operatorname{Hom}_{R_0}(R, R_0)/(a \operatorname{Hom}_{R_0}(R, R_0))$ as well.

□

Since we saw directly that the adjoint ideal was equal to the conductor ideal in the case of a nodal curve, this result gives another, less ad hoc, proof that the effective divisors equivalent to D are all defined by pullbacks of forms of degree m that contain Δ as constructed in Proposition 15.7.

Differentials. Let C_0° be the intersection of C_0 with the open set $\mathbb{A}^2 \cong U \subset \mathbb{P}^2$ where $Z \neq 0$, and let $C^\circ \subset C$ be the preimage of C_0°.

Theorem 15.16. *If C_0 meets the line L at infinity only in smooth points of C_0 other than $(0, 1, 0)$, then the complete canonical series on the normalization $\nu : C \to C_0$ is cut out by differentials of the form*

$$\frac{e(x, y)\, dx}{f_y}$$

where $e(x, y)$ is a polynomial of degree $\leq d - 3$ contained in the conductor ideal $\mathfrak{f}_{C^\circ/C_0^\circ}$.

As in the case of nodal curves, one can use Theorem 15.16 to re-embed a plane curve of geometric genus g as a canonical curve in $\mathbb{P}^{g-1}$:

Corollary 15.17. *The canonical ideal of the normalization C of a plane curve C_0 of degree d is the ideal of polynomial relations among the forms of degree $d - 3$ in the conductor ideal $\mathfrak{f}_{C/C_0}$.* □

Proof of Theorem 15.16. We proceed in four steps. First, because $(0,1,0)$ does not lie on C, the function x defines a ramified d-sheeted cover of C to $\mathbb{P}^1$. Because C_0 meets L only in smooth points and the differential dx has a pole of order 2 at the point at infinity in $\mathbb{P}^1$, dx has polar locus twice the divisor $\nu^{-1}(C_0 \cap L)$. It follows that the differential $\varphi_0 := dx/f_y$ is regular, with a zero of order $d - 3$, along the divisor of C lying over $C_0 \cap L$.

Second, the function on C_0° defined by x is a finite map to $\mathbb{A}^1$, and thus the field of rational functions $\kappa(C) = \kappa(C_0^\circ)$ is a finite separable extension of $\mathbb{C}(x)$. By [Eisenbud 1995, Section 16.5], the module of differentials $\omega_{\kappa(C)/\mathbb{C}}$ is generated over $\kappa(C_0^\circ)$ by dx. Thus every rational differential form on C can be expressed as a rational function times dx. Since $\varphi_0 := dx/f_y$ vanishes to order $d - 3$ along $C_0 \cap L$, the regular differential forms on C must be of the form $e(x, y)\varphi_0$ where $e(x, y)$ is a rational function of degree $\leq d - 3$. (The set of rational forms that occur in this way is called the *Dedekind complementary module*.)

Third, a sophisticated form of Hurwitz's theorem, to be explained in Chapter 16, shows that the sheaf ω_C of regular differential forms on C can be expressed as $\nu^! \pi^! \mathcal{H}om_{\mathbb{P}^1}(\pi_* \nu_*(\mathcal{O}_C)\omega_{\mathbb{P}^1})$, where π is the map $C_0 \to \mathbb{P}^1$ defined by x. Locally, this is expressed more simply as

$$\mathcal{H}om_{\mathbb{P}^1}(\nu_*(\mathcal{O}_C), \omega_{\mathbb{P}^1}),$$

where we use the action of $\mathcal{O}_C$ on $\nu_*(\mathcal{O}_C)$. Since the maps involved are finite, we will identify $\mathcal{O}_C$ with $\nu_*(\mathcal{O}_C)$ and write

$$\omega_C = \mathcal{H}om_{\mathbb{P}^1}(\mathcal{O}_C, \omega_{\mathbb{P}^1}),$$

Since $\mathcal{O}_{C_0} \subset \mathcal{O}_C$, this sheaf is naturally contained in

$$\mathcal{H}om_{\mathbb{P}^1}(\mathcal{O}_{C_0}, \omega_{\mathbb{P}^1}).$$

As will be explained in Chapter 16,

$$\omega_{C_0} := \mathcal{H}om_{\mathbb{P}^1}(\mathcal{O}_{C_0}, \omega_{\mathbb{P}^1}) = \mathcal{H}om_{\mathbb{P}^1}(\mathcal{O}_{C_0}, \mathcal{O}_{\mathbb{P}^1})(-2)$$

is properly called the dualizing module of the singular curve C_0. We will show in Theorem 16.17 that $\mathfrak{f}_{C/C_0}$ is the annihilator of the quotient of two $\mathcal{O}_{C_0}$ modules

$$\mathfrak{f}_{C/C_0} = \text{ann}_{C_0} \frac{\mathcal{H}om_{\mathbb{P}^1}(\mathcal{O}_{C_0}, \omega_{\mathbb{P}^1})}{\mathcal{H}om_{\mathbb{P}^1}(\mathcal{O}_C, \omega_{\mathbb{P}^1})}.$$

We will also see that $\mathcal{H}om_{\mathbb{P}^1}(\mathcal{O}_{C_0}, \omega_{\mathbb{P}^1})$ is an invertible sheaf on C_0, and since the quotient is supported only at the finite set of singularities of C_0, it follows that every regular differential form on C can be expressed as an element of $\mathfrak{f}_{C/C_0}$ times some element of ω_{C_0}.

To prove the theorem, it now suffices to show that $\varphi_0 = dx/f_y$ generates ω_{C_0} as a module over $\mathcal{O}_{C_0}$.

Passing to the field of rational functions $\kappa := \kappa(C)$, and noting that $\kappa(C) = \kappa(C_0)$ we use the well-known result from Galois theory that

$$\text{Hom}_{\kappa(\mathbb{P}^1)}(\kappa, \kappa(\mathbb{P}^1))$$

is a 1-dimensional vector space over κ, generated by the trace map. Tr. Moreover, because $\mathcal{O}_{C_0}$ is integral over $\mathcal{O}_{\mathbb{P}^1}$, which is normal,

$$\text{Tr}(\mathcal{O}_{C_0}) \subset \mathcal{O}_{\mathbb{P}^1}.$$

For the fourth and last step we need a result from commutative algebra:

Theorem 15.18. *If C_0° is an affine plane curve defined by the equation $f(x, y) = 0$ and such that $\mathbb{C}[x, y]/(f)$ is finite over $\mathbb{C}[x]$, then $\text{Hom}_{\mathbb{C}[x]}(\mathbb{C}[x, y]/(f), \mathbb{C}[x])$ is generated by $(1/f_y)T$.*

See [Kunz 2005, Theorem 15.1] for a proof using valuations, and [Eisenbud and Ulrich 2019, Theorem A.1] for a proof in a more general context.

Given Theorem 15.18, we can complete the proof of Theorem 15.16. Via the isomorphism $\mathbb{C}(x) \cong \mathbb{C}(x)\,dx$ sending 1 to dx, the trace map is identified (up to scalar) with the map $(1 \mapsto dx) \in \text{Hom}_{\mathbb{C}(x)}(\kappa, \mathbb{C}(x)\,dx)$. Thus by Theorem 15.18 the canonical module of C_0° is identified with $\mathcal{O}_{C_0^\circ}\phi_0$ as required. $\qquad\square$

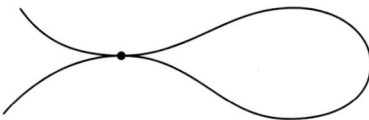

Figure 15.3. A tacnode: two smooth branches tangent to one another.

In Examples 15.19–15.21, we will consider a singular curve defined in the affine plane by an equation $f(x, y) = 0$ such that $\mathbb{C}[x, y]/(f)$ is finite over $\mathbb{C}[x]$, and write $\varphi_0 = dx/f_y$ for a local generator of the module of differentials.

Example 15.19 (nodes and cusps). We have already seen that in case C_0 has a node at a point q there are two points of C lying over q, and the multiplicities of φ_0 at these two points are $m_1 = m_2 = 1$; the adjoint ideal is thus the maximal ideal $\mathcal{I}_q$ at q. When q is a cusp of C_0 — that is, C_0 is analytically isomorphic to the zero locus of $y^2 - x^3$ in a neighborhood of the point $q = (0,0)$ — there is only one point $r \in C$ lying over q. In an analytic neighborhood of the cusp, C_0 can be parametrized by $x = t^2$, $y = t^3$ and it follows that the differential

$$\varphi_0 = \frac{dx}{f_y} = \frac{2t\, dt}{2t^3} = \frac{dt}{t^2}$$

has a pole of order 2 at r. Since the pullback to C of any polynomial g vanishing at the cusp q vanishes to order at least two at r, the adjoint ideal is again the maximal ideal at q. We can also see this by computing the conductor ideal as the annihilator of $\mathbb{C}[\![t]\!]/\mathbb{C}[\![t^2, t^3]\!]$.

Example 15.20 (tacnodes). Next, consider the case where C_0 has a tacnode at q (Figure 15.3); that is, a plane-curve singularity with two smooth branches simply tangent to one another, analytically isomorphic to the zero locus of $y^2 - x^4$ in a neighborhood of $q = (0,0)$. The two branches are parametrized locally analytically by $x = t$, $y = t^2$ and $x = t$, $y = -t^2$.

At the two points $r_1, r_2 \in C$ lying over q, we have

$$\varphi_0 = \pm \frac{dt}{2t^2}$$

and each of these differentials has a double pole at each of r_1, r_2. The adjoint ideal is thus the ideal of functions vanishing at q and having derivative 0 in the direction of the common tangent line to the branches.

Example 15.21 (ordinary n-fold points). Consider the case where C_0 has an ordinary n-fold point at q, where n smooth branches intersect pairwise transversely. There are n points r_i of C lying over q. The polynomial f_y vanishes to order $n - 1$ at q, so dx/f_y has a pole of order $n - 1$ at each r_i. It follows that for $e(x, y)\, dx/f_y$ to be regular, e must vanish to order $n - 1$ at each r_i.

We can see that the conductor ideal is the full $(n-1)$-rst power of (x, y) by using the normalization map

$$\nu^* : \frac{\mathbb{C}[x, y]}{\prod_{i=1}^{n}(x - \alpha_i y)} \to \prod_{i=1}^{n} \frac{\mathbb{C}[x, y]e_i}{(x - \alpha_i y)}$$

where the α_i are distinct elements of $\mathbb{C}$ and the e_i are orthogonal idempotents. The element $1 = \sum_i e_i$ goes to 0 in the cokernel of ν^* and e_i is annihilated by $x - \alpha_i y$, so the quotient is annihilated by each of the n elements

$$g_j := \prod_{i \neq j}(x - \alpha_i y).$$

As forms on $\mathbb{P}^1$, all the g_i except g_j vanish at the point $(\alpha_j, 1)$, so the g_i are linearly independent. Since $(x, y)^{n-1}$ is minimally generated by n elements, $(x, y)^{n-1} = (g_1, \ldots, g_n)$.

A consequence of this computation is that the δ invariant of the ordinary multiple point is $\binom{n}{2}$, the dimension of $k[x, y]/(x, y)^{n-1}$, as we proved just after Theorem 2.39 and in Proposition 15.9.

Example 15.22 (spatial triple points). Spatial triple points provide a contrast to the last example. A spatial triple point is a singularity consisting of three smooth branches, with linearly independent tangent lines, meeting in a point p so that its Zariski tangent space is 3-dimensional; the simplest example is the origin as a point on the union of the three coordinate axes in $\mathbb{A}^3$.

In this case the conductor is the annihilator of the cokernel of

$$\nu^* : R := \frac{\mathbb{C}[x, y, z]}{(xy, xz, yz)} \to \frac{\mathbb{C}[x, y, z]}{(x, y)} \times \frac{\mathbb{C}[x, y, z]}{(x, z)} \times \frac{\mathbb{C}[x, y, z]}{(y, z)} =: \bar{R}.$$

Since $x\bar{R} = x\mathbb{C}[x, y, z]/(y, z)$ is in the image of $\mathbb{C}[x, y, z]/(xy, xz, yz)$, and similarly with y and z, we see that the conductor is the maximal ideal (x, y, z). However, the δ invariant, the length of the quotient $(\bar{R})/R$, is 2: for a function f in $\bar{R}$ to be in R, it is necessary and sufficient that f take the same value at the three points above the singular point, and this amounts to two linear conditions on f.

15.3. Exercises

In Exercise 4.1, we saw how to use the description of the canonical series on a smooth plane curve to determine its gonality. Now that we have an analogous description of the canonical series on (the normalization of) a nodal plane curve, we can deduce a similar statement about the gonality of such a curve. Here are the first two cases:

Exercise 15.1. Let C_0 be a plane curve of degree $d \geq 4$ with one node p and no other singularities, and let C be its normalization. Show that C admits a unique map $C \to \mathbb{P}^1$ of degree $d - 2$, but does not admit a map $C \to \mathbb{P}^1$ of degree $d - 3$ or less. ◆

Exercise 15.2. Let C_0 be a plane curve of degree $d \geq 5$ with two nodes p and p' and no other singularities, and let C be its normalization. Show that C admits two maps $C \to \mathbb{P}^1$ of degree $d - 2$, but does not admit a map $C \to \mathbb{P}^1$ of degree $d - 3$ or less. ◆

Exercise 15.3. Generalizing the examples above, show that if a nodal plane curve of degree d has $\delta \leq d + 3$ nodes, then its gonality is $d - 2$, and moreover every g^1_d on the curve is given by projection from one of the nodes. You may use the following result (see [Eisenbud et al. 1996, p. 302] for a proof):

Proposition. *A set of $n \leq 2d + 2$ distinct points in the plane fails to impose independent conditions on curves of degree d if and only if either $d + 2$ of the points are collinear or $n = 2d + 2$ and all the points lie on a conic.* ◆

Exercise 15.4. Suppose that C is a smooth curve and $\nu : C \to C_0$ is a map to a plane curve with only nodes as singularities. Let $D = D_+ - D_-$ be a divisor on C. Modify the technique of Section 15.1.2 to compute the complete linear series $|D|$ without assuming that D is disjoint from the preimages of the singular points. ◆

Exercise 15.5. Let C_0 be a plane quartic curve with two nodes q_1, q_2 and let $\nu : C \to C_0$ be its normalization. By the adjunction formula, C has genus 1. For an arbitrary point $o \in C$ not lying over a node of C_0, give a geometric description of the group law on C with o as origin. ◆

Exercise 15.6. Let p be a point in $\mathbb{P}^2$ and let C_0 be a curve that, in a neighborhood of p, consists of 3 smooth branches that are pairwise simply tangent (C_0 could be given locally analytically by the equation $y(y - x^2)(y + x^2) = 0$, for example.) Use Corollary 15.11 to show that the δ invariant of C_0 at p is 6. ◆

Exercise 15.7. Find the adjoint ideals of these plane curve singularities:

(1) a *triple tacnode*, also known in classical language as a *triple point with an infinitely near triple point*: three smooth branches, pairwise simply tangent;

(2) a triple point with an infinitely near double point: three smooth branches, two of which are simply tangent, with the third transverse;

(3) a unibranch triple point, such as the zero locus of $y^3 - x^4$. ◆

Here is a general description in case the individual branches of C_0 at p are each smooth:

Exercise 15.8. Let $\nu : C \to C_0$ be the normalization of a plane curve C_0 and $p \in C_0$ a singular point. Denote the branches of C_0 at p by $B_1, \dots, B_k$, and let r_i be the point in B_i lying over p. If the individual branches B_i of C_0 at p are each smooth, and we set

$$m_i = \sum_{j \neq i} \text{mult}_p(B_i, B_j),$$

where mult_p is defined as in Section 2.6, then the adjoint ideal of C_0 at p is the ideal of functions g such that $\text{ord}_{r_i} \nu^* g \geq m_i$. ◆

Chapter 16

Linkage and the canonical sheaves of singular curves

16.1. Introduction

In this chapter, curves are purely 1-dimensional projective schemes, not always reduced or irreducible.

Linkage is an equivalence relation on varieties and schemes of a given dimension embedded in a common space. It was a key element in the classification of curves in $\mathbb{P}^3$ for which Max Noether and Georges-Henri Halphen received the Steiner prize of the Prussian Academy of Sciences in 1880, and it was a necessary ingredient in the work of Clebsch, Brill, Noether and Macaulay toward a version of the Riemann–Roch theorem couched in terms of the algebra of plane curves near the end of the nineteenth century. It was put on a firm modern footing in [Peskine and Szpiro 1974], and this foundation was used for further progress in projective geometry by Hartshorne, Rao and others. In this chapter we will explain some of these developments, starting with a simple example, and including the algebra necessary for a formulation in the natural generality of purely 1-dimensional schemes.

As we have seen, any plane curve is arithmetically Cohen–Macaulay, and its arithmetic genus is determined by its degree. Similarly, a curve in $\mathbb{P}^3$ that is a complete intersection of surfaces of degrees d, e is arithmetically Cohen–Macaulay (Theorem 0.1) and has arithmetic genus determined by d, e. Next simplest, perhaps, is a curve C that is *directly linked* to a complete intersection, which means roughly that its union $X = C \cup D$ with a complete intersection curve D is again a complete intersection (see Definition 16.3 for the general definition). We will see that, once again, such a curve C is arithmetically Cohen–Macaulay, and its genus is determined by the degrees of the equations of X and the degree and genus of D. Allowing sequences of direct links we define an equivalence relation called *linkage* or *liaison*, and curves in the linkage class of a complete intersection are said to be *licci*.

A famous theorem of Hartshorne and Rao [Prabhakar Rao 1978/79] shows that the linkage class of a curve $C \subset \mathbb{P}^3$ is classified by the finite-dimensional graded module

$$D(C) := H^1_*(\mathcal{J}_C) := \bigoplus_{m \in \mathbb{Z}} H^1(\mathcal{J}_C(m)),$$

called the *deficiency module* or *Hartshorne–Rao module* of C. (This is a module over the homogeneous coordinate ring $S = H^0_*(\mathcal{O}_{\mathbb{P}^3})$.) The correspondence is explicit: from a finite-dimensional graded module over S one can actually construct curves.

In the first sections of this chapter we will examine the equivalence relation on curves in $\mathbb{P}^3$ that is defined by linkage. Much of the story extends to the case of singular curves. This extension requires an understanding of the dualizing sheaves of singular curves, to which we turn in Section 16.5. We conclude the chapter with an analysis of the adjoint ideal, completing a result from Chapter 15, and allowing us to formulate the Riemann–Roch theorem for general curves and coherent sheaves.

Aside from the classification result above, linkage is useful in analyzing Hilbert schemes. We will exploit this systematically in cases of low degree and genera in Chapter 19, and we begin this chapter with what is perhaps the simplest example, computing the dimension of the component of the Hilbert scheme $\mathrm{Hilb}_{3m+1}(\mathbb{P}^3)$ that is the closure of the open subset $\mathcal{H}^\circ$ parametrizing twisted cubics (see Proposition 7.11 for another proof).

16.2. Linkage of twisted cubics

The simplest example of linkage is that of the union of a twisted cubic and one of its secant lines, pictured in Figure 16.1, and we will start with that.

Any twisted cubic curve $C \subset \mathbb{P}^3$ lies on a nonsingular quadric in class $(1,2)$. Adding a line L of class $(1,0)$ we get a divisor of class $(2,2)$, the class of the complete intersection of two quadrics. Since L is also a complete intersection, C is licci.

We can make the relation of L and C explicit as follows: The ideal of C is minimally generated by the three 2×2 minors of the matrix

$$\begin{pmatrix} x_0 & x_1 & x_2 \\ x_1 & x_2 & x_3 \end{pmatrix}.$$

Figure 16.1. A quadratic cone (red) intersecting a smooth quadric (yellow) in the union of a vertical line and a twisted cubic (credit: Herwig Hauser).

The minor $Q_{1,2}$ involving the first two columns and the minor $Q_{2,3}$ involving the last two columns both vanish on the line $L : x_1 = x_2 = 0$, which meets the twisted cubic in the two points $x_0 = x_1 = x_2 = 0$ and $x_1 = x_2 = x_3 = 0$. Thus L is a secant line to C. A general linear combination Q of $Q_{1,2}$ and $Q_{2,3}$ defines a smooth quadric, which is thus isomorphic to $\mathbb{P}^1 \times \mathbb{P}^1$. The curve C necessarily lies in the divisor class $(1, 2)$ (or, symmetrically, $(2, 1)$), and the line in class $(1, 0)$ (respectively, $(0, 1)$), summing to the complete intersection $(2, 2)$ of Q with (say) $Q_{1,2}$. See Figure 16.1.

Conversely, if two irreducible quadrics Q_1, Q_2 both contain a twisted cubic C then, by Bézout's theorem, $Q_1 \cap Q_2$ is the union of C with a line. If at least one of the quadrics is smooth, we are in the situation above.

This suggests that we set up an incidence correspondence between twisted cubics and their secant lines. Let $\mathbb{P}^9$ denote the projective space of quadrics in $\mathbb{P}^3$, and consider

$$\Phi = \left\{ (C, L, Q, Q') \in \mathcal{H}^\circ \times \mathbb{G}(1, 3) \times \mathbb{P}^9 \times \mathbb{P}^9 \mid Q \cap Q' = C \cup L \right\}.$$

We'll analyze Φ by considering the projection maps to $\mathcal{H}^\circ$ and $\mathbb{G}(1, 3)$; that is, by looking at the diagram

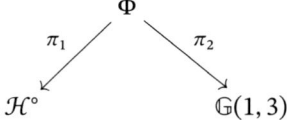

Consider the projection map $\pi_2 : \Phi \to \mathbb{G}(1,3)$ on the second factor. By what we just said, the fiber over any point $L \in \mathbb{G}(1,3)$ is an open subset of $\mathbb{P}^6 \times \mathbb{P}^6$, where $\mathbb{P}^6$ is the space of quadrics containing L. Since $\dim \mathbb{G}(1,3) = 4$ we see that Φ is irreducible of dimension $4 + 2 \times 6 = 16$. On the other hand, the map $\pi_1 : \Phi \to \mathcal{H}^\circ$ is surjective, with fiber over a curve C an open subset of $\mathbb{P}^2 \times \mathbb{P}^2$, where $\mathbb{P}^2$ is the projective space of quadrics containing C; we conclude that $\mathcal{H}^\circ$ is irreducible of dimension 12, in accord with our computation in Proposition 7.11 of the space of twisted cubics as $\mathrm{PGL}_4 / \mathrm{PGL}_2$.

16.3. Linkage of smooth curves in $\mathbb{P}^3$

If the union of two smooth curves in $\mathbb{P}^3$ is a complete intersection then the degrees and genera of the curves are related:

Theorem 16.1. *Let $C_1, C_2 \subset \mathbb{P}^3$ be distinct smooth irreducible curves whose union is the complete intersection of two surfaces S, T of degrees s, t, with S smooth. Then $\deg C_1 + \deg C_2 = st$ and*

$$g(C_1) - g(C_2) = \frac{s+t-4}{2}(\deg C_1 - \deg C_2).$$

In words, the difference between the genera of C_1 and C_2 is proportional to the difference in their degrees, with constant of proportionality $(s+t-4)/2$. In the example of a complete intersection of two quadrics described above, the multiplier $(s+t-4)/2$ is zero, and indeed the line and the twisted cubic have the same genus. The relation of degrees and genera is true more generally, as we shall see in the next section, but the special case is already useful.

Proof. The sum of degrees of C_1 and C_2 is the degree of $C_1 \cup C_2 = S \cap T$, which, by Bézout's theorem, is st.

By the adjunction formula in $\mathbb{P}^3$ the canonical divisor of S has class $K_S = (s-4)H$. Thus, from the adjunction formula on the surface S we get

$$g(C_i) = \frac{C_i^2 + C_i \cdot K_S}{2} + 1 = \frac{C_i^2 + (s-4)\deg C_i}{2} + 1.$$

Subtracting,

$$g(C_1) - g(C_2) = \frac{C_1^2 - C_2^2 + (s-4)(\deg C_1 - \deg C_2)}{2}.$$

Because $C_1 + C_2$ is in the class tH on S we have

$$C_1^2 - C_2^2 = (C_1 + C_2)(C_1 - C_2) = t(\deg C_1 - \deg C_2).$$

Combining the last two displays yields the second formula of the theorem. $\square$

Remark 16.2. Linkage is closely related to linear equivalence. Here is a special case: suppose that S is a smooth surface in $\mathbb{P}^3$, and $C \subset S$ is a curve. If T is a sufficiently general surface of degree t containing C then the curve C' that is the link of C with respect to S, T lies in the class $tH - C$. If we link again with respect to another surface T' of degree t' we thus arrive at $C'' = C + (t - t')H$. Thus if $t = t'$ we get a curve in the same linear equivalence class as C. Moreover, since every rational function on S is the restriction to S of the ratio of two forms of the same degree on $\mathbb{P}^3$, the set of curves on S that can be obtained from C by two linkages with surfaces T, T' of the same degree is exactly the linear series $|C|$ on S. This idea is generalized in the notion of a *basic double link*; see Exercise 16.7.

16.4. Linkage of purely 1-dimensional schemes in $\mathbb{P}^3$

To say that the union X of distinct reduced irreducible curves $C \cup C'$ is a complete intersection means that the ideal I_X equals $I_C \cap I_{C'}$. Since the latter contains $I_C I_{C'}$, the ideal quotient $(I_X : I_C) := \{F \mid FI_C \subset I_X\}$ contains $I_{C'}$.

On the other hand, if $F \notin I_{C'}$ and we choose $G \in I_C \setminus I_{C'}$, then $FG \notin I_{C'}$, so $F \notin (I_X : I_C)$, and thus $(I_X : I_C) = I_{C'}$. This relationship underlies the formulas connecting the degrees and genera of C, C', which hold for arbitrary purely 1-dimensional subschemes of $\mathbb{P}^3$, as we shall see in Theorem 16.5.

Definition 16.3. Let C, C' be purely 1-dimensional subschemes of $\mathbb{P}^3$. We say that C' is *directly linked* to C if there is a complete intersection X containing C, C' and satisfying $(I_X : I_C) = I_{C'}$. We say that C' is *linked* to C if they are connected by a chain of such direct linkages, and we say that C' is *evenly linked* to C if the chain involves an even number of direct linkages.

Note that in this setting, C and C' can have components in common. For example, the subscheme $C \subset \mathbb{P}^3$ defined by the square of the ideal $\mathcal{J}_{L/\mathbb{P}^3}$ of a line $L \subset \mathbb{P}^3$ is linked to the reduced line L in the complete intersection of two quadrics. This makes sense, since C is a flat limit of twisted cubics. C is an example of a rope; see Exercise 16.11 for more.

As in the smooth case treated above, direct linkage is a symmetric relation:

Proposition 16.4. *Let $C_1 \subset \mathbb{P}^3$ be a purely 1-dimensional subscheme with saturated homogeneous ideal I_1 and suppose that C_1 is contained in a complete intersection of hypersurfaces $X := S \cap T$. The ideal $I_2 = (I_X : I_1)$ is a saturated ideal, defining a purely 1-dimensional subscheme and $I_1 = (I_X : I_2)$ as well.*

Proof. The ideal I_X is unmixed of codimension 2, since X is a complete intersection [Eisenbud 1995, Proposition 18.13]. It follows that $I_2 = (I_X : I_1)$ is also unmixed of codimension 2, and therefore saturated. Thus it suffices to prove that $I_1 = (I_X : I_2)$ after localizing at a codimension 2 prime P that contains I_X.

Write R for the localization at P of the homogeneous coordinate ring of X. Because I_X is a complete intersection, the ring R is zero-dimensional and Gorenstein. By [Eisenbud 1995, Propositions 21.1 and 21.5], every finitely generated

R-module is reflexive. Since $I_{C_2}R = \mathrm{ann}_R(I_{C_1}R) \cong \mathrm{Hom}_R(R/I_{C_1}R, R)$, the proposition follows. □

16.5. Degree and genus of linked curves

The degrees and arithmetic genera of directly linked schemes are related exactly as in the pilot case of Section 16.3.

Theorem 16.5. *If $C_1, C_2 \subset \mathbb{P}^3$ are purely 1-dimensional schemes that are directly linked by surfaces S, T of degrees s, t, then $\deg C_1 + \deg C_2 = st$ and*

$$p_a(C_1) - p_a(C_2) = \frac{s+t-4}{2}(\deg C_1 - \deg C_2).$$

Since we have left the realm of smooth curves and surfaces, we will need a more sophisticated duality theory, and we postpone the proof to explain the necessary ideas.

Dualizing sheaves for singular curves. In Chapter 2 we said that the canonical sheaf of a smooth curve — the sheaf of differential forms — was the most important invertible sheaf after the structure sheaf. In the general setting of Cohen–Macaulay schemes, the analogue of the canonical sheaf is known as the dualizing sheaf. The general definition of the dualizing sheaf of a pure-dimensional projective scheme is not very illuminating; what is useful is how it is constructed and its cohomological properties relating to duality. However, having a definition may be comforting.

Definition 16.6. Let X be a projective scheme of pure dimension d over $\mathbb{C}$. The *dualizing sheaf* for X is a coherent sheaf ω_X together with a *residue map* $\eta : H^d(\omega_X) \to \mathbb{C}$ such that for every coherent sheaf $\mathcal{F}$ the composite map

$$H^d(\mathcal{F}) \times \mathrm{Hom}(\mathcal{F}, \omega_X) \to H^d(\omega_X) \xrightarrow{\eta} \mathbb{C}$$

is a perfect pairing.

If $\mathcal{F} = \mathcal{L}$ is an invertible sheaf on a projective curve C then $\mathrm{Hom}(\mathcal{L}, \omega_X) = \mathcal{L}^{-1} \otimes \omega$, so we recover Serre duality: $H^1(\mathcal{L})$ is the dual of $H^0(\mathcal{L}^{-1} \otimes \omega_C)$.

It follows from the definition that the pair (ω_X, η) is unique up to canonical isomorphism if it exists: The module $H^0_*(\omega_X) = \bigoplus_{n \in \mathbb{Z}}(\mathrm{Hom}_X(\mathcal{O}_X(n), \omega_X))$ is determined as the graded vector space dual of $H^d_*(\mathcal{O}_X)$, and the choice of η simply fixes the isomorphism. It may not be apparent that such a sheaf exists, but we will give a construction in Section 16.6.

Several properties of the dualizing sheaf on a purely 1-dimensional scheme are the same as in the smooth case, and follow easily from the definition:

Proposition 16.7. *Let C be a purely 1-dimensional projective scheme.*

(1) $\mathcal{H}om_C(\omega_C, \omega_C) = \mathcal{O}_C$; thus if C is integral then the generic rank of ω_C is 1.

(2) *For any invertible sheaf $\mathcal{L}$ on C we have*

$$H^1(\mathcal{L}^{-1}) = H^0(\mathcal{L} \otimes \omega_C) \quad \text{and} \quad H^0(\mathcal{L}^{-1}) = H^1(\mathcal{L} \otimes \omega_C).$$

In particular, $\chi(\mathcal{L}^{-1}) = -\chi(\mathcal{L} \otimes \omega_C)$.

Proof. (1) We will show that the natural map $\mathcal{O}_C \to \mathcal{H}om_C(\omega_C, \omega_C)$ is an isomorphism. Since the map is globally defined, it suffices to prove that it is an isomorphism locally.

Choose a Noether normalization of C, that is, a finite map $f : C \to \mathbb{P}^1$. We shall see in Theorem 16.8 below that $\omega_C \cong \mathcal{H}om_{\mathbb{P}^1}(\mathcal{O}_C, \omega_{\mathbb{P}^1})$, regarded as a sheaf on C (in Theorem 16.8 this is the sheaf $f^! \omega_{\mathbb{P}^1}$). Since C is purely 1-dimensional, $\mathcal{O}_C$ is torsion-free as an $\mathcal{O}_{\mathbb{P}^1}$-module, and is thus locally free. Also, since $\mathbb{P}^1$ is smooth, $\omega_{\mathbb{P}^1}$ is locally isomorphic to $\mathcal{O}_{\mathbb{P}^1}$. But if B is any commutative ring and A is a B-algebra that is finitely generated and free as a B-module, then the composite map

$$A \to \operatorname{Hom}_A(\operatorname{Hom}_B(A, B), \operatorname{Hom}_B(A, B)) \cong \operatorname{Hom}_B(\operatorname{Hom}_B(A, B), B)$$

sending an element a to the multiplication by a and thence to the map $f \mapsto f(a)$ may be identified with the isomorphism of A to its double dual as a B-module; this completes the argument. (See Exercise 16.8 for a generalization of this last step.)

(2) The definition of ω_C shows that, if $\mathcal{L}$ is an invertible sheaf, then

$$H^1(\mathcal{L}^{-1}) = \operatorname{Hom}_C(\mathcal{L}^{-1}, \omega_C)) = H^0(\mathcal{H}om_C(\mathcal{L}^{-1}, \omega_C)) = H^0(\mathcal{L} \otimes_C \omega_C).$$

Using part (1), we have

$$
\begin{aligned}
H^0(\mathcal{L}^{-1}) &= H^0(\mathcal{L}^{-1} \otimes_C \mathcal{O}_C) \\
&= H^0(\mathcal{L}^{-1} \otimes_C \mathcal{H}om_C(\omega_C, \omega_C)) \\
&= \operatorname{Hom}_C(\mathcal{L} \otimes_C \omega_C, \omega_C)) \\
&= H^1(\mathcal{L} \otimes_C \omega_C)
\end{aligned}
$$
□

16.6. The construction of dualizing sheaves

Dualizing sheaves do exist on any purely 1-dimensional projective scheme, and more generally on any projective Cohen–Macaulay scheme. We have already seen constructions in three cases:

- If X is a smooth scheme of dimension d over $\mathbb{C}$ then $\omega_X = \bigwedge^d \Omega_{X/\mathbb{C}}$ is a dualizing sheaf [Hartshorne 1977, Section III.7; [1978, p. 648, 708]].
- If $f : X \to Y$ is a map of smooth curves, then $\omega_X = f^*(\omega_Y)(\operatorname{ram}_{X/Y})$, where ram denotes the ramification divisor.
- If $X \subset Y$ is a Cartier divisor on a surface, then $\omega_X = \omega_Y(X)|_X$.

How can such different looking formulas all be correct? Grothendieck provided a general scheme that unifies them and gives many more. To understand what is needed for the general case, we first consider a setting generalizing Hurwitz's theorem. Suppose that $X \to Y$ is a finite map of projective schemes, and that $\mathcal{F}$ is a coherent sheaf on X.

If we restrict ourselves to open affine subsets $U := \operatorname{Spec} A \subset X$ mapping to $V := \operatorname{Spec} B \subset Y$ via the map of rings $f^* : B \to A$, then $F := \mathcal{F}_U$ is an A-module. Moreover, $f_*F := f_*(\mathcal{F})(V)$ is just F regarded as a B-module via the map f^*.

For any B-module M the module $\operatorname{Hom}_B(A, M)$ has a natural A-module structure, where $(a\phi)(m)$ is defined to be $\phi(am)$. The functor

$$f^!(-) := \operatorname{Hom}_B(A, -) : \operatorname{mod}_B \to \operatorname{mod}_A$$

defined in this way is the right adjoint of the functor $f_*(-) : \operatorname{mod}_A \to \operatorname{mod}_B$, which means that there is a natural isomorphism of functors

$$\operatorname{Hom}_B(f_*F, -) \cong \operatorname{Hom}_A(F, \operatorname{Hom}_B(A, -)) = \operatorname{Hom}_A(F, f^!(-)).$$

Thus if Y has dualizing sheaf ω_Y and we set $o_Y := \omega_Y(V)$, then

$$\operatorname{Hom}_B(f_*F, o_Y) \cong \operatorname{Hom}_A(F, f^!o_Y).$$

Also, there is a natural transformation η from $f_*f^!$ to the identity functor given by the formula

$$\eta : f_*f^!(M) = f_*(\operatorname{Hom}_B, (A, M)) = \operatorname{Hom}_B, (A, M) \to M, \quad \eta(\phi) = \phi(1),$$

for any A-module M. The transformation η called the *counit* of the adjoint pair $(f_*, f^!)$.

We also write $f^!(-)$ for the sheafification of the functor $\operatorname{Hom}_B(A, -)$. Again $f^!$ is right adjoint to f_* on coherent sheaves, and again there are natural maps $\eta : f_*f^!(\mathcal{F}) \to \mathcal{F}$.

Theorem 16.8. *Let $f : X \to Y$ be a finite map of d-dimensional projective schemes. If Y has a dualizing sheaf ω_Y, with residue map $\eta_Y : H^d(\omega_Y) \to \mathbb{C}$, then*

$$\omega_X := f^!\omega_Y,$$

with residue map

$$\rho_X : H^d(f^!\omega_X) = H^d(f_*f^!\omega_X) \xrightarrow{H^d(\eta)} H^d(\omega_Y) \xrightarrow{\rho_Y} \mathbb{C},$$

where η is the counit of the adjoint pair $(f_*, f^!)$, is a dualizing sheaf on X.

Proof. Let $\mathcal{F}$ be a coherent sheaf on X. Since f is finite, $H^d(\mathcal{F}) = H^d(f_*(\mathcal{F}))$. Thus, since $f^!(-)$ is a right adjoint of f_* there are natural isomorphisms

$$H^d(\mathcal{F}) = H^d(f_*\mathcal{F}) \cong \operatorname{Hom}_Y(f_*\mathcal{F}, \omega_Y)^\vee \cong \operatorname{Hom}_X(\mathcal{F}, f^!\omega_Y)^\vee,$$

the first being induced by ρ_Y. One can check that the composite isomorphism is the one induced by ρ_X, so $\omega_X = f^!(\omega_Y)$, completing the proof. $\qquad\square$

Cheerful Fact 16.9. Given this theorem, it seems natural to look for an adjoint functor $f^!$ for a wider class of morphisms f, but… in most cases, for example when f is the inclusion of a divisor on a smooth surface, no such functor exists on the category of coherent sheaves! An adjoint functor $f^!$ does exist on the derived category, where it is the right adjoint to Rf_*, leading to a theory of dualizing complexes.

Fortunately for the reader who is mostly interested in curves, this level of complication is unnecessary, and there is an intermediate level of generality that suffices for all the purposes of this book and more:

Theorem 16.10. *Suppose that $f : X \to Y$ is a finite map of projective schemes. If Y is Gorenstein with dualizing module ω_Y, then*

$$f^!(\omega_Y) \cong \mathcal{E}xt_Y^{\dim Y - \dim X}(\mathcal{O}_X, \omega_Y)$$

is a dualizing module for X.

The hypothesis is satisfied for any Y that is smooth, or even locally a complete intersection. The reason this works is that the complex $f^!\omega_Y$ can be identified with its one nonvanishing cohomology module, $\mathcal{E}xt_Y^{\dim Y - \dim X}(\mathcal{O}_X, \omega_Y)$. See for example [Altman and Kleiman 1970] for a thorough and accessible exposition and the website [Schreiber et al. 2011–] for an extensive list of further references.

Proof of Theorem 16.5. Let X be the complete intersection of surfaces of degrees s, t containing C, and let $R_X = S/(F, G)$ be its homogeneous coordinate ring, where $S = \mathbb{C}[x_0, \ldots, x_3]$ is the homogeneous coordinate ring of $\mathbb{P}^3$. From the free resolution

$$0 \to S(-s-t) \xrightarrow{\binom{G}{-F}} S(-s) \oplus S(-t) \xrightarrow{(F\ G)} S \to R_X \to 0$$

and Theorem 16.10 we see that

$$\omega_X = \mathcal{E}xt_C^2(\mathcal{O}_X, \omega_{\mathbb{P}^3}) = \mathcal{E}xt^2(\mathcal{O}_X, \mathcal{O}_{\mathbb{P}^3}(-4)) = \mathcal{O}_X(s+t-4).$$

Note that for any ideals $J \subset I$ in a ring A we have $\operatorname{Hom}_A(A/I, A/J) \cong (J : I)/J$, where the isomorphism sends a homomorphism ϕ to the element $\phi(1)$. From Theorem 16.10 we have

$$\omega_C = \mathcal{H}om_X(\mathcal{O}_C, \omega_X) = \mathcal{H}om_X(\mathcal{O}_C, \mathcal{O}_X)(s+t-4) = \frac{\mathcal{I}_X : \mathcal{I}_C}{\mathcal{I}_X}(s+t-4),$$

where we have identified $\mathcal{O}_C$ with its pushforward under the inclusion map $C \to X$.

By Proposition 16.7 we have $\chi(\omega_C(m)) = -\chi(\mathcal{O}_C(-m))$. It follows that the leading coefficient of the Hilbert polynomial of ω_C is equal to $\deg C$, and thus

$$st = \deg \mathcal{O}_X = \deg \mathcal{O}_{C'} + \deg \mathcal{O}_C,$$

as required by the formula for the sum of the degrees.

From Theorem 16.8 (or Theorem 16.10) we see that $\chi(\mathcal{O}_X) = st(4-s-t)/2$. Since $\mathcal{O}_{C'} = \mathcal{O}_{\mathbb{P}^3}/(\mathcal{J}_X : \mathcal{J}_C)$ and $(\mathcal{J}_X : \mathcal{J}_C)/(\mathcal{J}_X) = \omega_C(4-s-t)$ we have

$$\frac{4-s-t}{2}(\deg C + \deg C') = \frac{4-s-t}{2}st$$

$$= \chi(\mathcal{O}_X)$$

$$= \chi(\mathcal{O}_{C'}) + \chi(\omega_C(4-s-t))$$

$$= \chi(\mathcal{O}_{C'}) - \chi(\mathcal{O}_C(s+t-4))$$

$$= \chi(\mathcal{O}_{C'}) - (s+t-4)\deg C - \chi(\mathcal{O}_C)$$

$$= (1-p_a(\mathcal{O}_{C'})) - (1-p_a(\mathcal{O}_C)) - (s+t-4)\deg C,$$

whence

$$p_a(\mathcal{O}_C) - p_a(\mathcal{O}_{C'}) = \frac{s+t-4}{2}(\deg C - \deg C'). \qquad \square$$

Linkage also behaves in a simple way with respect to deficiency modules:

Theorem 16.11. *If C, C' are purely 1-dimensional subschemes of $\mathbb{P}^3$ that are directly linked by a complete intersection of degrees s, t then*

$$D(C') = \mathrm{Hom}_{\mathbb{C}}(D(C), \mathbb{C})(-s-t+4)$$

as graded modules over the homogeneous coordinate ring of $\mathbb{P}^3$.

A more general form of this result appears as Proposition 2.5 in [Peskine and Szpiro 1974], with an attribution to Daniel Ferrand.

Proof. Suppose that the homogeneous ideal of C is generated by forms of degree a_i, $i = 1,\ldots,s$. Since C is locally Cohen–Macaulay, the local rings $\mathcal{O}_{C,p}$ have projective dimension 2 as modules over $\mathcal{O}_{\mathbb{P}^3,p}$, and $\mathcal{J}_{C,p}$ has projective dimension 1. Thus we have an exact sequence

$$0 \to \mathcal{E} \to \bigoplus_i \mathcal{O}_{\mathbb{P}^3}(-a_i) \to \mathcal{J}_C \to 0.$$

Since the first and second cohomology groups of the twists of $\mathcal{O}_{\mathbb{P}^3}$ vanish, we deduce an isomorphism

$$D(C) := \bigoplus_{m\in\mathbb{Z}} H^1(\mathcal{J}_C(m)) \cong \bigoplus_{m\in\mathbb{Z}} H^2(\mathcal{E}(m)).$$

Let X be the complete intersection of two hypersurfaces, of degrees s, t, containing C. From the inclusion we deduce a map of resolutions

$$
\begin{array}{ccccccccc}
0 & \longrightarrow & \mathcal{E} & \longrightarrow & \bigoplus_i \mathcal{O}_{\mathbb{P}^3}(-a_i) & \longrightarrow & \mathcal{O}_{\mathbb{P}^3} & \longrightarrow & \mathcal{O}_C & \longrightarrow & 0 \\
& & \uparrow & & \uparrow & & \uparrow{\scriptstyle =} & & \uparrow & & \\
0 & \longrightarrow & \mathcal{O}_{\mathbb{P}^3}(-s-t) & \longrightarrow & \mathcal{O}_{\mathbb{P}^3}(-s) \oplus \mathcal{O}_{\mathbb{P}^3}(-t) & \longrightarrow & \mathcal{O}_{\mathbb{P}^3} & \longrightarrow & \mathcal{O}_X & \longrightarrow & 0
\end{array}
$$

We dualize this diagram, form the mapping cone, and twist by $-s - t$. Note that $\mathrm{Hom}_{\mathbb{P}^3}(\mathcal{O}_C, \mathcal{O}_{\mathbb{P}^3}) = 0$. Also, since the vertical map $\mathcal{O}_{\mathbb{P}^3} \to \mathcal{O}_{\mathbb{P}^3}$ on the right is the identity we may cancel these terms in the mapping cone. Noting that $\omega_C = \mathcal{E}xt^2(\mathcal{O}_C, \mathcal{O}_{\mathbb{P}^3}(-4))$ the result is a diagram with exact rows:

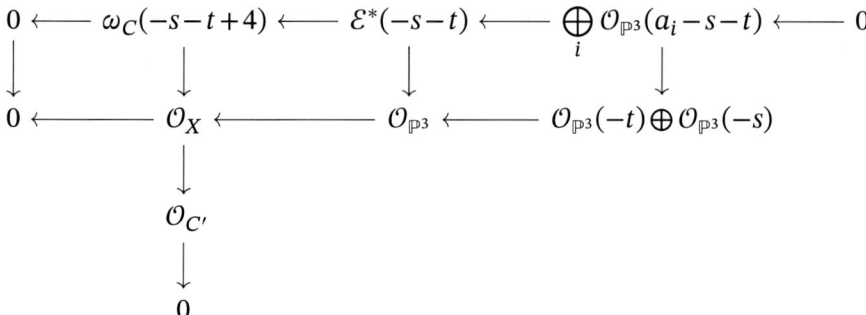

The map ϕ is a monomorphism because $(\mathcal{I}_X : \mathcal{I}_C)/\mathcal{I}_X \cong \omega_C(-s - t + 4)$, as explained above, so the column on the left is a short exact sequence. We can now write a resolution of $\mathcal{I}_{C'}$ as the mapping cone:

$$0 \leftarrow \mathcal{I}_{C'} \leftarrow \mathcal{O}_{\mathbb{P}^3}(-t) \oplus \mathcal{O}_{\mathbb{P}^3}(-s) \oplus \mathcal{E}^*(-s - t) \leftarrow \bigoplus_i \mathcal{O}_{\mathbb{P}^3}(a_i - s - t) \leftarrow 0.$$

From this we see that

$$H^1(\mathcal{I}_{C'}(m)) \cong H^1(\mathcal{E}^*(-s - t + m)) \cong \mathrm{Hom}_{\mathbb{C}}(H^2(\mathcal{E}(s + t - m - 4)), \mathbb{C}),$$

where the last equality is from Serre duality on $\mathbb{P}^3$. Summing over m we see that $D(C') \cong \mathrm{Hom}_{\mathbb{C}}(D(C)(s + t - 4), \mathbb{C})$, and since Serre duality is functorial, the isomorphism holds not only as graded vector spaces, but as graded S-modules. $\square$

Sometimes the following consequence is a useful way to compute the deficiency module:

Proposition 16.12. *If C is a purely 1-dimensional subscheme of $\mathbb{P}^3$ with homogeneous ideal $I = I_C$ then*

$$D(C) \cong \mathrm{Hom}_{\mathbb{C}}(\mathrm{Ext}^3(S/I, S), \mathbb{C})(-4), \mathbb{C})$$

as graded modules over the homogeneous coordinate ring S of $\mathbb{P}^3$.

Proof. We may choose a surjection $\psi : \bigoplus_i S(-a_i) \to I$, and choose the map $\phi : \bigoplus_i \mathcal{O}_{\mathbb{P}^3}(-a_i) \to \mathcal{I}_C$ in the proof of Theorem 16.11 to be the corresponding map of sheaves, so that $\mathcal{E}$ is the sheafification of the graded module $E = \ker \psi$.

Since I is a saturated ideal, the depth of S/I is at least 1, so pd $S/I \leq 3$, and I has a free resolution of the form

$$0 \to G \to F \to \bigoplus_i S(-a_i) \to S \to S/I \to 0.$$

where $G \to F$ is a free presentation of E. and there is an exact sequence

$$0 \to E^* \to F^* \to G^* \to \text{Ext}_S^3(S/I, S) \to 0.$$

By Theorem 18.10, the fact that C is Cohen–Macaulay implies that the projective dimension of each of its local rings is ≤ 2, and it follows that $\text{Ext}_S^3(S/I, S)$ has finite length. Writing $\widetilde{(\ \ )}$ for the sheafification functor, we have a short exact sequence of sheaves

$$0 \to \mathcal{E}^* \to \widetilde{F^*} \to \widetilde{G^*} \to 0.$$

From this we see that

$$\text{Ext}_S^3(S/I, S) = H_*^1(\mathcal{E}^*) = \text{Hom}_{\mathbb{C}}(H_*^2(\mathcal{E}(-4)), \mathbb{C}) = H_*^1(\mathcal{I})(-4),$$

proving the assertion. □

Proposition 16.12 is actually a special case of the local duality isomorphism between local cohomology and the dual of Ext; see for example [Eisenbud 2005, Theorem A.1.9].

16.7. The linkage equivalence relation

As an immediate consequence of Theorem 16.11 we have:

Corollary 16.13 (Hartshorne). *If two curves C, C' are linked by an even length chain of direct linkages, then $D(C)$ and $D(C')$ are isomorphic up to a shift in grading.* □

As we mentioned at the beginning of this chapter, the converse is also true: the Hartshorne–Rao modules, up to shift in grading, provide a complete invariant of linkage.

Cheerful Fact 16.14. Even more precise results are known (and the characteristic 0 hypothesis is largely unnecessary); here is a sample:

Theorem 16.15. *Let $S = \mathbb{C}[x_0, \ldots, x_3]$ be the homogeneous coordinate ring of $\mathbb{P}^3$, and let M be a graded S-module of finite length.*

(1) *There is a smooth curve C with $D(C) = M(m)$ for some integer m.*

(2) *There is a minimum value of m such that $M(m) = D(C_0)$ for some purely one-dimensional scheme C_0.*

Moreover, each linkage class has a relatively simple structure, known as the *Lazarsfeld–Rao property*. We say that C' is obtained from C by an *ascending double link* if $I_{C'} = fI_C + (g)$ for some regular sequence contained in I_C — see Exercise 16.7.

Theorem 16.16 [Ballico et al. 1991]. *Let $M = D(C_0)$ the Hartshorne–Rao module of a purely 1-dimensional subscheme of $\mathbb{P}^3$, and suppose that M is minimal in the sense that no $M(m)$ with $m > 0$ is the invariant of a purely 1-dimensional scheme.*

(1) *Any curve in $\mathbb{P}^3$ with $D(C) = M$ is a deformation of C_0 through curves with invariant M.*

(2) *Every curve in the even linkage class of C_0 is the result of a series of ascending double links followed by a deformation.*

In [Lazarsfeld and Rao 1983] it is shown that general curves in $\mathbb{P}^3$ that have reasonably large degree compared to their genus are minimal in the sense of Theorem 16.16.

16.8. Comparing the canonical sheaf with that of the normalization

In Chapter 15 we boasted in that we could effectively compute linear series on a smooth curve C given any plane curve C_0 with normalization C, and we showed how to do this when the plane curve has only nodes. To complete the discussion we need to compare the canonical sheaf ω_C of C with the dualizing sheaf ω_{C_0} of C_0; that is, we need a formula for the adjoint ideal of any curve singularity. A simplification occurs when the dualizing sheaf is invertible, that is, when the curve is Gorenstein (as is every plane curve).

Theorem 16.17. *If $\nu : C \to C_0$ is the normalization of a reduced connected projective curve then the adjoint ideal*

$$\mathfrak{A}_{C/C_0} := \mathrm{ann}_{\mathcal{O}_{C_0}} \frac{\omega_{C_0}}{\nu_* \omega_C}$$

is equal to the conductor ideal

$$\mathfrak{f}_{C/C_0} := \mathrm{ann}_{\mathcal{O}_{C_0}} \frac{\nu_* \mathcal{O}_C}{\mathcal{O}_{C_0}}.$$

Moreover, if C_0 is Gorenstein, then

$$\delta(C_0) = \mathrm{length}\, \frac{\nu_* \mathcal{O}_C}{\mathcal{O}_{C_0}} = \mathrm{length}\, \frac{\mathcal{O}_{C_0}}{\mathfrak{f}_{C/C_0}}.$$

As explained in Chapter 2, we can think of $\delta(C_0)$ as the number of nodes equivalent to the singularities of C_0. The formula for $\delta(C_0)$ in the theorem was first noted in Daniel Gorenstein's thesis[1] under Oscar Zariski. Hyman Bass [1963] explains that this is why Grothendieck named Cohen–Macaulay rings that have cyclic canonical modules after Gorenstein.

---

[1]Gorenstein is better remembered for his work on the classification of finite simple groups.

In Examples 15.19–15.21 we used the second equality in the formula for $\delta(C_0)$ in Theorem 16.17 to compute the δ invariant for several plane curve singularities. It fails for many space curve singularities; see Example 16.19 for a singularity that is not Gorenstein and behaves differently.

Proof of Theorem 16.17. Let $\rho : C_0 \to \mathbb{P}^1$ be a finite morphism. Both $\rho_*\nu_*\mathcal{O}_C$ and $\rho_*\mathcal{O}_{C_0}$ are torsion free coherent sheaves over $\mathcal{O}_{\mathbb{P}^1}$, and are thus locally free. Since ρ_* is left-exact, the inclusion $\mathcal{O}_{C_0} \subset \nu_*\mathcal{O}_C$ pushes forward to an inclusion

$$\alpha : \rho_*\mathcal{O}_{C_0} \hookrightarrow \rho_*\nu_*\mathcal{O}_C$$

and since $\mathcal{O}_C$ is equal to $\mathcal{O}_{C_0}$ generically on $\mathbb{P}^1$, the cokernel coker α has finite length; indeed, it is supported on the image in $\mathbb{P}^1$ of the singular locus of C_0. Since the maps ν and ρ are finite, we may harmlessly think of both $\mathcal{O}_{C_0}$ and $\mathcal{O}_C$ as coherent sheaves on $\mathbb{P}^1$, and we will simplify the notation by dropping ν_* and ρ_*. Taking duals into $\omega_{\mathbb{P}^1} = \mathcal{O}_{\mathbb{P}^1}(-2)$ and defining $\alpha^\vee := \mathrm{Hom}_{\mathbb{P}^1}(\alpha, \mathcal{O}_{\mathbb{P}^1})$ we get a map that fits into the long exact sequence of $\mathcal{E}xt_{\mathbb{P}^1}(-, \omega_{\mathbb{P}^1})$:

$$0 \to \mathrm{Hom}_{\mathbb{P}^1}(\mathrm{coker}\,\alpha, \omega_{\mathbb{P}^1}) \to \omega_C \xrightarrow{\alpha^\vee} \omega_{C_0}$$
$$\to \mathcal{E}xt^1_{\mathbb{P}^1}(\mathrm{coker}\,\alpha, \omega_{\mathbb{P}^1}) \to \mathcal{E}xt^1_{\mathbb{P}^1}(\mathcal{O}_C, \omega_{\mathbb{P}^1}) \to \cdots$$

We know that coker α has finite support, so $\mathrm{Hom}_{\mathbb{P}^1}(\mathrm{coker}\,\alpha, \omega_{\mathbb{P}^1})$ is trivial and $\mathcal{E}xt^1_{\mathbb{P}^1}(\mathrm{coker}\,\alpha, \omega_{\mathbb{P}^1})$ has the same length and the same annihilator as coker α. Because $\mathcal{O}_C$ is locally free as an $\mathcal{O}_{\mathbb{P}^1}$-module, the term $\mathcal{E}xt^1_{\mathbb{P}^1}(\mathcal{O}_C, \omega_{\mathbb{P}^1})$ vanishes, and we get the more manageable exact sequence

$$0 \to \omega_C \xrightarrow{\alpha^\vee} \omega_{C_0} \to \mathcal{E}xt^1_{\mathbb{P}^1}(\mathrm{coker}\,\alpha, \mathcal{O}_{\mathbb{P}^1}) \to 0.$$

It follows that the sheaves $\nu_*\mathcal{O}_C/\mathcal{O}_{C_0}$ and $\omega_{C_0}/\nu_*\omega_C$ have the same length $\delta(C_0)$. Note that the conductor $\mathfrak{f}_{C/C_0}$ (the annihilator of $\mathcal{O}_C/\mathcal{O}_{C_0}$ in $\mathcal{O}_{C_0}$) is at the same time an ideal sheaf of $\mathcal{O}_{C_0}$ and an ideal sheaf of $\mathcal{O}_C$ via the inclusion $\mathcal{O}_{C_0} \subset \mathcal{O}_C$. The argument above shows that $\mathfrak{f}_{C/C_0}$ is also the annihilator ideal of ω_{C_0}/ω_C. By definition, this is the adjoint ideal of C_0, proving the first statement of the theorem.

A further simplification occurs when $C_0 \subset \mathbb{P}^2$ is a plane curve, or more generally any Gorenstein curve. If the defining equation of C_0 is the form F of degree d, then there is a locally free resolution of $\mathcal{O}_{C_0}$ of the form

$$0 \to \mathcal{O}_{\mathbb{P}^2}(-d) \xrightarrow{F} \mathcal{O}_{\mathbb{P}^2} \to \mathcal{O}_{C_0} \to 0.$$

Thus $\omega_{C_0} \cong \mathcal{E}xt^1_{\mathbb{P}^2}(\mathcal{O}_{C_0}, \omega_{\mathbb{P}^2})$ is the cokernel of the map $\mathcal{O}_{\mathbb{P}^2}(-3) \to \mathcal{O}_{\mathbb{P}^2}(d-3)$ given by multiplication by F. It follows that ω_{C_0} is locally cyclic, and since ω_{C_0}/ω_C has finite support, ω_{C_0}/ω_C is globally cyclic, so $\omega_{C_0}/\omega_C \cong \mathcal{O}_{C_0}/\mathfrak{f}_{C/C_0}$. Since the length of ω_{C_0}/ω_C is equal to the length of $\mathcal{O}_C/\mathcal{O}_{C_0}$, the last statement of the theorem follows. $\qquad\square$

Example 16.18. Working locally, consider the germ of a node, represented by the ring $R_0 := k[\![x,y]\!]/(xy)$, and the projection to a line represented by the inclusion

$$P := k[\![t]\!] \subset R_0 : t \mapsto x + y.$$

The normalization of R_0 is the map $R_0 \to R := k[\![x]\!]e_1 \times k[\![y]\!]e_2$, where $e_1 = x/t$ and $e_2 = y/t$ are orthogonal idempotents. Writing

$$Q_1 = k(\!(t)\!) \subset Q := k(\!(x)\!) \times k(\!(y)\!)$$

for the map of total quotient rings, we know that, because the extension is separable, the trace map $\mathrm{Tr} := \mathrm{Tr}_{Q/Q_1} : Q \to Q_1$ generates $\mathrm{Hom}_{Q_1}(Q,Q_1)$ as a Q-vector space. Thus we may write the elements of $\omega_{R_0} = \mathrm{Hom}_P(R_0,P)$ and $\omega_R = \mathrm{Hom}_P(R,P)$ as multiples of Tr by elements of Q.

Since $R \cong Pe_1 \oplus Pe_2$ as a P-module, the module $\mathrm{Hom}_P(R,P)$ is generated by the two projections, and it is easy to check that these are the maps $(x/t)\,\mathrm{Tr}$ and $(y/t)\,\mathrm{Tr}$. One can also check easily that

$$g := \frac{x-y}{t^2}\,\mathrm{Tr} \in \mathrm{Hom}_P(R_0,R).$$

Since $xg = x/t$ and $yg = y/t$ in Q we have $xg = x/t\,\mathrm{Tr}$ and $yg = y/t\,\mathrm{Tr}$, the generators of $\mathrm{Hom}_P(R,P)$. The ring R_0, regarded as a P-module, is freely generated by 1 and x. Immediate computation shows that $g\,\mathrm{Tr}(1) = 0$ while $g\,\mathrm{Tr}(x) = 1$. Furthermore $(x-y)g = 1$ in Q, and thus $\frac{1}{2}(x-y)g\,\mathrm{Tr}$ takes 1 to 1 and x to t, proving that $g\,\mathrm{Tr}$ generates $\mathrm{Hom}_P(R_0,P)$. We also see directly from this that the adjoint ideal of ω_{R_0}/ω_R is the conductor ideal $(x,y)R_0 = (x,y)R = \mathfrak{f}_{R/R_0}$, as shown by Theorem 16.17.

Example 16.19. In Example 15.22 we showed that if C has a spatial triple point at 0, so that the completion of its local ring has the form

$$R_0 := k[\![x,y,z]\!]/(xy,xz,yz),$$

with normalization R, then it has δ invariant 2 but the conductor is $\mathfrak{f}_{R/R_0} = (x,y,z)R = (x,y,z)R_0$, and thus

$$\delta(C_0) \neq \mathrm{length}\,\frac{\mathcal{O}_{C_0}}{\mathfrak{f}_{C/C_0}}.$$

The S-module R_0, on the other hand, has free resolution

$$0 \longrightarrow S^2 \xrightarrow{\begin{pmatrix} 0 & z \\ y & -y \\ -x & 0 \end{pmatrix}} S^3 \xrightarrow{(xy\ xz\ yz)} S \longrightarrow R_0 \longrightarrow 0.$$

The canonical module of R_0 (which would be the germ of the canonical module of a global curve with such a singularity) is thus

$$\omega_{R_0} = \mathrm{Ext}^2(R_0,S) = \mathrm{coker}\begin{pmatrix} 0 & y & -x \\ z & -y & 0 \end{pmatrix},$$

a module requiring 2 generators. This shows that R_0 is not Gorenstein.

16.9. A general Riemann–Roch theorem

Using dualizing sheaves we can state a more general version of the Riemann–Roch theorem, applicable to any reduced, irreducible projective curve and any coherent sheaf thereupon. We will not make use of this generality later, so we only sketch the argument. We first need to extend the notion of degree of a sheaf:

Definition 16.20. If $\mathcal{F}$ is a sheaf of generic rank r on a projective reduced and irreducible curve C we define the *degree* of $\mathcal{F}$ as $\deg \mathcal{F} := \chi(\mathcal{F}) - r\chi(\mathcal{O}_C)$. Keeping in mind the definition of the arithmetic genus, we can write

$$(*) \qquad \chi(\mathcal{F}) = \deg \mathcal{F} + r\chi(\mathcal{O}_C) = \deg \mathcal{F} + r(1 - p_a(C)).$$

This formula would be pointless if there were no other way to compute $\deg \mathcal{F}$, but that is not the case:

Cheerful Fact 16.21. The degree of $\mathcal{F}$ coincides with that of a certain divisor class, the *first Chern class* of $\mathcal{F}$. See [Eisenbud and Harris 2016, Chapter 5] for more information.

More to the point, if $\mathcal{F}$ is a coherent sheaf on the reduced, irreducible projective curve C of generic rank r, and if $\mathcal{L}$ is an invertible sheaf on C, then the following assertions can be proved using nothing more than the additivity of the Euler characteristic:

(1) If $\mathcal{F}$ is generated by its global sections and $\sigma_1, \ldots, \sigma_r$ is a maximal generically independent collection of global sections of $\mathcal{F}$, then

$$M = \operatorname{coker}(\mathcal{O}_C^r \xrightarrow{(\sigma_1, \ldots, \sigma_r)} \mathcal{F})$$

has finite support, and

$$\deg \mathcal{F} = r\chi(\mathcal{O}_C) + \dim_{\mathbb{C}} H^0(M).$$

(2) $\deg(\mathcal{L} \otimes \mathcal{F}) = \deg \mathcal{F} + r \deg \mathcal{L}$.

Thus if $\mathcal{O}_C(1)$ is a very ample invertible sheaf on C and m is an integer that is large enough so that $\mathcal{F}(m)$ is generated by global sections, then the degree of $\mathcal{F}(m)$ and the degree of $\mathcal{O}_C(1)$ are computed by the formula in item (1) and $\deg \mathcal{F} = \deg \mathcal{F}(m) - m \operatorname{rank}(\mathcal{F}) \deg \mathcal{O}_C(1)$.

Using these facts and the dualizing property of ω_C we can reexpress the equality $(*)$ in the definition of $\deg \mathcal{F}$ as follows:

Theorem 16.22. *If C is a reduced and irreducible curve and $\mathcal{F}$ is a coherent sheaf on C, then*

$$h^0(\mathcal{F}) = \deg \mathcal{F} + \operatorname{rank}(\mathcal{F})(1 - p_a(C)) + h^0(\operatorname{Hom}_C(\mathcal{F}, \omega_C)).$$

16.10. Exercises

Exercise 16.1. Verify that the genus formula in Theorem 16.5 agrees with the usual calculation of degrees and genera for divisors on a quadric of classes (a, b) and $(d - a, d - b)$.

Exercise 16.2. Let $C_1, C_2 \subset \mathbb{P}^3$ be distinct smooth irreducible curves whose union is the complete intersection of two surfaces S, T of degrees s, t, with S smooth. Compute the intersection number $(C_1 \cdot C_2)$ in terms of the degrees and genera of C_1 and C_2.

Exercise 16.3. Let C be a reduced and irreducible projective curve, and let $\mathcal{E}$ be a locally free sheaf of rank r on C. Show that $\deg \mathcal{E} = \deg \bigwedge^r (\mathcal{E})$. ◆

Exercise 16.4. Let C be the disjoint union of 3 skew lines (see Figure 16.2).

(1) Prove that C lies on a unique quadric, and that $H^2(\mathcal{I}_C) = 0$.

(2) Compute the Hartshorne–Rao module $D(C)$.

(3) Show that if Γ is the union of 3 points in $\mathbb{P}^3$ then $H^1\mathcal{I}(\Gamma) = 0$ if and only if the three points are collinear.

(4) Using the exact sequence in cohomology coming from the short exact sequence

$$0 \to \mathcal{I}_C \xrightarrow{\ell} \mathcal{I}_C(1) \to \mathcal{I}_\Gamma(1) \to 0,$$

where ℓ is a linear form, show that the map of vector spaces

$$H^1(\mathcal{I}_C) \xrightarrow{\ell} H^1(\mathcal{I}_C(1))$$

has rank < 2 if and only if ℓ vanishes on 3 collinear points on the three lines (including the case when ℓ vanishes identically on one of the lines). Conclude that if a different union C' of 3 skew lines is linked to C, then C' lies on the same quadric as C.

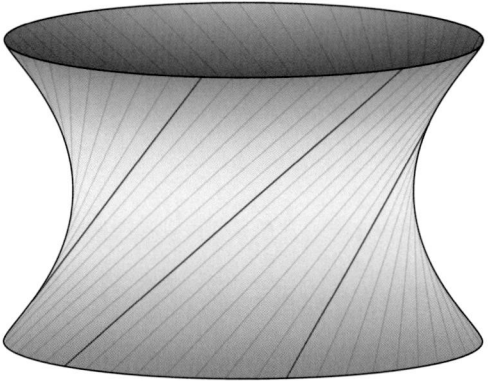

Figure 16.2. Any three skew lines in space lie on a unique (and necessarily smooth) quadric surface, and all belong to the same ruling.

See [Migliore 1986] for more examples of this type.

Exercise 16.5. Compute the Hilbert function of the Hartshorne–Rao module of a curve of type (a, b) on a smooth quadric surface.   ◆

Exercise 16.6 (linkage addition [Schwartau 1982]). Suppose that I, J are saturated ideals defining purely 1-dimensional subschemes of $\mathbb{P}^3$ and that f, g is a regular sequence with $f \in I$ and $g \in J$. Prove that $gI \cap fJ = (fg)$, and conclude that if I, J are saturated codimension 2 ideals defining purely 1-dimensional schemes C, C' in $\mathbb{P}^3$ then $(gI + fJ)$ is a saturated ideal defining a purely 1-dimensional scheme C'' with $D(C'') = D(C)(-\deg g) \oplus D(C')(-\deg f)$.   ◆

Exercise 16.7 (basic double links). The special case of the construction in Exercise 16.6 in which C' is trivial is already interesting.

(1) Show that if I is a saturated ideal of codimension 2 defining a purely 1-dimensional scheme C in $\mathbb{P}^3$ and (f, g) is a regular sequence with $g \in I$, then $fI + (g)$ defines a scheme C' with $D(C') = D(C)(-\deg f)$.

(2) Show directly that, with notation as above, C' is directly linked to C in two steps. Since the degrees of the generators of $D(C')$ are more positive, this is sometimes called an *ascending double link*. Geometrically it amounts to taking the union of C with some components that are complete intersections.

Exercise 16.8. Here is a more general form of the last step in the proof of Proposition 16.7(1). Suppose that $B \to A$ is a homomorphism of rings, X is an A-module and Y is a B-module. Show that there is a natural transformation

$$\phi : \mathrm{Hom}_A(X, \mathrm{Hom}_B(A, Y)) \cong \mathrm{Hom}_B(X, Y)$$

and that if $X = \mathrm{Hom}_B(A, Y)$, then the map

$$A \to \mathrm{Hom}_A(\mathrm{Hom}_B(A, Y), \mathrm{Hom}_B(A, Y))$$

taking an element $a \in A$ to multiplication by a on the A-module $\mathrm{Hom}_B(A, Y)$ is sent by ϕ to the evaluation map $\alpha \mapsto \alpha(a)$ for $\alpha \in \mathrm{Hom}_B(A, Y)$.

Ropes and ribbons. The simplest way to construct well-behaved nonreduced curves is to take neighborhoods of smooth ones. Ropes and ribbons are examples of this sort:

Definition 16.23. The *rope defined from a curve* $C \subset \mathbb{P}^n$ is the scheme $V(I_C^2)$ defined by the square of the ideal C.

Exercise 16.9. If C is the rope defined from a line $L \subset \mathbb{P}^3$ then the Hilbert function $h_C(m)$ and Hilbert polynomial $p_C(m)$ are both equal to $3m+1$. Thus C has degree 3 and arithmetic genus 0. Note that the degree can also be computed as the degree of a general hyperplane section, since this is defined by the square of the ideal of a point in $\mathbb{P}^2$.   ◆

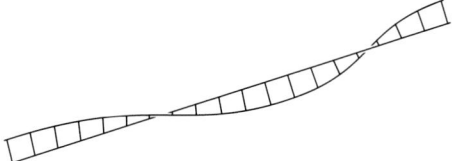

Figure 16.3. A ribbon supported on a line.

Exercise 16.10. To see why the rope in Exercise 16.9 should look like a twisted cubic, show that it is the flat limit of a twisted cubic as follows: Let $X \subset \mathbb{P}^3$ be the twisted cubic with parametrization $x_i = s^i t^{3-i}$. Consider the one-parameter subgroup of PGL_4 given in homogeneous coordinates $x_0, \ldots, x_3$ on $\mathbb{P}^3$ by

$$A_t : (x_0, \ldots, x_3) \mapsto (tX_0, X_1, X_2, tX_3).$$

Show that the flat limit, as $t \to 0$, of the twisted cubics $A_t(C)$ is the rope $V(X_0^2, X_0 X_1, X_1^2)$. ♦

Exercise 16.11. We saw in Section 16.1 that a twisted cubic curve is linked to a line by the complete intersection of two quadrics. Show that the same is true for the rope of Exercise 16.9.

If C is the rope defined from a line in $\mathbb{P}^2$, then the Zariski tangent space to C at any point is 2-dimensional; that is, it looks like a ribbon. More generally:

Definition 16.24. By a *ribbon* $X \subset \mathbb{P}^n$ we mean a scheme of pure dimension 1 and multiplicity 2 whose support is a smooth, irreducible curve $C \subset \mathbb{P}^n$ and whose Zariski tangent space at every point is 2-dimensional (Figure 16.3).

Exercise 16.12. Suppose that $C \subset \mathbb{P}^n$ is a ribbon. Show that C is contained in the rope defined from C_{red}, and show that the degree of C is twice that of C_{red}. ♦

Exercise 16.13. Unlike ropes, there are many different ribbons C with the same smooth curve C_{red}, and they can have different arithmetic genera. Suppose that $C \subset \mathbb{P}^3$ is a ribbon such that $X = C_{\mathrm{red}}$ is the line $V(x_0, x_1)$. Since C is contained in the rope defined from X we must have $(x_0^2, x_0 x_1, x_1^2) \subset I(C)$. The tangent space to C at a point $(0, 0, s, t)$ meets the line $X' = V(x_2, x_3)$ at some point $(F(s, t), G(s, t), 0, 0)$, so F and G define a morphism $\mathbb{P}^1 \to \mathbb{P}^1$; thus they are homogeneous polynomials of the same degree d. It follows that $I(C)$ also contains the element $x_0 G(x_2, x_3) - x_1 F(x_2, x_3)$. Show that the ideal of C is obtained by adding this form to the ideal of the rope, that is,

$$I_C = (X_0^2, X_0 X_1, X_1^2, F(X_3, X_4)X_0 + G(X_3, X_4)X_1).$$

In case $d = 1$, show that C lies on a smooth quadric.

General adjunction. The next two exercises illustrate Theorem 16.10:

Exercise 16.14. Show that if $C \to D$ is a map of smooth curves with ramification index e at $p \in C$, and t is a local analytic parameter at p, then locally analytically at p the sheaf $\mathcal{H}om_C(\mathcal{O}_C, \omega_D)$ is $\mathcal{O}_C(e)$.

Exercise 16.15. Show that if $C \subset S$ is a Cartier divisor on a surface S with canonical sheaf ω_S, then $\mathcal{E}xt^1(\mathcal{O}_C, \omega_S) \cong \mathcal{O}_C \otimes \mathcal{O}_S(C)$, and thus

$$K_C = (K_S + C) \cap C.$$

Chapter 17

Scrolls and the curves they contain

If you've never read T. S. Eliot's *Old Possum's Book of Practical Cats*, you could do worse than start with "The naming of cats", where the poet tells us that

> The Naming of Cats is a difficult matter [...]
> a cat must have THREE DIFFERENT NAMES

— one for daily use, one complex and unique, and one ineffable. So it is with the class of varieties known as *rational normal scrolls*. Although they're some of the simplest subvarieties in projective space — the first examples are shown in Figure 17.1 — they have a deeper aspect and myriad applications (for instance, in describing the embeddings of curves of low degree and genus), and can be examined from three rather different points of view, each useful in a different context. We will take up each in turn.[1]

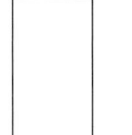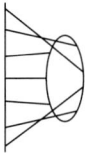

Figure 17.1. The smallest scrolls: from the left, $S(0,1) \cong \mathbb{P}^2$; $S(0,2)$, the cone over a conic; $S(1,1) \subset \mathbb{P}^3$, the union of lines joining corresponding points of two skew lines, isomorphic to a smooth quadric; and $S(1,2) \subset \mathbb{P}^4$, the union of the lines joining corresponding points on a line and a conic.

---

[1]However, we claim no secret knowledge comparable to the feline name "that no human research can discover– But THE CAT HIMSELF KNOWS." The full text of Eliot's poem is available at https://poets.org/poem/naming-cats.

In this chapter we will refer to rational normal scrolls simply as scrolls. We first give a classical geometric construction, then an algebraic description that allows one to "find" the scrolls containing a given variety, and then a more modern geometric definition that makes it easy to understand the divisors on a scroll. Finally, we turn to some of the applications to the embeddings of curves. We will focus on the case of 2-dimensional scrolls because this is the one that occurs most often in our applications.

17.1. Some classical geometry

To construct a scroll of dimension 2 in $\mathbb{P}^r$, we start by choosing integers $a_1 \geq 0$ and $a_2 \geq 1$ with $a_1 + a_2 = r - 1$, and consider a pair of complementary linear subspaces $\mathbb{P}^{a_1}$ and $\mathbb{P}^{a_2} \subset \mathbb{P}^r = \mathbb{P}(V)$ — that is, we express an $(r+1)$-dimensional vector space V as a direct sum $V = V_1 \oplus V_2$ of subspaces $V_1, V_2 \subset V$ of dimensions $a_1 + 1$ and $a_2 + 1$. For simplicity we will always assume, without loss of generality, that $a_1 \leq a_2$.

Next, for $i = 1, 2$, we take $\phi_i : \mathbb{P}^1 \to \mathbb{P}^{a_i}$ to be the parametrization of the rational normal curve of degree a_i given by a basis of homogeneous polynomials of degree a_i (if $a_i = 0$ this is just the constant map from $\mathbb{P}^1$ to a point). Finally, we define the scroll $S(a_1, a_2)$ to be the union of the lines

$$S(a_1, a_2) := \bigcup_{p \in \mathbb{P}^1} \overline{\phi_1(p)\,\phi_2(p)}.$$

We call the curve C_{a_1} a *directrix* of the scroll; if $a_1 < a_2$ this is unique. The family of lines $\overline{\phi_1(p)\phi_2(p)}$ is called a *ruling* of the scroll, and this is unique except when $a_2 = 1$, the case of $S(0,1) = \mathbb{P}^2$, and the case of a smooth quadric surface in $S(1,1) \subset \mathbb{P}^3$.

In the degenerate case $a_1 = 0$, the surface $S(0, a_2)$ is the cone in $\mathbb{P}^{a_2+1}$ over a rational normal curve of degree a_2. Since $S(0, a_2)$ is singular when $a_2 \geq 2$, it is useful to consider the surface

$$\tilde{S}(0, a_2) := \left\{ (t, q) \in \mathbb{P}^1 \times \mathbb{P}^r \mid q \in \overline{\phi_1(t), \phi_2(t)} \right\}.$$

This is the blowup of the cone $S(0, a_2)$ at its vertex; like the surfaces $S(a_1, a_2)$ with $a_1 > 0$ it is a $\mathbb{P}^1$ bundle over $\mathbb{P}^1$ and thus is smooth. As we shall see, $\tilde{S}(0, a_2)$ is isomorphic to the scroll $S(1, a_2 + 1)$. It is not hard to prove directly that $S(a_1, a_2)$ is an algebraic variety, and we shall soon write down its defining equations.

From the description above we can immediately deduce the dimension and degree of a scroll:

Proposition 17.1. (1) $S(a_1, a_2)$ *is a nondegenerate surface.*

(2) $S(a_1, a_2)$ *has degree $a_1 + a_2$, and codimension $a_1 + a_2 - 1$.*

(3) $S(a_1, a_2)$ *is nonsingular if $a_1 > 0$.*

Proof. The rational normal curves separately span the spaces $\mathbb{P}^{a_i}$, so a hyperplane containing both of them would contain $\overline{\mathbb{P}^{a_1}, \mathbb{P}^{a_2}} = \mathbb{P}^r$, proving nondegeneracy.

It is clear from our description that S is 2-dimensional, and thus of codimension $a_1 + a_2 + 1 - 2 = a_1 + a_2 - 1$.

To compute the degree, we choose a general hyperplane H containing $\mathbb{P}^{a_1}$. The intersection $H \cap C_2$ consists of a_2 reduced points. Thus the intersection $H \cap S$ consists of C_1 and the a_2 reduced lines connecting the points of $H \cap C_2$ with their corresponding points on C_1; this union has degree $a_1 + a_2$.

If $0 < a_1$ we also see from this argument that, given any point $p \in S(a_1, a_2)$, there is a hyperplane section that is nonsingular at p, and thus $S(a_1, a_2)$ is nonsingular at p. $\qquad\square$

A completely parallel construction creates scrolls of any dimension s. Start with a series of integers $0 \le a_1 \le \cdots \le a_s$; set $r + 1 = \sum_{i=1}^{s}(a_i + 1)$, and decompose $\mathbb{C}^{r+1}$ as

$$\mathbb{C}^{r+1} = \bigoplus_{i=1}^{s} \mathbb{C}^{a_i+1}.$$

Let $\mathbb{P}^{a_i} \subset \mathbb{P}^r$ be the subspaces corresponding to the summands, choose for each i a map $\phi_i : \mathbb{P}^1 \to \mathbb{P}^{a_i}$ given by a basis of homogeneous polynomials of degree a_i, and define the scroll $S \subset \mathbb{P}^r$ by

$$S = S(a_1, \dots, a_s) := \bigcup_{p \in \mathbb{P}^1} \overline{\phi_1(p), \phi_2(p), \dots, \phi_s(p)}.$$

For example, $S(a_1)$ is the rational normal curve of degree a_1. In general, the variety S is nondegenerate of codimension $r - s$ and degree $\sum a_i = r - s + 1$. The proof is similar to the one we gave for $s = 2$.

To put this construction in context, we recall that by Corollary 3.9 Any irreducible, nondegenerate variety X of codimension c in $\mathbb{P}^r$ has degree $\ge c + 1$.

Thus scrolls are *varieties of minimal degree*. The reader already knows that the rational normal curves of degree a in $\mathbb{P}^a$ are the only irreducible, nondegenerate curves of degree a and codimension $a - 1$. A celebrated theorem of del Pezzo (for surfaces) and Bertini (in general) generalizes this statement; a proof may be found in [Eisenbud and Harris 1987d].

Theorem 17.2. *An irreducible, nondegenerate variety $X \subset \mathbb{P}^r$ satisfying $\deg X = \operatorname{codim} X + 1$ is either a scroll, a quadric hypersurface, the Veronese surface in $\mathbb{P}^5$ or a cone over the Veronese surface.* $\qquad\square$

One interesting way to view the construction of a scroll is that we chose subvarieties $C_i \subset \mathbb{P}^{a_i}$ and a one-to-one correspondence between them, that is, a subscheme $\Gamma \subset \prod C_i$ that projects isomorphically onto each C_i; the scroll is then the union of the planes spanned by sets of points $p_i \in C_i$ that are "in correspondence". There are other interesting varieties constructed starting with other choices of subvarieties C_i and subschemes — not necessarily reduced — of $\prod C_i$. See [Eisenbud and Sammartano 2019] for an exploration of this idea.

Despite the choices made in the definition on page 291, we're entitled to talk about *the* scroll $S(a_1, \ldots, a_s)$:

Proposition 17.3. *Up to a linear automorphism of the ambient projective space* $\mathbb{P}$, *the scroll* $S(a_1, \ldots, a_s)$ *is independent of the choices made in its definition.*

Proof. In the construction of $S(a_1, \ldots, a_s)$ we chose

(1) independent subspaces $\mathbb{P}^{a_i} \subset \mathbb{P}$,

(2) a rational normal curve in each subspace, and

(3) an isomorphism between these curves.

Elementary linear algebra shows that there are automorphisms of $\mathbb{P}$ carrying any choice of linearly independent subspaces to any other choice. Further, since the rational normal curve of degree a is unique up to an automorphism of $\mathbb{P}^a$, the two choices in (2) differ by a linear automorphism. Finally, any automorphism of $C_{a_i} \cong \mathbb{P}^1$ extends to an automorphism of $\mathbb{P}^{a_i} = |\mathcal{O}_{\mathbb{P}^1}(a_i)|$, and this extends to an automorphism of $\mathbb{P}$ fixing all the $\mathbb{P}^{a_j}$, for $j \neq i$ pointwise, showing that $S(a_1, \ldots, a_s)$ is independent, up to an automorphism of the ambient space, of the choice in (3) as well. $\square$

17.2. 1-generic matrices and the equations of scrolls

Suppose that a scheme X is embedded in $\mathbb{P}^r$ by a complete linear series, and that $\mathcal{O}_X(1)$ can be "factored" as a tensor product $\mathcal{L} \otimes \mathcal{M}$ of invertible sheaves on X. If we pick bases of p elements $\{\ell_i\} \subset H^0(\mathcal{L})$ and q elements $\{m_i\} \subset H^0(\mathcal{M})$ then the multiplication map

$$\mu : H^0(\mathcal{L}) \otimes H^0(\mathcal{M}) \to H^0(\mathcal{O}_X(1)) = H^0(\mathcal{O}_{\mathbb{P}^r}(1))$$

gives rise to a $p \times q$ matrix M_μ of linear forms on $\mathbb{P}^r$ whose i, j entry is $\mu(\ell_i m_j)$. More abstractly, this is a linear space of matrices obtained from the "adjunction" isomorphism $\mathrm{Hom}(A \otimes B, C) \cong \mathrm{Hom}(A, \mathrm{Hom}(B, C))$.

Regarding the sections of invertible sheaves as rational functions on X, we see from the commutativity of multiplication that the 2×2 minors of

$$\det \begin{pmatrix} \ell_{i_1} m_{j_1} & \ell_{i_1} m_{j_2} \\ \ell_{i_2} m_{j_1} & \ell_{i_2} m_{j_2} \end{pmatrix},$$

vanish on X — that is, the ideal of 2×2 minors $I_2(M_\mu)$ is contained in the homogeneous ideal of X. We will show that the ideal of a scroll is generated by such minors.

Example 17.4. We have already seen this phenomenon in the case of the rational normal curve $C_a \subset \mathbb{P}^a$ if $X = \mathbb{P}^1$. Here C_a is embedded by the complete linear series $|\mathcal{O}_{\mathbb{P}^1}(a)|$, and we can write $\mathcal{O}_{\mathbb{P}^1}(a) = \mathcal{O}_{\mathbb{P}^1}(1) \otimes \mathcal{O}_{\mathbb{P}^1}(a-1)$. If we take bases $s^i t^j$ in each of $H^0(\mathcal{O}_{\mathbb{P}^1}(1))$, $H^0(\mathcal{O}_{\mathbb{P}^1}(a-1))$ and $H^0(\mathcal{O}_{\mathbb{P}^1}(a))$ we get the $2 \times a$ matrix

$$M_\mu := \begin{pmatrix} x_0 & x_1 & \cdots & x_{a-1} \\ x_1 & \cdots & x_{a-1} & x_a \end{pmatrix}.$$

When restricted to $\mathbb{P}^1$, this becomes

$$M_a = \begin{matrix} s \\ t \end{matrix} \begin{pmatrix} \overset{s^{a-1}}{s^a} & \overset{s^{a-2}t}{s^{a-1}t} & \overset{\cdots}{\cdots} & \overset{t^{a-1}}{st^{a-1}} \\ s^{a-1}t & s^{a-2}t^2 & \cdots & t^a \end{pmatrix}$$

where we have written s, t for the basis of $H^0(\mathcal{O}_{\mathbb{P}^1}(1))$, and bordered the matrix with the corresponding bases of $H^0(\mathcal{O}_{\mathbb{P}^1}(1))$ and $H^0(\mathcal{O}_{\mathbb{P}^1}(a-1))$, and it is obvious that the minors of M_a are 0.

By a *generalized row* of M_a we mean a $\mathbb{C}$-linear combination of the given rows of M_a. Note that the points at which the 2×2 minors of M_a vanish are the points at which the evaluations of the two rows are linearly dependent; that is, the points at which some generalized row of M_a vanishes identically.

Definition 17.5. A matrix of linear forms M is *1-generic* if every generalized row of M consists of $\mathbb{C}$-linearly independent forms.

Proposition 17.6. *Let X be an irreducible, reduced variety. If $\mathcal{L}, \mathcal{M}$ are invertible sheaves on X then the matrix M_μ coming from the map $\mu : H^0(\mathcal{L}) \otimes H^0(\mathcal{M}) \to H^0(\mathcal{L} \otimes \mathcal{M})$ is 1-generic.*

Proof. The entries of the generalized row of M_μ corresponding to $s \in H^0(\mathcal{L})$ are a basis of $s \cdot H^0(\mathcal{M}) \cong H^0(\mathcal{M})$, and are thus linearly independent. $\square$

Example 17.7. For example, the matrix

$$M = \begin{pmatrix} x & y \\ z & x \end{pmatrix}$$

over $\mathbb{C}[x, y, z]$ is 1-generic, since if a row and column transformation produced a 0 the determinant would be a product of linear forms, while $\det M = x^2 - yz$ is irreducible. (M corresponds to the multiplication $H^0(\mathcal{O}_{\mathbb{P}^1}(1)) \otimes H^0(\mathcal{O}_{\mathbb{P}^1}(1)) \to H^0(\mathcal{O}_{\mathbb{P}^1}(2))$.)

On the other hand, the matrix

$$M' = \begin{pmatrix} x & y \\ -y & x \end{pmatrix}$$

over $\mathbb{C}[x, y]$ is not 1-generic, since

$$\begin{pmatrix} 1 & 0 \\ -i & 1 \end{pmatrix} M' \begin{pmatrix} 1 & 0 \\ i & 1 \end{pmatrix} = \begin{pmatrix} x + iy & 0 \\ 0 & x - iy \end{pmatrix}.$$

The matrix M' would be 1-generic if we restricted scalars to $\mathbb{R}$; thus the definition depends on the field.

Lemma 17.8. *There exist 1-generic p×q matrices of linear forms in r+1 variables over $\mathbb{C}$ if and only if $r \geq p + q$.*

If M is 1-generic and involves more than p+q−1 variables, then the restriction of M to a general hyperplane is still 1-generic.

Proof. Consider a map of vector spaces $\mu : A \otimes B \to C$, where we regard C as a space of linear forms. With notation as above, if $\ell_i \otimes m_j \in \ker \mu$, then the i, j entry of M_μ is 0 and similarly for any *pure* tensor $\ell \otimes m \in A \otimes B$. Thus M_μ is 1-generic if and only if the linear subspace $\ker \mu \subset A \otimes B$ is disjoint from the set of pure tensors. Under the isomorphism $A \otimes B \cong \mathrm{Hom}(A^*, B)$ the set of pure tensors corresponds to matrices of rank 1, and this set has codimension $(p-1)(q-1) = pq-p-q+1$ in $\mathrm{Hom}(A^*, B)$. (Proof: a rank 1 matrix corresponds to the choice of a 1-quotient of A^* and a 1-dimensional subspace of B, thus a point in $\mathbb{P}^q \times \mathbb{P}^p$.) Thus for μ to correspond to a 1-generic matrix, we must have $\mathrm{codim} \ker \mu \geq q + p - 1$. Since $\mathrm{codim} \ker \mu = \dim \mathrm{im} \mu \subset C$ we see that any 1-generic matrix must involve at least $q + p + 1$ variables.

Furthermore, the restriction of M_μ to a hyperplane corresponds to the composite homomorphism $A \otimes B \to C \to C/\langle x\rangle$, or equivalently to the addition of one element to $\ker \mu$, and thus if M_μ is 1-generic and involves $> q + p - 1$ variables, then the restriction to a general hyperplane is again 1-generic. $\square$

We will see that 1-generic $2 \times r$ matrices correspond to scrolls. The beginning of the story is the calculation of the codimension of the ideal $I_2(M)$ generated by the 2×2 minors of M:

Lemma 17.9. *If M is a 1-generic $2 \times b$ matrix of linear forms in $\mathbb{C}[x_0, \ldots, x_r]$ then $V(I_2(M))$ is irreducible of codimension $b - 1$.*

Proof. The algebraic set $V := V(I_2(M))$ is the set of points on which a generalized row ρ_λ, $\lambda \in \mathbb{P}^1$ of M vanishes. Thus the map
$$W := \{(p, \lambda) \in \mathbb{P}^r \times \mathbb{P}^1 \mid p \in V(\rho_\lambda)\} \to V$$
is surjective. All the fibers of the second projection $\pi_2 : W \to \mathbb{P}^1$ are isomorphic to $\mathbb{P}^{r-b}$, so W is irreducible of dimension $r-b+1$, and thus V is irreducible. Each fiber $\pi_2^{-1}(\lambda)$ injects into V, and the images of these fibers are distinct since otherwise the entries of M would all be contained in a single generalized row, contradicting Lemma 17.8. It follows that V also has dimension $r - b + 1$, as required. $\square$

We have seen in Proposition 3.4 that the ideal of minors of
$$M_a := \begin{pmatrix} x_0 & x_1 & \cdots & x_{a-1} \\ x_1 & x_2 & \cdots & x_a \end{pmatrix}$$

generates the ideal of the rational normal curve of degree a in $\mathbb{P}^a$, and is thus prime. More generally:

Theorem 17.10. *Let $I = I_2(M)$ be the ideal generated by the 2×2 minors of a 1-generic, $2 \times a$ matrix M of linear forms in $S = \mathbb{C}[x_0, \dots, x_n]$.*

(1) *The ideal I is prime, and $V(I)$ either is smooth, or is a cone over a smooth variety.*

(2) *If $a = r$ then $V(I)$ is a rational normal curve. More generally, the variety $V = V(I) \subset \mathbb{P}^r$ has degree a and codimension $a - 1$ and is thus a variety of minimal degree.*

Proof. By Lemma 17.9, $V(I)$ has codimension $a - 1$. If the span of the linear forms in M is not the whole space of linear forms on $\mathbb{P}^r$, then $V(I)$ is a cone, so we may assume that the entries of M generate the maximal ideal.

Let $p \in V(I)$ be a point, so that there is a generalized row of M — without loss of generality the second row — whose entries all vanish at p. Since not all the linear forms of S can vanish at a point of $\mathbb{P}^r$, we may make column transformations to reduce to the case where $\ell_{1,j}$ also vanishes at p for all $j \neq 1$. Since the entries of each generalized row are linearly independent, the ideal generated by the entries of the second row define a plane of Λ codimension a containing p.

Over the local ring $\mathcal{O}_{\mathbb{P}^r, p}$ the element $\ell_{1,1}$ becomes a unit. Thus, locally at p, the elements $m_j := \ell_{2,j} + \ell_{1,1}^{-1} \ell_{2,1} \ell_{1,j}$ for $j = 2, \dots, a$ are in the ideal generated by the minors. Since the $\ell_{2,j}$ are independent regular parameters in $\mathcal{O}_{\mathbb{P}^r, p}$ and $\ell_{1,1}^{-1} \ell_{2,1} \ell_{1,j}$ is in the square of the maximal ideal of $\mathcal{O}_{\mathbb{P}^r, p}$ the $a - 1$ elements m_j are also regular parameters. Since $\operatorname{codim} I = a - 1$, it follows that I is prime and defines a smooth variety, as claimed.

If $a = r$ then $V(I)$ is a smooth, nondegenerate curve. In Chapter 18 we will construct the Eagon–Northcott complex $EN(M)$. By Theorem 18.18, the complex $EN(M)$ is a free resolution, and if M' is the ideal of the rational normal curve, then $EN(M')$ is again a resolution. Since the Hilbert function of $V(I)$ can be computed from the free resolution, it follows that $V(I)$ has the same Hilbert function as $V(I_2(M))$, and thus $V(I)$ is a rational normal curve.

If $a < r$ then by Lemma 17.6 there is linear form ℓ such that M remains 1-generic modulo ℓ. Since $I_2(M)$ is prime it has the same degree and codimension as $I_2(M) + (x)/(x) \subset S/(x)$, so by induction its degree is a. $\square$

We can use Theorem 17.10 to prove a classic result of Castelnuovo, characterizing large sets of points on a rational normal curve. It is a key element in Theorem 17.20 below, which classifies Castelnuovo curves of high degree,

Corollary 17.11 (Castelnuovo's $2r + 3$ lemma). *If $\Gamma \subset \mathbb{P}^r$ is a set of $d \geq 2r + 3$ distinct points in linearly general position, then Γ is contained in a rational normal curve if and only if Γ imposes only $2r + 1$ conditions on quadrics.*

For a classical proof of Corollary 17.11 see for example [Griffiths and Harris 1978, p. 531].

To understand the approach taken in the proof below, recall that in the case of a twisted cubic, whose equations are the minors of the matrix

$$\begin{pmatrix} x_0 & x_1 & x_2 \\ x_1 & x_2 & x_3 \end{pmatrix}$$

the first column defines a secant line to the twisted cubic, and the two minors involving the first column are a regular sequence defining the union of the twisted cubic and that secant line (see Figure 16.1). In general a rational normal curve may be defined by the 1-generic matrix associated to the product $H^0(\mathcal{O}_{\mathbb{P}^1}(1)) \otimes H^0(\mathcal{O}_{\mathbb{P}^1}(r-1)) \to H^0(\mathcal{O}_{\mathbb{P}^1}(r))$, and it follows from that description that the columns of the matrix define the $(r - 1)$-secant $(r - 2)$-planes to the curve. In the proof below we reconstruct the matrix starting from such a secant plane.

Proof. Suppose first that Γ is contained in a rational normal curve C of degree r in $\mathbb{P}^r$. If Q is a quadric vanishing on Γ then by Bézout's theorem $C \subset Q$. Since C is arithmetically Cohen–Macaulay, the dimension of the space of quadrics vanishing on C is

$$h^0(\mathcal{O}_{\mathbb{P}^r}(2)) - h^0(\mathcal{O}_C(2r)) = h^0(\mathcal{O}_{\mathbb{P}^r}(2)) - (2r + 1)$$

so Γ imposes $2r + 1$ conditions on quadrics.

Conversely, suppose that Γ imposes only $2r + 1$ conditions on quadrics. By Theorem 17.10 it suffices to construct a 1-generic $2 \times r$ matrix M whose minors vanish on Γ. Note that the dimension of the space of quadrics on $\mathbb{P}^r$, which is $\binom{r+2}{2}$, is equal to the sum $\binom{r}{2} + 2r + 1$, so the vector space of quadrics containing Γ has dimension $\binom{r}{2}$.

Write $\Gamma = \{p_1, \ldots, p_d\}$, and let $\Lambda = V(a_1, b_1)$ be the $(r - 2)$-dimensional linear space spanned by $p_1, \ldots, p_{r-1}$.

The number of conditions Λ imposes on quadrics is $\binom{r}{2}$, but the $r-1$ points of $\Gamma \cap \Lambda$ already impose $r - 1$ conditions, so the dimension of the space of quadrics containing $\Lambda \cup \Gamma$ is $\binom{r}{2} - (\binom{r}{2} - (r - 1)) = r - 1$. These quadrics are contained in the ideal (a_1, b_1), so a basis for them may be written as the 2×2 minors that involve the first column of a matrix

$$M' := \begin{pmatrix} a_1 & a_2 & \ldots & a_r \\ b_1 & b_2 & \ldots & b_r \end{pmatrix}.$$

We will show first that M' is 1-generic.

If M' were not 1-generic then we could perform row and column operations that do not change the span of the first column or the span of the minors involving the first column to arrive at a matrix with an entry equal to 0. The minor involving the first column and that column is nonzero, because the quadrics defined by these minors are linearly independent. But a reducible quadric can contain only $2r$ linearly independent points, a contradiction proving that M' is 1-generic.

It now suffices to show that all the minors of M vanish at all the points of Γ. At a point in $p \in \Gamma$ that is not in Λ, at least one of a_1, b_1 is nonzero. Since each minor of M involving the first column vanishes at p, each pair of scalars $(a_i(p), b_i(p))$ is a multiple of $(a_1(p), b_1(p))$. Thus all the minors of M vanish at p, and thus vanish on at least $\geq 2r + 3 - (r - 1) = r + 4$ points of Γ.

Because $a_2, \ldots, a_r$ are linearly independent, we may perform column operations to ensure that for $i = 2, \ldots, r$ the linear form a_i vanishes at all of $\{p_2, \ldots, p_r\}$ except possibly at p_i. Thus the minor of M involving columns i, j vanishes at each of $p_2, \ldots, p_r$ except possibly p_i, p_j, thus at $r - 3$ additional points, for a total of $2r + 1$ points of Γ. Since Γ imposes only $2r + 1$ conditions on quadrics, the minors of M vanish on all of Γ, as required. $\qquad\square$

The number $2r + 3$ in Corollary 17.11 is sharp: if C is a canonical curve of genus $r + 2$ in $\mathbb{P}^{r+1}$, then, since C is arithmetically Cohen–Macaulay (Corollary 10.7), Corollary 3.18 implies that the points of a hyperplane section lie on the same number of independent quadrics as does C, and so, by Corollary 10.9 such points lie on the same number of quadrics as do $2r + 2$ points on a rational normal curve.

Cheerful Fact 17.12. In [Harris 1982a] a similar argument is used to show that if Γ is a set of at least $2r + 1 + 2d$ points in uniform position imposing only $2r + d$ conditions on quadrics, then Γ lies on a rational normal scroll of dimension d. We do not know whether the requirement of uniform position can be weakened to linearly general position, as in the case $d = 1$.

Corollary 17.13. *Let* $a_1, \ldots, a_d$ *be nonnegative integers, and let*

$$r = d - 1 + \sum_{i=1}^{d} a_i.$$

The ideal of $S(a_1, \ldots, a_d) \subset \mathbb{P}^r$ *is generated by the* 2×2 *minors of the matrix*

$$M = \begin{pmatrix} x_{1,0} & x_{1,1} & \cdots & x_{1,a_1-1} & | & x_{2,0} & \cdots & x_{2,a_2-1} & | & \cdots & | & x_{r,0} & \cdots & x_{r,a_r-1} \\ x_{1,1} & x_{1,2} & \cdots & x_{1,a_1} & | & x_{2,1} & \cdots & x_{2,a_2} & | & \cdots & | & x_{r,1} & \cdots & x_{r,a_r} \end{pmatrix}$$

Moreover, the scroll admits a linear projection to each C_{a_i}.

Proof. We may think of the matrix M as consisting of d blocks, M_{a_i}. These blocks are 1-generic by Proposition 17.6. Since they involve distinct variables, it follows that M is 1-generic. Thus by Theorem 17.10, the ideal $I_2(M)$ is prime and of codimension $d = \sum a_i - 1$, as is the ideal of the scroll. Thus it suffices to show that the minors of M vanish on the scroll.

Let C_i be the rational normal curve in the subspace $\mathbb{P}^{a_i} \subset \mathbb{P}^r$. As always, the set $V(M)$ is the union of the linear spaces on which generalized rows of M vanish; and each such space is the space spanned by the points in the curves C_{a_i} corresponding to the part of that row in the block M_{a_i} — that is, $V(I_2(M))$ is the union of the spans of sets of corresponding points on the C_{a_i}, as required. $\square$

In Theorem 18.3 we will show that every 1-generic $2 \times (r-d)$ matrix of linear forms in r variables can be transformed by row and column transformations and a linear change of variables to one of the type shown in Corollary 17.13, and thus the minors of any 1-generic matrix define a scroll.

17.3. Scrolls as images of projective bundles

Recall that if X is a scheme and $\mathcal{E}$ is a locally free sheaf on X then $\mathbb{P}(\mathcal{E}) = \text{Proj}(\text{Sym } \mathcal{E})$ is a *projective space bundle*, and comes equipped with a projection $\pi : \mathbb{P}(\mathcal{E}) \to X$ and a *tautological invertible sheaf* $\mathcal{O}_{\mathbb{P}(\mathcal{E})}(1)$ that is a quotient of $\pi^*(\mathcal{E})$ such that $\pi_*(\mathcal{O}_{\mathbb{P}(\mathcal{E})}(1)) = \mathcal{E}$, and thus with

$$H^0(\mathcal{O}_{\mathbb{P}(\mathcal{E})}(1)) = H^0(\mathcal{E}).$$

Our third description of 2-dimensional scrolls is that they are the images of projective space bundles

$$\mathbb{P}(\mathcal{O}_{\mathbb{P}^1}(a_1) \oplus \mathcal{O}_{\mathbb{P}^1}(a_2))$$

under the map given by the complete series associated to the tautological line bundle. When both a_1 and a_2 are strictly positive, we will see that this is an embedding, and in any case we will focus on the projective bundle itself, which is always smooth. The case of higher-dimensional scrolls is similar. We follow [Hartshorne 1977, Chapter V], but restrict to the 2-dimensional rational case.

Theorem 17.14 [Hartshorne 1977, Proposition V.2.2 and V.2.3]. *Suppose that $0 \le a_1 \le a_2$ are integers and let*

$$\mathcal{E} = \mathcal{O}_{\mathbb{P}^1}(a_1) \oplus \mathcal{O}_{\mathbb{P}^1}(a_2).$$

Let $X = \mathbb{P}(\mathcal{E})$ be the corresponding $\mathbb{P}^1$ bundle over $\mathbb{P}^1$ and let $\pi : X \to \mathbb{P}^1$ be the natural projection. Set $\mathcal{L} = \mathcal{O}_{\mathbb{P}(\mathcal{E})}(1)$, the tautological 1-quotient of $\pi^*(\mathcal{E})$.

The complete linear series $|\mathcal{L}|$ is basepoint free, and is very ample if $0 < a_1$. Let $\phi : X \to \mathbb{P}^{a_1+a_2+1} = \mathbb{P}^r$ be the corresponding morphism. The image of ϕ is the scroll $S(a_1, a_2)$. More explicitly:

(1) *If $C_1, C_2 \subset X$ are the curves defined by the projections $\mathcal{E} \to \mathcal{O}_{\mathbb{P}^1}(a_i)$, then $\phi(C_1)$ and $\phi(C_2)$ are defined by the vanishing of the sections of $\mathcal{O}_{\mathbb{P}^1}(a_2)$ and*

$\mathcal{O}_{\mathbb{P}^1}(a_1)$ respectively. Thus $C_i \cong \mathbb{P}^1$, and the restriction of ϕ to C_i embeds it in $\mathbb{P}^{a_i}$ as the rational normal curve of degree a_i.

(2) The restriction of $\mathcal{L}$ to the fiber $\mathbb{P}^1$ of π is $\mathcal{O}_{\mathbb{P}^1}(1)$.

(3) The fibers of π meet each C_i in a point. The images of C_1, C_2 are contained in disjoint subspaces of $\mathbb{P}^r$, and the fibers of π are mapped to lines joining the corresponding points of the C_i. □

Corollary 17.15 [Hartshorne 1977, Section V.2]. *Let* $0 \leq a_1 \leq a_2$ *be integers. The divisor class group of the scroll*

$$X := S(a_1, a_2) \subset \mathbb{P}^r = \mathbb{P}^{a_1 + a_2 + 1}$$

is generated by the class of the hyperplane section and the class of a ruling. If $a_1 = 0$, *then the blowup of* X *at its singular point is* $S(1, a_2 + 1) \subset \mathbb{P}^{r+2}$, *and* $C_1 \subset S(1, a_2 + 1)$ *is the exceptional divisor. The blowup map* $S(1, a_2 + 1) \to S(0, a_2)$ *corresponds to the isomorphism*

$$\mathbb{P}(\mathcal{O}_{\mathbb{P}^1}(1) \oplus \mathcal{O}_{\mathbb{P}^1}(a_2 + 1)) \to \mathbb{P}(\mathcal{O}_{\mathbb{P}^1} \oplus \mathcal{O}_{\mathbb{P}^1}(a_2))$$

induced by tensoring with $\pi^*(\mathcal{O}_{\mathbb{P}^1}(-1))$.

Proposition 17.16. *Suppose that* $0 < a_1 \leq a_2$ *and* $\mathcal{E} = \mathcal{O}_{\mathbb{P}^1}(a_1) \oplus \mathcal{O}_{\mathbb{P}^1}(a_2)$. *Let* $\pi : X := \mathbb{P}(\mathcal{E}) \to \mathbb{P}^1$ *be the projection. The sections* $\sigma : \mathbb{P}^1 \to C \subset X$ *of* π — *that is, maps* σ *such that* $\pi\sigma$ *is the identity* — *correspond to surjections* $\mathcal{E} \to \mathcal{L}$ *for some line bundle* $\mathcal{L}$ *on* $\mathbb{P}^1$. *Considering* X *as embedded in* $\mathbb{P}^{a_1 + a_2 + 1}$, *we have* $\mathcal{L} = \sigma^* \mathcal{O}_C(1)$, *so the degree of* C *is equal to the degree of* $\mathcal{L}$.

Thus π *admits a section of degree* e *as a curve in* $\mathbb{P}^{a_1 + a_2 + 1}$ *if and only if* $e = a_1$ *or* $e \geq a_2$.

Proof. Apply [Hartshorne 1977, II.7.12]. □

17.4. Curves on a 2-dimensional scroll

Finding a scroll containing a given curve. One reason we are interested in scrolls is for the study of the curves contained in them. We may begin with the curve, and try to find a scroll containing it; or we may begin with the scroll and ask what curves it contains. We start with the first approach:

Proposition 17.17. *Suppose that* $C \subset \mathbb{P}^r$ *is a linearly normal reduced and irreducible curve, and* D *is a Cartier divisor on* C *such that* $|D|$ *is basepoint free. If the linear span of* D, *regarded as a subscheme of* $\mathbb{P}^r$, *is* t-*dimensional with* $t \leq r - 2$, *then* C *lies on a scroll of dimension* $t + 1$.

Proof. Choose a basepoint-free pencil $\mathbb{C}^2 \cong V \subset H^0(\mathcal{O}_C(D))$, and let H be a hyperplane section of C. Since the span of D is t-dimensional, the vector space $W := H^0(\mathcal{O}_C(H - D))$ has dimension $r - t$, and the natural 1-generic mapping

$V \otimes W \to H^0(\mathcal{O}_C(H))$ corresponds, as in Theorem 17.10 and Theorem 18.3, to the desired scroll.   □

Corollary 17.18. *Suppose that either*

(1) $C \subset \mathbb{P}^r$ *is a linearly normal hyperelliptic curve, and* $\{D_\lambda \mid \lambda \in \mathbb{P}^1\}$ *are the divisors of the* g_2^1 *on C, or*

(2) $C \subset \mathbb{P}^{g-1}$ *is a trigonal canonical curve, and* $\{D_\lambda \mid \lambda \in \mathbb{P}^1\}$ *are the divisors of a g_3^1.*

The union of the lines spanned by the D_λ is a scroll $S(a_1, a_2)$, and $a := \max\{a_1, a_2\}$ is the maximal integer such that aD_λ is a special divisor.

Proof. Let $\mathcal{L}$ be the invertible sheaf $\mathcal{O}_C(D_\lambda)$ corresponding to the g_2^1 in case 1 or the g_3^1 in case 2 and let s_λ be the section vanishing on D_λ. Setting $\mathcal{M} = \mathcal{L}^{-1} \otimes \mathcal{O}_C(1)$, we see that $s_\lambda \cdot H^0(\mathcal{M}) \subset H^0(\mathcal{O}_C(1))$ is the space of linear forms vanishing on D_λ. In both cases, this space is a line: this is obvious in case 1, and follows from the geometric Riemann–Roch theorem in case (2).

These forms make up the generalized rows of the matrix M_μ corresponding to the multiplication $\mu : H^0(\mathcal{L}) \otimes H^0(\mathcal{M}) \to H^0(\mathcal{O}_C(1))$, we see that the union of the lines is a scroll $S(a_1, a_2)$ cut out by $I_2(M_\mu)$.

The proof is completed by the more general Proposition 17.19.   □

Proposition 17.19. *Let $S(a_1, a_2) \subset \mathbb{P}^{a_1+a_2+1}$ be a scroll with $a_1 \leq a_2$. The maximal number of rulings contained in a proper subspace of $\mathbb{P}^{a_1+a_2+1}$ is a_2.*

Equivalently, if the scroll is defined from a multiplication map $V \otimes H^0(\mathcal{L}_2) \to H^0(\mathcal{O}_X(1))$, where V is a basepoint-free pencil in $H^0(\mathcal{L}_1)$, then a_2 is the maximal integer such that $\mathcal{L}_1^{-a_2}\mathcal{O}_X(1)$ is effective.

Proof. A hyperplane containing C_{a_1} meets C_{a_2} in a_2 points, and thus contains a_2 rulings of the scroll. If H were a hyperplane containing more than a_2 rulings, then H would meet each curve C_{a_i} in more than a_i points, and thus H would contain both these curves, so that $S(a_1, a_2) \subset H$. Since $S(a_1, a_2)$ is nondegenerate, this is impossible.

The second characterization follows because the rulings of the scroll are the divisors of elements of V.   □

Theorem 17.20. *If $C \subset \mathbb{P}^r$ is a reduced and irreducible curve of degree $d \geq 2r+1$ with arithmetic genus equal to the Castelnuovo bound $\pi(r, d)$, then C lies on a 2-dimensional scroll or $r = 5$ and C lies on the Veronese surface.*

For the classes in which these Castelnuovo curves lie, see [Harris 1982a, Theorem 3.11] or [Griffiths and Harris 1978, p. 533].

Proof. From the proof of Castelnuovo's theorem (Theorem 10.4) we see that a general hyperplane section $H \cap C$ is a set of $d \geq 2(r-1) + 3$ points in linearly general position in H imposing $2(r-1)+1$ conditions on quadrics. Moreover the curve is arithmetically Cohen–Macaulay, so the dimension of the space of quadrics containing the hyperplane section is the same as the dimension of the space of quadrics containing C. By Theorem 17.11 the hyperplane section lies on a rational normal curve whose ideal is generated by the quadrics containing the points; thus the quadrics containing C intersect in a surface of minimal degree containing C.   □

Both scrolls and the Veronese surface occur (Exercises 17.6 and 17.9).

Finding curves on a given scroll. We now turn to the reverse approach: given a 2-dimensional scroll, what are the curves that lie on it? In Example 2.42 we studied the special case of curves on a smooth quadric surface such as that shown in Figure 17.2. The key is to understand the divisor class group and the canonical class.

Notation: Throughout this section we consider the vector bundle

$$\mathcal{E} = \mathcal{O}_{\mathbb{P}^1}(a_1) \oplus \mathcal{O}_{\mathbb{P}^1}(a_2)$$

with $0 \le a_1,\ 1 \le a_2$ and $a_1 \le a_2$. The scroll $S(a_1, a_2)$ lies in $\mathbb{P}^r$, where $r = a_1+a_2+1$. This scroll is the image of $X = \mathbb{P}(\mathcal{E})$ by a map that is an isomorphism if $0 < a_1$ and is the blowdown of C_1 if $a_1 = 0$. We write $\pi : X \to \mathbb{P}^1$ for the projection, and $C_{a_i} \subset X$ for the directrix of degree a_i. The degree of $S(a_1, a_2)$ is $d := a_1 + a_2 = r - 1$.

The case of a smooth scroll. Now assume in addition that $1 \le a_1$, so that $S(a_1, a_2)$ is smooth.

Theorem 17.21. (1) *The Picard group of $X = S(a_1, a_2)$ is $\operatorname{Pic} X \cong \mathbb{Z}^2$, freely generated by the class F of a ruling and the class H of a hyperplane section.*

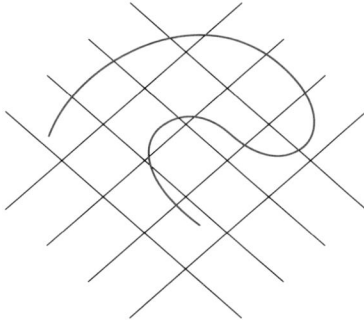

Figure 17.2. A curve of type (2,3) on a smooth quadric meets one ruling twice and the other 3 times.

(2) *The intersection form on* $\mathrm{Pic}(X)$ *is given by*

$$
\begin{array}{c c c}
\cdot & F & H \\
\begin{array}{c} F \\ H \end{array} & \begin{pmatrix} 0 & 1 \\ 1 & d \end{pmatrix} &
\end{array}
$$

(3) *The canonical class of the scroll is* $K := -2H + (d-2)F$, *so* $K^2 = 8$.

(4) *The degree of a curve in the class* $pH + qF$ *is* $pd + q$, *while its arithmetic genus is* $\binom{p}{2}d + pq - p - q + 1$.

(5) *The class of* C_{a_1} *is* $H - a_2F$, *and the class of* C_{a_2} *is* $H - a_1F$.

(6) *If* $C \subset \mathbb{P}^{g-1}$ *is a trigonal canonical curve and* X *is the scroll swept out by the trisecants of* C, *then the class of* C *is* $3H + (4-g)F = H - K$.

Proof. We have

$$
H^2 = \deg X = a_1 + a_2 = d, \quad H.F = \deg F = 1, \quad F^2 = 0,
$$

where the last equality follows because any two fibers are linearly equivalent and disjoint. It follows that H and F are linearly independent.

We next show that $\mathrm{Pic}\, X$ is generated by H and F. If D is any divisor, then $D' = D - (F \cdot D)H$ meets F in degree 0, and it now suffices to show that $D' \sim aF$ for some integer a. Since $\mathcal{O}_X(D')|_F = \mathcal{O}_F$ for any fiber F, we see that $\pi_*(\mathcal{O}_X(D'))$ is a torsion-free sheaf on $\mathbb{P}^1$ whose fiber at each point is 1-dimensional, so $\mathcal{L}$ is a line bundle on $\mathbb{P}^1$. Possibly replacing D' by $-D'$, and using the fact that $\pi_*(\mathcal{O}_X(-D')) = (\pi_*(\mathcal{O}_X(D'))^{-1}$, we may assume that $\mathcal{L}$ is globally generated, and it follows that the natural map of line bundles $\pi^*\mathcal{L} = \pi^*\pi_*\mathcal{O}_X(D') \rightarrow \mathcal{O}_X(D')$ is a surjection, whence $\mathcal{O}_X(D') \cong \pi^*\mathcal{L}$. Thus if $q = \deg \mathcal{L}$ then $D' \sim qF$. Note that we could also recover q as $H \cdot D'$. This completes the proof of parts (1) and (2).

To compute the canonical class $K_X = pH + qF$ we use the adjunction formula on the rational curves H and F. Thus $-2 = (F + K) \cdot F = p$ and

$$
-2 = (H + K) \cdot H = d + pd + q = d + (-2)d + q
$$

whence $q = d - 2$ as required for part (3).

Part (4) is a direct computation from the adjunction formula.

For part (5) we observe that a hyperplane containing C_{a_1} meets X in C_{a_1} plus a_2 rulings; thus $C_{a_1} \sim H - a_2F$. Similar reasoning holds for C_{a_2}.

If $C \subset S(a_1, a_2) \subset \mathbb{P}^{g-1}$ is a trigonal canonical curve of genus g, then the degree d of the scroll must be $g - 2$. Moreover $C.F = 3$ and $C.(C+K) = 2g - 2$. These equations have the unique solution $C \sim H - K = 3H + (4-g)F$. $\quad\square$

Now we can say exactly which classes on the scroll contain curves:

Theorem 17.22. *Again, suppose that* $0 < a_1$. *There are reduced curves in the class* $D = pH + qF$ *if and only if one of the following holds:*

(1) $D \sim qF$; *that is,* $p = 0, q > 0$.

(2) $D \sim C_{a_1}$; *that is,* $p = 1, q = -a_2$.

(3) $p \geq 0$ *and* $D \cdot C_{a_1} > 0$; *that is,* $q \geq -pa_1$.

In case (3) *the linear series* $|D|$ *is basepoint free. When, in addition,* $a_2 > a_1$ *or* $q > -pa_1$ *the class* $|D|$ *contains irreducible smooth curves.*

Note that in case (1) we have $D^2 = 0$, because any two fibers of π are disjoint; in case (2) we have $D^2 = a_1 - a_2 \leq 0$ and in case (3) we have $D^2 \geq 0$, and $D^2 = 0$ only if $d = 2, q = -pa_1$. In particular, no irreducible curve on X other than C_{a_1} can have negative self-intersection.

Proof. The existence of smooth curves of types (1) and (2) is obvious; the following result will show that the ones of type (3) move in a basepoint free linear series. By Bertini's theorem (Theorem 0.4), such a series must contain smooth curves unless the associated map factors through a curve, in which case $D^2 = dp^2 - 2pq = p(pa_1 + pa_2 - 2q) = 0$, which implies that $a_2 = a_1$ and $q = -pa_1$.

Theorem 17.23, below, thus completes the proof. $\qquad\square$

Theorem 17.23. *Again, suppose that* $0 < a_1$. *Suppose that* D *is a divisor on the scroll* X *as above. If* $D \sim pH + qF$ *is effective, then* $p \geq 0$ *and*

$$H^0(\mathcal{O}_X(D)) = H^0(\mathcal{O}_{\mathbb{P}^1}(q) \otimes \mathrm{Sym}^p \mathcal{E}) = \bigoplus_{0 \leq i \leq p} H^0\big(\mathcal{O}_{\mathbb{P}^1}(q + (p-i)a_1 + ia_2)\big).$$

The linear series $|D|$ *is basepoint free if and only if every summand in the last expression is nonzero. Thus, numerically,*

$$h^0(\mathcal{O}_X(D)) = \sum_{i \mid q+(p-i)a_1+ia_2 \geq 0} 1 + (q + (p-i)a_1 + ia_2),$$

and $|D|$ *is basepoint free if and only if* $p \geq 0$ *and* $q \geq -pa_1$.

Proof. First, If $q < -pa_1$, then

$$D \cdot C_{a_1} = (pH + qF) \cdot (H - a_2 F) = p(a_1 + a_2) - pa_1 + q = pa_1 + q < 0,$$

so any effective divisor in the class of D must have a component in common with C_{a_1}.

Let $\pi : X \to \mathbb{P}^1$ be the structure map of the projective bundle $X = \mathbb{P}_{\mathbb{P}^1}(\mathcal{E})$. We have $H^0(\mathcal{O}_X(pH + qF)) = H^0(\pi_*(\mathcal{O}_X(pH + qF)))$. Also, we may write $\mathcal{O}_X(pH + qF)$ as $\mathcal{O}_X(p) \otimes \pi^* \mathcal{O}_{\mathbb{P}^1}(q)$, and since $\mathcal{O}_{\mathbb{P}^1}(q)$ is a line bundle we see that

$$\pi_*\big(\mathcal{O}_X(p) \otimes \pi^* \mathcal{O}_{\mathbb{P}^1}(q)\big) = \pi_*\big(\mathcal{O}_X(p)\big) \otimes \mathcal{O}_{\mathbb{P}^1}(q).$$

The projective bundle $\mathbb{P}(\mathcal{E})$ is by definition Proj of the symmetric algebra $\mathbb{P}(\mathcal{E}) := \text{Proj}(\text{Sym}\,\mathcal{E})$. Over any open subset U of the base $\mathbb{P}^1$ over which $\mathcal{E}$ is free, this is $\pi^{-1}(U) = U \times \mathbb{P}^1$, and it follows that $\pi_*(\mathcal{O}_{\mathbb{P}(\mathcal{E})}(p)) = \text{Sym}^P \mathcal{E}$. Thus

$$\pi_*(\mathcal{O}_X(pH + qF)) = \pi_*(\mathcal{O}_X(p) \otimes \pi^*\mathcal{O}_{\mathbb{P}^1}(q))$$
$$= \pi_*(\mathcal{O}_X(p)) \otimes \mathcal{O}_{\mathbb{P}^1}(q) = \text{Sym}^P \mathcal{E} \otimes \mathcal{O}_{\mathbb{P}^1}(q)$$
$$= \Big(\bigoplus_{0 \le i \le p} \mathcal{O}_{\mathbb{P}^1}((p-i)a_1 + ia_2) \Big) \otimes \mathcal{O}_{\mathbb{P}^1}(q),$$

and the first formula follows.

The term $H^0(\mathcal{O}_{\mathbb{P}^1}(q + (p-i)a_1 + ia_2))$ is nonzero for all i if and only if $H^0(\mathcal{O}_{\mathbb{P}^1}(q + pa_1))$ is nonzero, which holds if and only if $q \ge -pa_1$. If $\sigma = \sum \sigma_i$ is a section of $\mathcal{O}_X(D)$ written according to the decomposition above, then the restriction of σ to the rational normal curve $C_{a_1} = \mathbb{P}(\mathcal{O}_{\mathbb{P}^1}(a_1))$ is the component σ_0, and similarly for C_{a_2} and σ_p. Thus when all the summands are nonzero there are sections vanishing on C_{a_1} but not C_{a_2}, and vice versa, so the system is basepoint free. $\qquad\square$

The case of a singular scroll $S(0, a_2)$. We now assume that $a_1 = 0$. We will use a general fact about projective bundles:

Proposition 17.24. *If X is a scheme, $\mathcal{E}$ a locally free sheaf on X, $\mathcal{L}$ an invertible sheaf on X and $\pi : \mathbb{P}(\mathcal{E}) \to X$ the natural projection. Under the isomorphism*

$$\mathbb{P}(\mathcal{L} \otimes \mathcal{E}) \cong \mathbb{P}(\mathcal{E})$$

the invertible sheaf $\mathcal{O}_{\mathbb{P}(\mathcal{E} \otimes \mathcal{L})}(1)$ corresponds to the invertible sheaf

$$\mathcal{O}_{\mathbb{P}(\mathcal{E})}(1) \otimes \pi^*\mathcal{L}.$$

Moreover, the singular scroll $S(0, a_2)$, which is the cone over a rational normal curve of degree a_2, is the image of $S(1, a_2 + 1)$ under the map corresponding to the complete linear series

$$\big|\mathcal{O}_{S(1,a_2+1)}(1) \otimes \pi^*(\mathcal{O}_{\mathbb{P}^1}(-1))\big|,$$

which blows down the line C_1.

Note that $S(1, a_2 + 1)$ is isomorphic to the surface $\tilde{S}(0, a_2)$ of Section 17.1.

Proof. The isomorphism $\mathbb{P}(\mathcal{E}) \to \mathbb{P}(\mathcal{E} \otimes \mathcal{L})$ corresponds to the surjection

$$\pi^*(\mathcal{E} \otimes \mathcal{L}) \cong \pi^*(\mathcal{E}) \otimes \pi^*(\mathcal{L}) \to \mathcal{O}_{\mathbb{P}(\mathcal{E})}(1) \otimes \pi^*(\mathcal{L}),$$

and the inverse map is formed similarly.

When $\mathcal{E} = \mathcal{O}_{\mathbb{P}^1}(a_1) \oplus \mathcal{O}_{\mathbb{P}^1}(a_2 + 1)$ and $a_1 = 1$ the invertible sheaf

$$\mathcal{O}_{S(a_1,a_2+1)}(1) \otimes \pi^*(\mathcal{O}_{\mathbb{P}^1}(-1))$$

corresponds to the divisor class $H-F$, which meets C_{a_1} in degree 0, so the image of C_{a_1} under the corresponding linear series is a point. However the description of $H^0(\mathcal{E})$ in Theorem 17.23 shows that the restriction to F is still the complete linear series $|\mathcal{O}_{\mathbb{P}^1}(1)|$, and the restriction to $C_{a_2+1} \cong \mathbb{P}^1$ is $|\mathcal{O}_{\mathbb{P}^1}(a_2 + 1)|$. Thus the displayed linear series in the proposition is basepoint free and does define a map as claimed. $\quad\square$

Theorem 17.25. *Suppose that C is a smooth curve lying on the scroll $S(0,d)$, with $1 \leq d$. Let H, F be the (Weil) divisor classes of the hyperplane section and ruling, respectively.*

(1) *$C \sim mH$ or $C \sim mH + F$ for some $m \geq 0$.*

(2) *If $C \sim mH$ then $\deg C = md$ and the genus of C is $g = \binom{m}{2}d - m + 1$.*

(3) *If $C \sim mH + F$ then $\deg C = md + 1$ and the genus of C is $g = \binom{m}{2}d$.*

In particular, $\deg C$ is congruent to 0 or 1 modulo d.

Proof. $d = 1$: In this case, $S(0,1) \cong \mathbb{P}^2$, with a distinguished point $\phi_1(\mathbb{P}^1)$ and $mH + F = (m+1)H$. The formulas for degree and genus are familiar from the theory of curves in the plane.

We now assume $2 \leq d$, so $S(0,d)$ is a cone over the rational normal curve of degree d. Let $\pi : S(1, d+1) = \tilde{S}(0,d) \rightarrow S(0,d)$ be the blowup of the vertex of the cone, and let E, L be the exceptional divisor and ruling on $S(1, d+1)$, so that $E = C_1$ and $E^2 = -d$. The proper transform of the hyperplane class H on $S(0,d)$ has intersection number 1 with L and 0 with E, and is thus of class $E + dL$. Let $\tilde{C} \cong C$ be the proper transform of C on $S(1,d)$. Since C is smooth, $\tilde{C}$ meets E at most once, whence the first assertion.

The degree assertions of items 2,3 are immediate, and the genus assertions follow by direct computation from the adjunction formula on $S(1, d+1)$. $\quad\square$

In these terms we can also see which classes on a singular scroll correspond to hyperelliptic or canonical trigonal curves in the way described by Corollary 17.18:

Corollary 17.26. (1) *For a smooth canonical curve $C \subset \mathbb{P}^{g-1}$ to lie on a singular scroll it is necessary that $g \leq 4$. If $g = 4$ then C is a complete intersection of the cone over a conic with a cubic surface not passing through the vertex of the cone; thus C is trigonal, as is also the case for $g = 3$.*

(2) *If C is a smooth hyperelliptic curve of genus $g \geq 2$, and the union of the secant lines to C corresponding to the hyperelliptic involution sweep out a singular scroll $S(0,d)$ then $C \sim 2H$ or $2H + F$ on $\tilde{S}(0,d) = S(1, d+1)$ in which cases the degree of C is $2d$ or $2d + 1$ and the genus of C is $d - 1$ or d respectively. Conversely, any curve in these classes is hyperelliptic and the lines on the singular scroll $S(0,d) \subset \mathbb{P}^d$ meet the image of C in the pairs of points that are conjugate under the hyperelliptic involution.*

Proof. (1) If $C \subset \mathbb{P}^{g-1}$ is a smooth nonhyperelliptic canonical curve lying on a singular 2-dimensional scroll in $\mathbb{P}^{g-1}$ then the scroll is $S(0, g - 2)$. By Theorem 17.25, $\deg C = 2g - 2$ must be congruent to 0 or 1 modulo $g - 2$, and this is the case only for $g = 3, 4$, the cases of a plane quartic or the complete intersection of a quadric cone with a cubic surface not passing through the vertex of the cone.

(2) By hypothesis the strict transform $\tilde{C} \subset \tilde{S} = S(1, d + 1)$ has class $2H + \epsilon F$, and by Theorem 17.25, ϵ is 0 or 1. The formulas for degree and genus follow are given in that theorem. $\qquad\square$

17.5. Exercises

Exercise 17.1. Suppose $1 \le a_1 < a_2$. Show that the projection of $S(a_1, a_2)$ from any point on C_{a_1} is $S(a_1 - 1, a_2)$, and that the projection from any point on C_{a_2} is $S(a_1, a_2 - 1)$. (If $a_1 = a_2$, projection of $S(a_1, a_2)$ from any point is $S(a_1 - 1, a_2)$.)

This gives another proof that the blowup of the cone $S(0, a_2)$ over the rational normal curve of degree a_2 is $S(1, a_2 + 1)$.

Exercise 17.2. Show that a matrix M of linear forms is 1-generic if and only if, even after arbitrary row and column transformations, its entries are all nonzero.

Exercise 17.3. Let M be a 1-generic $p \times q$ matrix of linear forms, with $p \le q$. Show that the codimension of $I_p(M)$ is $q - p + 1$.   ◆

Exercise 17.4. Show that if $X \subset \mathbb{P}^r$ is a projective variety whose homogeneous ideal I contains m independent quadrics, then the ideal of the general hyperplane section of X in $\mathbb{P}^{n-1}$ contains at least m independent quadrics. Use this to prove that if X has codimension c then $m \le \binom{c+1}{2}$, and that equality holds if and only if X is a variety of minimal degree (that is, $\deg X = c + 1$.)

Exercise 17.5. If $X \subset \mathbb{P}^r$ is an irreducible, nondegenerate projective variety of codimension c whose homogeneous ideal contains $\binom{c}{2}$ independent quadratic forms, then X is a variety of degree c.   ◆

Exercise 17.6. Show that every reduced, irreducible curve on the Veronese surface is a Castelnuovo curve.

Exercise 17.7. (1) Use the Leray spectral sequence to compute the cohomology $H^i(\mathcal{O}_X(D))$ on a scroll $X \subset \mathbb{P}^r$.

(2) Using this, show that X is arithmetically Cohen–Macaulay.   ◆

Exercise 17.8. Show that if $X \subset \mathbb{P}^r$ is an arithmetically Cohen–Macaulay surface and $D \subset X$ is a curve, then D is arithmetically Cohen–Macaulay if and only if $H^1_*(\mathcal{J}_{D/X}) = 0$. Use this and Exercise 17.7 to determine the possible divisor classes of arithmetically Cohen–Macaulay curves on a 2-dimensional scroll.

Exercise 17.9. By Theorem 10.4 Castelnuovo curves are arithmetically Cohen–Macaulay. Show that all the arithmetically Cohen–Macaulay curves on a 2-dimensional scroll are Castelnuovo curves. ◆

Exercise 17.10. State and prove an analogue of Proposition 17.19 for curves on scrolls of dimension > 2. ◆

Exercise 17.11. Improve the result of Exercise 17.1 by showing that, if $0 \leq a_1 \leq a_2$, then the projection of $S(a_1, a_2)$ from any point not on C_{a_1} (or not the vertex, in case $a_1 = 0$) is $S(a_1, a_2 - 1)$.

Exercise 17.12. (1) If $C \subset S(0, a_2) \subset \mathbb{P}^{a_2+1}$ is a smooth curve with class m times the hyperplane section (i.e., corresponding to a curve on $S(1, a_2 + 1)$ with class $mH - mF$), show that

$$\deg C = ma_2 \quad \text{and} \quad g(C) = \binom{m}{2}a_2 - m + 1.$$

(2) If $C \subset S(0, a_2) \subset \mathbb{P}^{a_2+1}$ is a smooth curve with class m times the hyperplane section plus a ruling (i.e., corresponding to a curve on $S(1, a_2 + 1)$ with class $mH - (m-1)F$), show that

$$\deg C = ma_2 + 1 \quad \text{and} \quad g(C) = \binom{m}{2}a_2.$$

Exercise 17.13. Show that 3 general quadrics meet in 8 points in $\mathbb{P}^3$ that do not lie on a twisted cubic, and deduce that the bound $d \geq 2r + 3$ in Theorem 17.11 is sharp.

Exercise 17.14 (trigonal curves of genus 5). In our study of canonical curves of genus 5 (second case, page 175) we analyzed trigonal curves $C \subset \mathbb{P}^4$ of genus 5, showing that there are three independent quadrics in I_C that are either a complete intersection (and generate I_C or meet in a nondegenerate surface). From Corollary 17.18 we know that in the second case the surface is a scroll. Complete the analysis by deciding between the two types of scrolls in $\mathbb{P}^4$: $S(1, 2)$ and $S(0, 3)$. What is the divisor class of the curve on the scroll? ◆

Exercise 17.15. If C is a smooth trigonal curve, canonically embedded in $\mathbb{P}^{g-1}$, then C lies on a 2-dimensional scroll $S(a_1, a_2)$, as in Theorem 17.21. Show that

$$a_2 - a_1 \leq \frac{g + 2}{3}.$$

(In this situation the difference $|a_2 - a_1|$ is called the *Maroni invariant*.) ◆

Chapter 18

Free resolutions and canonical curves

In Chapter 16 we related the free resolution of the homogeneous coordinate ring of a curve C in $\mathbb{P}^3$ to the Hartshorne–Rao module $H^1_*(I_C)$ of the curve. These are extrinsic invariants — they depend on the choice of an embedding of the curve. In this chapter we will develop the language and machinery to understand Green's conjecture, which posits a subtle relationship between the intrinsic geometry of a curve and the free resolution of the canonical image of that curve.

The first two sections of this chapter constitute a quick review of some homological commutative algebra (a more complete exposition can be found in [Eisenbud 1995, Part III]). We explain how the condition that a curve in $\mathbb{P}^r$ is arithmetically Cohen–Macaulay manifests itself in the free resolution of the homogeneous ideal of the curve, and we show how the uniqueness of minimal free resolutions leads to the classification of the matrices whose minors define scrolls. In Section 18.3 we introduce the Eagon–Northcott complexes and give a full proof of their properties. We also explain their relation to the free resolutions of rational normal scrolls. This is followed in Section 18.4 by an introductory account of Green's conjecture.

18.1. Free resolutions

The basic results described here depend on Nakayama's lemma and hold in two parallel contexts: modules over a local ring, and graded modules over a polynomial ring whose variables have positive degree, in which case we restrict attention to graded modules and homogeneous ideals. We will work both with local rings and with polynomial rings whose variables have degree 1, as convenient, and leave the formulation of the corresponding results in the other context to the reader.

Let M be a finitely generated graded module over $S := \mathbb{C}[x_0, \ldots x_r]$. A *free resolution* of M is an exact complex of graded free modules, with maps of degree 0:

$$(\mathbb{F}, \phi): \quad 0 \leftarrow M \xleftarrow{\epsilon} \bigoplus_j S(-j)^{\beta_{0,j}} \xleftarrow{\phi_1} \cdots \xleftarrow{\phi_t} \bigoplus_j S(-j)^{\beta_{t,j}} \leftarrow 0.$$

Here $S(-j)$ denotes the graded free module of rank 1 with generator in degree j. The map to M is not considered part of the free resolution. The resolution is called *minimal* if a minimal set of generators of $F_i := \bigoplus_j S(-j)^{\beta_{i,j}}$ maps to a minimal set of generators of the kernel of the following map, or equivalently (by Nakayama's lemma) the maps in $\mathbb{F} \otimes_S \mathbb{C}$ are all 0. In this case $\beta_{i,j} = \dim_\mathbb{C} \mathrm{Tor}_i^S(M, \mathbb{C})$. The numbers $\beta_{i,j}$ are called the *Betti numbers* of M.

Minimal free resolutions of a given module M are all isomorphic, and thus provide interesting invariants of M. The Betti numbers of a graded module over the polynomial ring determine the Hilbert function of the module as an alternating sum of the Hilbert functions of the free modules in the resolutions — this was Hilbert's original motivation for proving the syzygy theorem. As we shall see in the last section of this chapter, the Betti numbers are a much finer invariant.

To construct the minimal free resolution of a finitely generated graded module M, suppose that a minimal homogeneous set of generators of M contains $\beta_{0,j}$ generators of degree j for each j; the choice of generators defines a degree 0 map ϵ from $\bigoplus_j S(-j)^{\beta_{0,j}}$ onto M. We proceed to do the same with the kernel of ϵ to construct ϕ_1, and continue similarly to construct $\phi_2 \ldots$. Hilbert's basis theorem and syzygy theorem together imply that for some $t \leq r + 1$ we will find that ϕ_t has kernel equal to 0; that is, every finitely generated S-module has a finite free resolution of length $t \leq r + 1$ [Eisenbud 1995, Corollary 19.7]. The minimal such t is called the *projective dimension* of M, and is written pd M. Computing $\mathrm{Ext}(M, -)$ from such a resolution, Nakayama's lemma implies that,

$$\mathrm{pd}\, M = \max\{s \mid \mathrm{Ext}_S^s(M, k) \neq 0\}.$$

It follows from the Auslander–Buchsbaum formula [Eisenbud 1995, Theorem 19.9] that $t \geq \mathrm{codim}\, \mathrm{ann}_S(M)$, the codimension of the support of M.

Cheerful Fact 18.1. Fundamental results of Auslander, Buchsbaum and Serre say that a local ring R is *regular* — that is, the Krull dimension of R is equal to the minimal number of generators of is maximal ideal — if and only if the minimal free resolution of the residue field is finite, in which case every module has a finite free resolution.

Example 18.2. Koszul complexes were first defined, despite the name, in [Cayley 1848], and are found as examples in [Hilbert 1890]. They can be used to resolve S/I when $I = (f_1, \ldots, f_t)$ is a complete intersection, that is, $f_1, \ldots, f_t$ is a regular sequence. For $t = 2, 3$, if $\deg f_i = d_i$, these have the form

$$S \xleftarrow{(f_1\ f_2)} S(-d_1) \oplus S(-d_2) \xleftarrow{\binom{-f_2}{f_1}} S(-d_1 - d_2) \longleftarrow 0$$

and

$$S \xleftarrow{(f_1\ f_2\ f_3)} F_1 \xleftarrow{\begin{pmatrix} 0 & f_3 & -f_2 \\ -f_3 & 0 & f_1 \\ f_2 & -f_1 & 0 \end{pmatrix}} F_2 \xleftarrow{\begin{pmatrix} f_1 \\ f_2 \\ f_3 \end{pmatrix}} F_3 \longleftarrow 0$$

where

$$F_1 = \bigoplus_{j=1}^{3} S(-d_j), \quad F_2 = \bigoplus_{1 \le i < j \le 3} S(-d_i - d_j), \quad F_3 = S(-d_1 - d_2 - d_3).$$

Any Koszul complex K is *symmetric* in the sense that $\mathrm{Hom}(K, S)$ is isomorphic to S up to a shift.

The classification of 1-generic $2 \times f$ matrices. We can use the uniqueness of minimal free resolutions to give a simple proof of Kronecker's classification of 1-generic $2 \times f$ matrices, which was announced in Chapter 17.

Theorem 18.3. *Every 1-generic $2 \times f$ matrix of linear forms can be transformed by row and column operations and a linear change of variables to one of the type shown in Corollary 17.13, and thus the minors of any 1-generic $2 \times f$ matrix define a scroll.*

To prove this result we first reinterpret the 1-generic condition: An $a \times b$ matrix of linear forms in c variables can be interpreted as a $\mathbb{C}$-linear map of vector spaces $A \otimes B \to C$, where A, B and C have dimensions a, b and c respectively. Such a map can be viewed in several ways, for example as a map $C^* \otimes B \to A^*$ — in other words, a $c \times b$ matrix in a variables — or equivalently an a-dimensional family of $c \times b$ scalar matrices (and similarly for other permutations of A, B, C). This bit of trivial formalism pays off in the following observation:

Proposition 18.4. *An $a \times b$ matrix M of linear forms over $S = \mathbb{C}[x_0, \ldots, x_r]$ is 1-generic if and only if the corresponding family N of $b \times c$ scalar matrices over $\mathbb{P}^{a-1}$ has constant rank b; that is, for any point $x \in \mathbb{P}^{a-1}$, the rank of N evaluated at x is b.*

Proof. By Lemma 17.8, $c \ge a + b - 1 \ge b$. The b rows of N correspond to the b columns of M, while the columns of N are indexed by the c variables in M. The evaluation of N at x corresponds to a generalized row of M; and the image of $N(x)$ is the span of the variables in that generalized column of M. The matrix M is 1-generic if the dimension of that span is b for every generalized column. □

Thus the classification of 1-generic $2 \times f$ matrices of linear forms in $r + 1$ variables of Theorem 18.3 is equivalent to the classification of *matrix pencils* of constant maximal rank — that is, $f \times (r + 1)$ matrices of linear forms over $\mathbb{P}^1$ with constant rank f. Such "matrix pencils" were first classified by Kronecker; see [Gantmacher 1959, Chapter 12] for an exposition, and [Eisenbud and Harris 1987d] for a geometric approach.

Proof of Theorem 18.3. Let M be a 1-generic $2 \times b$ matrix in $r + 1$ variables. We may assume that the span of the entries is equal to the vector space of linear forms in $\mathbb{P}^r$. The associated $(r + 1) \times b$ matrix

$$N : \mathcal{O}_{\mathbb{P}^1}(-1)^b \to \mathcal{O}_{\mathbb{P}^1}^{r+1}$$

of linear forms over $\mathbb{P}^1$ has constant rank b. For every point $x \in \mathbb{P}^1$ the scalar matrix $N(x)$ is a split inclusion, and thus coker N is a vector bundle on $\mathbb{P}^1$ that is generated by its global sections, and thus, necessarily of the form $\sum_{i=1}^{d} \mathcal{O}_{\mathbb{P}^1}(a_i)$ for some integers $a_i \geq 0$.

Regarding N as a matrix over the homogeneous coordinate ring $S := \mathbb{C}[s, t]$ of $\mathbb{P}^1$, and taking global sections of all nonnegative twists, we get

$$0 \to S(-1)^b \xrightarrow{N} S^{r+1} \to \bigoplus_{i=1}^{d}\left(\bigoplus_{j=0}^{\infty} H^0(\mathcal{O}_{\mathbb{P}^1}(j + a_i))\right) \to 0,$$

which is a minimal free resolution because $H^1(\mathcal{O}_{\mathbb{P}^1}(j)) = 0$ for all $j \geq -1$. It follows that the map N is the direct sum of the minimal S-free resolutions of the modules

$$\bigoplus_{j=0}^{\infty} H^0(\mathcal{O}_{\mathbb{P}^1}(j + a_i)) = (s, t)^{a_i},$$

where we take the generators to be in degree 0. As explained in Example 18.14, the minimal presentation of $(s, t)^{a_i}$ is the $(a_i + 1) \times a_i$ matrix

$$\phi_a = \begin{pmatrix} s & 0 & 0 & \cdots & 0 \\ -t & s & 0 & \cdots & 0 \\ 0 & -t & s & \cdots & 0 \\ 0 & 0 & -t & \cdots & 0 \\ \vdots & \vdots & & \ddots & \vdots \\ 0 & 0 & 0 & \ddots & s \\ 0 & 0 & 0 & \cdots & -t \end{pmatrix}.$$

Thus, by a change of bases and variables, the matrix N can be transformed into the direct sum of such matrices. The direct sum decomposition of N corresponds to a block decomposition of M. Translating this back to a $2 \times (r + d - 1)$ matrix M, we have transformed M into a matrix of the desired form. □

How to look at a resolution. As is apparent even in Example 18.2, free resolutions can be bulky to describe; the *Betti table* is a compact representation of
the numerical information in the resolution introduced by Bayer and Stillman
to improve the output of their symbolic computation system *Macaulay*. Suppose that F is a minimal free resolution of a graded module M as constructed
in the beginning of the previous section. Since we choose a minimal set of
generators at each stage, the matrices of the ϕ_i have entries in the maximal
ideal $(x_0, \ldots, x_r)$, and thus if $\beta_{i+1,j} \neq 0$ there must be a number $\ell < j$ such
that $\beta_{i,\ell} \neq 0$. For this reason it is convenient to tabulate the Betti numbers so
that $\beta_{i,j}$ is in the i-th column and $(j-i)$-th row:

j	$i = 0$	1	$\cdots$	n
$\vdots$	$\vdots$	$\vdots$		$\vdots$
0	$\beta_{0,0}$	$\beta_{1,1}$	$\cdots$	$\beta_{n,n}$
1	$\beta_{0,1}$	$\beta_{1,2}$	$\cdots$	$\beta_{n,n+1}$
$\vdots$	$\vdots$	$\vdots$		$\vdots$

Example 18.5. The differentials in the Koszul complex $K(x_0, \ldots, x_r)$ that resolves the module $S/(x_0, \ldots, x_r)$ are all linear, so $\beta_{i,j}(K) = 0$ unless $i = j$; thus
the Betti table of K has the form

j	$i = 0$	1	2	$\ldots$	r+1
0	1	$r+1$	$\binom{r+1}{2}$	$\cdots$	1

Example 18.6. The Koszul complex that resolves the homogeneous coordinate
ring $S/(Q, F_1, F_2)$ of the complete intersection of 2 quadrics and a cubic in $\mathbb{P}^3$
has the form

$$S \leftarrow S(-2) \oplus S(-3)^2 \leftarrow S(-5)^2 \oplus S(-6) \leftarrow S(-8) \leftarrow 0,$$

which has Betti table

j	$i = 0$	1	2	3
0	1	–	–	–
1	–	1	–	–
2	–	2	–	–
3	–	–	2	–
4	–	–	1	–
5	–	–	–	1

where a dash represents 0.

When is a finite free complex a resolution? How does a free resolution
over $S := \mathbb{C}[x_0, \ldots x_r]$ "know" to end no later than the $(r+1)$-st step? The main
theorem of [Buchsbaum and Eisenbud 1973] (Theorem 18.9 below) describes
a sense in which the maps in the resolution change as the resolution continues.
See [Eisenbud 1995, Theorem 20.9] for further exposition.

A central role in the theorem is played by the ideals of minors of the differentials in the complex: If $\phi : F \to G$ is a map of finitely generated free

R-modules with cokernel M, then $I_t(\phi)$ denotes the ideal generated by all the $t \times t$ minors (subdeterminants) of a matrix representing ϕ; this is independent of the choice of bases used to represent ϕ as a matrix. The ideal $I_{\operatorname{rank} G-j}(\phi)$ depends only on M; it called the j-th *Fitting ideal* of coker ϕ and usually written Fitt$_j M$. Some basic properties of these ideals are given in Exercise 18.12, where the reader may show that the annihilator of M has the same radical as Fitt$_0 M$. Actually more is true:

Cheerful Fact 18.7. If $\phi : F \to G$ is an exact sequence of finitely generated R-modules and we set $I = \operatorname{ann} \operatorname{coker} \phi$, then $I^{\operatorname{rank} G} \subset \operatorname{Fitt}_0 \phi \subset I$. For this and other such inequalities, see [Buchsbaum and Eisenbud 1977b].

The *rank* of $\phi : F \to G$ is by definition the largest size of a nonvanishing minor of a matrix representing ϕ, or equivalently the largest k such that the exterior power

$$\bigwedge^k \phi : \bigwedge^k F \to \bigwedge^k G$$

is nonzero.

We define $I(\phi) := I_{\operatorname{rank} \phi}(\phi)$; if the source or target of ϕ is 0 then, since an empty product is equal to 1, we set $I(\phi) = R$. The ideal $I(\phi)$ plays a special role: If $I(\phi)$ contains a nonzerodivisor, then the construction of $I(\phi)$ commutes with localization, and the cokernel of ϕ is locally free indexfree!locally if and only if $I(\phi) = R$.

Lemma 18.8. *If R is a local ring and $\phi : F \to G$ is a map of free R modules with G finitely generated, then coker ϕ is free if and only if $I(\phi) = R$, the unit ideal.*

Proof. If $H := \operatorname{coker} \phi$ is free, then we may write $G = G' \oplus H$ and $F = F' \oplus G'$, where ϕ is the projection to G'. Thus ϕ is the direct sum of a unit matrix and zero, whence $I(\phi) = R$.

For the converse, set $r := \operatorname{rank} \phi$. If $I(\phi) = R$ then, since R is local, some $r \times r$ submatrix ϕ_1 of ϕ has unit determinant, and after row and column operations we can write $\phi \cong \phi_1 \oplus \phi_2$. If ϕ_2 were nonzero the rank of ϕ would be greater than r, so $\phi_2 = 0$, and coker ϕ is a free module isomorphic to the target of ϕ_2. $\square$

Recall that in Section 3.3 we defined the *grade* of an ideal I to be the length of a maximal regular sequence contained in I, or ∞ if $I = R$. If R is a Cohen–Macaulay ring such as $\mathbb{C}[x_0, \ldots, x_r]$ then the grade of any proper ideal is equal to its codimension, so grade becomes a geometric notion. Readers less familiar with commutative algebra will lose little if they stick with the case when R is regular, or even the case when R is $\mathbb{C}[x_0, \ldots, x_r]$, and replace "grade" by "codimension". This suffices, for example, for the applications of the Eagon–Northcott complex described below.

Theorem 18.9 [Buchsbaum and Eisenbud 1973]. *Let R be a Noetherian ring and let*

$$\mathbb{F}: \quad F_0 \xleftarrow{\phi_1} F_1 \leftarrow \cdots \leftarrow F_{n-1} \xleftarrow{\phi_n} F_n \leftarrow 0$$

be a finite complex of free S-modules. Set $r_i := \operatorname{rank} \phi_i$. The complex $\mathbb{F}$ is acyclic (that is, $H_i(\mathbb{F}) = 0$ for all $i > 0$) if and only if, for all i,

$$\operatorname{rank} F_i = r_i + r_{i+1} \quad and \quad \operatorname{grade} I_{r_i}(\phi_i) \geq i. \qquad \square$$

A familiar case occurs when $n = 1$ and R is a domain. In this case the theorem says that a map $F_1 \to F_0$ is a monomorphism if and only if it becomes a monomorphism after tensoring with the field of rational functions K, which follows from the flatness of localization and the fact that F_1 is torsion-free, so that $F_1 \subset F_1 \otimes K$. More generally, if R is Noetherian, then a map of finitely generated free modules $\phi : F \to G$ is injective if and only if $I_{\operatorname{rank} F}(\phi)$ contains a nonzerodivisor.

To see the relevance of the grade hypothesis, suppose for a moment that $R = \mathbb{C}[x_0, \ldots, x_r]$ and that the $\mathbb{F}$ is a finite free resolution of an R-module. The conclusion that the grade of $I(\phi_{r+2})$ is at least $r + 2$ can only be satisfied if $I(\phi_{r+2}) = R$ (so that its grade is ∞ by convention). This is equivalent to $\operatorname{coker} \phi_{r+2} = \ker \phi_r$ being free. Thus the theorem "explains" why a minimal free resolution has length $\leq r + 1$.

18.2. Depth and the Cohen–Macaulay property

If M is a graded $\mathbb{C}[x_0, \ldots x_r]$-module then an *M-regular sequence* is a sequence of homogeneous polynomials $f_1, \ldots, f_m \in (x_0, \ldots, x_r)$ such that f_1 is a non-zerodivisor on M, f_2 is a nonzerodivisor on $M/(f_1 M)$, and so on. The maximal length of such a sequence is called the *depth* of M, or more properly the depth of $(x_0, \ldots, x_r)$ on M. The lengths of all maximal M-regular sequences are the same:

Theorem 18.10 (Auslander–Buchsbaum). *If M is a finitely generated graded module over $S := \mathbb{C}[x_0, \ldots x_r]$, then the length of every M-regular sequence is*

$$m = r + 1 - \operatorname{pd} M,$$

and m is the smallest integer m such that $\operatorname{Ext}_S^m(S/(x_0, \ldots x_r), M) \neq 0$. $\qquad \square$

The depth of a module M is bounded above by $\dim M$, the Krull dimension. The reason is that if the dimension of M is d, and $f_1 \in (x_0, \ldots x_r)$ is a nonzero-divisor on M, then $\dim M/(f_1)M = \dim M - 1$. Thus by induction, if $f_1, \ldots, f_d$ is M-regular then $M/(f_1, \ldots, f_d)M$ has dimension 0, which is equivalent to its being Artinian. Thus any $f_{d+1} \in (x_0, \ldots x_r)$ acts as a nilpotent endomorphism of $M/(f_1, \ldots, f_d)M$.

It follows from these facts that the depth of an S-module M is equal to the dimension of M if and only if the projective dimension of M is equal to the codimension of M; in this case we say that M is a *Cohen–Macaulay module*.

As we asserted in Chapter 16, a curve $C \subset \mathbb{P}^3$ is linked to a complete intersection if and only if $H^1_*(\mathcal{J}_C) := \bigoplus_{m \in \mathbb{Z}} H^1(\mathcal{J}_C(m)) = 0$, in which case we said that C was arithmetically Cohen–Macaulay. This is equivalent to the condition that the homogeneous coordinate ring R_C of C is Cohen–Macaulay. In general, we say that a projective scheme $X \subset \mathbb{P}^r$ is arithmetically Cohen–Macaulay if its homogeneous coordinate ring R_X is Cohen–Macaulay, and this is true if and only if $\operatorname{pd} R_X = \operatorname{codim} X$. A scheme $X \subset \mathbb{P}^r$ is said to be Cohen–Macaulay if all its local rings are Cohen–Macaulay; this is weaker than saying that X is ACM. For example, every smooth scheme is Cohen–Macaulay, but most curves in $\mathbb{P}^3$ are not ACM.

The Gorenstein property. An important homological condition on a scheme X is that ω_X be an invertible sheaf; when this holds, we say that X is *quasi-Gorenstein*. When, in addition, X is Cohen–Macaulay we say that X is *Gorenstein*. As with the Cohen–Macaulay property, the Gorenstein property is interpreted locally. Thus any scheme that is locally a complete intersection, such as any smooth scheme, is Gorenstein. Since the restriction of $\mathcal{O}_{\mathbb{P}^r}(1)$ to a subvariety is always invertible, saying that a scheme X is canonically embedded implies that X is quasi-Gorenstein.

In Chapter 16, we expressed ω_X for a subscheme $X \subset Y$ of a scheme Y as $\operatorname{Ext}^{\operatorname{codim} X}_{\mathcal{O}_Y}(\mathcal{O}_X, \omega_Y)$. Since $\omega_{\mathbb{P}^r} = \mathcal{O}_{\mathbb{P}^r}(-r-1)$, we take $\omega_S = S(-r-1)$ where S is the homogeneous coordinate ring of $\mathbb{P}^r$. If $X \subset \mathbb{P}^r$ is a scheme with homogeneous coordinate ring R_X, we define ω_{R_X} to be $\operatorname{Ext}^{r-1}_S(R_C, S(-r-1))$, a graded module whose sheafification is ω_C. We say that a projective scheme X is *arithmetically Gorenstein* if it is ACM and $\omega_{R_X} \cong R_X(a)$ for some integer $a = a(X)$. Since canonically embedded curves are Castelnuovo curves they are ACM and thus arithmetically Gorenstein.

If $X \subset \mathbb{P}^r$ has codimension c and is arithmetically Cohen–Macaulay, so that $\operatorname{pd} X = c$, then by computing $\operatorname{Ext}^c_S(R_X, S(-r-1))$ from the minimal free resolution

$$(\mathbb{F}, \phi): \quad 0 \leftarrow R_X \leftarrow S \xleftarrow{\phi_1} F_1 \leftarrow \cdots F_{c-1} \xleftarrow{\phi_c} F_c \leftarrow 0$$

we see that ω_{R_X} is (up to twist) the cokernel of the dual $\phi_c^* = \operatorname{Hom}(\phi_c, S)$ of the last map ϕ_c in the resolution. Theorem 18.9 implies that the complex $(\mathbb{F}^*, \phi^*)$ which is the dual of the resolution $(\mathbb{F}, \phi)$ is again acyclic, so it is the minimal free resolution of ω_{R_X}. Thus ω_{R_X} is a Cohen–Macaulay module. Just as the Cohen–Macaulay property of R_X implies that $R_X = H^0_*(\mathcal{O}_X)$, it follows that $\omega_{R_X} = H^0_*(\omega_X)$.

If $C \subset \mathbb{P}^{g-1}$ is a canonical curve, so that $\omega_C = \mathcal{O}_C(1)$, then we derive

$$\omega_{R_C} = \mathrm{coker}(\phi^*_{g-2})(-g) = R_C(1),$$

so $\mathrm{coker}\,\phi^*_{g-2} = R_C(r)$. Thus $\mathrm{Hom}(\mathbb{F}, S(-g-1))$ is a minimal free resolution of R_C, and is therefore isomorphic to $\mathbb{F}$; that is, $\mathbb{F}$ is self-dual, up to shift. We have seen a different example of such a duality already in the Koszul complex (a complete intersection is arithmetically Gorenstein).

Taking into account that in a resolution $(\mathbb{F}, \phi)$ each summand $S(-j)$ of F_{i+1} can only map to summands $S(-l)$ of F_i with $\ell < j$, and similarly for the dual, we see that the Betti table of the minimal free resolution of a canonical curve of genus $g \geq 4$ must have the form

j | $i = 0$	1	2	$\cdots$	$g-3$	$g-2$
0	1	–	–	$\cdots$	–	–
1	–	b_1	b_2	$\cdots$	b_{g-3}	–
2	–	b_{g-3}	b_{g-4}	$\cdots$	b_1	–
3	–	–	–	$\cdots$	–	1

It turns out that the b_i depend on the particular canonical curve, but since the Hilbert function of the curve is the alternating sum of the Hilbert functions in the resolution, the differences $b_i - b_{g-i-1}$ are independent of the curve. This together with the self-duality implies that the whole Betti table of a canonical curve is determined by the numbers $b_1, \ldots, b_{g-2}$.

The geometry of linear series on C influences these Betti numbers. For example, if $|\mathcal{L}|$ is a g_d^1 on C then the multiplication map

$$H^0(\mathcal{L}) \otimes H^0(\mathcal{L}^{-1} \otimes \omega_C) \to H^0(\omega_C) = H^0(\mathcal{O}_C(1))$$

corresponds to a $2 \times h^1(\mathcal{L})$ matrix of linear forms on $\mathbb{P}^{g-1}$ as in Chapter 17, and the ideal I generated by the minors of this matrix is contained in the homogeneous ideal of C. In fact, by Proposition 18.21 below, the whole resolution of I is a subcomplex of the resolution of R_C. Resolutions of such determinantal ideals can be described explicitly, and we turn now to this description.

18.3. The Eagon–Northcott complex

Let $\phi : F \to G$ be a map of finitely generated free modules. The Eagon–Northcott complex $EN(\phi)$ [Eagon and Northcott 1962] associated with ϕ is a generalization of the Koszul complex, which is the case rank $G = 1$. Like the Koszul complex, it is tautological: its existence depends only on the properties of commutative rings; and like the Koszul complex it is exact or not depending on a property of the matrix ϕ related to regular sequences. It is part of a family of complexes described in [Eisenbud 1995, Appendix A2], and, from a more conceptual and general point of view, in [Weyman 2003].

We are interested in $EN(\phi)$ because its shape influences the shape of the free resolutions of canonical curves in an interesting way, leading to Green's conjecture. This conjecture, one of the central open problems in the theory of algebraic curves, is described in the last section of this chapter. We will also use the Eagon–Northcott complex, in a special case, to give a proof of the classification of matrix pencils and an analysis of the ideals of ACM curves in $\mathbb{P}^3$.

Two special cases of $EN(\phi)$ are of interest in their own right: when G has rank 1 and when rank G is arbitrary but rank $F = \text{rank } G + 1$. We will describe these first. The general case is notationally more complicated, but the ideas necessary for describing it and proving its properties are all present in these two special cases.

The case rank $G = 1$: the Koszul complex. In this case the complex $EN(\phi) = K(\phi)$ is just a Koszul complex. Let $\phi : F = R^f \to R$ be a homomorphism from a free module to a ring R. We may write $EN(\phi)$ in the form

$$ S \xleftarrow{\delta_1} F \xleftarrow{\delta_2} \textstyle\bigwedge^2 F \xleftarrow{\delta_3} \cdots \xleftarrow{\delta_f} \textstyle\bigwedge^f F \leftarrow 0, $$

where $\delta_1 = \phi$.

To define the complex, we must construct the differentials δ_i and prove that $\delta_i \delta_{i+1} = 0$. Since the modules are free, it suffices to do this for the dual maps

$$ \delta_{i+1}^* = \partial_i : \textstyle\bigwedge^i F^* \to \textstyle\bigwedge^{i+1} F^*, $$

and it turns out that this is in a sense even more natural.

It is convenient to think of R as an $S := \mathbb{Z}[x_1, \ldots, x_f]$-algebra by the map sending x_i to ϕ_i; we define the Koszul complex of ϕ over R by tensoring the Koszul complex of $(x_1, \ldots, x_f)$ with R.

Thus for the definition we take the map ϕ to be

$$ \phi : S^f \xrightarrow{(x_1 \ldots x_f)} S. $$

First of all, the map ∂_i (like the map δ_i) is *linear*: the image of a basis vector of $\bigwedge^i F^*$ is a sum of variables times basis vectors of $\bigwedge^{i+1} F^*$. We may write S as $\text{Sym}(V)$, where V is the free $\mathbb{Z}$-module generated by $x_1, \ldots, x_f$, and we may think of F as the module $V \otimes S$ with the map ϕ sending $V \otimes 1 \subset F$ by the identity to $V = S_1 \subset S$ — the *tautological map*. Let $t \in V \otimes V^* \subset S \otimes \bigwedge V^*$ be the *trace element* represented in terms of any basis $\{x_i\}$ of V and dual basis $\{\hat{e}_i\}$ of V^* as $t = \sum x_i \otimes \hat{e}_i$. Because $\bigwedge V^*$ is an anti-commutative algebra, we have $t^2 = 0$.

We define the map

$$ \partial_i : S \otimes_{\mathbb{C}} \textstyle\bigwedge^i V^* = \textstyle\bigwedge^i F^* \to \textstyle\bigwedge^{i+1} F^* = S \otimes_{\mathbb{C}} \textstyle\bigwedge^{i+1} V^* $$

to be multiplication by t, and thus $\partial_{i+1}\partial_i$ is multiplication by $t^2 = 0$.

Having defined the complex $K(\phi) = EN(\phi)$ in the case rank $G = 1$, we next ask what conditions on ϕ make it acyclic (that is, a free resolution of coker δ_1).

Theorem 18.11. *Suppose that R is a ring and $\phi : F \to R$ is a map from a free R-module of rank f. The complex $K(\phi)$ is acyclic if and only if the ideal $I := I_1(\phi)$ has grade $\geq f$.*

Proof. Theorem 18.9 directly implies that if $K(\phi)$ is acyclic then rank $\delta_f = 1$ and grade $I(\delta_f) \geq f$. Since $I(\delta_f) = I$, this proves one implication.

Now assume that grade $I \geq f$. We first prove that $K(\phi)$ is split exact when $I = R$; that is, $\bigwedge^i F \cong \ker \delta_i \oplus \operatorname{im} \delta_i$ and $\ker \delta_i = \operatorname{im} \delta_{i+1}$ for every i, or equivalently $\bigwedge^i F^* \cong \ker \partial_i \oplus \operatorname{im} \partial_i$ and $\ker \partial_i = \operatorname{im} \partial_{i-1}$ for every i. The condition $I = R$ implies that δ_1 is a split surjection, or equivalently that ∂_0 is a split injection. In this case we may write $F^* = R \oplus F'^*$ in such a way that ∂_0 is the injection into the first summand, and we may choose a basis of F^* such that the last $f - 1$ basis elements are a basis for F'^*. Specializing the sequence $x_1, x_2, \ldots, x_f$ to the sequence $1, 0, \ldots, 0$, the differential of $K(\phi)^*$ becomes multiplication by $1 \otimes e_1$.

The module $\bigwedge^i F^*$ decomposes as

$$\textstyle\bigwedge^i F^* = \big(Re_1 \otimes_R \bigwedge^{i-1} F^*\big) \oplus \bigwedge^i F'^*.$$

Because $e_1 \wedge e_1 = 0$ the differential $\partial_i : \bigwedge^i F^* \to \bigwedge^{i+1} F^*$ has the form

$$
\begin{array}{ccc}
Re_1 \otimes \bigwedge^{i-1} F'^* & \xrightarrow{\ 0\ } & Re_1 \otimes \bigwedge^i F'^* \\
\end{array}
$$

$$
\textstyle\bigwedge^i F^* = \qquad \oplus \qquad\qquad \overset{\cong}{\nearrow} \qquad\qquad \oplus \qquad\quad = \bigwedge^{i+1} F^*
$$

$$
\begin{array}{ccc}
\bigwedge^i F'^* & \xrightarrow{\ 0\ } & \bigwedge^{i+1} F'^*
\end{array}
$$

Thus we see that $K(\phi)$ is split exact when ϕ is a split surjection.

We now assume only that grade $I \geq f$. From what we just proved we see that if we localize R by inverting any element of I the complex $K(\phi)$ becomes split exact. Since grade $f \geq 1$, we can find such a nonzerodivisor in I, and inverting it does not change the ranks of the maps ϕ_i. Because ranks of free modules are additive in direct sums, it is obvious that in the split exact case the condition on the ranks of the ϕ_i is satisfied; more precisely, $\operatorname{rank}(\delta_i) = \binom{f-1}{i-1}$. We also see that after inverting a nonzerodivisor in $I(\delta_1)$ we have $I(\delta_i) = R$; equivalently,

$$I \subset \sqrt{I(\delta_i)}.$$

(In fact $I(\delta_i) = I^{\binom{f-1}{i-1}}$, though this requires a separate argument.) Thus if grade $I = f$ then grade $I(\delta_i) \geq i$ for all i, so $K(\phi)$ is acyclic. $\qquad\square$

The case rank F = rank G + 1: the Hilbert–Burch complex. In the case g = rank G and f = rank F = $g + 1$ the Eagon–Northcott complex has the form

$$EN(\phi) : \quad 0 \to G^* \otimes \textstyle\bigwedge^f F \xrightarrow{\delta_2} \bigwedge^{f-1} F \xrightarrow{\delta_1} \bigwedge^g G.$$

Here $\delta_1 = \bigwedge^g \phi$, so that the entries of a matrix for δ_1 are the $g \times g$ minors of ϕ.

We choose an identification $\bigwedge^f F = S$, called an *orientation* of F, and get a perfect pairing

$$\textstyle\bigwedge^g F \times F \to \bigwedge^f F = S$$

so that we may identify $\bigwedge^g F$ with F^*. With this identification, we define δ_2 as

$$\delta_2 : G^* \xrightarrow{\phi^*} F^* = \textstyle\bigwedge^g F.$$

We also choose an orientation $\bigwedge^g G = S$, in terms of which the image of δ_1 is the ideal generated by the $(f-1) \times (f-1)$ minors of ϕ.

We first claim that $EN(\phi)$ is a complex; that is, $\delta_1 \delta_2 = 0$. As with the Koszul complex, it is convenient to dualize and consider the maps

$$EN(\phi)^* : \quad 0 \to \textstyle\bigwedge^g G \to \bigwedge^g F^* = F \xrightarrow{\phi} G.$$

The fact that this composition is 0 is often taught as Cramer's rule for solving a system of homogeneous equations represented by a $g \times (g + 1)$ matrix of rank g. The solutions — that is, the elements of $\ker \phi$ — are multiples of the column $\Delta_1, \ldots, \Delta_g$ where the Δ_j is $(-1)^j$ times the determinant of the matrix obtained from ϕ by leaving out the j-th column. The composition $\delta_1 \delta_2$ is zero because the composition is a column matrix whose i-th entry is the $(g + 1) \times (g + 1)$ determinant of the matrix obtained from ϕ by repeating the i-th row, and is thus zero.

Theorem 18.12. *Suppose that R is a ring and $\phi : F \to G$ is a map of free R-modules, where G has rank g and F has rank $f = g + 1$. The complex $EN(\phi)$ is acyclic if and only if the ideal $I := I_g(\phi)$ has grade ≥ 2.*

Example 18.13. We have seen that the ideal of the twisted cubic is generated by the 2×2 minors of the matrix

$$\phi := \begin{pmatrix} x_0 & x_1 & x_2 \\ x_1 & x_2 & x_3 \end{pmatrix}$$

and it follows that the free resolution of its homogeneous coordinate ring is the Eagon–Northcott complex

$$0 \to S^2(-3) \xrightarrow{\begin{pmatrix} x_0\, x_1 \\ x_1\, x_2 \\ x_2\, x_3 \end{pmatrix}} S^3(-2) \xrightarrow{\phi \wedge \phi} S$$

Example 18.14. In Section 18.1 we asserted that the $(a + 1) \times a$ matrix ϕ_a given on page 312 is the minimal presentation of the ideal $(s^a, s^{a-1}t, \ldots, t^a) \subset R := \mathbb{C}[s,t]$. It is not hard to check this directly, but in any case it's easy to see that its $a \times a$ minors of ϕ_a generate this ideal, which has grade 2, so the Eagon–Northcott resolution $EN(\phi)$ has the form

$$R \xleftarrow{\wedge^a \phi_a} R^{a+1} \xleftarrow{\phi_a} R^a \leftarrow 0,$$

verifying the assertion.

Proof. Once having shown that $EN(\phi)$ is a complex, as we did above, the proof of the equivalence in the theorem follows the same pattern as the proof given above for the Koszul complex:

If $EN(\phi)$ is acyclic, then by Theorem 18.9 the $g \times g$ minors of $\phi = \delta_2$ must have grade ≥ 2. For the converse, suppose first that $I_g(\phi)$ is the unit ideal. We may split F as $S \oplus G$ with $\Delta_1 = 1$ and $\Delta_j = 0$ for $j > 1$, and then $EN(\phi)^*$ has the form $0 \to G^* \to F^* \to S$ with maps as follows:

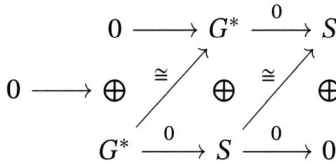

Thus $EN(\phi)$ is split exact in this case.

We next show that if $I_g(\phi)$ has grade 2 then both conditions of Theorem 18.9 are satisfied. From what we just proved we see that if we localize S by inverting any element of I then the complex $EN(\phi)$ becomes split exact. Since we have assumed that $I_1(\delta_1) = I_g(\phi)$ has grade at least 2, we can choose the element to be a nonzerodivisor and therefore, before localizing, $\mathrm{rank}(\delta_1) = 1$ and $\mathrm{rank}\,\delta_2 = g$. In this case it follows from the definition that $I_1(\delta_1) = I_g(\phi) = I_g(\delta_2)$ so, by Theorem 18.9, the complex is exact. $\square$

The Hilbert–Burch theorem. In a regular local ring any ideal whose primary components all have codimension 1 is principal (divisors are all Cartier) and thus of projective dimension 0. What about ideals of projective dimension 1? The answer is the content of the *Hilbert–Burch* theorem, proven in 1890 by David Hilbert in the case of homogeneous ideals in $\mathbb{C}[x_0, x_1]$ and in general by Lindsay Burch [1967]. We can deduce it as an application of the Eagon–Northcott complex in the case $f = g + 1$:

Corollary 18.15 (Hilbert–Burch theorem). *Suppose that R is a local ring. Any ideal $I \subset R$ of projective dimension 1 has the form aI' where I' is an ideal of grade 2 generated by the $g \times g$ minors of a $g \times (g + 1)$ matrix and a is a nonzerodivisor of R; and conversely any ideal of this form has projective dimension 1.*

In particular, if $C \subset \mathbb{P}^3$ is an ACM curve whose homogeneous ideal I is generated by f elements, then I is minimally generated by the $(f - 1) \times (f - 1)$ minors of the syzygy matrix of I.

Proof. If $C \subset \mathbb{P}^3$ is ACM, then the projective dimension of the homogeneous coordinate ring R_C is 2 by the Auslander–Buchsbaum formula, and thus the ideal of C has projective dimension 1.

Now suppose that $I \subset R$ is an ideal of projective dimension 1 in any local ring, and suppose that I is generated by f elements, so that we have a surjection $F := R^f \to I$. The module R/I has free resolution of the form

$$\mathbb{F}: \quad R \xleftarrow{\alpha} R^f \xleftarrow{\phi} G \leftarrow 0,$$

where $I = I_1(\alpha)$, so by Theorem 18.9 the free module G has rank $g = f - 1$, the $g \times g$ minors of ϕ generate an ideal I' of grade ≥ 2, and the ideal I has grade at least 1. Theorem 18.9 implies that both the Eagon–Northcott complex $EN(\phi)$ and its dual are acyclic.

The dual of the complex $\mathbb{F}$ will not be acyclic unless grade $I = 2$, but there is at least a comparison map

$$
\begin{array}{ccccccccc}
\mathbb{F}^*: & G^* & \xleftarrow{\phi^*} & F^* & \xleftarrow{\alpha^*} & R & \longleftarrow & 0 \\
& \downarrow{=} & & \downarrow{=} & & \downarrow{a} & & \\
EN(\phi)^*: & G^* & \xleftarrow{\phi^*} & F^* & \xleftarrow{\wedge^g \phi^*} & R & \longleftarrow & 0
\end{array}
$$

It follows that $I = aI'$, and since I has grade at least 1, a must be a nonzerodivisor.

Conversely, if $\phi : R^f \to R^g$ is a map with $f = g + 1$ and grade $I_g(\phi) \geq 2$, then the acyclicity of $EN(\phi)$ shows that $I_g(\phi)$ has projective dimension 1; and if a is a nonzerodivisor, then $I := aI_g(\phi) \cong I_g(\phi)$ as R-modules, so I has projective dimension 1 as well. $\qquad\square$

This argument applies, for example, to the case of a nonhyperelliptic curve of genus 3 and degree 6 in $\mathbb{P}^3$, discussed on page 121.

The general case of the Eagon–Northcott complex. With these two special cases in hand, we are ready for the general case.

Definition 18.16. If $\phi : F \to G$ is a map of free S-modules with $f := \operatorname{rank} F \geq g := \operatorname{rank} G$, then the *Eagon–Northcott complex of ϕ* is a complex of free S-modules

$$EN(\phi): \quad S \xleftarrow{\delta_1} \textstyle\bigwedge^g F \xleftarrow{\delta_2} G^* \otimes \textstyle\bigwedge^{g+1} F \xleftarrow{\delta_3} (\operatorname{Sym}^2 G)^* \otimes \textstyle\bigwedge^{g+2} F$$

$$\xleftarrow{\delta_4} \cdots \xleftarrow{\delta_{f-g+1}} (\operatorname{Sym}^{f-g} G)^* \otimes \textstyle\bigwedge^f F \leftarrow 0$$

with the following properties:

(1) After identifying $\bigwedge^g G$ with S, the map δ_1 is identified with $\bigwedge^g \phi$.

(2) For $i > 1$ the dual of the differential δ_i,

$$\partial_{i-1} := \delta_i^* : \operatorname{Sym}^{i-2} G \otimes \textstyle\bigwedge^{g+i-2} F^* \to \operatorname{Sym}^{i-1} G \otimes \textstyle\bigwedge^{g+i-1} F^*$$

is multiplication by the trace element $\sum_{i=1}^f \phi(e_i) \otimes \hat{e}_i$, where $\{e_i\}$ and $\{\hat{e}_i\}$ are dual bases for F and F^*, respectively.

Proof that $EN(\phi)$ is a complex. The proof that $\delta_1 \delta_2 = 0$ is almost the same as in the case $f = g + 1$ because a basis element

$$b := e_{i_1} \wedge \cdots \wedge e_{i_{g+1}} \in \textstyle\bigwedge^{g+1} F$$

can be thought of as coming from a rank $g + 1$ summand of F, and the value of $\delta_2 \delta_1 b$ is the same as it would be if F were replaced by this summand. Thus from the case $f = g + 1$ we see that $\delta_1 \delta_2 b = 0$, and thus $\delta_2 \delta_1 = 0$.

On the other hand if $i \geq 2$ then, just as in the case $f = 1$, the map $\partial_{i+1}\partial_i$ is multiplication by

$$\left(\sum_{i=1}^f \phi(e_i) \otimes \hat{e}_i \right)^2,$$

which we may think of as the square of an element of degree 1 in the exterior algebra of the free $\operatorname{Sym}(G)$-module $\bigwedge(\operatorname{Sym}(G) \otimes F^*) = \operatorname{Sym}(G) \otimes \bigwedge F^*$, and hence this square is 0. $\qquad\square$

Example 18.17. If ϕ is a matrix of linear forms, then the first map of $EN(\phi)$ is represented by the row of $g \times g$ minors of ϕ, which are forms of degree g, but all the rest of the maps are represented by matrices of linear forms. Thus, for example, the Betti table of the Eagon–Northcott complex of a $2 \times f$ matrix of linear forms is

j	$i = 0$	1	2	$\cdots$	$f-1$
0	1	–	–	$\cdots$	–
1	–	$\binom{f}{2}$	$2\binom{f}{3}$	$\cdots$	$(f-1)\binom{f}{f}$

Theorem 18.18. *Suppose that R is a ring and $\phi : F \to G$ is a map of free R-modules, where G has rank g and F has rank $f \geq g$. The complex $EN(\phi)$ is acyclic if and only if the ideal $I := I_g(\phi)$ has grade $\geq f - g + 1$.*

Proof. The dual of the last differential, δ_{f-g+1}, of $EN(\phi)$ is

$$\partial_{f-g} : \operatorname{Sym}^{f-g-1} G \otimes \textstyle\bigwedge^{f-1} F^* \to \operatorname{Sym}^{f-g}(G) \otimes \textstyle\bigwedge^f F^*.$$

After choosing a generator of $\bigwedge^f F^*$ we may identify $\bigwedge^f F^*$ with S and $\bigwedge^{f-1} F^*$ with F, and the map ∂_{f-g} is then identified with the multiplication map

$$\operatorname{Sym}^{f-g-1} G \otimes F \xrightarrow{1 \times \phi} \operatorname{Sym}^{f-g} G,$$

whose cokernel is $\operatorname{Sym}^{f-g}(\operatorname{coker} \phi)$ by the right exactness of the symmetric algebra functor [Eisenbud 1995, Proposition A2.2].

Because Sym is a multilinear functor, the support of $\mathrm{Sym}^{f-g}(\mathrm{coker}\,\phi)$ is contained in the support of coker ϕ, but in fact they are equal: if a localization of $(\mathrm{coker}\,\phi)_P$ over a local ring S_P is nonzero, then by Nakayama's lemma it surjects onto $S_P/P_P = \kappa(P)$. By the right exactness of the symmetric algebra functor $\mathrm{Sym}^{f-g}(\mathrm{coker}\,\phi)_P$ surjects onto $(\mathrm{Sym}^{f-g}(\kappa(P)))_P = \kappa(P) \neq 0$.

By Theorem 18.9 we see from this that if $EN(\phi)$ is acyclic, then the support of coker ϕ has grade $\geq f - g + 1$. This support is defined by the radical of $I_g(\phi)$, so grade $I_g(\phi) \geq f - g + 1$ as required.

Conversely, to show that $EN(\phi)$ is acyclic under the given hypothesis we first treat the case where R is local and $I_g(\phi) = S$, and prove in this case that $EN(\phi)^*$ is split exact. This is the most complicated part of the proof, but, given our understanding of the Koszul complex, it is purely formal.

In this case the cokernel of ϕ is 0, so we may split F and assume that $F = G \oplus F'$, the map ϕ being the projection onto the first summand. Using the splitting of F we may write the exterior powers of F as

$$\textstyle\bigwedge^i F = \bigoplus_{j=0}^{i} \bigwedge^j G \otimes \bigwedge^{i-j} F'.$$

In terms of the splitting of $\bigwedge^g F$, the map δ_1 is the projection onto the summand with $j = g$, so we must show that in this case $EN(\phi)$ is a split exact resolution of $\bigwedge^g G$ together with its augmentation map, a projection onto $\bigwedge^g G$.

It is simplest to prove the corresponding statement for the dual complex. Assuming that the summand $G \subset F$ has basis $e_1, \ldots, e_g$, the dual differential ∂_i takes the form $\sum_{i=1}^{g} e_i \otimes \hat{e}_i$. For $i \geq 1$ the terms of $EN(\phi)^*$ decompose as

$$EN_i(\phi)^* = \mathrm{Sym}^{i-1} G \otimes \textstyle\bigwedge^{g+i-1} F^* = \bigoplus_{j=0}^{g} \mathrm{Sym}^{i-1} G \otimes \bigwedge^j G^* \otimes \bigwedge^{g+i-1-j} F'^*.$$

The map $\partial_i = \delta_i^*$ is a direct sum of the maps

$$\mathrm{Sym}^{i-1} G \otimes \textstyle\bigwedge^j G^* \otimes \bigwedge^{g+i-1-j} F'^* \to \mathrm{Sym}^i G \otimes \bigwedge^{j+1} G^* \otimes \bigwedge^{g+i-1-j} F'^*$$

given by the tensor product of multiplication by $\sum_{k=1}^{g} e_k \otimes \hat{e}_k$ on the first two tensor factors and the identity on the last tensor factor, $\bigwedge^{g+i-1-j} F'^*$.

Thus, setting $\ell = g + i - 1 - j$, we see that $EN(\phi)^*$ is the direct sum for $0 \leq \ell \leq f - g$ of the free module $\bigwedge^\ell F'^*$ tensored with the complex

$$(*_\ell) \qquad \cdots \to \mathrm{Sym}^{\ell+j-g} G \otimes \textstyle\bigwedge^j G^* \to \mathrm{Sym}^{\ell+j-g+1} G \otimes \bigwedge^{j+1} G^* \cdots \to .$$

Let $R := \mathrm{Sym}\, G = S[e_1, \ldots, e_g]$, write $\phi' : R^g \to R$ for the map sending the i-th basis element of R^g to the element $e_i \in R$, and let $K(\phi') = EN(\phi')$ be the Koszul complex. The R-dual $\mathrm{Hom}(K(\phi'), R)$ decomposes as

$$R \otimes_S \textstyle\bigwedge G^* = \mathrm{Sym}(G) \otimes_S \bigwedge G^* = \bigoplus_{p,q} \mathrm{Sym}^p G \otimes \bigwedge^q G^*$$

and has differential equal to multiplication by the trace element $\sum_{k=1}^{g} e_k \otimes \hat{e}_k$. Each complex $(*_\varrho)$ is a direct summand of this complex.

We proved in Theorem 18.11 that $K(R)$ is a free resolution of $R/(e_1, \ldots, e_g) = S$ via the isomorphism $\mathrm{Sym}^0 G \otimes \bigwedge^0 G^* \to S$. Thus, as a complex of S-modules, $K(R)$ is exact with homology S in degree 0, and since it is free as an S-module, it becomes a split exact complex if we add this isomorphism.

It follows that each of the complexes $(*_\varrho)$ is split exact except for the one-term complex $(*_0)$, which is

$$0 \to \mathrm{Sym}^0 G \otimes \bigwedge^g G^* \to 0,$$

and we see finally that $EN(\phi)^*$ is a direct sum of split exact complexes plus the one-term complex

$$0 \to \mathrm{Sym}^0 G \otimes \bigwedge^g G^* \otimes \bigwedge^0 F'^* \to 0.$$

Adjoining the isomorphism

$$\partial_1 : \bigwedge^g G^* \to \mathrm{Sym}^0 G \otimes \bigwedge^g G^* \otimes \bigwedge^0 F'^*,$$

we see at last that $EN(\phi)$ is split exact in this case, as required.

The argument above shows that the complex $EN(\phi)$ becomes split exact after localizing at any prime P not containing the ideal I of $g \times g$ minors of ϕ.

To establish the rank condition in Theorem 18.9 we first consider the generic case when $I_g(\phi)$ contains a nonzerodivisor x. Inverting x does not change the rank of any map, and after inverting x the complex $EN(\phi)$ becomes split exact, so we see that

$$\mathrm{rank}\, \delta_i + \mathrm{rank}\, \delta_{i+1} \le \mathrm{rank}\, EN(\phi)_i.$$

Now suppose that $I_g(\phi)$ has grade $\ge f - g + 1$. Since this number is at least 1, it follows from the preceding argument that the rank condition of Theorem 18.9 is satisfied. Moreover after localizing at any prime not containing $I_g(\phi)$ the ideals $I(\delta_i)$ become equal to the unit ideal. Thus $I_g(\phi) \subset \sqrt{I(\delta_i)}$, so that all the ideals have grade at least $\ge f - g + 1$, verifying the grade condition of Theorem 18.9 and completing the proof of Theorem 18.18.   □

Corollary 18.19. *If $X \subset \mathbb{P}^r$ is the rational normal scroll defined by $I_2(\phi)$, then the resolution of the homogeneous coordinate ring of X is $EN(\phi)$.*

Proof. The map ϕ is represented by a 1-generic matrix of size $2 \times (1 + \mathrm{codim}\, X)$, and thus $I_2(\phi)$ has grade equal to $\mathrm{codim}\, X = (1 + \mathrm{codim}\, X) - 2 + 1$.   □

Corollary 18.20. *With notation as in Theorem 18.18, suppose that $(R, \mathfrak{m}, k)$ is a local Cohen–Macaulay ring and that $k \otimes_R \phi = 0$. If $I_g(\phi)$ has codimension $\ge f - g + 1$ then it has codimension exactly $f - g + 1$, and the $\binom{f}{g}$ forms that are the $g \times g$ minors of a matrix for ϕ are linearly independent over k in $I_g(\phi)/(\mathfrak{m} I_g(\phi))$, and $R/I_g(\phi)$ is Cohen–Macaulay.*

Proof. By Theorem 18.18 the projective dimension of $R/I_g(\phi)$ is less than or equal to $f - g + 1$. Since the projective dimension of a module is at least the codimension of its annihilator, the equality follows, and the Auslander–Buchsbaum formula implies that $R/I_g(\phi)$ is Cohen–Macaulay. The linear independence of the minors of ϕ follows because $EN(\phi)$ is a resolution and the differential δ_2, which exhibits the relations on the $g \times g$ minors, has image in the maximal ideal time $EN(\phi)_1$. $\square$

In general when $X \subset Y \subset \mathbb{P}^r$, so that $I_X \supset I_Y$, it may be hard to see which syzygies of I_X come from syzygies of I_Y. But when the degrees of the syzygies of I_Y are smaller than those of I_X, the situation is simpler. Here is the special case we will use:

Proposition 18.21. *Let $C \subset \mathbb{P}^r$ be a nondegenerate curve. If $C \subset X \subset \mathbb{P}^r$, where X is a rational normal scroll, then the Eagon–Northcott complex that is the minimal free resolution of I_X is termwise a direct summand of the minimal free resolution of I_C. Thus each number in the Betti table of the resolution of I_C is no less than the corresponding number in the Betti table of the resolution of I_X.*

Proof. Let EN be the minimal resolution of I_X, and let $\mathbb{F}$ be the minimal resolution of I_C. The inclusion $I_X \subset I_C$ induces a map $\psi : EN \to \mathbb{F}$, unique up to homotopy. Since the minimal generators of I_X are quadratic, and I_C contains no linear forms, $\phi_0 : EN_0 \to \mathbb{F}_0$ is a split monomorphism.

By Corollary 18.19 the minimal free resolution of the homogeneous coordinate ring of X is $EN(\phi)$ for a suitable $2 \times (1 + \operatorname{codim} X)$ matrix ϕ. By induction, we may assume that ψ_{i-1} is a split monomorphism. The free module $EN(\phi)_i$ is generated in degree $i + 1$, while the free module $\mathbb{F}_i$ is generated in degrees $\geq i + 1$. It follows that the relations represented by EN_i, extended by 0, are among the minimal generators of the relations represented by $\mathbb{F}_i$, completing the proof. $\square$

18.4. Green's conjecture

To describe relationship between the geometry of a curve and the Betti table of its homogeneous coordinate ring in its canonical embedding conjectured by Mark Green, we begin at the beginning of the resolution, with the number of quadratic and cubic generators of the homogeneous ideal.

Corollary 10.9 implies that the dimension of the vector space of forms of degree d vanishing on a canonical curve is independent of the curve; for example, if C is a curve of genus $g \geq 3$ that is not hyperelliptic, $\dim(I_C)_2 = \binom{g-2}{2}$. Since there are no linear forms in the ideal this implies that in the Betti table of the homogeneous coordinate ring of every canonically embedded curve of genus g we have $\beta_{1,2} = \binom{g-2}{2}$.

However, $\beta_{2,2}$, the number of cubic generators required, may vary, and reflects the geometry of the curve. We have seen in Chapter 9 that if C is a canonically embedded curve of genus 5 in $\mathbb{P}^4$ that is not trigonal, then C is the complete intersection of three quadrics, and thus the Betti table of the homogeneous coordinate ring R_C is

j	$i = 0$	1	2	3
0	1	–	–	–
1	–	3	–	–
2	–	–	3	–
3	–	–	–	1

On the other hand the ideal of a trigonal curve C' of genus 5 requires two cubic generators in addition to three quadrics. Since the Hilbert functions are the same in the two cases, and there were no linear relations on the quadrics in the nontrigonal case, there must be 2 linear relations on the quadrics in the trigonal case. Since the homogeneous coordinate ring of a canonical curve is Gorenstein, the free resolution is symmetric, and is therefore the Betti table for C' is

j	$i = 0$	1	2	3
0	1	–	–	–
1	–	3	2	–
2	–	2	3	–
3	–	–	–	1

Using the knowledge of scrolls from Chapter 17 we can refine this observation. By the geometric Riemann–Roch theorem, C' has a 1-dimensional family of trisecant lines, and any quadric containing C' must contain all these. As we have seen in Chapter 17, these lines sweep out the 2-dimensional rational normal scroll defined by the 1-generic $2 \times (g - 2)$ matrix M corresponding to the decomposition of $\mathcal{O}_{C'}(1)$ into a tensor product of the invertible sheaf $\mathcal{L}$ associated to the g_3^1 and the residual sheaf $\omega_{C'} \otimes \mathcal{L}^{-1}$. The latter has $g - 2$ sections, and we see from Section 17.2 that the scroll itself lies on the $\binom{g-2}{2}$ quadrics defined by the minors of M. The exactness of the Eagon–Northcott complex associated to this matrix shows that there are no relations of degree 0 on these minors — that is, they are linearly independent over the ground field. It follows that they generate the vector space of all quadrics containing C'. We have seen that the minimal resolution of the ideal of the scroll is, term by term, a summand of the minimal resolution of $I_{C'}$, and indeed we see the Betti table

j	$i = 0$	1	2	3
0	1	–	–	–
1	–	3	2	–

of the scroll in the first two rows of the Betti table for C' above.

Another example of a canonical curve whose ideal requires cubic generators occurs in genus 6 when C is isomorphic to a plane quintic curve. The canonical series of the plane quintic is $5 - 3 = 2$ times the hyperplane series, and it follows that the canonical image of C lies on the Veronese surface in $\mathbb{P}^5$. As explained in the general construction at the beginning of Section 17.2, the Veronese surface is contained in (in fact, equal to) the intersection of the quadrics defined by the 2×2 minors of a generic symmetric matrix, coming from the multiplication map

$$H^0(\mathcal{O}_{\mathbb{P}^2}(1)) \otimes H^0(\mathcal{O}_{\mathbb{P}^2}(1)) \to H^0(\mathcal{O}_{\mathbb{P}^2}(2)) = H^0(\mathcal{O}_{\mathbb{P}^5}(1))$$

and there are $6 = \binom{g-2}{2}$ independent quadrics in this ideal, so these are all the quadrics in I_C. Thus I_C requires cubic generators. One can show that if C is not trigonal and not isomorphic to a plane quintic, then the resolution of R_C has Betti table

j	$i = 0$	1	2	3	4
0	1	–	–	–	–
1	–	6	5	–	–
2	–	–	5	6	–
3	–	–	–	–	1

whereas if C is either trigonal or isomorphic to a plane quintic, the Betti table of R_C is

j	$i = 0$	1	2	3	4
0	1	–	–	–	–
1	–	6	8	3	–
2	–	3	8	6	–
3	–	–	–	–	1

and the first two rows are the Betti tables of either the rational normal scroll swept out by the trisecant lines (in the trigonal case) or the Veronese surface (in the plane quintic case).

One might fear that these examples are the start of a long series of types of curves whose canonical image is not cut out by quadrics, but this is not so:

Theorem 18.22 (Petri). *The ideal of a canonical curve of genus ≥ 5 is generated by the $\binom{g-2}{2}$-dimensional space of quadrics it contains unless the curve is either trigonal or isomorphic to a plane quintic; in the latter cases, the ideal of the curve is generated by quadrics and cubics.*

For modern treatments of Petri's theorem see [Schreyer 1991] or [Arbarello and Sernesi 1978].

As we have just seen, the Betti tables for curves of genus 6 do not distinguish between the two exceptional types of curves of genus 6, and this is a clue to the general case. The two exceptions in Petri's theorem are unified by the notion of the Clifford index:

Definition 18.23. The *Clifford index* Cliff $\mathcal{L}$ of an invertible sheaf $\mathcal{L}$ on a curve C is $d - 2r$, where $d := \deg \mathcal{L}$ and $r := h^0(\mathcal{L}) - 1$. The Clifford index Cliff C of a curve C of genus ≥ 2 is the minimum of the Clifford indices of the invertible sheaves $\mathcal{L}$ with $h^0(\mathcal{L}) \geq 2$ and $h^1(\mathcal{L}) \geq 2$.

See Exercise 18.1 for some equivalent formulations.

Clifford's theorem (Corollaries 2.33 and 10.14) asserts that, for any curve C, Cliff $C \geq 0$, and Cliff $C = 0$ if and only if C is hyperelliptic. If C is not hyperelliptic, then Cliff $C = 1$ if and only if C is either trigonal or isomorphic to a plane quintic, the two exceptional cases in Theorem 18.22. Corollary 12.3 shows that the Clifford index of any smooth curve of genus $g \geq 2$ is $\leq \lceil \frac{g-2}{2} \rceil$, with equality for a general curve. An invertible sheaf $\mathcal{L}$ of maximal Clifford index often has only 2 sections, though there is an infinite sequence of examples where this "Clifford dimension" is greater beginning with plane quintics and complete intersections of two cubics in $\mathbb{P}^3$ [Eisenbud et al. 1989].

Putting together the construction of Section 17.2 with Proposition 18.21 allows us to say something about the meaning of the Betti table of a canonical curve.

Theorem 18.24. *Let $C \subset \mathbb{P}^{g-1}$ be a reduced, irreducible canonical curve. If C has an invertible sheaf $\mathcal{L}$ of degree $d \leq g - 1$ with $h^0(\mathcal{L}) = 2$ then there is a 1-generic $2 \times (g + 1 - d)$ matrix of linear forms whose minors define a scroll of codimension $g - d$ containing C; and thus an Eagon–Northcott complex of length $g - d$ is a subcomplex of the minimal free resolution of R_C. In particular, the Betti table of R_C is termwise $\geq$ that of the homogeneous coordinate ring of the scroll.*

Proof. With notation as in the theorem, $h^1(\mathcal{L}) = h^0(\mathcal{L}^{-1} \otimes \omega_C) = g - d + 1$, and the scroll corresponding to the product

$$H^0(\mathcal{L}) \otimes H^0(\mathcal{L}^{-1} \otimes \omega_C) \to H^0(\omega_C)$$

has codimension $g - d$. $\square$

For a curve as in the theorem, the existence of the g_d^1 on C, together with the symmetry of the resolution of the Gorenstein ring R_C, implies that the Betti table of R_C has the form

j $\backslash$ $i=0$	1	2	...	$d-3$	$d-2$	...	$g-d$	$g-d+1$	...	$g-3$	$g-2$
0	1	–	–	...	–	–	–	–	–	–	–
1	–	*	*	...	*	*	...	*	?	...	?	?
2	–	?	?	...	?	*	...	*	*	...	*	*
3	–	–	–	...	–	–	–	–	–	–	1

where we have assumed for illustration that $d - 2 < g - d$. As before, a dash indicates a place that is definitely 0; asterisks indicate some that are definitely nonzero. The entries of row 1 are greater than or equal to the corresponding entries of the Betti table of the scroll. Note that Cliff $\mathcal{L} = d - 2$ in this case.

We can summarize this by saying that if the curve C has an invertible sheaf $\mathcal{L}$ of degree d with exactly 2 sections (which is thus of Clifford index $c = d - 2$) then $\beta_{g-d,1} \neq 0$ and, by symmetry, $\beta_{c,c+2} \neq 0$. As with the case of the plane quintics above, one can make a similar argument for *any* invertible sheaf of Clifford index c. Thus:

Corollary 18.25. *If* $\operatorname{Cliff} C \leq c$ *then* $\beta_{c,c+2}(S/I_C) \neq 0$.

Starting from examples such as the case of genus 6, Mark Green made a bold conjecture that is still open as of this writing:

Conjecture 18.26 (Green's conjecture). *If C is a smooth canonical curve of genus g over a field of characteristic 0 and R_C is the homogeneous coordinate ring of C in its canonical embedding, then the Clifford index of C is $\leq c$ if and only if $\beta_{c,c+2}(R_C) \neq 0$.*

The conjecture was made for curves over a field of characteristic 0, and is known in many cases, though it is also known to fail in small finite characteristics (see [Bopp and Schreyer 2021] for an amended conjecture that may hold in all characteristics). All the possibilities of Betti tables for canonical curves of genus up to 8 (in characteristic 0) were analyzed in [Schreyer 1986], and the project was carried further to genus 9 in [Sagraloff 2018]; these computations establish the conjecture for these values of g. Claire Voisin [2002; 2005] proved the conjecture for an open set of curves of each Clifford index; simpler proofs then appeared in [Aprodu et al. 2019; Kemeny 2021; Rathmann 2022].

See the excellent survey [Farkas 2017] for more information.

Perhaps the most definitive of the partial results related to Green's conjecture amounts to a complete characterization of curves of maximal Clifford index in the odd genus case; it is a theorem combining work from [Hirschowitz and Ramanan 1998] and [Voisin 2005]. The reader may compare the theorem to the special case of genus 5 at the beginning of this section.

Theorem 18.27. *Let $g = 2c + 1$ be an odd integer. The curves of Clifford index strictly less than c form an effective divisor Δ in M_g, and the Betti tables of the homogeneous coordinate rings of the canonical embeddings of curves outside this divisor all have the form*

j	$i = 0$	1	2	...	$c-2$	$c-1$	c	$c+1$	...	$2c-1$	$2c$	$2c+1$
0	1	–	–	...	–	–	–	–	...	–	–	–
1	–	*	*	...	*	*	–	–	...	–	–	–
2	–	–	–	...	–	–	*	*	...	*	*	–
3	–	–	–	...	–	–	–	–	...	–	–	1

as predicted by Green's conjecture.

One can at least understand a local equation for the divisor Δ directly. Let C be a curve of genus $2c + 1$ and Clifford index c, and consider a small affine neighborhood U of C in the Hilbert scheme of smooth canonical curves of this genus, The corresponding family of homogeneous coordinate rings over U will have Betti table of the form

j	$i = 0$	1	2	...	$c-2$	$c-1$	c	$c+1$	...	$2c-1$	$2c$	$2c+1$
0	1	–	–	...	–	–	–	–	...	–	–	–
1	–	*	*	...	*	*	$\mathcal{E}$		...	...	–	–	–
2	–	–	–	...	...	$\mathcal{E}^*$	*	*	...	*	*	–
3	–	–	–	...	–	–	–	–	...	–	–	1

where now the stars represent sheaves on U. By restricting U we may assume that $\mathcal{E}$ is a vector bundle on $U - \{p\}$.

The specialization of this resolution over a point $p \in U$ is in general *not* the minimal resolution of the corresponding homogeneous coordinate ring; but the minimal resolution can be obtained by eliminating all nonzero scalar entries. This will yield a resolution of the form exhibited in Theorem 18.27 if and only if the map $\mathcal{E} \to \mathcal{E}^*$ in the middle of the resolution specializes to an isomorphism. Since $\mathcal{E}$ and $\mathcal{E}^*$ have the same rank, the map fails to be an isomorphism on the divisor Δ defined locally by the vanishing of the determinant.

The divisor Δ plays another important role in the history of our subject: it is a crucial ingredient in the original proof in [Harris and Mumford 1982] of the fact that M_g is of general type for large odd values of g!

18.5. Exercises

Exercise 18.1. Show that if C is a curve of genus $g \geq 2$ and $\mathcal{L}$ is an invertible sheaf on C then:

(1) Cliff $\mathcal{L} = g + 1 - (h^0(\mathcal{L}) + h^1(\mathcal{L}))$.

(2) Cliff $C = \min\{$Cliff $\mathcal{L} \mid \deg \mathcal{L} \leq g - 1$ and $h^0(\mathcal{L}) \geq 2\}$.   ◆

Exercise 18.2. Let $q : F \to G$, with $F = S^2(-1)$ and $G = S^2$, be described by the matrix $\left(\begin{smallmatrix} x_0 & x_1 \\ x_2 & x_3 \end{smallmatrix}\right)$, and let $Q \subset \mathbb{P}^3$ be the quadric defined by the determinant of q.

(1) Show that the sheafification of the graded module $M := \operatorname{coker} q$ is $\mathcal{O}_Q(1, 0)$ and the sheafification of $\operatorname{coker} q^*$ is $\mathcal{O}_Q(0, 1)$ by computing the vanishing locus of the two sections corresponding to the generators of the module.

(2) Show that the sheafification of $\operatorname{Sym}^a(M)$ is $\mathcal{O}_Q(a, 0)$. Conclude that if $a \leq b$ then the relative ideal sheaf $\mathcal{I}_{C/Q} = \mathcal{O}(-a, -b)$ of a curve C of type (a, b) is the sheafification of the module $\operatorname{Sym}^{b-a}(M)(-b)$.

(3) Let $S = \mathbb{C}[x_0, \ldots, x_3]$. Show that the minimal S-free resolution of $\operatorname{Sym}^a(M)$ has the form $0 \to \bigwedge^2 F \otimes \operatorname{Sym}^{a-2} G(-2) \to F \otimes \operatorname{Sym}^{a-1} G(-1) \to \operatorname{Sym}^a G$.   ◆

(4) Show that if $C \subset \mathbb{P}^3$ is a curve of type (a, b) with $a \leq b$ then there is a free resolution of a module whose sheafification is $\mathcal{I}_{C/\mathbb{P}^3}$ that is a mapping cone of the map of complexes

$$
\begin{array}{ccccccc}
0 & \longrightarrow & 0 & \longrightarrow & \bigwedge^2 F & \xrightarrow{\bigwedge^2 q} & \bigwedge^2 G \\
\uparrow & & \uparrow & & \uparrow & & \uparrow \\
0 \longrightarrow \bigwedge^2 F \otimes \mathrm{Sym}^{b-a-2} G(-b-2) & \longrightarrow & F \otimes \mathrm{Sym}^{b-a-1} G(-b-1) & \longrightarrow & \mathrm{Sym}^{b-a} G(-b)
\end{array}
$$

(5) Using the construction above, compute the Betti table of the homogeneous coordinate ring of a curve of type (a, b) on the smooth quadric in $\mathbb{P}^3$.

(6) Conclude that the Hartshorne–Rao module of a curve of type (a, b) with $a \leq b$ is the cokernel of a map

$$\bigwedge^2 F^* \otimes \mathrm{Sym}^{b-a-2}(G^*)(b+2) \xleftarrow{F^*} \otimes \mathrm{Sym}^{b-a-1} G^*(b+1)$$

and thus, after choosing a basis of $F = S^2$, may be identified as the sheafification of the cokernel of

$$\mathrm{Sym}^{b-a-2}(G^*)(b+2) \xleftarrow{F} \otimes \mathrm{Sym}^{b-a-1} G^*(b+1)$$

where the map is the action of F on $\bigwedge G^*$ via the map $q : F \to G$.

Exercise 18.3. Let $S = k[x_0, \ldots, x_r]$, and let

$$\mathbb{F}: \quad F_0 \xleftarrow{\phi_1} F_1 \leftarrow \cdots \leftarrow F_{n-1} \xleftarrow{\phi_n} F_n \leftarrow 0$$

be a finite complex of free S-modules. Set

$$X_i = \{ p \in \mathbb{A}^{n+1} \mid H_i(\mathbb{F} \otimes \kappa(p)) \neq 0 \}.$$

Use Theorem 18.9 to prove that $\mathbb{F}$ is acyclic if and only if $\mathrm{codim}\, X_i \geq i$ for all $i > 0$ and that $X_0 \supseteq X_1 \supseteq \cdots \supseteq X_n$.   ◆

Exercise 18.4. The *depth lemma* states that if

$$0 \to A \to B \to C \to 0$$

is an exact sequence of nonzero finitely generated modules over a local ring R then

$$\mathrm{depth}\, C \geq \min\{\mathrm{depth}\, B, \mathrm{depth}\, A - 1\},$$

$$\mathrm{depth}\, A \geq \min\{\mathrm{depth}\, B, \mathrm{depth}\, C + 1\}.$$

Prove this in the special case when R is regular using the characterization of depth via projective dimension.

Exercise 18.5. Prove that over a local ring every projective module is free, and show that $I(\phi)$ defines the nonfree locus of coker ϕ.

Exercise 18.6. Suppose that $X \subset \mathbb{P}^r$ is arithmetically Gorenstein. Prove that if ω_{R_X} is isomorphic to $R_X(a)$ for some a.   ◆

Exercise 18.7. Find a degree 6 embedding of a curve of genus 3 that is not arithmetically Cohen–Macaulay, and another that is.   ◆

Exercise 18.8. Referring to the Betti table of a canonical curve just after Theorem 18.24, give a formula for the differences $b_i - b_{g-i-1}$ that depends only on i and g.   ◆

Exercise 18.9. Show that if $I \subset S := \mathbb{C}[x_0, \dots, x_r]$ is a codimension 2 ideal, then S/I is Cohen–Macaulay if and only if the minimal S-free resolution of S/I has the form

$$0 \to S^{n-1} \to S^n \to S$$

for some n. Show that S/I is Gorenstein if and only if I is a complete intersection.   ◆

Exercise 18.10. Show that sets of points are always ACM schemes. Which sets of 4 distinct points in $\mathbb{P}^2$ are arithmetically Gorenstein? Which sets of 5 points in $\mathbb{P}^2$? Which sets of 6 points in $\mathbb{P}^2$?

Exercise 18.11. Give an example of a set of points in $\mathbb{P}^3$ that is arithmetically Gorenstein but not a complete intersection.   ◆

Exercise 18.12. Let $\phi : F \to G \to M \to 0$ be an exact sequence of finitely generated R-modules, with F and G free.

(1) Show that the annihilator of coker ϕ has the same radical as the ideal of minors $I_{\text{rank }G}(\phi)$.
(2) Show that the cokernel of ϕ is locally free if and only if $I_{\text{rank }\phi}(\phi) = R$.

Chapter 19

Hilbert Schemes

In earlier chapters, we described some degree d embeddings of curves of genus g in projective spaces $\mathbb{P}^r$ for small g, r, d. In this chapter, we will try to describe the *restricted Hilbert scheme* $\mathcal{H}^\circ_{g,3,d}$, defined to be the open subscheme of the Hilbert scheme $\mathcal{H}_{g,3,d} := \mathrm{Hilb}_{dm-g+1}(\mathbb{P}^3)$ parametrizing smooth, irreducible and nondegenerate curves of degree d and genus g in $\mathbb{P}^3$.

Three basic questions about the schemes $\mathcal{H}^\circ_{g,r,d}$ are:

- Is $\mathcal{H}^\circ_{g,r,d}$ irreducible?
- What is its dimension (or the dimensions of its components)?
- Where is it smooth, and where is it singular?

Of course, there are many more questions one could ask about the geometry of $\mathcal{H}^\circ_{g,r,d}$, many of which are open. For example, what is the closure $\overline{\mathcal{H}^\circ_{g,r,d}} \subset \mathcal{H}_{g,r,d}$ in the whole Hilbert scheme? (In other words, when is a subscheme $X \subset \mathbb{P}^r$ with Hilbert polynomial $dm - g + 1$ *smoothable*, in the sense that it is the flat limit of a family of smooth curves?) In general, no one knows!

19.1. Degree 3

The first case to consider is that of the Hilbert scheme $\mathcal{H}_{0,3,3}$. The corresponding restricted Hilbert scheme $\mathcal{H}^\circ_{0,3,3}$, parametrizing twisted cubics, is one we have encountered and described already: in Proposition 7.11 we showed that $\mathcal{H}^\circ_{0,3,3}$ is irreducible of dimension 12, and we gave another proof, based on linkage, in Chapter 16. By Exercise 19.1, the normal bundle of a twisted cubic C is $\mathcal{N}_{C/\mathbb{P}^3} = \mathcal{O}_{\mathbb{P}^1}(5) \oplus \mathcal{O}_{\mathbb{P}^1}(5)$ so by Theorem 7.8 the tangent space to the Hilbert scheme at C is $H^0(\mathcal{N}_{C/\mathbb{P}^3}) = \mathbb{C}^{12}$. Thus $\mathcal{H}_{0,3,3}$ is smooth at this point.

The other component of $\mathcal{H}_{0,3,3}$**.** We might expect that the closure $\overline{\mathcal{H}_{0,3,3}^{\circ}}$ would be all of $\mathcal{H}_{0,3,3}$, but this is not the case: $\mathcal{H}_{0,3,3}$ has two irreducible components.

One component is the closure of $\mathcal{H}_{0,3,3}^{\circ}$. To describe the "extraneous" second component, suppose we start with a plane cubic curve $C_0 \subset \mathbb{P}^2 \subset \mathbb{P}^3$. This is of course a curve of degree 3, but its Hilbert polynomial is $p(m) = 3m$ rather than $3m + 1$, reflecting the fact that the genus of C_0 is 1, not 0.

But that can be fixed: adding a point $p \in \mathbb{P}^3 \setminus C_0$ to C_0 has the effect of increasing the Hilbert polynomial by 1, so that the Hilbert polynomial of $C := C_0 \cup \{p\}$ is $3m + 1$. Thus C corresponds to a point of $\mathcal{H}_{0,3,3}$. In fact, the locus in $\mathcal{H}_{0,3,3}$ of curves C of this form is open, and its closure $\mathcal{H}'$, which includes plane cubics with an embedded point, is a second irreducible component of $\mathcal{H}_{0,3,3}$.

Theorem 19.1. *$\mathcal{H}_{0,3,3}$ has two irreducible components (Figure 19.1). One has generic point corresponding to a twisted cubic, and the other has generic point corresponding to the union of a smooth plane cubic and a point outside the plane. They have dimensions 12 and 15 respectively.*

Proof. Let C' be the purely 1-dimensional scheme defined by the intersection of the 1-dimensional primary ideals in the decomposition of I_C. If the Hilbert polynomials of C and C' are $p(m)$ and $p'(m)$ then $p(m) \geq p'(m)$ for all large m; equality for large m would imply that $C' = C$.

Whatever 0-dimensional components C' may have do not contribute to the degree (= leading coefficient of the Hilbert polynomial) so $\deg C' = \deg C = 3$. Thus the curve C' is either irreducible or the union of two or three irreducible components. In the first case C' is either nondegenerate, in which case it is a twisted cubic by Theorem 1.7 and $C' = C$; or a plane cubic. If C' is a plane cubic, then it has Hilbert function $3m$, so $\mathcal{I}_{C'}/\mathcal{I}_C$ corresponds to a point in $\mathbb{P}^3$, either embedded in C' or not. Such a union is specified by the choice of the plane, the cubic in it, and the point,[1] and thus has dimension $3 + 9 + 3 = 15$.

On the other hand, if C is not planar and not irreducible, then C' consists of 3 lines or the union of a (planar) conic and a line not in the plane. In this case each connected component has arithmetic genus 0 and thus Hilbert polynomial with constant term 1; so the curve must be connected. All such curves can be realized as divisors of type $(1, 2)$ on a quadric, and thus have Hilbert function equal to that of C, whence again $C' = C$. $\square$

---

[1] If the point is an embedded point, specifying C' and p does not determine C — if p is a smooth point of C', for example, there is a one-parameter family of curves C with support C' and an embedded point at p; the tangent space to C at p can be any plane containing the tangent line to C' at p — but these are all flat limits of disjoint unions $C' \sqcup \{p\}$, so they don't contribute a separate component of $\mathcal{H}_{0,3,3}$.

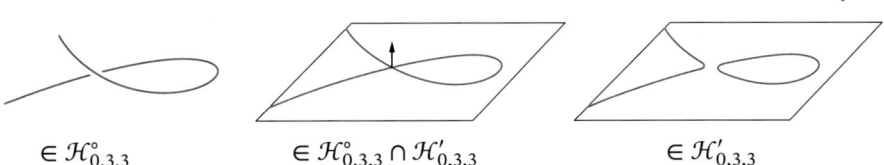

$\in \mathcal{H}^{\circ}_{0,3,3}$ $\quad\quad\quad$ $\in \mathcal{H}^{\circ}_{0,3,3} \cap \mathcal{H}'_{0,3,3}$ $\quad\quad\quad$ $\in \mathcal{H}'_{0,3,3}$

Figure 19.1. $\mathcal{H}_{0,3,3}$ has two components, whose generic members are shown on the left and right, with the generic member of the intersection shown in the middle. On the left is a generic point of the "principal" component: a smooth twisted cubic. This degenerates to a singular plane cubic with an embedded point at the node. In the other component, the singular plane cubic becomes smooth, and the extra point is free to move anywhere in space.

Cheerful Fact 19.2. In [Piene and Schlessinger 1985] it is also shown that the two components of $\mathcal{H}_{0,3,3}$ are smooth and rational and that their intersection is smooth and rational of dimension 11.

The presence of components of the Hilbert scheme whose general member is not smooth, irreducible and nondegenerate is not exceptional, and in the following section we'll describe some of the "extraneous" components more generally. But one aspect of the geometry of $\mathcal{H}_{0,3,3}$ is special: the action of the group PGL_4 on $\mathcal{H}_{0,3,3}$ has only finitely many orbits (if you want to see the orbits in the principal component $\mathcal{H}^{\circ}_{0,3,3}$, see [Harris 1982a]). It's not known if there are other examples of this, beyond Grassmannians, Hilbert schemes of quadric hypersurfaces and Hilbert schemes of triples of points. Later (page 345) we will discuss the opposite situation: the possibility of "rigid curves," corresponding to points in the Hilbert scheme $\mathcal{H}^{\circ}_{dm-g+1}$ with $g > 0$ whose PGL_{r+1} orbit is open.

19.2. Extraneous components

Let $\mathcal{H}_d(\mathbb{P}^r)$ be the Hilbert scheme of subschemes of $\mathbb{P}^r$ with Hilbert polynomial of degree 0, say equal to the constant d. There is an open subset $\mathcal{H}^{\circ}_d(\mathbb{P}^r)$ of dimension dr whose points correspond to reduced d-tuples of points in $\mathbb{P}^r$. This open subset is isomorphic to the complement of the diagonals in the d-th symmetric power of $\mathbb{P}^r$. We call the closure of this open set the *principal component* of $\mathcal{H}_d(\mathbb{P}^r)$.

The Hilbert scheme $\mathcal{H}_d(\mathbb{P}^2)$ is smooth and irreducible of dimension $2d$, and a similar result holds for 0-dimensional schemes on any smooth surface, but Iarrobino [1985] showed that, for any $r \geq 3$ and any sufficiently large d, there are components of $\mathcal{H}_d(\mathbb{P}^r)$ having dimension strictly larger than dr; see Exercise 19.17. There are also examples of extraneous components of dimension $< dr$; see [Cartwright et al. 2009]. No one knows how many irreducible components the Hilbert scheme $\mathcal{H} = \mathcal{H}_d(\mathbb{P}^r)$ has, or what their dimensions might be.

This in turn infects the Hilbert schemes of curves. For example, the Hilbert scheme $\mathcal{H}_{g,r,d}$ has a component whose general point corresponds to a union of a plane curve of degree d and $\binom{d-1}{2}-g$ points; moreover, if Γ is any irreducible component of the Hilbert scheme of zero-dimensional subschemes of degree $\binom{d-1}{2}-g$ in $\mathbb{P}^3$, there is a component of $\mathcal{H}_{g,3,d}$ whose general point corresponds to a union of a plane curve of degree d and the subscheme corresponding to a general point of Γ. Further, we can replace the plane curves in this construction with any component of the Hilbert scheme of curves of degree d and genus $g' > g$. There can also be components of $\mathcal{H}_{g,3,d}$ whose general point corresponds to a subscheme of $\mathbb{P}^3$ with a spatial embedded point — see [Chen and Nollet 2012].

Bottom line: it's a mess. For many g, d the Hilbert scheme $\mathcal{H}_{g,3,d}$ has many components. In most cases no one knows how many, or what their dimensions are, which is why we most often focus on the restricted Hilbert scheme.

19.3. Degree 4

By Clifford's theorem, an irreducible nondegenerate curve of degree 4 in $\mathbb{P}^3$ must have genus 0 or 1; we consider these cases in turn.

Genus 0. We can deal with rational quartics by a slight variant of the first method we used to deal with twisted cubics. A rational curve of degree 4 is the image of a map $\phi_F : \mathbb{P}^1 \to \mathbb{P}^3$ given by a four-tuple $F = (F_0, F_1, F_2, F_3)$ with $F_i \in H^0(\mathcal{O}_{\mathbb{P}^1}(4))$. The space of all such four-tuples up to scalars is a projective space of dimension $4 \times 5 - 1 = 19$; let $U \subset \mathbb{P}^{19}$ be the open subset of four-tuples such that the map ϕ is a nondegenerate embedding. By the universal property of the Hilbert scheme there is a surjective map $\pi : U \to \mathcal{H}^\circ_{0,3,4}$, whose fiber over a point C is the space of maps with image C. Since any two such maps differ by an automorphism of $\mathbb{P}^1$ — that is, an element of PGL_2 — the fibers of π are three-dimensional, proving that $\mathcal{H}^\circ_{0,3,4}$ is irreducible of dimension 16.

The same analysis can be used on rational curves of any degree d in any projective space $\mathbb{P}^r$:

Proposition 19.3. *The open set $\mathcal{H}^\circ_{0,r,d}$ parametrizing smooth, irreducible nondegenerate rational curves $C \subset \mathbb{P}^r$ is irreducible of dimension $(r+1)(d+1)-4$; in case $r = 3$ in particular it has dimension $4d$.*

Proof. The space U of nondegenerate embeddings $\mathbb{P}^1 \to \mathbb{P}^r$ of degree d is an open subset of the projective space $\mathbb{P}^{(r+1)(d+1)-1}$ of $(r+1)$-tuples of homogeneous polynomials of degree d on $\mathbb{P}^1$ modulo scalars; and the fibers of the corresponding map $U \to \mathcal{H}^\circ_{0,r,d}$ are copies of PGL_2. $\square$

In contrast to the case of twisted cubics, smooth rational curves in $\mathbb{P}^r$ of the same degree may have different normal bundles. This gives an interesting stratification of the restricted Hilbert scheme of rational curves; see [Coskun and Riedl 2018] for a discussion.

Genus 1. As we saw in Section 4.5, a quartic curve $C \subset \mathbb{P}^3$ of genus 1 is the intersection of two quadric surfaces, and by Lasker's theorem, every quadric containing C is a linear combination of those two. Conversely, the intersection of two general quadrics in $\mathbb{P}^3$ is a quartic curve of genus 1, as follows from Bézout's theorem and the adjunction formula. We can thus construct a family of quartics of genus 1: let $V = H^0(\mathcal{O}_{\mathbb{P}^3}(2))$ be the 10-dimensional vector space of homogeneous quadric polynomials in $\mathbb{P}^3$ and $G(2, V)$ the Grassmannian of 2-planes in V, and consider the incidence correspondence

$$\Gamma = \left\{ (\Lambda, p) \in G(2, V) \times \mathbb{P}^3 \mid F(p) = 0 \ \forall \ F \in \Lambda \right\}.$$

The fiber of Γ over a point $\Lambda \in G(2, V)$ is thus the base locus of the pencil of quadrics represented by Λ; let $B \subset G(2, V)$ be the Zariski open subset over which the fiber is smooth, irreducible and nondegenerate of dimension 1. By the universal property of Hilbert schemes, the family $\pi_1 : \Gamma_B \to U$ induces a map $\phi : B \to \mathcal{H}^{\circ}_{1,3,4}$ that is one-to-one on points; it follows that the reduced subscheme of $\mathcal{H}^{\circ}_{1,3,4}$ is birational to an open subset of the Grassmannian $G(2, 10)$. We conclude that $\mathcal{H}^{\circ}_{1,3,4}$ is irreducible of dimension 16. Exercise 19.6 shows that this map is actually an isomorphism.

The argument here — where we constructed a family $\mathcal{C} \to B$ of curves of given type, and then invoked the universal property of the Hilbert scheme to get a map $B \to \mathcal{H}$ — is typical in analyses of Hilbert schemes.

19.4. Degree 5

Let $C \subset \mathbb{P}^3$ be a smooth, irreducible, nondegenerate quintic curve of genus g. By Clifford's theorem the bundle $\mathcal{O}_C(1)$ must be nonspecial, so by the Riemann–Roch theorem we must have $0 \leq g \leq 2$. We have already seen that the space $\mathcal{H}^{\circ}_{0,3,5}$ of rational quintic curves is irreducible of dimension 20. We will treat the case $g = 2$ in detail, and leave the case $g = 1$ as Exercise 19.10. (Degree 5 will be covered in a different way in Section 19.11.)

Genus 2. We have considered curves of genus 2 in Section 6.3. Recall that a curve of genus 2 and degree 5 in $\mathbb{P}^3$ is contained in the intersection of a unique quadric surface Q and a cubic surface S not containing Q. The intersection $Q \cap S$ has degree 6, and is thus the union of C and a line. If Q is smooth then, in terms of the isomorphism $Q \cong \mathbb{P}^1 \times \mathbb{P}^1$, the curve C is a divisor of type $(2, 3)$ on the quadric Q. Note that conversely if $L \subset \mathbb{P}^3$ is a line and Q and $S \subset \mathbb{P}^3$ are general quadric and cubic surfaces containing L, and if we write

$$Q \cap S = L \cup C$$

then the curve C is a curve of type $(2,3)$ on the quadric Q and hence, by the adjunction formula, a quintic of genus 2.

This suggests two ways of describing the family $\mathcal{H}^\circ_{2,3,5}$ of such curves. First, we can use the fact that C is linked to a line to make an incidence correspondence

$$\Psi = \{(C, L, Q, S) \in \mathcal{H}^\circ \times \mathbb{G}(1,3) \times \mathbb{P}^9 \times \mathbb{P}^{19} \mid Q \cap S = C \cup L\},$$

where the $\mathbb{P}^9$ (respectively, $\mathbb{P}^{19}$) is the space of quadric (respectively, cubic) surfaces in $\mathbb{P}^3$. Given a line $L \in \mathbb{G}(1,3)$, the space of quadrics containing L is a $\mathbb{P}^6$, and the space of cubics containing L is a $\mathbb{P}^{15}$; thus the fiber of the projection $\pi_2 : \Psi \to \mathbb{G}(1,3)$ over L is an open subset of $\mathbb{P}^6 \times \mathbb{P}^{15}$, and we see that Ψ is irreducible of dimension $4 + 6 + 15 = 25$.

On the other hand, the fiber of Ψ over a point $C \in \mathcal{H}^\circ_{2,3,5}$ is an open subset of the $\mathbb{P}^5$ of cubics containing C; and we conclude that $\mathcal{H}^\circ_{2,3,5}$ is irreducible of dimension 20.

Another approach to describing the restricted Hilbert scheme $\mathcal{H}^\circ_{2,3,5}$ would be to use the fact that the quadric surface Q containing a quintic curve $C \subset \mathbb{P}^3$ of genus 2 is unique. We thus have a map

$$\mathcal{H}^\circ \to \mathbb{P}^9,$$

whose fiber over a point $Q \in \mathbb{P}^9$ is the space of quintic curves of genus 2 on Q.

The general fiber of this map, the space of quintic curves of genus 2 on a smooth quadric Q is reducible: it consists of the disjoint union of the open subsets of smooth elements in the two linear series of curves of type $(2,3)$ and $(3,2)$ on Q, each of which is a $\mathbb{P}^{11}$, while over a quadric cone the fiber is irreducible, since the cone has a unique family of lines. We can conclude immediately that $\mathcal{H}^\circ_{2,3,5}$ is of pure dimension 20; to conclude that it is irreducible we use the fact that in the family of all smooth quadric surfaces, the monodromy exchanges the two rulings (Example 11.2).

19.5. Degree 6

Again the Clifford and Riemann–Roch theorems suffice to compute the possible genera of a curve of degree 6. To start with, if the line bundle $\mathcal{O}_C(1)$ is nonspecial, then by the Riemann–Roch theorem we have $g \leq 3$. Suppose on the other hand that $\mathcal{O}_C(1)$ is special. Since $h^0(\mathcal{O}_C(1)) \geq 4$, we have equality in Clifford's theorem, and either C is hyperelliptic and $\mathcal{O}_C(1)$ is a multiple of the g^1_2 or C is a canonically embedded curve of genus 4. The first case cannot occur, since no special series on a hyperelliptic curve is very ample; thus C must be a canonical curve of genus 4. Thus a smooth irreducible, nondegenerate curve of degree 6 in $\mathbb{P}^3$ has genus at most 4.

The cases of genera 0, 1 and 2 are covered under Proposition 19.5 below, leaving us with $g = 3$ and 4. In both cases we can describe the ideal of the curve.

Genus 4. As we've seen in Theorem 9.1, a canonical curve of genus 4 is the complete intersection of a (unique) quadric Q and a cubic surface S. We thus have a map

$$\alpha : \mathcal{H}^{\circ}_{4,3,6} \to \mathbb{P}^9$$

sending a curve C to the quadric Q containing it. Moreover, the fibers of this map are open subsets of the projective space $\mathbb{P}V$, where V is the quotient of the space of all cubic polynomials modulo cubics containing Q,

$$V = \frac{H^0(\mathcal{O}_{\mathbb{P}^3}(3))}{H^0(\mathcal{I}_{Q/\mathbb{P}^3}(3))}.$$

Since this vector space has dimension 16, the fibers of α are irreducible of dimension 15, and we see that the space $\mathcal{H}^{\circ}_{4,3,6}$ is irreducible of dimension 24.

See Exercise 19.12 for the generalization to arbitrary smooth complete intersections in $\mathbb{P}^3$.

Genus 3. We leave this to the reader in Exercises 19.13 and 19.14.

19.6. Degree 7

Using the tools above, we invite the reader to show that each component of the restricted Hilbert schemes of curves of degree 7 in $\mathbb{P}^3$ has expected dimension 28. See Exercise 19.16 for an outline.

19.7. The expected dimension of $\mathcal{H}^{\circ}_{g,r,d}$

The sharp-eyed reader will have noticed that, in every case analyzed so far, the Hilbert scheme $\mathcal{H}^{\circ}_{g,3,d}$ parametrizing smooth curves of degree d and genus g in $\mathbb{P}^3$ has dimension $4d$. This is the expected dimension, in a sense we will now make precise:

Let $C \subset \mathbb{P}^r$ be a smooth curve of genus g and degree d. In Section 7.3 we computed the tangent space to $\mathcal{H}^{\circ}_{g,r,d}$ at the point $[C]$ as $H^0(\mathcal{N}_{C/\mathbb{P}^r})$, so the dimension $h^0(\mathcal{N}_{C/\mathbb{P}^r})$ is an upper bound for $\dim \mathcal{H}^{\circ}_{g,r,d}$. We can compute a lower bound as well:

Cheerful Fact 19.4. The completion of the local ring

$$(R, \mathfrak{m}) = \mathbb{C}[\![x_1, \dots, x_t]\!]/J$$

of $\mathcal{H}^\circ_{g,r,d}$ at the point $[C]$ representing a smooth curve can, in principle, be computed by deformation theory. Though this can actually be carried out in small cases, it is hard to get much qualitative information from the process except for two numbers: the dimension of the Zariski tangent space $(\mathfrak{m}/\mathfrak{m}^2)^*$, which we have computed as $t := h^0(\mathcal{N}_{C/\mathbb{P}^r})$; and an upper bound for the number of generators of the ideal J, which is $h^1(\mathcal{N}_{C/\mathbb{P}^r})$. See [Fantechi and Göttsche 2005, Corollaries 6.2.5, 6.4.11 and Proposition 6.5.2], where a similar result is given for any locally complete intersection subscheme of $C \subset \mathbb{P}^r$.

Thus from the principal ideal theorem [Eisenbud 1995, Theorem 10.2] it follows that if C is a smooth curve, then

$$\chi(\mathcal{N}_{C/\mathbb{P}^r}) \leq \dim(\mathcal{H}^\circ_{g,r,d}) \leq h^0(\mathcal{N}_{C/\mathbb{P}^r}) \text{ locally at } [C].$$

If the upper bound is achieved, then $[C]$ is a smooth point of $\mathcal{H}^\circ_{g,r,d}$, and if the lower bound is achieved then the local ring of $\mathcal{H}^\circ_{g,r,d}$ at $[C]$ is a complete intersection.

Using the deformation bound (Cheerful Fact 19.4) we can at least control the Hilbert scheme at a point representing a nonspecial embedding.

Theorem 19.5. *For any smooth curve $C \subset \mathbb{P}^r$ of genus g and degree d*

$$\chi(\mathcal{N}_{C/\mathbb{P}^r}) = (r+1)d - (r-3)(g-1),$$

which is a lower bound for the dimension of the Hilbert scheme locally at $[C]$. If $\mathcal{O}_C(1)$ is nonspecial, then $H^1(\mathcal{N}_{C/\mathbb{P}^r}) = 0$ so $\mathcal{H}^\circ_{g,r,d}$ is smooth and of dimension $(r+1)d - (r-3)(g-1)$ at $[C]$. If $d > 2g-2$, then $\mathcal{H}^\circ_{g,r,d}$ is irreducible as well.

Note that in the case of $\mathbb{P}^3$ we have $\chi(\mathcal{N}_{C/\mathbb{P}^r}) = 4d$, and we have seen that when $d \leq 7$ this is always equal to the dimension of the restricted Hilbert scheme, whether or not the embedding is special. In view of this, we define the *expected dimension* of $\mathcal{H}^\circ_{g,r,d}$ to be

$$h(g, r, d) := (r+1)d - (r-3)(g-1).$$

We will give yet another argument for calling this the expected dimension in Section 19.11.

Proof. From the exact sequence

$$0 \to \mathcal{T}_C \to \mathcal{T}_{\mathbb{P}^r}|_C \to \mathcal{N}_{C/\mathbb{P}^r} \to 0$$

we see that $\chi(\mathcal{N}_{C/\mathbb{P}^r}) = \chi(\mathcal{T}_{\mathbb{P}^r}|_C) - \chi(\mathcal{T}_C)$ and that if $H^1(\mathcal{T}_{\mathbb{P}^r}|_C) = 0$ then $\chi(\mathcal{N}_{C/\mathbb{P}^r}) = h^0(\mathcal{N}_{C/\mathbb{P}^r})$. Since $\mathcal{T}_C = \omega_C^{-1}$, the Riemann–Roch theorem gives $\chi(\mathcal{T}_C) = -3g + 3$.

To compute $\chi(\mathcal{T}_{\mathbb{P}^r}|_C)$ we restrict the Euler sequence

$$0 \to \mathcal{O}_{\mathbb{P}^r} \to \mathcal{O}_{\mathbb{P}^r}(1)^{r+1} \to \mathcal{T}_{\mathbb{P}^r} \to 0$$

to C. Using the Riemann–Roch theorem again we deduce that

$$\chi(\mathcal{T}_{\mathbb{P}^r}) = (r+1)\chi(\mathcal{O}_C(1)) - \chi(\mathcal{O}_C) = (r+1)(d-g+1) - (1-g) = (r+1)d + r(1-g)$$

From the restriction of the sequence above we also see that if $\mathcal{O}_C(1)$ is nonspecial then $H^1(\mathcal{T}_{\mathbb{P}^r}|_C) = 0$.

Putting these values together, we get $\chi(\mathcal{N}_{C/\mathbb{P}^r}) = (r+1)d - (r-3)(g-1)$ as required.

Whenever $\dim \mathcal{H}^\circ_{g,r,d} = h^0(\mathcal{N}_{C/\mathbb{P}^r})$ the Hilbert scheme is smooth at C by Theorem 7.8. From the deformation theory argument in 19.4 we know that this dimension equality is true for any nonspecial embedding since in this case $H^1(\mathcal{N}_{C/\mathbb{P}^r}) = 0$, so the upper and lower estimates for the dimension coincide. If $d > 2g - 2$, then every curve in $\mathcal{H}^\circ_{g,r,d}$ is nonspecial.

We can also prove the dimension statement for nonspecial embeddings invoking only the existence of the relative Picard scheme from Chapter 5: If $\mathcal{O}_C(1)$ is nonspecial, then the nonspecial invertible sheaves of degree d on curves of genus g form an open subset of the $(3g - 3 + g)$-dimensional relative Picard scheme. The dimension of the space of sections is $d - g + 1$, so the family of r-dimensional linear series associated to each invertible sheaf is the dimension of the Grassmannian, $\dim G(r+1, d-g+1) = (r+1)(d-g-r)$, and the choice of a basis of the linear series, up to scalars, adds $(r+1)^2 - 1 = \dim \mathrm{PGL}_{r+1}$. Thus the dimension of $\mathcal{H}^\circ_{g,r,d}$ near $[C]$ is

$$3g - 3 + g + (r+1)(d-g-r) + (r+1)^2 - 1 = (r+1)d - (r-3)(g-1)$$

as required.

The irreducibility of the Hilbert scheme in the case $d > 2g - 2$ also follows from this argument together with the existence of a connected family containing all curves of genus g, for example over the Hilbert scheme of tricanonical curves. $\qquad\square$

19.8. Some open problems

Brill–Noether in low codimension. If $C \subset \mathbb{P}^r$ is a nonspecial embedding, then the set of isomorphism classes of the curves represented in the component of $\mathcal{H}^\circ_{g,r,d}$ containing $[C]$ is open in M_g. We shall see in Section 19.11 that every component of $\mathcal{H}^\circ_{g,r,d}$ dominating M_g in this sense has dimension $h(g,r,d)$. What about "smaller" components? Observations suggest that components of $\mathcal{H}^\circ$ whose images in M_g have low codimension still have the expected dimension $h(g,r,d)$: among the examples we know of components of the Hilbert scheme whose dimension is strictly greater than the expected $h(g,r,d)$, there are none whose image in M_g has codimension less than $g - 5$.

Conjecture 19.6. *If $\mathcal{K} \subset \mathcal{H}^{\circ}_{g,r,d}$ is any component of a restricted Hilbert scheme, and the image of $\mathcal{K}$ in M_g has codimension $\leq g - 5$, then* $\dim \mathcal{K} = h(g,r,d)$.

For some examples see [Keem 1994, Theorem 3.4].

Maximally special curves. How special can a linear series on a special curve be?

To make such a question precise, let $\widetilde{M}^r_{g,d} \subset M_g$ be the closure of the image of the map $\phi : \mathcal{H}^{\circ}_{g,r,d} \to M_g$ sending a curve to its isomorphism class. We ask,

(1) What is the smallest possible dimension of $\mathcal{H}^{\circ}_{g,r,d}$?

(2) What is the smallest possible dimension of $\widetilde{M}^r_{g,d}$? and

(3) Modifying the last question slightly, let $M^r_{g,d} \subset M_g$ be the closure of the locus of curves C that possess a g^r_d (in other words, we are dropping the condition that the g^r_d be very ample). What is the smallest possible dimension of $M^r_{g,d}$?

One might guess that the most special curves, from the point of view of questions 2 and 3, are hyperelliptic curves. A hyperelliptic curve is determined by a set of $2g + 2$ points in $\mathbb{P}^1$, modulo the action of PGL_2, so the locus in M_g of hyperelliptic curves has dimension $2g - 1$. Smooth plane curves are a better guess — the locus in M_g of smooth plane curves has dimension asymptotic to g (Exercise 19.21) — but there are still a lot of them.

Can we do better? Consider the locus of smooth complete intersections of two surfaces of degree m in $\mathbb{P}^3$. (Exercise 19.23 suggests why we are choosing complete intersections of surfaces of the same degree.) As we saw in Exercise 19.7, these comprise an open subset $\mathcal{H}^{\circ}_{ci}$ of the Hilbert scheme of curves of degree $d = m^2$, and genus g given by the relation

$$2g - 2 = \deg K_C = m^2(2m - 4),$$

or, asymptotically,

$$g \sim m^3.$$

Moreover, the dimension of this component of the Hilbert scheme is easy to compute: as we saw in Exercise 19.8, it is isomorphic to an open subset of the Grassmannian $G(2, \binom{m+3}{3})$, and so has dimension

$$2\left(\binom{m+3}{3} - 2\right) \sim \frac{m^3}{3}.$$

To compute the dimension of the fibers of the map from this Hilbert scheme to M_g we use the facts that if $C \subset \mathbb{P}^r$ is a complete intersection curve of genus $g > 1$ then the canonical sheaf K_C is a positive power of $\mathcal{O}_C(1)$, and C is arithmetically Cohen–Macaulay by Exercise 3.16. For a given abstract curve C there are only finitely many invertible sheaves having the canonical sheaf as a power; and since an arithmetically Cohen–Macaulay curve is necessarily embedded by a complete linear series, there are only finitely many embeddings of a given curve as a complete intersection, up to PGL_{r+1}. Thus the fibers of $\mathcal{H}^\circ_{ci}$ over M_g have dimension $\dim(\mathrm{PGL}_{r+1}) = r^2 + 2r$.

This construction gives a sequence of components of the restricted Hilbert scheme $\mathcal{H}^\circ_{g,3,d}$ whose images in M_g have dimension tending asymptotically to $g/3$.

More generally, we can consider complete intersections of $r - 1$ hypersurfaces of degree m in $\mathbb{P}^r$. Such curves have genus $g = m^{r-1}((r-1)m-r-1)/2+1$ in a similar fashion we can calculate that their images in M_g have dimension asymptotically approaching $2g/r! = (m-1)(r-1)/r!$ as $m \to \infty$, as we ask you to verify in Exercise 19.23.

These components have the smallest images in M_g of any we know. To pose a precise question: if we fix r, can we find a sequence of components $\mathcal{H}_n$ of restricted Hilbert schemes $\mathcal{H}^\circ_{g_n,r,d_n}$ of curves in $\mathbb{P}^r$ such that

$$\lim \frac{\dim \mathcal{H}_n}{g_n} = 0?$$

Rigid curves? In the last section, we considered components of the restricted Hilbert scheme whose image in M_g was "as small as possible". Let's go now all the way to the extreme, and ask: is there a component of the restricted Hilbert scheme $\mathcal{H}^\circ_{g,r,d}$ whose image in M_g is a single point? (Of course M_0 itself is a single point, so we exclude genus 0.) We can give three flavors of this question, in order of ascending preposterousness. Let $C \subset \mathbb{P}^r$ be a smooth irreducible nondegenerate curve.

(1) We call C *moduli-rigid* if it lies in a component of the restricted Hilbert scheme whose image in M_g is just the point $[C] \in M_g$ — in other words, if the linear series $|\mathcal{O}_C(1)|$ does not deform to any nearby curves.

(2) We call C *rigid* if it lies in a component $\mathcal{H}^\circ_{g,r,d}$ of the restricted Hilbert scheme such that PGL_{r+1} acts transitively on $\mathcal{H}^\circ_{g,r,d}$. This is saying that C is moduli rigid, plus the line bundle $\mathcal{O}_C(1)$ does not deform to any other g^r_d on C.

(3) We call C *deformation-rigid* if the curve $C \subset \mathbb{P}^r$ has no nontrivial infinitesimal deformations other than those induced by PGL_{r+1} — in other words, every global section of the normal bundle $\mathcal{N}_{C/\mathbb{P}^r}$ is the image of the restriction of a vector field on $\mathbb{P}^r$.

Do any rigid curves exist? In $\mathbb{P}^3$ we have a positive lower bound $h(g,r,d) \geq 4d$ on the dimension of the restricted Hilbert scheme, which shows that there are no rigid curves in $\mathbb{P}^3$. But this argument does nothing for the general case: when $r \geq 4$, the number $h(g, r, d)$ is sometimes negative (for example already when $r = 4, d \geq 36$ and g the maximal genus allowed by the Castelnuovo bound, though there are certainly no rigid curves in this case, since the curves move in a linear series on the scroll).

Indeed, the recent paper [Keem et al. 2019] shows that no moduli-rigid curves exist in certain ranges. More generally, the existence of irrational rigid curves seems outlandish; we don't know anyone who thinks there are such things. But then, why can't we prove that they don't exist?

19.9. Degree 8, genus 9

So far, with the reader's presumed assistance, we have examined all the $\mathcal{H}^\circ_{g,3,d}$ with $d \leq 7$ and found only irreducible varieties whose components have the expected dimension $4d = h(g, 3, d)$. But it turns out that this is not typical.

We start with an example of a component of $\mathcal{H}^\circ_{9,3,8}$ whose dimension is strictly greater than $4d$. Let C be a curve in this Hilbert scheme, and consider the restriction map

$$\rho_2 : H^0(\mathcal{O}_{\mathbb{P}^3}(2)) \to H^0(\mathcal{O}_C(2)).$$

The source of ρ_2 has dimension 10. The Riemann–Roch theorem admits two possibilities for the dimension of the target.

$$h^0(\mathcal{O}_C(2)) = \begin{cases} 9 & \text{if } \mathcal{O}_C(2) \cong K_C, \\ 8 & \text{if } \mathcal{O}_C(2) \not\cong K_C. \end{cases}$$

However, if $h^0(\mathcal{O}_C(2))$ were 8 then C would lie on two distinct quadrics Q and Q'. Since C is nondegenerate, it cannot lie on a reducible quadric; thus Q and Q' would be irreducible, violating Bézout's theorem. We deduce that $\mathcal{O}_C(2) \cong K_C$, and thus that C lies on a unique quadric surface Q.

Similarly, since $\deg C > 2 \cdot 3$, the curve C cannot lie on any cubic not containing Q. Moving on to quartics, we look again at the restriction map

$$\rho_4 : H^0(\mathcal{O}_{\mathbb{P}^3}(4)) \to H^0(\mathcal{O}_C(4)).$$

The dimensions here are, respectively, 35 and $4 \cdot 8 - 9 + 1 = 24$; and we deduce that C lies on at least an 11-dimensional vector space of quartic surfaces. On the other hand, only a 10-dimensional vector subspace of these vanish on Q. Thus C lies on a quartic surface not containing Q. It follows from Bézout's theorem and Lasker's theorem that $C = Q \cap X$ and moreover, by Lasker's theorem, the homogeneous ideal of C is generated by the forms defining Q and X. Thus $\ker(\rho_4)$ has dimension exactly 11, and X is unique modulo quartics vanishing on Q.

From these facts it is easy to compute the dimension of $\mathcal{H}^\circ_{9,3,8}$: Associating to C the unique quadric on which it lies gives a map $\mathcal{H}^\circ_{9,3,8} \to \mathbb{P}^9$ with dense image, and each fiber is an open subset of the projective space $\mathbb{P}V$, where V is the 25-dimensional vector space

$$V = \frac{H^0(\mathcal{O}_{\mathbb{P}^3}(4))}{H^0(\mathcal{I}_{Q/\mathbb{P}^3}(4))}.$$

By Exercise 7.5, $\mathcal{H}^\circ_{9,3,8}$ is generically smooth, as well.

In sum, we have proven:

Proposition 19.7. *The scheme $\mathcal{H}^\circ_{9,3,8}$ is generically smooth and irreducible of dimension 33 — one larger than the expected 4d.*

This is a special case of Exercise 19.12. See Exercise 19.20 for a rich set of examples of components whose dimension is $> h(g,r,d)$.

19.10. Degree 9, genus 10

For an example of a restricted Hilbert scheme that is reducible, consider $\mathcal{H}^\circ_{10,3,9}$, the scheme of smooth irreducible curves of degree 9 and genus 10 in $\mathbb{P}^3$.

Proposition 19.8. *The scheme $\mathcal{H}^\circ_{10,3,9}$ parametrizing smooth, irreducible, non-degenerate curves $C \subset \mathbb{P}^3$ of degree 9 and genus 10 has two irreducible components, each generically smooth of dimension 36, the expected dimension. One consists of the complete intersections of two cubics. The other consists of curves of type $(3,6)$ or $(6,3)$ on a smooth quadric surface.*

Proof. To describe a smooth irreducible nondegenerate curve C of degree 9 and genus 10 in $\mathbb{P}^3$ we look at the restriction maps $\rho_m : H^0(\mathcal{O}_{\mathbb{P}^3}(m)) \rightarrow H^0(\mathcal{O}_C(m))$. The Riemann–Roch theorem admits these possibilities:

$$h^0(\mathcal{O}_C(2)) = \begin{cases} 10 & \text{if } \mathcal{O}_C(2) \cong K_C \text{ (type 1)}, \\ 9 & \text{if } \mathcal{O}_C(2) \not\cong K_C \text{ (type 2)}. \end{cases}$$

Unlike the situation in degree 8, both occur.

<u>Type 1.</u> Suppose first that C does not lie on any quadric surface (so that C is of type 1), and consider the map $\rho_3 : H^0(\mathcal{O}_{\mathbb{P}^3}(3)) \rightarrow H^0(\mathcal{O}_C(3))$. By the Riemann–Roch theorem, the dimension of the target is $3 \cdot 9 - 10 + 1 = 18$, from which we conclude that C lies on at least a pencil of cubic surfaces. Since C lies on no quadrics, all of these cubic surfaces must be irreducible, and it follows by Bézout's theorem that the intersection of two such surfaces is exactly C. At this point, Lasker's theorem assures us that C lies on exactly two cubics.

By Exercise 19.8 the space $\mathcal{H}^\circ_1$ of curves of this type is an open subset of the Grassmannian $G(2,20)$ of pencils of cubic surfaces, which is irreducible of dimension $36 = 4d$. By Exercise 7.5, $\mathcal{H}^\circ_1$ is generically smooth.

<u>Type 2.</u> Next, suppose that C has type 2 — that is, C does lie on a quadric surface $Q \subset \mathbb{P}^3$; let $\mathcal{H}^\circ_2 \subset \mathcal{H}^\circ_{10,3,9}$ be the locus of such curves. First, suppose

that the quadric Q is singular. Since C is irreducible and nondegenerate, Q must be the cone over a conic, that is, the scroll $S(0,2)$. Since $\deg C = 9$ we must have $C \sim 4H + F$, where H is the hyperplane and F the ruling on Q. By Theorem 17.25 the genus of such a curve is 12, not 10, a contradiction proving that the quadric Q is smooth.

If C has class (a,b) on Q then $a+b = 9, (a-1)(b-1) = 10$ has the solutions $(a,b) = (3,6)$ or $(6,3)$.

We can show that $\mathcal{H}_2^\circ$ is generically smooth by computing its tangent space $H^0(\mathcal{N}_{C/\mathbb{P}^3})$. We start with the normal bundle sequence

$$0 \to \mathcal{N}_{Q/\mathbb{P}^3}|_C \to \mathcal{N}_{C/\mathbb{P}^3} \to \mathcal{N}_{C/Q} \to 0.$$

By the adjunction formula, $\omega_C = \mathcal{O}_Q(1,4)|_C$. Thus both $\mathcal{N}_{Q/\mathbb{P}^3}|_C = \mathcal{O}_Q(2,2)|_C$ and $\mathcal{N}_{C/Q} = \mathcal{O}_Q(3,6)|_C$ are nonspecial, so

$$h^0(\mathcal{N}_{C/\mathbb{P}^3}) = h^0(\mathcal{N}_{\mathcal{O}_Q(2,2)}|_C) + h^0(\mathcal{O}_Q(3,6)|_C) = 10 - 1 + (4 \cdot 7 - 1) = 36$$

We outline two proofs that $\mathcal{H}_2^\circ$ is irreducible in Exercise 19.18. $\square$

These examples are far from exhausting the possibilities for the schemes $\mathcal{H}_{g,3,d}^\circ$. For example the scheme $\mathcal{H}_{14,3,24}^\circ$ has 3 components, one of which is everywhere nonreduced [Mumford 1962; Nasu 2008].

19.11. Estimating the dimension of the restricted Hilbert schemes using the Brill–Noether theorem

The Brill–Noether theorems lead to an understanding of at least one component of $\mathcal{H}_{g,r,d}^\circ$ when the Brill–Noether number is nonnegative:

Theorem 19.9. *Let* $g \geq 2, d$ *and* r *be nonnegative integers such that the Brill–Noether number* $\rho(g,r,d) = g - (r+1)(g-d+r) \geq 0$ *and* $r \geq 3$*. There is a unique component* $\mathcal{H}_0$ *of the restricted Hilbert scheme* $\mathcal{H}_{g,r,d}^\circ$ *dominating the moduli space* M_g*; and this component has the "expected dimension"*

$$\dim \mathcal{H}_0 = h(g,r,d).$$

The component $\mathcal{H}_0$ identified in Theorem 19.9 is called the *principal component* of the Hilbert scheme; there may be others as well, of possibly different dimension, and we do not know precisely for which d, g and r these occur. In case $\rho < 0$, the Brill–Noether theorem tells us that there is no component of $\mathcal{H}_{g,r,d}^\circ$ dominating M_g.

Proof. Consider the spaces

$$\mathcal{H}_{g,r,d}^\circ \to \mathcal{P}_{d,g} = \{(C,\mathcal{L}) \mid \mathcal{L} \in W_d^r(C)\} \to M_g.$$

Starting from the right, we compute dimensions:

- M_g is irreducible of dimension $3g - 3$. Let C be a general curve of genus g.
- The Brill–Noether theorem tells us that the variety $W_d^r(C)$ has dimension ρ, and the general point of $W_d^r(C)$ corresponds to a very ample line bundle with exactly $r + 1$ sections. If $\rho > 0$ then $W_d^r(C)$ is irreducible, while if $\rho = 0$ then the monodromy action on $W_d^r(C)$ is transitive.
- Therefore, over a general point of $W_d^r(C)$ the fiber of $\mathcal{H}_{g,r,d}^\circ$ is isomorphic to PGL_{r+1}. Thus there is a unique component of $\mathcal{H}_{g,r,d}^\circ$ dominating $W_d^r(C)$ and therefore a unique component of $\mathcal{H}_{g,r,d}^\circ$ dominating M_g.
- Adding up the dimensions of base and fiber, we see that the dimension of this unique component is $(3g-3)+\rho(g,r,d)+((r+1)^2-1)$, and arithmetic shows that this is $h(g,r,d)$. □

Cheerful Fact 19.10. Although when $\rho(g,r,d) < 0$ there is no component of $\mathcal{H}_{g,r,d}^\circ$ dominating moduli, one could hope that when ρ is not very negative one could understand a component dominating a subset of M_g of codimension $-\rho$, leading to a proof of Conjecture 19.6 in such cases. Indeed, this can be done when $\rho(g,r,d) \geq -2$: In [Eisenbud and Harris 1989], it is shown that if $\Sigma \subset M_g$ is any subvariety of codimension 1, then the curve C corresponding to a general point of Σ has no linear series with Brill–Noether number $\rho < -1$; and [Edidin 1993] proves the analogous (and much harder) result for subvarieties of codimension 2.

19.12. Exercises

Exercise 19.1. Let $C \cong \mathbb{P}^1 \subset \mathbb{P}^3$ be a twisted cubic. Show that the normal bundle $\mathcal{N}_{C/\mathbb{P}^3} \cong \mathcal{O}_{\mathbb{P}^1}(5)^{\oplus 2}$, the direct sum of two line bundles of degree 5. Use this to prove that the restricted Hilbert scheme $\mathcal{H}_{0,3,3}^\circ$ of twisted cubics is everywhere smooth. ◆

Exercise 19.2. Show that the locus Σ of schemes X consisting of a nodal plane cubic curve C with a spatial embedded point of multiplicity 1 at the node is dense in the intersection $\overline{\mathcal{H}_{0,3,3}^\circ} \cap \overline{\mathcal{H}'}$, where $\mathcal{H}'$ is the second component of the Hilbert scheme of twisted cubics described in Section 19.1.

Exercise 19.3. Compute the dimension of the following subsets of Hilb_{3m+1}:

(1) unions of a conic and a line meeting it in 1 point;

(2) the connected union of 3 lines not all contained in the same plane;

(3) nodal plane cubics together with an embedded point at the node that is not contained in the plane of the cubic. ◆

Exercise 19.4. Give an argument for Proposition 19.3 in case $d = 4$ using linkage. ◆

Exercise 19.5. Let $C \cong \mathbb{P}^1 \subset \mathbb{P}^3$ be a smooth rational curve of any degree d.

(1) Show that $h^1(\mathcal{N}_{C/\mathbb{P}^3}) = 0$; that is, the normal bundle of C is nonspecial.

(2) Using this, the Riemann–Roch formula for vector bundles on a curve and Proposition 19.3, show that the Hilbert scheme $\mathcal{H}$ is smooth at the point $[C]$. ◆

Exercise 19.6. Let $C = Q \cap Q' \subset \mathbb{P}^3$ be a smooth curve of degree 4 and genus 1. Identify the normal bundle $\mathcal{N}_{C/\mathbb{P}^3}$ of C, and use this to conclude that $\mathcal{H}^\circ_{1,3,4}$ is itself reduced, and even smooth, and thus isomorphic to an open subset of the Grassmannian $G(2, 10)$. ◆

Exercise 19.7. Let $m \geq n > 0$ be two positive integers. Show that the locus $U_{m,n} \subset \mathcal{H}^\circ$ of curves $C \subset \mathbb{P}^3$ that are smooth complete complete intersection intersections of surfaces of degrees m and n is an open subset of the Hilbert scheme. ◆

Exercise 19.8. Consider the locus $U_{n,n} \subset \mathcal{H}^\circ$ of curves $C \subset \mathbb{P}^3$ that are smooth complete intersections of two surfaces of degrees n. Show that $U_{n,n}$ is isomorphic to an open subset of the Grassmannian $G(2, H^0(\mathcal{O}_{\mathbb{P}^3}(n)))$.

Exercise 19.9. Let $C \subset \mathbb{P}^3$ be a smooth curve of degree 5 and genus 2, and assume that the quadric surface Q containing C is smooth. From the exact sequence

$$0 \to \mathcal{N}_{C/Q} \to \mathcal{N}_{C/\mathbb{P}^3} \to \mathcal{N}_{Q/\mathbb{P}^3}|_C \to 0,$$

calculate $h^0(\mathcal{N}_{C/\mathbb{P}^3})$ and deduce that $\mathcal{H}^\circ_{2,3,5}$ *is smooth at the point* $[C]$. Does this conclusion still hold if Q is singular?

Exercise 19.10. Show that a smooth, irreducible, nondegenerate curve $C \subset \mathbb{P}^3$ of degree 5 and genus 1 is residual to a rational quartic in the complete intersection of two cubics, and use the result of Section 19.3 (genus 0) to deduce that the space of genus 1 quintics is irreducible of dimension 20.

Exercise 19.11. (1) Show that all genera $g \leq 4$ occur among curves of degree 6 in $\mathbb{P}^3$; that is, there exists a smooth irreducible, nondegenerate curve of degree 6 and genus g in $\mathbb{P}^3$ for all $g \leq 4$.

(2) What is the largest possible genus of a smooth irreducible, nondegenerate curve $C \subset \mathbb{P}^3$ of degree $d = 7$? Can you do this with Clifford and Riemann–Roch, or do you need to invoke Castelnuovo?

Exercise 19.12. As before, let $U_{m,n} \subset \mathcal{H}^\circ$ be the locus of curves $C \subset \mathbb{P}^3$ that are smooth complete intersections of surfaces of complete intersection degrees m and n. In case $m > n$, show that $U_{m,n}$ is isomorphic to an open subset of a projective bundle over the projective space $\mathbb{P}(H^0(\mathcal{O}_{\mathbb{P}^3}(n))) \cong \mathbb{P}^{\binom{n+3}{3}-1}$ of surfaces of degree n, with fiber over the point $[S] \in \mathbb{P}(H^0(\mathcal{O}_{\mathbb{P}^3}(n)))$ the projective space $\mathbb{P}(H^0(\mathcal{O}_{\mathbb{P}^3}(m)))/H^0(\mathcal{I}_{S/\mathbb{P}^3}(m)) \cong \mathbb{P}^{\binom{m+3}{3}-\binom{m-n+3}{3}-1}$.

Note that by Exercise 7.5 the scheme $\mathcal{H}^\circ$ is smooth.

Exercise 19.13. Let C be a curve of degree 6 and genus 3, and assume that C does not lie on any quadric surface. Show that C is residual to a twisted cubic in the complete intersection of two cubic surfaces, and use this to deduce that the space of such curves is irreducible of dimension 24.

Exercise 19.14. Now let C again be a curve of degree 6 and genus 3 in $\mathbb{P}^3$, but now assume that C *does* lie on a quadric surface Q. Show that such a curve is a flat limit of curves of the type described in the last exercise, and conclude that $\mathcal{H}^\circ_{3,3,6}$ is irreducible of dimension 24. ◆

Exercise 19.15. By analyzing the geometry of linear series of degrees $2g - 1$ and $2g$ on a curve of genus g, extend Proposition 19.5 to the cases $d = 2g - 1$ and $2g$. What goes wrong if $d \leq 2g - 2$?

Exercise 19.16. Show that each component of the restricted Hilbert schemes of curves of degree 7 in $\mathbb{P}^3$ has expected dimension 28:

(1) $g \geq 6$: By Castelnuovo's theorem, the largest possible genus of a curve of degree 7 in $\mathbb{P}^3$ is 6, and by Theorem 17.20 or a comparison of the Hilbert polynomial with the Hilbert polynomial of $\mathbb{P}^3$, these lie on a quadric.

(2) $g = 5$: Show that there is a unique component, and it dominates the moduli space of curves of genus 5. Show that such a curve is residual to a conic in the complete intersection of two cubics.

(3) $g \leq 4$: Since these curves are nonspecial, there is a unique irreducible component.

Exercise 19.17 (Iarrobino). Show that ideals generated by a independent forms of degree d, together with all the forms of degree $d + 1$ in 3 variables form a family of 0-dimensional schemes of degree $d := \binom{3+d}{3} - a$ in $\mathbb{P}^3$, which can be interpreted as a subvariety of the Hilbert scheme $\mathrm{Hilb}_d(\mathbb{P}^3)$. Show that the dimension of this subvariety is $h := a(\binom{2+d}{2} - a)$. Find values of d and a such that $h > 3d$, and conclude that the Hilbert scheme $\mathrm{Hilb}_d(\mathbb{P}^3)$ has more than one component. ◆

The next three exercises refer to the discussion of the two types of curves of degree 9 and genus 10 in Section 19.10.

Exercise 19.18. In Section 19.10 we identified a locus $\mathcal{H}^\circ_2$ of degree 9, genus 10 curves lying on a quadric. One can show that $\mathcal{H}^\circ_2$ is irreducible by a monodromy argument using Example 11.2, but one can also prove it via a liaison: a curve in $\mathcal{H}^\circ_2$ is residual to a union of three skew lines in the intersection of a quadric and a sextic surface. Use this to establish that $\mathcal{H}^\circ_2$ is irreducible.

Exercise 19.19. In this and the next exercise we refer to the classification of curves of degree 9 and genus 10 in $\mathbb{P}^3$ given in Proposition 19.8. We used a dimension count to conclude that a general curve of type 1 could not be a specialization of a curve of type 2, and vice versa. Prove these assertions directly: specifically, argue that

(1) by upper-semicontinuity of $h^0(\mathcal{I}_{C/\mathbb{P}^3}(2))$, argue that a curve C not lying on a quadric cannot be the specialization of curves C_t lying on quadrics; and

(2) show that for a general curve of type $(3, 6)$ on a quadric, $K_C \not\cong \mathcal{O}_C(2)$, and deduce that a general curve of type 2 is not a specialization of curves of type 1.

Exercise 19.20. Let $\mathcal{H}$ be a component of the restricted Hilbert scheme parametrizing curves of degree d and genus g in $\mathbb{P}^3$ that dominates the moduli space M_g. For $s, t \gg d$, let $\mathcal{K}^\circ$ be the family of smooth curves residual to a curve $C \in \mathcal{H}$ in a complete intersection of surfaces of degrees s and t.

(1) Show that $\mathcal{K}^\circ$ is open and dense in a component of the Hilbert scheme of curves of degree $st - d$ and the appropriate genus.

(2) Calculate the dimension of $\mathcal{K}^\circ$, and in particular show that it is strictly greater than $h(g, r, d)$.

Exercise 19.21. (1) Let $C \subset \mathbb{P}^2$ be a smooth plane curve of degree $d \geq 4$. Show that the g_d^2 cut by lines on C is unique; that is, $W_d^2(C)$ consists of one point. ◆

(2) Using this, find the dimension of the locus of smooth plane curves in M_g.

Exercise 19.22. Consider the locus of curves $C \subset \mathbb{P}^3$ that are complete intersections of a quadric surface and a surface of degree m. Show that these comprise components of the restricted Hilbert scheme, and that their images in moduli have dimension asymptotically approaching g as $m \to \infty$.

Exercise 19.23. Consider the locus $\mathcal{H}^\circ_{ci}$ of smooth, irreducible, nondegenerate curves $C \subset \mathbb{P}^r$ that are complete intersections of $r - 1$ hypersurfaces of degree m, in the restricted Hilbert scheme $\mathcal{H}^\circ$.

(1) Show that $\mathcal{H}^\circ_{ci}$ is open in $\mathcal{H}^\circ$.

(2) Calculate the dimension of $\mathcal{H}^\circ_{ci}$ (and observe that it is irreducible).

(3) Show that the dimension of the image of $\mathcal{H}^\circ_{ci}$ in M_g is asymptotically $2g/r!$ as $m \to \infty$.

Appendix: A historical essay on some topics in algebraic geometry

by **Jeremy Gray**

(Open University, Milton Keynes, United Kingdom)

This essay offers a brief overview of aspects of the long history of plane curves and algebraic geometry.[1] It starts with some historical remarks about conic sections, moves on to look at some aspects of the early study of cubic and quartic curves and the emergence of real and complex projective space, then nods briefly at Bernhard Riemann's ideas about algebraic curves and Luigi Cremona's ideas about plane transformations before concluding with an introduction to the work of Alexander Brill and Max Noether in the late 19th century.

1. Greek mathematicians and conic sections

Quadratic problems were part of high-level teaching in ancient Mesopotamia around 1800 BCE. One such tablet (BM 13901) begins[2]

I have added up the area and the side of my square: 0;45.

The unstated task is to find the side of the square. The number 0;45 is $\frac{3}{4}$ in sexagesimal notation, so if we rewrite this as an equation for s, the unknown

---

[1] I am grateful to Andrea del Centina for letting me read the manuscript of his forthcoming book *The History of Projective Geometry*, which has been very helpful and will surely become the definitive work on the subject.

[2] This is tablet number 13901 in the British Museum collection.

side, we have

$$s^2 + s = \tfrac{3}{4}.$$

The tablet then goes through the steps of an algorithm much as we would for finding the positive root: $s = \tfrac{1}{2}$.

Other problems involve knowledge of the Pythagorean theorem, known also to Chinese and Indian mathematicians. But the first mathematicians to have studied curves more complicated than the straight line and the circle seem to have been those of ancient Greece. Surviving sources suggest that the study of conic sections, literally sections of a cone, may well have arisen with Menaechmus's work on doubling the cube around 350 BCE. This was a long-standing problem that was given a theological spin in the form of the Delian problem, which, according to one story, asked for an altar double the size of a given one, apparently to get rid of a plague. Finding this difficult, workers consulted Plato, who said that the real reason for the task was to reproach the Greeks for their neglect of geometry.[3] It may well also have stood out because Plato in the *Meno* dialogue made such a fuss of doubling the square. Hippocrates of Chios, who flourished around 470 BCE, had already reduced the cube doubling problem to that of finding two mean proportionals between 1 and 2; i.e., quantities x and y such that

$$(*) \qquad\qquad 1 : x = x : y = y : 2.$$

For Greek mathematicians, these quantities would have been line segments of those lengths.

Many solutions to the problem of finding two mean proportionals were proposed. Menaechmus, a younger associate of Plato and a pupil of Eudoxus, may have been the first to connect the problem of doubling the cube to the idea of conic sections, or perhaps we should say quadratic curves, because it has been argued that he considered equation $(*)$ as defining or generating curves that would have been drawn and understood pointwise. At all events, he expressed the solution to the Delian problem in terms of the intersection of a hyperbola and a parabola. [4]

About a century after Menaechmus' time, Diocles, and soon after him Apollonius made significant progress with the conic sections.[5] Diocles identified the curve that would focus the sun's rays to a point as the parabola, and knew precisely how to cut a cone so as to obtain it.[6]

---

[3]Theon of Smyrna, 2nd century CE, in (Thomas 1939, Vol. 1, 257).

[4]See Eutocius *Commentary on Archimedes' Sphere and Cylinder*, in (Cohen and Drabkin 1948, 62–66). Among Eudoxus's many achievements is to have provided the first theory of magnitudes that went beyond the rational numbers; it is now found in Book V of Euclid's *Elements*.

[5]The first places to go for an account of Greek mathematics are the analytical account in (Netz 2022) and the detailed if sometimes dated century-old account (Heath 1921).

[6]Apollonius's four books of his conics survive in Greek and three more in Arabic; the eighth and final volume is lost.

Apollonius produced the first *theory* of conic sections as sections of a cone. The books are very dry. Van der Waerden called him a virtuoso "in dealing with geometric algebra, and also a virtuoso in hiding his original line of thought", while going on to say that "his reasoning was crystal clear and elegant."[7] Apollonius began with the basic definitions of a cone (on a circle) and its three types of section: the hyperbola, parabola, and ellipse; the names are due to him. He showed that all the known sections that had been studied as sections of a cone with vertical angle a right angle could be obtained as sections of a suitable but otherwise arbitrary cone.

It's arguable that a system of coordinates was present in his work, inasmuch as he generally referred everything to a pair of distinguished lines in the plane of section of the cone, but it is buried in the heavy use of proportion theory between different line segments. Apollonius produced a theory of the principal diameters of conics, studied such problems as finding the tangents to a conic from an exterior point, and came close to observing the properties of cross-ratio that modern writers were to detect. He also had a theory about normals to conics from which it's a short step to obtaining their evolutes. The biggest weakness in his theory was that very often he had to treat the three kinds of conic section separately, although you can see him trying for greater generality, as, for example, when he introduced the concept of conjugate hyperbolas in Book I to make the proofs of theorems about conjugate diameters run equally elegantly and simply for hyperbolas as ellipses.

Mathematicians of the Islamic era (active from the 9th to the 15th century CE) also studied conic sections and the various problems they inherited from the Greeks. One of the most outstanding is 'Umar ibn Ibrāhīm al-Khāyyamī (usually known in the West as Omar Khayyām), who was born in Nishapur in northern Iran in the 11th century CE and spent much of his life in Isfahan, where he led a group of Islamic astronomers at the observatory there. In his book the *Risāla* he gave a complete account of how to solve cubic equations using pairs of conic sections where necessary.[8] Other Islamic mathematicians advanced the study of trigonometry, investigated the foundations of geometry (Euclid's parallel postulate (which al-Khāyyamī also investigated), invented methods for the numerical solution of polynomial equations, and studied various properties of conic sections, but there was little study of other kinds of curves.

2. The first appearance of complex numbers

In the 16th century, various Italian mathematicians, notably Gerolamo Cardano, Ludovico Ferrari, and Rafael Bombelli, succeeded in finding algebraic

[7]See (Van der Waerden 1961, 248).
[8]See (al-Khāyyamī 1950).

methods for solving cubic and quartic equations. This inevitably led them to have views about complex numbers. Cardano recognised in his (1545, Chapter 37) that they will appear in the formulae for the solution of these equations, even though he said "So progresses arithmetic subtlety, the end of which, as is said, is as refined as it is useless." Bombelli was more hospitable to the new numbers in his (1572), and gave rules for manipulating them, but they remained mysterious and suspect for many years.

At this point, the history of algebraic geometry broadly divides into two. One part concerns the theory of conic sections — algebraically, curves of degree two — and the other the theory of curves of higher degree. The first can regard Girard Desargues as a major innovator, the second his contemporary, René Descartes. Together, they lead to the concept of a projective space. Let us take the study of conic sections first.

3. Conic sections from the 17th to the 19th centuries

In 1639 Desargues published fifty copies of a short pamphlet known as the *Brouillon Project* (1639), on what we could call the projective theory of conics. He was perhaps hoping for useful critical feedback, but he never returned to rewrite his account. In the essay, all non-degenerate conics are treated on a par, and treated as perspective images of the circle. The idea that all (non-degenerate) conics are the projective image of a circle is not original with Desargues. It was stated by Francesco Maurolico, for example, in his book (1611), which it is likely Desargues knew, and in any case it is visually apparent to anyone who thinks of the conics as sections of a cone on a circular base. The trick is to make this insight work.

Desargues's key idea is that of properties of figures invariant under projection, such as the cross-ratio of four points on a line: if A, B, C, D are four points on a line, their cross-ratio can be defined to be $(AC.DB)/(AD.CB)$. The special case, when the cross-ratio is -1 and $AC/CB = -AD/DB$ and people said B and D separate A and C in the same ratio, arises frequently in the theory of the poles and polars of conic section.[9] Desargues also had a more complicated property of six points, which, like the four-point property, was invariant under a projection. He connected all this to the study of the "complete quadrilateral" (see Figure A.1) — take four points, no three collinear, and join them in all possible ways — and thence to the study of conics through those four points.

The *Brouillon Project* is fiercely unreadable.[10] It was also lost for a long time and known only through commentaries by later authors until in 1845 Michel

---

[9]Desargues did not use the concepts of pole and polar but of four or six points in involution.

[10]Happily, it has been the subject of sequence of detailed analyses in recent papers by Marie Anglade and Jean-Yves Briend; see (Anglade and Briend 2022) and the references to their earlier papers cited there.

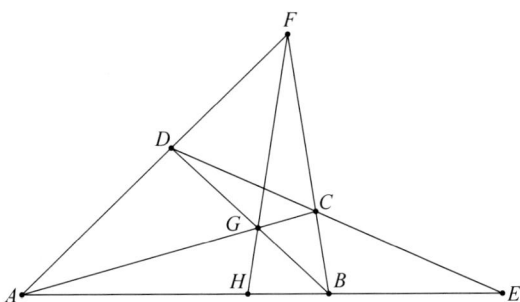

Figure A.1. A complete quadrilateral, consisting of the four points A, B, C, D and the three lines joining them in pairs.

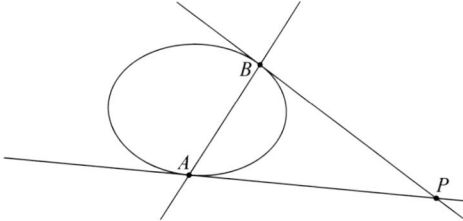

Figure A.2. The tangents from a point P outside the conic touch it at A and B; the line AB is the *polar* of P and P is the *pole* of the line AB. The construction can be extended to points inside the conic.

Chasles found a handwritten copy made by Philippe de la Hire.[11] In particular, de la Hire, a generation after Desargues, wrote some much more readable, and longer, works in Desargues's spirit, illuminating the role of cross-ratio in the theory of tangents that came close to a theory of duality. Desargues's famous theorem on two triangles in perspective was published separately in (Bosse 1648).

The best attention Desargues's little book got was from his younger contemporary Blaise Pascal, who evidently produced a virtually complete theory of conics around what he called the "mystical hexagram". Unhappily, much of it is lost, and known to us only from some notes on it made by Leibniz, but the idea is that while there is always a conic through five points something happens if you want a conic through six points: we call it Pascal's theorem. Then, if you let the sixth point collapse onto one of the other five you get a tangent to the conic through those five points. One way or another all the key properties of conics are wrapped up in this idea, or so Pascal seems to have shown, and much of the early 19th century work in France can be seen as attempts to recover such a theory, which would include such topics as duality, in the form of the pole and polar relationship with respect to a conic (see Figure A.2).[12]

---

[11] The only known copy of the original *Brouillon project* turned up in 1950.

[12] For a thorough analysis of Pascal's work and its context, see (Del Centina 2020).

Thereafter, truly projective geometry languished for much of the 18th century, despite some insightful contributions by Isaac Newton that we shall mention below (see p. 361) and a few others. Its revival is conventionally dated from the early work of Gaspard Monge, who as a young man seeking a job in the French army was interested in how to depict three dimensions on two. He devised a method of plan and elevation (projections onto a horizontal and a vertical plane) that he could couple to some simple algebra in a way that was easy to use; as a result he was offered a job at the military Academy in Mézières and his discovery made a military secret. During the French Revolution, Monge was influential in setting up the École Polytechnique, where he was an inspiring teacher of geometry, and this did much to revive the subject.

Among those so inspired there was Jean Victor Poncelet, who promoted a much more general theory of transformations with a view to unifying the theory of conics that he began to develop while a prisoner-of-war in Saratov during Napoleon's disastrous invasion of Russia. His *Traité des Propriétés Projectives des Figures* (1822) is a visionary book that relies on some rather mysterious arguments about ideal points of intersection (Augustin-Louis Cauchy, who reviewed the book, urged that they be regarded as points with complex coordinates; Poncelet never agreed, and reproduced the review between the Preface and the Introduction to his *Traité* presumably to show his disdain). As a result, some of the transformations it invokes necessarily require complex coordinates. The most famous result in the book is Poncelet's closure theorem, which has continued to attract attention to this day, but it is also notable for many other theorems involving pairs of conics.

In the 1820s, Poncelet had a dispute with Joseph Diaz Gergonne, the editor of the only journal at the time entirely devoted to mathematics, about what duality in the plane actually is.[13] Poncelet always saw it as pole and polar with respect to a conic, Gergonne saw it as a new and fundamental feature of projective geometry.[14] Gergonne's view led to confusion when it was applied to curves of degree three or more, a matter that began to be sorted out only with Julius Plücker's work, as we shall see below.

Another mathematician who was inspired by Monge was Michel Chasles, who used the projective invariance of the cross-ratio of four points to eliminate much of the weirdness of Poncelet's ideas. Independently, Jakob Steiner did very similar work in Germany. He can be credited with the first truly projective definition of a conic section. Even so, there remained an irritating feature of cross-ratio: it was given as a function of four lengths, but length is a property of Euclidean geometry, not projective geometry. If projective geometry is to

[13]Gergonne's journal was called the *Annales de Mathématiques Pures et Appliquées*. It ran from 1810 to 1832, and it was succeeded in by Liouville's *Journal de Mathématiques Pures at Appliquées* in 1836. Crelle's *Journal für die reine und angewandte Mathematik* was founded in 1826.

[14]See (Gergonne 1826).

be regarded as more fundamental than Euclidean geometry, because it rests on fewer assumptions or axioms, then the intrusion of Euclidean length in the definition of cross-ratio is at the very least unfortunate. Nor can one easily speak of there being a concept of projective space in the 1820s; rather, much of geometry at this time was about figures in the plane subject to a variety of projective transformations. The first truly foundational work on real and complex projective geometry that avoided deriving it from Euclidean geometry was the achievement of Karl von Staudt in his *Geometrie der Lage* (1847), a long and difficult work that influenced Felix Klein when he succeeded von Staudt as a Professor at Erlangen twenty-five years later.[15]

All in all, a surprising amount of work was done on the theory of conic sections at the start of the 19th century, and before it is dismissed as arcane it should be stressed that the subject was a proving ground for the development of projective geometry. Two approaches stand out: the search for entirely general methods that would treat all non-degenerate conics on a par; and the emergence of the property of duality. Matters are complicated by the existence of two separate traditions, usually called the synthetic and the analytic, supposedly divided into classical geometric methods and more algebraic ones that introduce coordinates. Recent historical work suggests that it was all a bit murky, and algebraic methods were also often used in synthetic geometry.[16] Several things promoted the use of synthetic methods. They can be elegant when algebraic methods are blunt; they correspond to the visual form of the conics; they provide a language for describing what is apparent or to be found in a problem. Against them is the obstinate fact that algebra is more general: it does not care if some quantities become negative, but what is a negative length? Once ways round this obstacle were found (by Poncelet and then Chasles) the way was open to a truly systematic synthetic theory of conics.

How then did this change, and purely synthetic projective geometry begin to wane? Very few of the original protagonists disdained algebra outright, and as the (projective) theory of conics reached completion in the 1820s and established its fundamental character, being more general than metrical Euclidean geometry, it also had its baroque aspects. But worse, it did not generalise at all readily to the study of curves of higher degree. For that, as even Newton had recognised, a hefty dose of algebra was required.

4. Curves of higher degree from the 17th to the early 19th century

A common way to think of curves in antiquity was pointwise: some length depends in a given way on some other length. Accordingly, at least in principle,

---

[15]See, for example, (Gray 2015).
[16]See (Lorenat 2016).

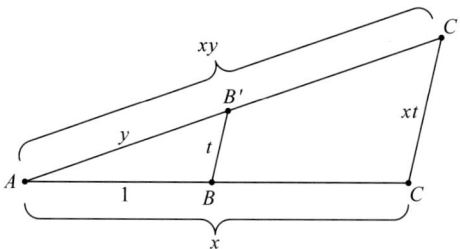

Figure A.3. If AB is of length 1, AC of length x, AB' of length y, and BB' and CC' are parallel then AC' is of length xy.

if you know the independent length (the ordinate, or x coordinate) you know the dependent length (the abscissa, or y coordinate).

By Descartes' time there was already some sophisticated algebra expressed in a formalism that hadn't quite shaken off the Greek insistence on seeing everything as geometrical magnitudes: lengths, areas, volumes, and, well, what exactly? The French mathematician François Viète in his *Isagoge* boldly spoke of magnitudes having the dimensions of side, square, cube, square-square, square-cube and so on.[17] It was possible to write polynomial equations in this language, as Viète did in the early 1600s. First Fermat in 1636, and then much more boldly Descartes in 1637, realised that you could extend the language to two variables and so describe curves in the plane.

Descartes's first achievement in his *La Géométrie* (1637) was to eliminate the dimensional aspect. A simple use of similar triangles allowed him to show that the product of two lengths could be seen as another length (not an area) so all geometrical quantities could be regarded as one-dimensional and the idea of dimension quietly dropped (see Figure A.3). He also replaced Viète's cumbersome algebra, which was written in capital letters with verbal abbreviations for the algebraic operations, with something much more like what we use today. He wrote x and y for the key variables (not A and E in the manner of Viète). Descartes was clear that he was using coordinates, although his x and y coordinates could have oblique axes.

Then came the real work. Almost all mathematical problems in his day were expressed in the language of geometry, except for some that we would call diophantine and were implicitly about integers and rational numbers. Accordingly, the answer had to be expressed geometrically. Descartes's idea was to give letters to all the lengths involved in a problem, use the statement of the problem to express relationships between the letters, and reduce the equations to a single equation. Then solve the equation and express the answer again in geometrical terms.

---

[17]See his collected works, (Viète 1646).

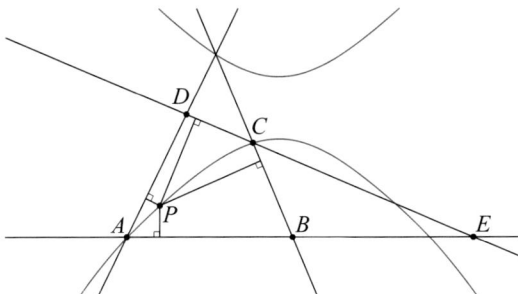

Figure A.4. The Pappus problem: the product of the distances of *P* from the lines *AB* and *BC* is proportional to the product of the distances of *P* from the lines *CD* and *DA*.

In this way he solved the famous Pappus problem (see Figure A.4): given four lines and four angles (nothing is lost if we take these angles to be right angles), find the locus of points *P* such that the product of the distances of *P* from the first two lines is proportional to the product of the distances of *P* from the last two lines. As he showed, and Pappus had known, the answer is a conic section, but Descartes went further and claimed that in this way he could solve the Pappus problem for any number of lines.

This was to infuriate Newton, who showed that Greek methods were indeed adequate to the original Pappus problem. The background here is that Newton was also engaged in demolishing Descartes's theory of planetary motion in favour of his own, which led him into the theory of conic sections and to pose and solve the problem of finding the conic (or conics) through *n* points and tangent to $5 - n$ lines, which he did in Book I, Section 5 of his *Principia Mathematica*.

It can be argued that the theory of specifically algebraic curves has its origin in Descartes's method for finding normals to a curve at a point, which involved considering all the circles through the given point and imposing the condition that the equation for the circle be such that it passes twice through the given point; this circle has its centre on the normal to the curve. On the basis of three examples, he claimed that the method always worked if the curve had an algebraic equation, and he described a system of linkages (sliding rulers) that he said could be adapted to draw any such curve. Curves like the cycloid that were patently not algebraic he sought to exclude from geometry — this exclusion was something else that annoyed Newton.

In the 1660s and 1670s, but only published the 1690s and again in 1704 as an Appendix to his *Opticks*, Newton worked on a classification of cubic curves. He claimed that there were 72 distinct types which could be derived by simplifying the general equation using what we would call affine or linear transformations, although at one point he used a simple birational transformation. For some reason, in writing up his work for publication he omitted four of the possible cubics, and they were speedily found by various other mathematicians. Newton also made the striking remark that this collapses to five types if projective transformations are allowed.[18]

A plane cubic curve is defined by an equation involving ten coefficients, or more precisely by the nine ratios between the coefficients, which suggests that it should be determined by nine points in the plane because their coordinates provide nine equations that should determine these nine ratios. Moreover, it seemed to mathematicians in the early 18th century that any set of nine points and therefore any set of nine equations in the nine unknowns should have a unique solution — there was no theory of linear equations at the time — and as a result even the best mathematicians got into difficulties. For example, the Scottish mathematician Colin MacLaurin knew in 1720 that there were problems with the idea that nine points in the plane should determine a cubic: any two cubics will meet in nine points, and so these nine points do not determine a unique cubic, and indeed there will be infinitely many cubics through these nine points.[19]   Maclaurin did not, however, know how to solve this puzzle. This apparent contradiction with the claim that nine points in the plane always determine a unique cubic became known as Cramer's paradox when the Swiss mathematician Gabriel Cramer addressed the problem in his book (1750, Chapter 3). Part of the confusion at the time was a lack of insight into systems of linear equations, and part was the lack of understanding about what are the implications of configurations that do not determine a unique curve.

More interestingly, Cramer also suggested that if the $\frac{1}{2}n(n+3)$ points needed to determine a curve of degree n contain tn points common to a curve of degree $t < n$ then the curve through the $\frac{1}{2}(n + 1)(n + 2)$ points breaks up into two or more curves, one of which passes through the tn points. We note that Cramer considered only irreducible curves: a circle and a line, for example, would be considered as two curves, not one.[20]

Cramer conveyed the (incorrect) claim about cubic curves, and more generally the analogous claim for curves of degree n — that they are determined by $\frac{1}{2}n(n + 3)$ general points — to his friend Leonhard Euler, who repeated them in the second volume of his *Introductio* (1748, § 81). Euler then wrote a paper (Euler 1750) in which he first spelled out the problem. It is a general proposition, he said, that k linear equations in k unknowns have a unique solution. Accordingly, 9 points in the plane will determine a unique cubic curve. But nine points common to two cubics plainly do not determine a unique cubic

---

[18]"The five divergent parabolas, by their shadows, generate all other curves of the second genus [i.e., cubic curves]." See (Newton 1704) and Talbot (1860, p. 25).

[19]MacLaurin also knew that a general cubic curve has nine flexes, and the line joining any two passes through a third.

[20]Curves were taken to be irreducible even in Salmon's book *Higher Plane Curves* (1852).

curve. To resolve this apparent contradiction, as he called it, he began with three equations in three unknowns and showed by example that there will not be a unique solution if one of these equations is contained in (that is, is a consequence of) the others. He drew the same conclusion about four linear equations in four unknowns, and stated more generally that there will not be a unique solution to a system of n equations in n unknowns if one or more are contained in all the others.

He then turned to the geometrical implications. In the case of conics, he showed that the containment condition corresponds to the case when four, or all five, of five given points lie on a line. For cubics, he remarked that one easily understands that not just one, but two or more of the nine equations specified by nine points may be contained in the others, and to obtain a unique cubic it will be necessary to specify that the curve passes through one or more additional points. However, he said, it was very difficult to see what the consequences were when this is the case, because there were so many points and coefficients that things become too complicated, although it was possible to draw conclusions in simple cases. For example, if the nine equations arise in part from four points lying on a line then the remaining five equations must determine a conic, which might itself consist of two lines. Euler concluded his paper with a few remarks about curves of higher degree.

The idea that algebraic curves of degrees k and m in the plane should meet in km points was something of a folklore result in the early 18th century, but it travelled without a proof for many years. Euler discussed it in the second volume of his *Introductio* (1748, Chapter 19) and noted that even in simple cases for the counting to work one would have to take care of multiple points (such as tangents), points 'at infinity' (consider a parabola and a line parallel to its axis), and allow the coordinates of intersection points to be complex (consider the intersection of two circles). The first person to find a persuasive way of tackling the problem was Étienne Bézout, who lived from 1739 to 1783, and made his living teaching mathematics at the French military and naval academies. He published the theorem that bears his name in a book of 1779; it is based on his theory of the resultant of two polynomial equations that he developed in a paper of 1764. His proof of the theorem is gappy and intuitive by any standards, but so much better than what had been done before that his immediate successors were willing to give him real credit for doing as much as he did. His results inspired later work by Cauchy and James Joseph Sylvester.

To state Bézout's theorem in general requires a notion of the multiplicity of an intersection of two plane curves without common components. Such a notion was developed by Max Noether, and refined by Francis Sowerby Macaulay. A modern proof would invoke Noether's Fundamental Theorem — see p. 372 —

and a computation of the Hilbert polynomial. This is a story that has been pursued, with successive new generalisations, right up to the present day.

As this activity suggests, throughout the 18th century mathematicians had become more and more comfortable with the idea of complex numbers in algebra, and with the idea of proving that a polynomial of degree n has n roots, possibly with repetitions (the so-called fundamental theorem of algebra).[21] Euler's attitude to complex numbers was that there was nothing to explain. Although they cannot be ordered, expressions of the form $a + bi$ behave arithmetically like numbers, so they can reasonably be considered numbers and "are usually called *imaginary quantities*, because they exist merely in the imagination."[22] Cauchy's view later was more explicit and very close to regarding the field of complex numbers as $\mathbb{R}[x]/(x^2 + 1)$. This works for finding a proof of the fundamental theorem of algebra, which he gave a proof of in his (1817a, b), but was not productive in contexts involving contour integration.

Credit for the first rigorous proof of the fundamental theorem is often given to Carl Friedrich Gauss for his paper (1799), although he did base his argument on the claim that an algebraic curve that enters a bounded region of the plane also leaves it, which is surely no easier to prove. Gauss went on to give three more proofs of the theorem, and soon any doubts about the algebraic nature of complex numbers were resolved.[23] Niels Abel and Carl Gustav Jacob Jacobi, in their work on elliptic functions, took a formal, non-geometrical attitude to complex numbers. Both men were algebraists at heart — formidable algebraists — and the exact nature the plane of complex numbers did not really interest them.

All things considered, however, progress in the study of cubic, quartic, and higher degree curves in the 18th century was piecemeal and slight, and the sheer enormity of the equations involved seems to have baffled even Euler. New methods for handling such curves would have to be found, such as came in with Plücker in the 1820s. His key idea was to study families of curves, using the symbolic notation devised by Gergonne.[24] If S_1 and S_2 stand for the equations of two curves of the same degree, then $S_1 + \lambda S_2$ is generally the equation of another curve of that degree. This may be the first occurrence of the idea of a linear series, with which this book is much concerned. Using this idea allowed Plücker to pull out geometrical properties of cubic and quartic plane curves while avoiding the algebraic complexities that had defeated Euler.

Plücker also resolved the paradox about dual curves in the theory of plane curves in his (1834) that had stumped Gergonne and Poncelet. The paradox arises because the dual of a smooth curve of degree d has degree $d(d-1)$ so the

---

[21] For a history of attempts on this theorem, see (Gilain 1991).

[22] See Euler, *Algebra* § 143, p. 43. Euler here rejected the idea that ultimately mathematical quantities must be exhibited in nature: three sheep, a length of $\sqrt{2}$, and so on.

[23] William Rowan Hamilton published his rigorous theory of ordered pairs of real numbers in his (1837).

[24] See (Gergonne 1826–1827), (Plücker 1835) and (Plücker 1839).

dual of the dual "ought" to have degree $d(d-1)(d(d-1)-1)$. However, the dual of the dual is the original curve. Plücker's response, which would strike us today as at best heuristic, was that any line through a double point, say, is counted in this way as a tangent, and this throws the count of genuine tangents off. More precisely, he showed that each node on a curve reduces the degree of the dual curve by 2, and each cusp reduces it by 3. Consequently, he claimed that a curve of degree d with δ double points and κ cusps has a dual curve of degree $d(d-1) - 2\delta - 3\kappa$.

Plücker also showed in his (1839, 207–227) that the nodes of C^*, the dual curve of C, corresponded to the bitangents of C, while the cusps of C^* corresponded to the flexes of C, and vice versa. To see how he resolved the duality paradox, consider the case of a non-singular cubic curve. It has no nodes, cusps, or bitangents, and an as yet indeterminate number, j of flexes. Its dual has degree 6 and j cusps. The dual of the dual curve therefore has degree $6 \cdot 5 - 3j$, which will equal 3 if $j = 9$. As we saw, the result that a non-degenerate cubic curve has nine flexes had been known since MacLaurin. For curves of higher degree, it is necessary to consider a curve of degree n that has δ double points, κ cusps, τ bitangents, and ι inflections. Suppose that its dual has degree n'. Plücker argued that

$$n' = n(n-1) - 2\delta - 3\kappa \quad \text{and} \quad \iota = 3n(n-2) - 6\delta - 8\kappa,$$

with another formula for τ, along with the corresponding formula for the dual curve and its inflection points and bitangents.

The number ι is most easily found using an argument due to Ludwig Otto Hesse, who showed in his (1844) that the set of flexes $\Gamma \subset G$ on a curve G of degree d with equation $F(x, y, z) = 0$ could be characterized as the intersection of G with the curve D defined by the vanishing of the *Hessian determinant*:

$$H = \det \begin{pmatrix} \partial^2 F/\partial x^2 & \partial^2 F/\partial x \partial y & \partial^2 F/\partial x \partial z \\ \partial^2 F/\partial x \partial y & \partial^2 F/\partial y^2 & \partial^2 F/\partial y \partial z \\ \partial^2 F/\partial x \partial z & \partial^2 F/\partial y \partial z & \partial^2 F/\partial z^2 \end{pmatrix}.$$

Since the entries of this matrix have degree $n - 2$, the degree of H is $3(n-2)$. For general F the curves G and D meet transversely, so Bézout's Theorem shows that a general curve of degree m has exactly $\iota = 3n(n-2)$ flexes.[25]

The most famous case Plücker established in his book (1839, 247), concerns a nonsingular plane curve C of degree $n = 4$ (no nodes, cusps, or other singular points) having τ bitangents and ι flexes. Its dual curve has degree 12 and will have τ double points and ι cusps, and the dual of the dual will have degree $12 \cdot 11 - 2\tau - 3h = 4$. Hesse's theorem shows that $h = 24$ (Plücker had an ad hoc argument to the same effect) and so $\tau = 28$, and we see that C has exactly 28 bitangents. Plücker showed that they can all be real (see Figure A.5).

---

[25]This result was first proved in a different way in (Plücker 1835, 264), as Hesse acknowledged.

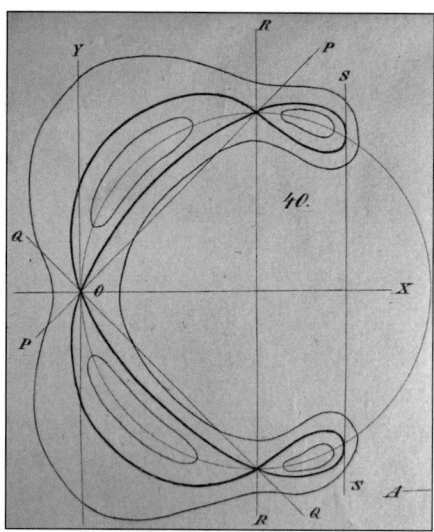

Figure A.5. Plücker's quartic curve has 28 bitangents. From (Plücker 1839).

In 1849, Plücker gave up mathematics for physics, particularly the study of cathode rays; he was awarded the Royal Society of London's Copley medal (its highest honour) for this work in 1866. It is sometimes said that he gave up geometry because he tired of the criticisms of Steiner, who had a secure position in Berlin and influenced the decisions of the *Journal für Mathematik*, and only returned to geometry after Steiner died in 1863. Now he took up the field of line geometry, which he transformed into a new branch of the subject.

Plücker's formulae work well for curves of degrees 3 and 4, but not so well thereafter because other types of singularity can appear. He listed all the solutions he could find to the equations that bear his name up to curves of degree 10, but he did not discuss the kinds of singularities a curve may have that are not of the type he had considered.[26] Progress was only made with a paper by Arthur Cayley (1866), who drew on Victor Puiseux's paper (1850) that analysed how curves are ramified at a singular point, showed how the branches are permuted in cycles, and how this is captured by the local power series expansions, which begin with fractional indices.[27]

Puiseux was one of a number of mathematicians in the circle around Cauchy. Cauchy had spent the 1830s and early 1840s following the Bourbon Court around Europe from a strange belief that the oath of allegiance he had sworn to the crown on becoming a professor compelled him to do so. As a result, by 1850 few people knew the work he had done in those years, and even he seems

---

[26]See (Plücker 1839, 214).

[27]Cayley's paper was corrected by Otto Stolz in his (1875), who found that the conclusions were correct but the proof "as he [Cayley] himself remarks", was not quite complete.

to have forgotten what he had done in complex variable theory back in the 1820s, which included a version of what we call the Cauchy integral theorem (Cauchy 1825) restricted to rectangular contours. Only on his return to Paris did he begin to think much more geometrically; previously his attitude to a many-valued 'function' was to cut the plane and study just a branch of it on what remains. His understanding of branch points was quite limited, and this left space for Puiseux to study integrals on arbitrary contours and what happens to integrals taken around branch points.

5. The birth of projective space

When did projective space come to be regarded as something more rigorous than Euclidean space with a "line at infinity"?[28] This might simply be a suitable set of coordinates. Plücker used coordinates for the plane with a line at infinity in his (1830) with a view to enabling homogeneous equations to correspond to curves; in this system the coordinates (p, q, r) denote the signed distances of a point from the three sides of a triangle of reference. However, in his study of curves (1835) he would first discuss them in the plane, and then as they went off to infinity; he didn't say that the line at infinity could be mapped by a projective transformation into the finite part of the plane. The way forward was indicated by August Möbius, who introduced barycentric coordinates in his (1827). Pick three points forming a triangle, say ABC, and attach weights, positive, zero, or negative (not all zero) to these points. The barycentre or centre of gravity of these three weighted points is a point P which can be said to have those three weights (or, better, their ratios) as its barycentric coordinates. If you put the points A, B, C at, say, $(0, 0), (1, 0), (0, 1)$ you get an easy way to relate points in what could be called the Cartesian and barycentric coordinate planes. The big plus is that the line at infinity, which is invisible in Cartesian coordinates, is a perfectly sensible line in barycentric coordinates. In this system, a line has an equation of the form $ax + by + cz = 0$, and by treating a, b, c as the coordinates of the line Möbius obtained a simple theory of conics and duality in the plane. (He also showed that there are dualities in P^3 that are not pole-polar dualities.)

If you drop the talk about weights, and keep the idea that barycentric coordinates are best thought of as ratios of three numbers, you have projective coordinates; this was one of the contributions of Hesse in the 1840s.[29] Mathematicians were still reluctant, however, to decide if this space was $\mathbb{P}^2_{\mathbb{R}}$ or, less likely, $\mathbb{P}^2_{\mathbb{C}}$.

---

[28] For an interesting set of essays analysing what projective space might be and how it came about, including an extensive analysis by J.-P. Friedelmeyer of the work of Poncelet, see (Biosemat-Martagon 2010), and the discussion in del Centina's *The History of Projective Geometry* (to appear).

[29] See, for example, (Hesse 1844).

Mathematicians found it hard to accept complex coordinates, even though Plücker used the term 'imaginary' (as in 'imaginary' points, 'imaginary' straight lines, 'imaginary' tangents) over two hundred times in his (1828–1831).[30] He employed the term confidently in his (1839) when, for example, he discussed how many of the 28 bitangents to a quartic curve can be real. In fact, the whole question of a complex space was obscure for a long time. As late as 1878 Cayley could write about complex curves as sets of points in $\mathbb{C} \times \mathbb{C}$ and remark "I was under the impression that the theory was a known one; but I have not found it set out anywhere in detail."[31] For quite some time attention was fixed on real curves in the real plane, which could conveniently sprout points with complex coordinates when they intersected other curves. This unstable situation could not last, but how it was to be swept away was not clear to mathematicians initially.

6. Riemann's theory of algebraic curves and its reception

It was, however, entirely clear to Riemann. The crucial issue in defining complex-valued functions of a complex variable, where a complex variable can be taken as an expression of the form $x + iy$, is defining what it is for such a function to be something more particular than a mapping from $\mathbb{R}^2$ to $\mathbb{R}^2$, and this comes down to defining what it is for such a function to be differentiable as a function of a complex variable. As is well-known, Riemann solved this problem in the opening pages of his inaugural dissertation (1851) by identifying — much more clearly than Cauchy — the role of what we call the Cauchy–Riemann equations. He proceeded to give a thoroughly geometric theory of complex functions, which he extended in his great paper on Abelian functions (1857). There he developed a strikingly topological account, which classified orientable, boundaryless surfaces by the number, $2p+1$, of cuts needs to disconnect them. He called this number the order of connectivity of the surface, when $p = 0$ he called the surface simply connected. The theory is too complicated to be described fully here, but briefly, he showed that the dimension of the space of "everywhere finite" differential forms on such a surface is p, and the integral of each such differential gives rise to a many-valued function, which can be expressed in the form $f(z, w) = 0$, where z and w are complex variables. He also deduced the Riemann inequality in this form (1857, § 5): the number of arbitrary constants in a function w that has m first-order poles on a surface of order of connectivity $2p + 1$ is $m - p + 1$ when $m \geq p + 1$. He gave a more detailed account that involves special cases that were interpreted by his student Roch to give us the Riemann–Roch theorem.

---

[30] Del Centina, *The History of Projective Geometry*, forthcoming.
[31] See (Cayley 1878, 32).

Riemann's health started to decline in the early 1860s, and he died in Italy in July 1866. By then, Rudolf Clebsch had decided to develop Riemann's ideas and to try to persuade people around him to take up the cause. In his (1864) he applied the theory of Abelian functions to the study of plane algebraic curves. Here he used coordinates that belonged either to $\mathbb{C}^2$ or $\mathbb{P}^2_{\mathbb{C}}$ and passed easily between them as required, so it would seem that he and the people he influenced had become comfortable with complex projective space in all but name.

In his papers (1865a, 43) and (1865b, 98) Clebsch sorted curves into different genera according to the value of the number p associated to them, but he did not speak of the genus of a curve.[32] If we allow ourselves to do so, we may say that Clebsch defined the genus of a plane algebraic curve with δ double points as[33]

$$p = \tfrac{1}{2}(n-1)(n-2) - \delta.$$

In his (1865b) and his book with Paul Gordan (1866) he extended this to encompass curves with κ cusps, so $p = \frac{1}{2}(n-1)(n-2) - \delta - \kappa$. This tells us incidentally that no singular points more complicated than those considered by Plücker in the late 1830s had been looked at. Clebsch's formula relied on being able to count the number of constants in an integral of an everywhere finite differential correctly, a result that requires the completeness of the adjoint series. Clebsch and Gordan offered in their book (1866, Chapter 3) a proof that the genus of a plane algebraic curve is invariant under a birational transformation, but it was valid only for the case of a curve with simple cusps and double points.

Clebsch became Riemann's successor in Göttingen in 1868, but he died of diphtheria in 1872 at the age of 39. His plans for the study of algebraic geometry now devolved upon Alexander Brill and Max Noether, with whose theory, modernised and made rigorous, this book is in part concerned.[34]

Alexander Wilhelm Brill was born in Darmstadt in 1842.[35] His uncle was the mathematician Christian Wiener, an expert in descriptive geometry. He entered the University of Giessen intending to study architecture but his mathematical ability brought him to the attention of Clebsch, who was then at Giessen and who encouraged him to go to Berlin, where Ernst Kummer, Leopold Kronecker, and Karl Weierstrass taught. This broadened Brill's horizons considerably, but he returned to Giessen in 1867 to take his Habilitation under Clebsch.

In Giessen, Brill met Max Noether. Noether had been born in Mannheim in 1844, but polio at the age of 14 left him paralysed in one leg and delayed his education. He had to be privately schooled, which gave him broad literary

---

[32]See (Lê 2020), who suggests that Felix Klein was the first to do so.

[33]See (Clebsch 1864, 192).

[34]A modern account of much of the material described above can be found in (Brieskorn and Knörrer 1986). (Coolidge 1940) remains a historically informative if not entirely rigorous account of many of these developments, more additional mathematical details are in (Coolidge 1931).

[35]See (Severi 1922), and Brill's obituary (Finsterwalder 1936).

and cultural interests. At university he initially intended to study astronomy, but then he switched to mathematics, first at Heidelberg and then at Giessen and Göttingen in 1868 and 1869. He habilitated in Heidelberg in 1870 with a thesis on surfaces possessing a family of rational curves. In 1875 he became a Professor at Erlangen, where he remained until his death in 1921.[36]

Noether worked not only on plane algebraic curves, including the analysis of their singular points, but the geometry of algebraic surfaces, algebraic curves in space, and Cremona transformations of the plane; van der Waerden justly remarked that "Algebraic geometry was created by Max Noether."[37] He drew largely on the work of Plücker on algebraic curves and their singular points, and was less interested in the computational side of the theory that Clebsch had emphasised. As a result, his obituarists — Guido Castelnuovo, Federigo Enriques, and Francesco Severi — noted that Noether's work was inclined to be qualitatively valuable even though he may not have paid sufficient attention to "contingent aspects of the formulae". However, they added that:[38]

> geometric intuition, which was always his guide, saved him from error, and the work, even if it was not perfect, and perhaps for that reason, was richly suggestive.

Brill chafed under the direction of Richard Baltzer, who was Clebsch's successor at Giessen, and moved to the new Polytechnic in Darmstadt in 1869, which was academically a step downwards. There, he did his work on the Cayley–Brill correspondence principle, and in 1875 he accepted Klein's offer of becoming a Professor at the Polytechnic in Munich, with responsibility for reforming high school education. Brill found Klein's inventiveness hard to match, but he enjoyed getting to know some of the other mathematicians there, including Philipp Seidel and the young Alfred Pringsheim. Among his pupils were Walther Dyck, Isaak Bacharach, and Ferdinand Lindemann, who went on to produce the two-volume geometry textbook *Vorlesungen über Geometrie* (known as Clebsch–Lindemann) before achieving fame as the mathematician who proved that π is transcendental.

7. First ideas about the resolution of singular points

In their work on algebraic curves, Brill and Noether found Riemann's use of the Dirichlet principle at the heart of his theory of complex analytic functions to be unsound, so they studied curves firmly in the complex plane $\mathbb{C}^2$ or the complex projective plane $\mathbb{P}_{\mathbb{C}}^2$. They were concerned to treat such curves in full generality, and this led them to confront the issue of arbitrary singularities of plane curves. Their preferred method was to argue that any curve with whatever singularities

---

[36] See his obituary (Brill 1923).

[37] See (Waerden 1971, 171) who immediately commented that the logical foundations are shaky.

[38] See (Castelnuovo, Enriques, and Severi 1925, 162)

could be reduced to one of the same genus but having only double points by a sequence of Cremona transformations, and then to deal in detail with these simpler curves. It was clear from the formula for the genus of a plane curve that if the genus of the curve is not a triangular number then the curve will always have singular points, so the process of desingularising cannot always lead to a nonsingular curve.

Luigi Cremona had outlined a general theory of transformations of the plane in his papers (1863) and (1865). He looked for geometrical transformations of the plane mapping a given figure one-to-one onto its image, and reciprocally; notably transformations that map straight lines to curves of order n, for some n. When $n = 2$ (the case of quadratic transformations) as he showed, almost all straight lines must have images that are conics passing through 3 fixed points.[39]

Noether in his (1871a) was the first to appreciate that Cremona transformations could be used to simplify singular points. He wrote a Cremona transformation of the plane algebraically as

$$[x, y, z] \mapsto [x', y', z'] = [\phi(x, y, z), \psi(x, y, z), \chi(x, y, z)],$$

where the functions ϕ, ψ and χ are homogeneous polynomials in x, y and z of the same degree. To simplify a singular point he used transformations defined as above where ϕ, ψ, and χ all vanish at the singular point. In the affine plane this becomes

$$(x, y) \mapsto (x_1, y_1) = \left(\frac{\varphi(x, y)}{\chi(x, y)}, \frac{\psi(x, y)}{\chi(x, y)} \right),$$

where φ, ψ, and χ are polynomial functions that vanish at the singular point. To see what this does, it is convenient to take the familiar example of the simplest Cremona transformation, other than a projective transformation. It is given algebraically by the quadratic transformation

$$[x, y, z] \mapsto \left[\frac{1}{x}, \frac{1}{y}, \frac{1}{z} \right] = [yz, zx, xy].$$

The union of the lines $x = 0, y = 0, z = 0$ is called the exceptional triangle, and the images of these lines are the points $[1, 0, 0], [0, 1, 0]$ and $[0, 0, 1]$ respectively. The transformation is many-valued at these points, each of which is "blown up" into a line but it is well-defined away from those three points and it is one-to-one away from the lines $x = 0, y = 0$, and $z = 0$.

Noether's argument that any plane algebraic curve can be reduced to one with only double points is sketchy at best. His use of Puiseux's work suggests that he appreciated that quadratic transformation might have to be used repeatedly to reduce a complicated singularity to a simpler one, but he did not fully appreciate what happens to the rest of the curve at the points where it

---

[39] Such examples had been studied before; Cremona cited papers by Steiner, Magnus, and Schiaparelli, but for higher values of n, which Cremona analysed in his (1865), everything was new.

crosses the exceptional lines. In his more rigorous account (1876), he gave a better definition of a singular point, and tried to redefine the genus of a plane algebraic curve and prove its invariance under birational transformations.[40]

8. The work of Brill and Noether

Noether began, in his (1873), with the question of when a plane curve E with equation $H = 0$ can be expressed in the form $H = AF + BG$, where $F = 0$ and $G = 0$ are the equations of two curves C and D. He rightly pointed out that people hitherto had assumed that it was necessary and sufficient that the curve E passes simply through the intersection points of the curves C and D, but in fact this is insufficient when the curves C and D have singular points or intersections with multiple tangents. His own account, which was based on counting the number of independent equations imposed on $H = 0$ by the equations $F = 0$ and $G = 0$, was a significant advance, but it too had flaws and he and other authors, notably Eugenio Bertini, offered improvements. Eventually, Noether offered this formulation of his Fundamental Theorem:[41]

Theorem. *$H(x, y)$ is representable in the form $AF + BG$ if and only if in some neighbourhood of each common point (x_0, y_0) of $F = 0$ and $G = 0$ there are power series $a(x-x_0)$ and $b(x-x_0)$ in $\mathbb{C}[y][[x]]$ whose coefficients are polynomials in y of orders $n - 1$ and $m - 1$ respectively, such that $H = aF + bG$.*

Only in the monumental joint work with Brill (Brill and Noether 1894, 352) was it shown that in the plane case there is nothing more to be said, the local conditions are indeed sufficient even in the projective case.

Today we would say that Noether's Fundamental Theorem is true for two polynomials F and G because F, G is a regular sequence and so the ideal they generate is, in later terminology due to Macaulay (1916), unmixed.[42]

Brill and Noether recognised that the Riemann–Roch theorem was of central importance to the theory of plane algebraic curves; indeed, they were the first to name it.[43] Clebsch had studied what we would call holomorphic differential forms on a curve C of degree d in his (1864, 193) by supposing that they are given by homogeneous polynomials of degree $d - 3$ that vanish at the double points of the curve. By Bézout's Theorem, such an intersection also has degree $d(d - 3) = 2p - 2$, as it should. The dimension of the family of divisors cut out on C in this way is the dimension of the projective space of forms of

---

[40]Noether also claimed the remarkable theorem that every such Cremona transformation is a composition of quadratic transformations and projective transformations. See his (1871b), with a small correction in his (1873).

[41]See (Noether 1887, 413).

[42]See (Eisenbud and Gray 2023) and (Eisenbud and Gray 2024) for a historical account of Macaulay's work.

[43]See (Brill and Noether 1874, § 5).

degree $d - 3$ in $\mathbb{C}[x_0, x_1, x_2]$, which is conveniently equal to $(d - 1)(d - 2)/2$. For this to be correct and account for all the canonical divisors, it has to be shown that every divisor E on the curve C that differs from an intersection $D \cap C$, where D is the curve defined by the equation $G = 0$, by the divisor of the restriction to C of a rational function P/Q (where P, Q are forms of the same degree), is again of the form $D' \cap C$ for some curve D'. This is the content of Noether's Fundamental Theorem. His Fundamental Theorem is essential when it comes to handling sets of points cut out on a fixed curve C by families of curves, counting the degree of an intersection of two curves at multiple points, dividing an arbitrary intersection into two subsets, and delivering a satisfactory version of the Riemann–Roch theorem.

For reasons of space, it is impossible to pursue the history of curves after the Riemann–Roch theorem here, except to note that some of the consequences are the subject of the present book to which this essay is appended.

9. Historical bibliography

Anglade, M. and J.-Y. Briend 2022. Nombrils, bruslans, autrement foyerz; la géométrie en action dans le *Brouillon project* de Girard Desargues, *Archive for History of Exact Sciences*, 76, 173–206.

Bézout, E. 1764. Sur le degré des équations résultantes de l'évanouissment des inconnues, *Mémoires de l'Académie Royale des Sciences*, 288–338.

Bézout, E. 1779. *Théorie générale des équations algébriques*, Ph.-D. Pierres, Paris, 1779. English translation by Eric Feron, Princeton University Press, 2006.

Biosemat-Martagon, L. 2010. *Eléments d'une biographie de l'Espace projectif*, Presses Universitaires de Nancy.

Bombelli, R. 1572. *L'algebra*, Bologna.

Bosse, A. 1648. *Maniere universelle de Mr. Desargues pour pratiquer la perspective, etc*, Paris.

Brieskorn, E. and H. Knörrer 1986. *Plane Algebraic Curves*, Birkhäuser.

Brill, A. 1923. Max Noether, *Jahrsbericht den Deutschen mathematiker Vereinigung* 32, 211–233.

Brill, A. and M. Noether 1874. Ueber die algebraischen Functionen und ihre Anwendung in der Geometrie, *Mathematische Annalen* 7, 269–310.

Brill A. and M. Noether 1894. Die Entwicklung der Theorie der algebraischen Functionen in alterer und neuerer Zeit, *Jahrsbericht den Deutschen mathematiker Vereinigung* 3, 107–566.

Cardano, G. 1545. *Artis Magnae sive de regulis algebraicis liber unus*, Nuremburg.

Castelnuovo, G., F. Enriques, and F. Severi, 1925. Max Noether, *Mathematische Annalen* 93, 161–181.

Cauchy, A.-L. 1817a. Sur les racines imaginaires des équations. *Nouv. Bull. Soc. Philom.* 5–9, in *Oeuvres Complètes* (2) 2, 210–216.

Cauchy, A.-L. 1817b. Seconde note sur les racines imaginaires des équations. *Nouv. Bull. Soc. Philom.* 161–164, in *Oeuvres Complètes* (2) 2, 217–222.

Cauchy, A.-L. 1825. Mémoire ser les intégrales définies prises entre des limites imaginaires, Imprimerie Royale, Paris, in *Oeuvres Complètes* ser. 2, vol. 2, 41–89.

Cayley, A. 1866. On the higher singularities of a plane curve, *Quarterly Journal of Pure and Applied Mathematics* 7, 212–223, in *Collected Mathematical Papers* V, no. 374, 520–582.

Cayley, A. 1878. On the geometrical representation of imaginary variables by a real correspondence of two planes, *Proceedings of the London Mathematical Society* 9, 31–39 in *The Collected Mathematical Papers of Arthur Cayley* X, no. 689, 316–323.

Chasles, M. 1837. *Aperçu historique sur l'origine et le développement des méthodes en géométrie*, Hayez, Bruxelles.

Clebsch, R. F. A. 1864. Ueber die Anwendung der Abelschen Functionen in der Geometrie, *Journal für die reine und angewandte Mathematik* 63, 189–243.

Clebsch, R. F. A. 1865a. Ueber die diejenigen ebenen Curven, deren Coordinaten rationale Functionen eines Parameters sind, *Journal für die reine und angewandte Mathematik* 64, 43–65.

Clebsch, R. F. A. 1865b. Ueber die Singularitäten algebraischer Curven, *Journal für die reine und angewandte Mathematik* 64, 98–100.

Clebsch, A. and P. Gordan 1866. *Theorie der Abelschen Functionen*, Teubner, Leipzig.

Cohen, M. R. and Drabkin, I. E. 1948. *A Source Book in Greek Science*, Harvard University Press.

Coolidge, J. L. 1931. *A Treatise on Algebraic Plane Curves* Oxford U. P., Dover reprint 2004.

Coolidge, J. L. 1940. *A history of geometrical methods*, Oxford U. P., Dover reprint 2003.

Cramer, G. 1750. *Introduction à l'analyse des lignes courbes algébriques*. Frères Cramer et Cl. Philibert, Geneva.

Cremona, L. 1863. Sulle trasformazioni geometriche delle figure piane (nota 1), *Giornale di Matematiche di Battaglini* 1, 305–311.

Cremona, L. 1865. Sulle trasformazioni geometriche delle figure piane (nota 2), *Giornale di Matematiche di Battaglini* 3, 269–280, 363–376.

Del Centina, A. 2020. Pascal's Mystic Hexagram, and a conjectural restoration of his lost Treatise on Conic Sections. *Archive for History of Exact Sciences* 74.5, 469–521.

Desargues G. 1639. *Brouillon project d'une atteinte aux evenements des rencontres du Cone avec un Plane*, Paris. J. V. Field and J. J. Gray (eds. and transl.)*The geometrical work of Girard Desargues*, 1987, Springer, New York.

Descartes, R. 1637. *La Géométrie* in *Discours de la Méthode, etc.* Leyden, English translation *The Geometry of René Descartes*, D. E. Smith and M. L. Latham, Open Court 1925, Dover reprint 1954.

Eisenbud, D. E. and J. J. Gray, 2023. F. S. Macaulay: From plane curves to Gorenstein rings, *Bulletin of the American Mathematical Society* (*new series*) 60.3, 371–406, https://doi.org/10.1090/bull/1787.

Eisenbud, D. E. and J. J. Gray, to appear. *F. S. Macaulay: from plane geometry to modern commutative algebra.*

Euler, L. 1748. *Introductio in Analysin Infinitorum*, two vols., *Opera Omnia* (1) Vols. 8 and 9, transl. *Introduction to Analysis of the Infinite*, Book I, Springer, 1988, Book II, Springer, 1990 (E101, E102).

Euler, L. 1750. Sur un contradiction apparente dans la doctrine des lignes courbes. *Mémoires de l'Académie des Sciences de Berlin* 4, 1750, 219–233, in *Opera Omnia* (1) 26, 33–45 (E147).

Euler, L. 1770. *Vollständige Einleitung zur Algebra* in *Opera Omnia* (1) 1, transl. *Elements of Algebra*, Rev. J. Hewlett, London, 1840, repr. Springer, 1972 (E387).

Finsterwalder, S. 1936 Alexander v. Brill. Ein Lebensbild, *Mathematische Annalen* 112, 653–663.

Gauss, C. F. 1799. Demonstratio nova theorematis omnem functionem algebraicam ... resolvi posse, Helmstadt, in *Werke* III, 1–30.

Gergonne, J. D. 1826. Philosophie mathématique. Considérations philosophiques sur les élémens de la science de l'étendue, *Annales de Mathématiques* 16, 209–231.

Gergonne, J. D. 1826–1827. Recherches sur quelques lois générales qui régissent les lignes et surfaces algébriques de tous les ordres. *Annales de Mathématiques* 17, 214–252.

Gilain, Ch. 1991. Sur l'histoire du théorème fondamental de l'algèbre: théorie des équations et calcul intégral, *Archive for History of Exact Sciences* 42, 91–136.

Gray, J. J. 2015. Klein and the Erlangen Programme, *Sophus Lie and Felix Klein: The Erlangen Program and its Impact in Mathematics and Physics*, Lizhen Ji and Athanase Papadopoulos (eds.), European Mathematical Society, 59–76.

Hamilton, W. R. 1837. Theory of conjugate functions, or algebraic couples; with a preliminary and elementary essay on algebra as the science of pure time, *Transactions of Royal Irish Academy* 17, 293–422.

Heath, Sir T. L. 1921. *A History of Greek mathematics*, 2 vols, Clarendon Press, Oxford, repr. Dover, 1981.

Hesse, O. 1844. Ueber die Wendepunkte der Curven dritter Ordnung *Journal für die reine und angewandte Mathematik* 28, 97–107, in *Gesammelte Werke*, 123–136.

Hesse, L. O. 1855. Ueber die Doppeltangenten der Curven vierter Ordnung. *Journal für die reine und angewandte Mathematik* 49, 243–264, in *Gesammelte Werke*, 319–344.

Hesse, L. O. 1897. *Gesammelte Werke*. Verlag der K. Akademie, München. Repr. Chelsea, New York 1972.

al-Khāyyamī, 'Umar 1950. In H. J. J. Winter and W. Arafat, The Algebra of Omar Khayyām, *Journal of the Royal Asiatic Society of Bengal*, 16, 27–78.

Lê, F. 2020. "Are the genre and the Geschlecht one and the same number?" An inquiry into Alfred Clebsch's Geschlecht, *Historia Mathematica* 53, 71–107.

Lorenat, J. 2016. Synthetic and analytic geometries in the publications of Jakob Steiner and Julius Plücker (1827–1829) *Archive for History of Exact Sciences* 70.4, 413–462.

Macaulay, F. S. 1916. *The algebraic theory of modular systems*, Cambridge Tracts in Mathematics, No. 19.

MacLaurin, C. 1720. *Geometria Organica, sive Descriptio Linearum Curvarum Universalis*, London.

Maurolico, F. 1611. *Theoremata de lumine et umbra*.

Möbius, A. F. 1827. *Der Barycentrische Calcul*, Barth, Leipzig.

Netz, R. 2022. *A new History of Greek Mathematics*, Cambridge University Press.

Newton, I. 1687. *Philosophiae Naturalis Principia Mathematica* London, 2nd ed. 1713, 3rd ed. 1726. English translation *Isaac Newton's Philosophiae Naturalis Principia Mathematica. The third edition, 1726, with variant readings*, I. B. Cohen and A. Whitman (transls. and eds.), Cambridge University Press, 1972, 1999.

Newton, I. 1704. *Enumeratio linearum tertii ordinis*, appendix in *Opticks*, London. English translation *Sir Isaac Newton's Enumeration of Lines of the third Order* C. R. M. Talbot, 1860.

Noether, M. 1871a. Ueber die algebraischen Functionen einer und zweier Variabeln, *Nachrichten von der Königlichen Gesellschaft der Wissenschaften und der Georg-Augusts Universität zu Göttingen* 267–278.

Noether, M. 1871b. Ueber Flächen, welche Schaaren rationaler Curven besitzen, *Mathematische Annalen* 3, 161–227 and 547–580.

Noether, M. 1873. Ueber einen Satz aus der Theorie der algebraischen Functionen, *Mathematische Annalen* 6, 351–359.

Noether, M. 1876. Ueber die singulären Werthsysteme einer algebraischen Function und die singulären Punkte einer algebraischen Curve, *Mathematische Annalen* 8, 166–182.

Noether, M. 1887. Ueber den Fundamentalsatz der Theorie der algebraischen Functionen, *Mathematische Annalen* 30, 410–417.

Plücker, J. 1829a, b. Recherches sur les courbes algébriques de tous les degrés, *Annales de Mathématiques Pures et Appliquées* 19, 97–106 and 129–137.

Plücker, J. 1830. Über ein neues Coordinatensystem, *Journal für die reine und angewandte Mathematik* 5, 1–36.

Plücker, J. 1828–1831. *Analytisch-geometrische Entwicklung*, 2 vols. Essen.

Plücker, J. 1834. Solution d'une question fondamentale concernant la théorie générale des courbes, *Journal für die reine und angewandte Mathematik* 12, 105–108.

Plücker, J. 1835. *System der analytischen Geometrie*, Berlin.

Plücker, J. 1839. *Theorie der algebraischen Curven*, Bonn.

Poncelet, J. V. 1822. *Traité des Propriétés Projectives des Figures*, Gauthier-Villars, Paris.

Puiseux, V. 1850. Recherches sur les fonctions algébriques, *Journal de Mathématiques Pures et Appliquées* 15, 365–480.

Riemann, B. 1851. Grundlagen für eine allgemeine Theorie der Functionen einer veränderlichen complexen Grösse (Inaugural dissertation), Göttingen, in *Mathematische Werke*, 3–45.

Riemann, B. 1857. Theorie der Abelschen Functionen, *Journal für die reine und angewandte Mathematik* 54, 115–155, in *Mathematische Werke* 88–144.

Riemann, B. 1990. *Bernhard Riemann's Gesammelte Mathematische Werke und Wissenschaftliche Nachlass*, R. Dedekind and H. Weber (eds.) with Nachträge, M. Noether and W. Wirtinger (eds.). 3rd ed. R. Narasimhan (ed.), Springer.

Severi, F. 1922. Alexander von Brill zum achtzigsten Geburtstag am 20, September 1922, *Jahrsbericht den Deutschen mathematiker Vereinigung* 31, 89–96.

Staudt, K. G. C. von, 1847. *Geometrie der Lage*, Nuremberg.

Steiner, J. 1832. *Systematischer Entwickelung der Abhängigkeit geometrischer Gestalten von einander*, Fincke.

Stolz, O. 1875. Ueber die singulären Punkte der algebraischen Functionen und Curven, *Mathematische Annalen* 8, 415–443.

Thomas, I. 1939. *Selections Illustrating the History of Greek Mathematics* (2nd ed. 1980), Heinemann.

Viète F. 1646. *Opera Mathematica ... Recognita Opera*, F. Schooten (ed.), Lugduni Batavorum.

Van der Waerden, B. L. 1961. *Science Awakening*, A. Dresden (transl.), Oxford University Press.

Waerden, B. L. van der, 1971. The Foundations of Algebraic Geometry from Severi to André Weil, *Archive for History of Exact Sciences* 7, 171–180, in *Zur algebraische Geometrie*, 1–10, Springer, 1983.

Hints to marked exercises

Exercises in Chapter 1

1.1. Use Bézout's theorem.

1.4. For the first part, use Bézout's theorem.

1.7. For parts (3) and (4), observe that if $\beta \neq \pm\alpha$ then $\Gamma_{\alpha,\beta}$ is locally the intersection of C with the line $L \subset \mathbb{A}^2$ given by $\alpha x + \beta y = 0$, but when $\beta = \pm\alpha$ the intersection $L \cap C$ has multiplicity 3 at p, and is not equal to $\Gamma_{\alpha,\beta}$.

Exercises in Chapter 2

2.1. For (1) and (3), use the Serre–Grothendieck vanishing theorem, plus the Riemann–Roch formula in the case of (3). For (2), compare the Hilbert polynomial of X with that of a hyperplane section of X.

2.2. For the degree formula, note that $\sum_{p\in C} \text{mult}_p(C, H)p$ is the divisor associated to the morphism $\widetilde{C} \to \mathbb{P}^r$ with image C.

2.3. For (3), let r be the third point of intersection of the line $\overline{pq}$ with the closure of $C°$, and consider the pencil of lines through r.)

2.6. For (1), find rational functions in x and y whose pullback to C has poles along $D = p + q + r_0 + r_3$ but nowhere else. For the last part, observe that the equation of the image corresponds to the kernel of the map $\text{Sym}^4 H^0(\mathcal{O}_C(D)) \to H^0(\mathcal{O}_C(4D))$.

2.11. First show that there are at most two points in the normalization of C lying over p.

2.13. The linear forms in $\mathbb{P}^{\binom{m+2}{2}-1}$ vanishing on the image of C_0 correspond to elements of the kernel of the map $H^0(\mathcal{O}_{\mathbb{P}^2}(m)) \to H^0(\mathcal{O}_C(m))$.

Exercises in Chapter 3

3.1. Use Proposition 1.11.

3.4. Find the rank of the restriction map $H^0(\mathcal{O}_{\mathbb{P}^N}(1)) \to H^0(\mathcal{O}_{\nu_d(C)}(1))$.

3.8. Use Bézout's theorem.

3.9. Look at the linear series cut on C by the two rulings of the quadric.

3.10. To any such embedding we can associate the g_3^1 given by one of the rulings of the quadric surface containing the image curve. Modulo PGL_2, there is a one-parameter family of g_3^1s on $\mathbb{P}^1$.

3.11. Use induction on n, starting with the case $n = 2$.

3.13. For any point $p \in C$, let $L_p \subset \mathcal{N}_{C/\mathbb{P}^3}$ be the line bundle of $\mathcal{N}_{C/\mathbb{P}^3}$ whose fiber over any point $q \neq p \in C$ is the one-dimensional subspace of $(\mathcal{N}_{C/\mathbb{P}^3})_q$ spanned by the line $\overline{pq}$. (This of course only defines a line subbundle of $\mathcal{N}_{C/\mathbb{P}^3}$ over $C \setminus \{p\}$, but there is a unique extension to a line subbundle of $\mathcal{N}_{C/\mathbb{P}^3}$ over all of C.) Show that for $p \neq p'$ we have

$$\mathcal{N}_{C/\mathbb{P}^3} = L_p \oplus L_{p'}.$$

3.14. For $p \in C$, define a line subbundle $L_p \subset \mathcal{N}_{C/\mathbb{P}^d}$ as in the preceding problem, and show that for $p_1, \ldots, p_{d-1}$ distinct points the bundles L_{p_i} are independent.

3.15. It is the projectivization of the action of $\mathrm{SL}(V)$ on $\mathrm{Sym}^{d-2} V$.

3.16. Since the $r-1$ surfaces intersect only in codimension $r-1$, they form a regular sequence; and the length of any maximal regular sequence in $\mathbb{C}[x_0, \ldots, x_r]$ is $r + 1$.

3.17. Let $p \in L \cap X$ be a point. If every secant to X through p lies entirely in X, then X is a cone over p; but since p was a general point, this would imply that X is a linear space, contradicting nondegeneracy. Hence the projection $\pi_p : X \to \mathbb{P}^{N-1}$ is a generically finite (rational) map from X to $X' := \pi_p(X)$, and thus $\dim X' = \dim X$ and $\mathrm{codim}\, X' = \mathrm{codim}\, X - 1$. The plane $\pi_p(L)$ meets X' in the images of the points of $L \cap X$ other than p, so $\deg X \geq \deg X' + 1$. By induction, $\deg X' \geq \mathrm{codim}\, X' + 1 = \mathrm{codim}\, X$, completing the argument.

Exercises in Chapter 4

4.1. Show that any $d - 2$ points impose independent conditions on forms of degree $d - 3$, and use Bertini's theorem to reduce to this case.

4.2. First, the space of sextics double at three noncollinear points visibly has dimension 19 (take the points to be the coordinate points and count monomials). Then, since this includes the triangle with vertices at the three points plus arbitrary cubics, the subspace of those double at the fourth point will have codimension 3.

Exercises in Chapter 5

5.1. Use the fact that the intersection of two affine open subsets of a quasi-projective scheme is again affine.

5.3. For (2), change coordinates to $z_i := x_i + y_i$, $w_i := x_i - y_i$.

5.5. If $\mathcal{L}$ is an invertible sheaf of degree d with $h^0(\mathcal{L}) > d - g + 1$, show that for general $p, q \in C$ we have $h^0(\mathcal{L}(p - q)) = h^0(\mathcal{L}) - 1$.

5.6. For (2), use the fact that μ_2 is the blowup of $J(C)$ at a point.

5.7. Show that for general points $p, q \in C$, there exists a unique pair $(p', q') \neq (p, q)$ with $p - q \sim p' - q'$.

5.9. A general divisor of degree $\geq g + 1$ is nonspecial, so

$$h^0(\mathcal{O}_C(D)) = (g + 1) - g + 1 = 2.$$

To say that the linear series $|D|$ has no basepoints means that D is not equivalent to a divisor of the form $D' + p$, where $h^0(D')$ has degree g and two independent sections. By the Riemann–Roch theorem, $h^0(D') = 1 + h^0(K - D')$, so the condition on D' means that $K - D'$ is effective. But $K - D'$ is a general divisor of degree $g - 2$, and the set of classes of effective divisors, the image of C_{g-2}, has dimension only $g - 2$. Thus the set of divisors of the form $D' + p$ with $h^0(D') \geq 2$ has dimension only $g - 2 + 1$ so a general divisor D of degree $g + 1$ is basepoint free and defines a map $C \to \mathbb{P}^1$.

By Hurwitz's theorem the total ramification of such a map is $2g + 2$. The divisors with points of ramification index 2 or more are linearly equivalent to divisors of the form $D' + 3p$, and the divisors that have two ramification points mapping to the same point are equivalent to divisors of the form $D'' + 2p + 2q$. Each of these sets of divisors fills only a $(g-1)$-dimensional family of equivalence classes, the images of $C_{g-2} \times C$ or $C_{g-3} \times C_2$ respectively; so not every divisor D is of this form.

Exercises in Chapter 6

6.1. In either case the complement $\overline{C}^{\circ} \setminus C^{\circ}$ consists of a single point, with two points of C mapping to it; now use the genus formula in either $\mathbb{P}^2$ or $\mathbb{P}^1 \times \mathbb{P}^1$.

6.2. Such a cover is specified by giving $2g + 2$ transpositions, not all equal, whose product is a nontrivial 3-cycle, modulo simultaneous conjugation. We have already worked out the number of such tuples whose product is the identity; just subtract.

6.3. Topologically, such covers are in 1-1 correspondence with subgroups of index 2 in $\pi_1(C)$; and such a subgroup is necessarily the preimage of a subgroup of index 2 in the abelianization $H_1(C, \mathbb{Z}) \cong \mathbb{Z}^{2g}$.

6.4. If $f : X \to C$ is an unramified double cover, consider the direct image $f_*(\mathcal{O}_X)$. This is a locally free sheaf of rank 2 on C, on which the group $\mathbb{Z}/2$ acts; the +1-eigenspace is the structure sheaf $\mathcal{O}_C$, and the -1-eigenspace is an invertible sheaf $\mathcal{L}$ on C such that $\mathcal{L}^2 \cong \mathcal{O}_C$.

6.5. By our analysis, to specify such a cover, we have to specify the monodromy around representative loops generating $H_1(E, \mathbb{Z}) \cong \mathbb{Z}^2$; thus there are four possibilities.

6.6. Choose any line $M \subset Q$ of the opposite ruling, and look at the linear forms H, H' on $\mathbb{P}^3$ vanishing on $L \cup M$ and $L' \cup M$.

6.7. If $h^0(\mathcal{L}) = 0$, we have $h^0(\mathcal{L}(p_k)) = 1$ for any ramification point p_k; show that the unique effective divisor in $|\mathcal{L}(p_k)|$ must be the sum of two ramification points.

6.8. For (1) — which implies that the map $\phi_\mathcal{L}$ is an immersion — observe that $h^0(\mathcal{L} \otimes K_C^{-1}) = 1$, meaning p and q are unique. Part (2) says that the images of the differential $d\phi_\mathcal{L}$ at p and q are distinct.

Exercises in Chapter 7

7.1. The Hilbert polynomials satisfy $p_X = p_Y + p_Z$, which follows from the vanishing of $h^1(\mathcal{J}_{Y \cup Z}(m))$ for large m; the Hilbert functions satisfy $h_X \leq h_Y + h_Z$. (When $h_X(m) = h_Y(m) + h_Z(m)$, we say that Y and Z *impose independent conditions* on $|\mathcal{O}_{\mathbb{P}^n}(m)|$.)

7.2. Use the exact sequence $0 \to \mathcal{J}_{Y \cup Z} \to \mathcal{J}_Y \oplus \mathcal{J}_Z \to \mathcal{J}_{Y \cap Z} \to 0$.

7.3. Any cubic vanishing on X vanishes identically on H.

7.4. As in the twisted cubic case, the group PGL_{r+1} acts transitively on $\mathcal{H}^\circ$ with stabilizer PGL_2.

7.5. The normal bundle is $\mathcal{N} = \mathcal{O}_C(d) + \mathcal{O}_C(e)$. To prove smoothness, use Exercise 3.16 to compute $H^0(\mathcal{N})$.

Exercises in Chapter 8

8.4. Show that the canonical sheaf K_x is nontrivial.

Exercises in Chapter 9

9.1. Recall that the g_3^1s on C are cut by the rulings of the quadric Q.

9.2. Try blowing up the plane at the nodes. Look at Section 15.1.1 if you get stuck.

9.3. For the first part: if the curve $C \subset \mathbb{P}^3$ lay on a quadric, what would be its class? For the second, if $S \cap T = C \cup D$, calculate the degree and genus of D. See Chapter 16 if you get stuck.

9.4. Show that each of the divisors E of the g_2^1 span a line in $\mathbb{P}^3$.

9.5. Consider the incidence correspondence

$$\Gamma := \{(Q, L) \in \mathbb{P}^9 \times \mathbb{G}(1, 3) \mid L \subset Q\}$$

where $\mathbb{P}^9$ is the space of quadrics $Q \subset \mathbb{P}^3$.

9.6. By projection, show that C would have to be double at the vertex of the cone.

9.7. Show that the linear series $|K_C - p|$ is very ample if and only if C is not trigonal.

9.8. Consider separately the cases where Γ contains a fat point (that is, the scheme defined by the square of the maximal ideal at a point) or is curvilinear (that is, has Zariski tangent space of dimension at most 1 everywhere).

Exercises in Chapter 10

10.16. Show that the items of Theorem 10.15 are true in this situation.

10.3. For (1), show that $h^0(D - K_C) = 0$, and conclude that the map $\phi_D \times \phi_E$ is birational onto its image; using the genus formula, conclude the image curve is smooth. For (2), observe that if L is a line through p meeting C more than once elsewhere, the line L must be contained in Q.

10.5. If $d > mr$, then the hypersurfaces of degree m containing Γ are exactly the hypersurfaces of degree m containing D, and we know how many there are. If $d \le mr$, then the hypersurfaces of degree m containing Γ cut out on D the linear series $|\mathcal{O}_D(m)(-\Gamma)| = |\mathcal{O}_{\mathbb{P}^1}(mr - d)|$ and again we know the dimension.

10.6. Show that if $h^0(\mathcal{O}_C(m + 1)) - h^0(\mathcal{O}_C(m)) \le \deg C$, and that if equality holds for all $m \ge m_0$, then $\mathcal{O}_C(m)$ is nonspecial for all $m \ge m_0$.

10.7. If $|D'|$ were another g_d^r on C, show that $h^0(mD + D') > h^0((m + 1)D)$.

10.8. For (1), we can calculate the dimension of the Hilbert scheme $\mathcal{H}^\circ$ of Castelnuovo curves by calculating the dimension of the incidence correspondence of pairs (Q, C) with Q a quadric surface in $\mathbb{P}^3$ and $C \subset Q$ a curve of type (k, k); then invoke Exercise 10.7 to say the fibers of $\mathcal{H}^\circ \to M_g$ are isomorphic to PGL_4. For (2), we already know the dimension of the locus of hyperelliptic curves is $2g - 1$ ($2g + 2$-tuples of points in $\mathbb{P}^1$ mod PGL_2); the point is just that Castelnuovo curves are rarer than hyperelliptic ones.

Exercises in Chapter 11

11.1. Say $k = 2$. If the general hyperplane section of X were reducible, there would be distinguished subsets of the intersection of X with a general $(r - 2)$-plane Λ (the intersections of Λ with the components of $H \cap X$, for H a general hyperplane containing Λ); but we know the monodromy on $X \cap \Lambda$ is the full symmetric group.

11.2. Choose a basepoint of the pencil, say $p = [1, i, 0, 0]$. The lines of the two rulings of Q_t passing through p are $Y - iX = Z - \pm i\sqrt{t}W = 0$, which are exchanged under the monodromy as t goes around 0.

11.4. We need to know that the dual hypersurfaces $C_i^* \subset (\mathbb{P}^r)^*$ are all distinct; given this, we can exhibit loops that induce a given permutation of the points of $H \cap C_i$ while fixing the points of $H \cap C_j$ for $j \neq i$. To see that the dual hypersurfaces $C_i^* \subset (\mathbb{P}^r)^*$ are all distinct, invoke the duality theorem $(C_i^*)^* = C_i$ (see for example [Eisenbud and Harris 2016]).

11.5. For (1), we can fix the curve E and prove the a priori stronger statement that the monodromy on the points of $D \cap E$ as D varies is the symmetric group; this follows from Theorem 11.3 applied to the Veronese embedding $\nu_d(E)$. Alternatively, we can exhibit a transposition by finding a pair (D, E) such that $D \cap E$ consists of $de - 2$ simple points and one double point, and prove double transitivity with the usual incidence correspondence argument.

11.6. The line joining any two flexes of a plane cubic E contains a third.

11.7. Show that a transitive subgroup of S_n generated by transpositions is all of S_n.

Exercises in Chapter 12

12.2. For (2), we know that $\phi_F(E)$ lies on at least 5 quadrics by the usual restriction sequence; if it lay on 6 or more it would be a rational normal curve. (Alternatively, see Section 4.6.)

12.3. Use the description of the canonical model of C together with the geometric Riemann–Roch theorem.

12.5. Take the image of C under the map associated to the g_6^2 cut out by conics through two of the three collinear nodes and the one remaining node.

12.7. The curve in question is the normalization of a plane sextic curve with one double point, consisting of two smooth branches with contact of order 4 with each other and contact of order 3 with their common tangent line. The exercise asks you to both prove that such a curve exists, and that the g_4^1 cut out by lines through the double point is the unique g_4^1 on C.

Exercises in Chapter 13

13.2. Since we know that f has finite order, we can take the quotient $B = C/\langle f \rangle$ of C by the cyclic group $\langle f \rangle$; apply the Riemann–Hurwitz formula to the quotient map $C \to B$.

13.3. Use the Hodge index theorem 13.10 and the adjunction formula to show that the maximum possible genus of such a curve C is obtained when C is linearly equivalent to a sum of fibers of the two projections, in which case the inequality becomes an equality.

13.4. Since we know that $\mathrm{Aut}\, C$ has finite order, we can take the quotient $B = C/\mathrm{Aut}\, C$ of C; again, apply the Riemann–Hurwitz formula to the quotient map $C \to B$. (Warning: the idea is the same as in Exercise 13.2, but the execution is substantially more complicated.)

13.5. For $\Lambda \in \Sigma_\beta(\mathcal{V})$, consider bases of Λ such that $\Lambda \cap V_i$ is the span of basis vectors.

13.8. Let each pair of points come together, and use the result for the tangent lines, a special case of Theorem 13.22.

13.9. It's enough to look at the case $d = 2$ and $r = 1$.

13.11. The inflection points are the points $p \in E$ such that $\mathcal{O}_E(np) \cong \mathcal{O}_E(1)$.

Exercises in Chapter 14

14.1. Using the canonical embedding to realize C as a plane quartic and project from the point of intersection of the tangent lines at p and q.

14.2. Show, using dimension counts:

(1) Λ does not contain any tangent line to C.

(2) Λ does not meet any secant line to C other than the lines $\overline{p_i q_i}$.

(3) Λ does not meet the 3-plane $\overline{\mathbb{T}_{p_i} C, \mathbb{T}_{q_i} C}$ in a line.

14.3. Let the p_i, q_i approach each other, reducing to the case of tangent lines. Then use Theorem 13.22. For a direct proof see [Griffiths and Harris 1980, Lemma, p. 259].

14.4. Let C_0 be a curve with a node p, and $C \xrightarrow{\nu} C_0$ its partial normalization at p. Denote by $q, r \in \widetilde{C}$ the points lying over p. If $\mathcal{L}$ is an invertible sheaf on C, and $\mathcal{M} := \nu^*(\mathcal{L})$ the pullback of $\mathcal{L}$ to $\widetilde{C}$, then $\mathcal{M}$ is an invertible sheaf on $\widetilde{C}$. Its fibers over q and r are both identified with the fiber $\mathcal{L}_p$ of $\mathcal{L}$ at p, and hence with each other. Conversely, given an invertible sheaf $\mathcal{M}$ on $\widetilde{C}$ and an identification of the fibers $\mathcal{M}_q$ and $\mathcal{M}_r$, we can form an invertible sheaf $\mathcal{L}$ on C by taking the subsheaf of $\nu_*\mathcal{M}$ whose sections agree at q and r, in terms of the identification.

Exercises in Chapter 15

15.1. If $D = q_1 + \cdots + q_{d-2}$ had $r(D) \geq 1$, the points $q_1 + \cdots + q_{d-2}$ and p would fail to impose independent conditions on plane curves of degree $d - 3$ and hence lie on a line.

15.2. If $D = q_1 + \cdots + q_{d-2}$ had $r(D) \geq 1$, the points $q_1 + \cdots + q_{d-2}$ and p, p' would fail to impose independent conditions on plane curves of degree $d - 3$ and hence by Proposition 15.3 below $d - 1$ of them would lie on a line.

15.3. A g^1_{d-2} is a set of points that, together with the nodes, impose dependent conditions on forms of degree $d - 3$.

15.4. Be careful to subtract the right multiples of the points that are preimages of the singular points.

15.5. To add two points s and $t \in C$, choose a conic curve D passing though s, t, q_1 and q_2, and let u and v be the remaining points of $C_0 \cap D$; then take the conic D' passing though u, v, q_1, q_2 and o. The sum $s + t$ will then be the remaining point of $D' \cap C_0$.

15.6. Show that in addition to the triple point at p, the curve C_0 has one infinitely near point of multiplicity 3.

15.7. For (1), the adjoint ideal is the ideal of functions vanishing to order 4 on each branch (so that the general member of the ideal with have zero locus consisting of two smooth branches simply tangent to the branches of the triple point). For (3), the adjoint is simply the square of the maximal ideal.

15.8. The computation can be done locally analytically. Let $R = \widehat{\mathcal{O}_{C_0,p}}$ be the completion of the local ring of C_0 at r_i. The integral closure is then the product of rings $R_i = \widehat{\mathcal{O}_{B_i}} \cong k[\![t_i]\!]$, with $R_i = R/P_i$ as P_i runs over the minimal primes of R. The multiplicity m_i is the colength of the ideal $\sum_{j\neq i} P_j \subset R$.

Exercises in Chapter 16

16.3. First show that any locally free sheaf on C is an iterated extension of invertible sheaves.

16.5. The ideal sheaf of the curve on the quadric Q is an extension of the ideal sheaf of the quadric in $\mathbb{P}^3$ with the ideal sheaf of the curve on the quadric, which is

$$\mathcal{O}_Q(-a,-b) = \pi_1^*(\mathcal{O}_{\mathbb{P}^1}(-a)) \otimes \pi_2^*(\mathcal{O}_{\mathbb{P}^1}(-b)),$$

where π_1, π_2 are the projections to $\mathbb{P}^1$. Use the Künneth formula

$$H^1(\mathcal{O}_Q(p,q)) = H^1(\mathcal{O}_{\mathbb{P}^1}(p)) \otimes H^0(\mathcal{O}_{\mathbb{P}^1}(q)) \oplus H^0(\mathcal{O}_{\mathbb{P}^1}(p)) \otimes H^1(\mathcal{O}_{\mathbb{P}^1}(q))$$

to compute the necessary cohomology.

16.6. Use the exact sequence

$$0 \to (fg) \to gI \oplus fJ \to gI + fJ \to 0$$

and the corresponding exact sequence of quotients by these ideals.

16.9. Count the monomials of each degree in square of the ideal of a line.

16.10. Use the description of $I(X)$ as the ideal of 2×2 minors of

$$\begin{pmatrix} x_0 & x_1 & x_2 \\ x_1 & x_2 & x_3 \end{pmatrix}.$$

16.12. Look at hyperplane sections of C.

Exercises in Chapter 17

17.3. Imitate the proof of Lemma 17.9.

17.5. Let $Y = X \cap H$ be a general hyperplane section of X and consider the restriction map $H^0(\mathcal{J}_{X/\mathbb{P}^r}(2)) \to H^0(\mathcal{J}_{Y/\mathbb{P}^{n-1}}(2))$; repeat $n - c$ times.

17.7. See [Hartshorne 1977, Section V.1]. Part (2) also follows from the length of the Eagon–Northcott complex described in Section 18.3.

17.9. See [Harris 1982a, Section 3c].

17.10. a_1 is again the maximal number of rulings. What's the cleanest statement after that? Does one have to project the scroll to $S(a_2, \dots)$ to see the rest?

17.14. $S(0, 3)$ does not occur, because by Theorem 17.25 the smooth curves on $S(0, 3)$ are either hypersurface sections, and thus of degree 3 for some integer a, or in the rational equivalence class of a hypersurface section plus one line, of degree $3a + 1$, and thus not of degree 8. One can see this directly too: if the scroll were the cone over a twisted cubic, then the lines on the cone would cut out the g_3^1 on C. Since a pair of lines on the cone span only a 2-plane, the sum D of the two divisors would be a divisor of degree 6 and thus $K - D$ would be a g_2^1, so the curve would be hyperelliptic.

In the case when C lies on $S := S(1, 2)$, we may write the class of C in terms of the hyperplane class H and the class F of a ruling, $C \sim pH + qF$, and we see from the projection of S to $\mathbb{P}^1$ that C admits a degree p covering of $\mathbb{P}^1$.

Since we have assumed that C is not hyperelliptic, we must have $p \geq 3$. By Theorem 17.22 we must have $q \geq -p$. Since $\deg C = C \cdot H = 3p + q = 8$, we must have either $C \sim 3H - F$ or $C \sim 4H - 4F$. By the adjunction formula, in the second case,

$$2g(C) - 2 = 8 = (C + K_S) \cdot C = (4H - 4F) + (-2H + F)) \cdot (4H - 4F) = 4$$

whereas a similar computation in the first case yields 8; thus $C \sim 3H - F$.

The key thing to note here is that the curve C has intersection number 3 with the lines of the ruling of S, meaning that C is a trigonal curve. (We also see that the g_3^1 on C is unique: if $D = p + q + r$ is a divisor moving in a pencil, the points p, q and r must lie on a line; since the surface S is the intersection of quadrics, those lines must lie on S and so must be lines of the ruling.) We see also that the locus $W_4^1(C)$ has two components: there are pencils with a basepoint, that is, consisting of the g_3^1 plus a basepoint; and there are the residual series $K_C - g_3^1 - p$. Each of these components of $W_4^1(C)$ is a copy of the curve C itself, and they meet in two points, corresponding to the points of intersection of C with the directrix of the scroll S.

17.15. Use intersection theory on the scroll.

Exercises in Chapter 18

18.1. Use the Riemann–Roch theorem.

18.2. For (3), use multilinear algebra (as in [Eisenbud 1995]) to define the maps, and imitate the proof of Theorem 18.18 to prove the exactness of the given resolution of $\text{Sym}^a(M)$. See also [Eisenbud 1995, Appendix A2.6].

18.3. Elementary linear algebra shows that, if k is a field, then a complex

$$k^p \xleftarrow{\phi} k^q \xleftarrow{\psi} k^r$$

is exact at k^q if and only if $\text{rank}\,\phi + \text{rank}\,\psi = q$.

18.6. Nakayama's lemma can be used to prove that projectives are free in some cases.

18.7. Show that the 3×3 minors of a general 4×3 matrix of linear forms defines a Cohen–Macaulay curve of genus 3. Show that a curve of type $(2, 4)$ on a smooth quadric in $\mathbb{P}^3$ is not arithmetically Cohen–Macaulay.

18.8. Use the formula for the Hilbert function of the canonical curve.

18.9. To show that the rank of the syzygy matrix is $n - 1$, tensor with the field of rational functions.

18.11. Consider the hyperplane section of a trigonal canonical curve of genus 5.

Exercises in Chapter 19

19.1. By restricting the presentation matrix of I_C to C — that is, substituting the forms of degree 3 in 2 variables for the variables in the presentation matrix — we get a presentation

$$R(-9)^2 \xrightarrow{A} R(-6)^3 \to I_C/I_C^2 \to 0.$$

The kernel of the dual of A is the normal bundle; as a module over $\mathbb{C}[s,t]$ it has 2 linear generators, the columns of the matrix

$$\begin{pmatrix} t & 0 \\ -s & t \\ 0 & -s \end{pmatrix}.$$

19.3. The dimensions are 11, 10 and 11.

19.4. A rational quartic curve is residual to the union of two skew lines in the complete intersection of a quadric and a cubic.

19.5. On a smooth rational curve any locally free sheaf that is generated by global sections is nonspecial.

19.6. $\mathcal{N}_{C/\mathbb{P}^3} \cong \mathcal{O}_C(2)^{\oplus 2}$.

19.7. If $C = S \cap T$, show that every deformation of C lies on a deformation of S and a deformation of T.

19.14. Let L, Q and F denote a general linear form, a general quadratic form and a general cubic form, and consider the pencil of surfaces $S_t = V(tF+LQ) \subset \mathbb{P}^3$ specializing from the cubic surface $V(F)$ the to reducible cubic $V(LQ)$.

19.17. The ideals in question are sandwiched between two successive powers of the maximal ideal of a point in $\mathbb{P}^3$.

19.21. For (1), see Corollary 4.8.

Bibliography

A. Altman and S. Kleiman (1970), *Introduction to Grothendieck duality theory*, Lecture Notes in Mathematics **146**, Springer, Berlin, 1970. DOI: 10.1007/BFb0060932.

A. Álvarez, F. Sancho, and P. Sancho (2008), "Homogeneous Hilbert scheme", *Proc. Amer. Math. Soc.* **136**:3 (2008), 781–790. DOI: 10.1090/S0002-9939-07-09169-1.

M. Aprodu, G. Farkas, ş. Papadima, C. Raicu, and J. Weyman (2019), "Koszul modules and Green's conjecture", *Invent. Math.* **218**:3 (2019), 657–720. DOI: 10.1007/s00222-019-00894-1.

E. Arbarello and E. Sernesi (1978), "Petri's approach to the study of the ideal associated to a special divisor", *Invent. Math.* **49**:2 (1978), 99–119. DOI: 10.1007/BF01403081.

E. Arbarello, M. Cornalba, P. A. Griffiths, and J. Harris (1985), *Geometry of algebraic curves*, vol. I, Grundlehren der Mathematischen Wissenschaften **267**, Springer, New York, 1985. DOI: 10.1007/978-1-4757-5323-3.

E. Arbarello, M. Cornalba, and P. A. Griffiths (2011), *Geometry of algebraic curves*, vol. II, Grundlehren der mathematischen Wissenschaften **268**, Springer, 2011. DOI: 10.1007/978-3-540-69392-5.

A. Atanasov, E. Larson, and D. Yang (2019), "Interpolation for normal bundles of general curves", *Mem. Amer. Math. Soc.* **257**:1234 (2019), v+105. DOI: 10.1090/memo/1234.

M. F. Atiyah and I. G. Macdonald (1969), *Introduction to commutative algebra*, Addison-Wesley, Reading, MA, 1969. DOI: 10.1201/9780429493638.

I. Bacharach (1886), "Über den Cayley'schen Schnittpunktsatz", *Math. Ann.* **26** (1886), 275–299. DOI: 10.1007/BF01444338.

E. Ballico and P. Ellia (1983), "Generic curves of small genus in $\mathbf{P}^3$ are of maximal rank", *Math. Ann.* **264**:2 (1983), 211–225. DOI: 10.1007/BF01457526.

E. Ballico, G. Bolondi, and J. C. Migliore (1991), "The Lazarsfeld–Rao problem for liaison classes of two-codimensional subschemes of $\mathbf{P}^n$", *Amer. J. Math.* **113**:1 (1991), 117–128. DOI: 10.2307/2374823.

H. Bass (1963), "On the ubiquity of Gorenstein rings", *Math. Zeitschr.* **82** (1963), 8–28. DOI: 10.1007/BF01112819.

D. Bayer and D. Eisenbud (1991), "Graph curves", *Adv. Math.* **86**:1 (1991), 1–40. DOI: 10.1016/0001-8708(91)90034-5.

D. Bayer and D. Eisenbud (1995), "Ribbons and their canonical embeddings", *Trans. Amer. Math. Soc.* **347**:3 (1995), 719–756. DOI: 10.2307/2154871.

A. Beauville (1996), *Complex algebraic surfaces*, 2nd ed., London Mathematical Society Student Texts **34**, Cambridge University Press, Cambridge, 1996. DOI: 10.1017/CBO9780511623936.

C. Bopp and F.-O. Schreyer (2021), "A version of Green's conjecture in positive characteristic", *Exp. Math.* **30**:4 (2021), 475–480. DOI: 10.1080/10586458.2019.1576082.

E. Brieskorn and H. Knörrer (1986), *Plane algebraic curves*, Birkhäuser, Basel, 1986. DOI: 10.1007/978-3-0348-5097-1.

A. Brill and M. Nöther (1874), "Ueber die algebraischen Functionen und ihre Anwendung in der Geometrie", *Math. Ann.* **7**:2 (1874), 269–310. DOI: 10.1007/BF02104804, https://eudml.org/doc/156642.

A. Bruno and A. Verra (2005), "M_{15} is rationally connected", pp. 51–65 in *Projective varieties with unexpected properties*, Walter de Gruyter, Berlin, 2005. DOI: 10.1515/9783110199703.51.

W. Bruns and J. Herzog (1993), *Cohen–Macaulay rings*, Cambridge Studies in Advanced Mathematics **39**, Cambridge University Press, Cambridge, 1993. DOI: 10.1017/CBO9780511608681.

D. A. Buchsbaum and D. Eisenbud (1973), "What makes a complex exact?", *J. Algebra* **25** (1973), 259–268. DOI: 10.1016/0021-8693(73)90044-6.

D. A. Buchsbaum and D. Eisenbud (1977a), "Algebra structures for finite free resolutions, and some structure theorems for ideals of codimension 3", *Amer. J. Math.* **99**:3 (1977), 447–485. DOI: 10.2307/2373926.

D. A. Buchsbaum and D. Eisenbud (1977b), "What annihilates a module?", *J. Algebra* **47**:2 (1977), 231–243. DOI: 10.1016/0021-8693(77)90223-X.

L. Burch (1967), "A note on ideals of homological dimension one in local domains", *Proc. Cambridge Philos. Soc.* **63** (1967), 661–662. DOI: 10.1017/s0305004100041633.

A. Calabri, D. Paccagnan, and E. Stagnaro (2014), "Plane algebraic curves with many cusps, with an appendix by Eugenii Shustin", *Ann. Mat. Pura Appl.* (4) **193**:3 (2014), 909–921. DOI: 10.1007/s10231-012-0306-6.

D. A. Cartwright, D. Erman, M. Velasco, and B. Viray (2009), "Hilbert schemes of 8 points", *Algebra Number Theory* **3**:7 (2009), 763–795. DOI: 10.2140/ant.2009.3.763.

G. Castelnuovo (1889), "Numero delle involuzioni razionali giacenti sopra una curva di dato genere", *Rom. Acc. L. Rend.* (4) **5**:2 (1889), 130–133. http://emeroteca.braidense.it/beic_attacc/sfoglia_articolo.php?IDTestata=927&IDT=34&IDV=388&IDA=19791.

A. Cayley (1848), "On the theory of elimination", *Cambridge and Dublin Math. J.* **3** (1848), 116–120. https://rcin.org.pl/dlibra/publication/143346/edition/119075.

D. Chen and S. Nollet (2012), "Detaching embedded points", *Algebra Number Theory* **6**:4 (2012), 731–756. DOI: 10.2140/ant.2012.6.731.

C. H. Clemens (2003), *A scrapbook of complex curve theory*, 2nd ed., Graduate Studies in Mathematics **55**, American Mathematical Society, Providence, RI, 2003. DOI: 10.1090/gsm/055.

C. H. Clemens and P. A. Griffiths (1972), "The intermediate Jacobian of the cubic threefold", *Ann. of Math.* (2) **95** (1972), 281–356. DOI: 10.2307/1970801.

M. Coppens (1983), "Some sufficient conditions for the gonality of a smooth curve", *J. Pure Appl. Algebra* **30**:1 (1983), 5–21. DOI: 10.1016/0022-4049(83)90035-X.

I. Coskun and E. Riedl (2018), "Normal bundles of rational curves in projective space", *Math. Z.* **288**:3-4 (2018), 803–827. DOI: 10.1007/s00209-017-1914-z.

P. Deligne and D. Mumford (1969), "The irreducibility of the space of curves of given genus", *Inst. Hautes Études Sci. Publ. Math.* **36** (1969), 75–109. DOI: 10.1007/BF02684599, http://www.numdam.org/item?id=PMIHES_1969__36__75_0.

A. Deopurkar and A. Patel (2015), "The Picard rank conjecture for the Hurwitz spaces of degree up to five", *Algebra Number Theory* **9**:2 (2015), 459–492. DOI: 10.2140/ant.2015.9.459.

R. Donagi (1980), "Group law on the intersection of two quadrics", *Ann. Scuola Norm. Sup. Pisa Cl. Sci.* (4) **7**:2 (1980), 217–239. http://www.numdam.org/item?id=ASNSP_1980_4_7_2_217_0.

J. A. Eagon and D. G. Northcott (1962), "Ideals defined by matrices and a certain complex associated with them", *Proc. Roy. Soc. London Ser. A* **269** (1962), 188–204. DOI: 10.1098/rspa.1962.0170.

D. Edidin (1993), "Brill-Noether theory in codimension-two", *J. Algebraic Geom.* **2**:1 (1993), 25–67.

D. Eisenbud (1980), "Transcanonical embeddings of hyperelliptic curves", *J. Pure Appl. Algebra* **19** (1980), 77–83. DOI: 10.1016/0022-4049(80)90094-8.

D. Eisenbud (1995), *Commutative algebra with a view toward algebraic geometry*, Graduate Texts in Mathematics **150**, Springer, New York, 1995. DOI: 10.1007/978-1-4612-5350-1.

D. Eisenbud (2005), *The geometry of syzygies: a second course in commutative algebra and algebraic geometry*, Graduate Texts in Mathematics **229**, Springer, New York, 2005. DOI: 10.1007/b137572.

D. Eisenbud and E. G. Evans, Jr. (1973), "Every algebraic set in n-space is the intersection of n hypersurfaces", *Invent. Math.* **19** (1973), 107–112. DOI: 10.1007/BF01418923.

D. Eisenbud and J. Gray (2023), "F. S. Macaulay: from plane curves to Gorenstein rings", *Bull. Amer. Math. Soc.* (*N.S.*) **60**:3 (2023), 371–406. DOI: 10.1090/bull/1787.

D. Eisenbud and J. Harris (1983), "Divisors on general curves and cuspidal rational curves", *Invent. Math.* **74**:3 (1983), 371–418. DOI: 10.1007/BF01394242.

D. Eisenbud and J. Harris (1987a), "Existence, decomposition, and limits of certain Weierstrass points", *Invent. Math.* **87**:3 (1987), 495–515. DOI: 10.1007/BF01389240, https://doi.org/10.1007/BF01389240.

D. Eisenbud and J. Harris (1987b), "Irreducibility and monodromy of some families of linear series", *Ann. Sci. Éc. Norm. Supér.* (4) **20**:1 (1987), 65–87. DOI: 10.24033/asens.1524.

D. Eisenbud and J. Harris (1987c), "The Kodaira dimension of the moduli space of curves of genus ≥ 23", *Invent. Math.* **90**:2 (1987), 359–387. DOI: 10.1007/BF01388710.

D. Eisenbud and J. Harris (1987d), "On varieties of minimal degree (a centennial account)", pp. 3–13 in *Algebraic geometry* (Brunswick, ME, 1985), Proc. Sympos. Pure Math. **46**, Amer. Math. Soc., Providence, RI, 1987.

D. Eisenbud and J. Harris (1989), "Irreducibility of some families of linear series with Brill–Noether number −1", *Ann. Sci. École Norm. Sup.* (4) **22**:1 (1989), 33–53. DOI: 10.24033/asens.1574.

D. Eisenbud and J. Harris (1992), "Finite projective schemes in linearly general position", *J. Algebraic Geom.* **1**:1 (1992), 15–30.

D. Eisenbud and J. Harris (2000), *The geometry of schemes*, Graduate Texts in Mathematics **197**, Springer, 2000. DOI: doi.org/10.1007/b97680.

D. Eisenbud and J. Harris (2016), *3264 and all that: a second course in algebraic geometry*, Cambridge University Press, Cambridge, 2016. DOI: 10.1017/CBO9781139062046.

D. Eisenbud and W. Neumann (1985), *Three-dimensional link theory and invariants of plane curve singularities*, Annals of Mathematics Studies **110**, Princeton University Press, Princeton, NJ, 1985.

D. Eisenbud and A. Sammartano (2019), "Correspondence scrolls", *Acta Math. Vietnam.* **44**:1 (2019), 101–116. DOI: 10.1007/s40306-018-0296-6.

D. Eisenbud and F. Schreyer (2022), "Hyperelliptic curves and Ulrich sheaves on the complete intersection of two quadrics", arxiv preprint, 2022. https://arxiv.org/abs/2212.07227.

D. Eisenbud and B. Ulrich (2019), "Duality and socle generators for residual intersections", *J. Reine Angew. Math.* **756** (2019), 183–226. DOI: 10.1515/crelle-2017-0045.

D. Eisenbud, H. Lange, G. Martens, and F.-O. Schreyer (1989), "The Clifford dimension of a projective curve", *Compositio Math.* **72**:2 (1989), 173–204. http://www.numdam.org/item?id=CM_1989__72_2_173_0.

D. Eisenbud, M. Green, and J. Harris (1996), "Cayley–Bacharach theorems and conjectures", *Bull. Amer. Math. Soc. (N.S.)* **33**:3 (1996), 295–324. DOI: 10.1090/S0273-0979-96-00666-0.

T. Ekedahl, S. Lando, M. Shapiro, and A. Vainshtein (2001), "Hurwitz numbers and intersections on moduli spaces of curves", *Invent. Math.* **146**:2 (2001), 297–327. DOI: 10.1007/s002220100164.

B. Fantechi and L. Göttsche (2005), "Local properties and Hilbert schemes of points", pp. 139–178 in *Fundamental algebraic geometry*, Math. Surveys Monogr. **123**, Amer. Math. Soc., Providence, RI, 2005.

G. Farkas (2017), "Progress on syzygies of algebraic curves", pp. 107–138 in *Moduli of curves*, Lect. Notes Unione Mat. Ital. **21**, Springer, 2017. https://arxiv.org/abs/1703.08056.

G. Farkas, D. Jensen, and S. Payne (2023), "The Kodaira dimensions of $\overline{M}_2 2$ and $\overline{M}_2 3$", preprint, 2023. http://web.ma.utexas.edu/users/sampayne/pdf/M23.pdf.

O. Forster (1981), *Lectures on Riemann surfaces*, Graduate Texts in Mathematics **81**, Springer, 1981. DOI: 10.1007/978-1-4612-5961-9.

W. Fulton (1969), *Algebraic curves: an introduction to algebraic geometry*, Addison-Wesley, 1969. https://dept.math.lsa.umich.edu/~wfulton/CurveBook.pdf. Notes written with the collaboration of Richard Weiss. Reprinted 1989 (Advanced Book Classics) and 2008.

W. Fulton (1984), *Intersection theory*, Ergebnisse der Mathematik und ihrer Grenzgebiete (3) **2**, Springer, Berlin, 1984. DOI: 10.1007/978-1-4612-1700-8.

W. Fulton and J. Harris (1991), *Representation theory: a first course*, Graduate Texts in Mathematics **129**, Springer, 1991. DOI: 10.1007/978-1-4612-0979-9.

W. Fulton and R. Lazarsfeld (1981), "On the connectedness of degeneracy loci and special divisors", *Acta Math.* **146**:3-4 (1981), 271–283. DOI: 10.1007/BF02392466.

F. R. Gantmacher (1959), *The theory of matrices*, vol. 2, Chelsea, New York, 1959.

S. I. Gelfand and Y. I. Manin (2003), *Methods of homological algebra*, 2nd ed., Springer, 2003. DOI: 10.1007/978-3-662-12492-5.

D. Gieseker (1982), "Stable curves and special divisors: Petri's conjecture", *Invent. Math.* **66**:2 (1982), 251–275. DOI: 10.1007/BF01389394.

R. Godement (1973), *Topologie algébrique et théorie des faisceaux*, 3ème ed., Publications de l'Institut de Mathématique de l'Université de Strasbourg **13**, Hermann, Paris, 1973.

G. Gotzmann (1978), "Eine Bedingung für die Flachheit und das Hilbertpolynom eines graduierten Ringes", *Math. Z.* **158**:1 (1978), 61–70. DOI: 10.1007/BF01214566.

M. Green (1989), "Restrictions of linear series to hyperplanes, and some results of Macaulay and Gotzmann", pp. 76–86 in *Algebraic curves and projective geometry* (Trento, 1988), Lecture Notes in Math. **1389**, Springer, Berlin, 1989. DOI: 10.1007/BFb0085925.

P. A. Griffiths (1989), *Introduction to algebraic curves*, Translations of Mathematical Monographs **76**, American Mathematical Society, Providence, RI, 1989. DOI: 10.1090/surv/039. Translated from the Chinese by Kuniko Weltin.

P. Griffiths and J. Harris (1978), *Principles of algebraic geometry*, Wiley-Interscience, New York, 1978. DOI: 10.1002/9781118032527. Reprinted 1994.

P. Griffiths and J. Harris (1980), "On the variety of special linear systems on a general algebraic curve", *Duke Math. J.* **47**:1 (1980), 233–272. http://projecteuclid.org/euclid.dmj/1077313873.

L. Gruson and C. Peskine (1982), "Genre des courbes de l'espace projectif, II", *Ann. Sci. École Norm. Sup.* (4) **15**:3 (1982), 401–418. DOI: 10.24033/asens.1431, http://www.numdam.org/item?id=ASENS_1982_4_15_3_401_0.

R. C. Gunning (1966), *Lectures on Riemann surfaces*, Princeton University Press, Princeton, N.J., 1966.

W. J. Haboush (1975), "Reductive groups are geometrically reductive", *Ann. of Math.* (2) **102**:1 (1975), 67–83. DOI: 10.2307/1970974.

J. Harris (1979), "Galois groups of enumerative problems", *Duke Math. J.* **46**:4 (1979), 685–724. DOI: 10.1215/S0012-7094-79-04635-0.

J. Harris (1982a), *Curves in projective space*, Séminaire de Mathématiques Supérieures **85**, Presses de l'Université de Montréal, 1982. With the collaboration of David Eisenbud.

J. Harris (1982b), "Theta-characteristics on algebraic curves", *Trans. Amer. Math. Soc.* **271**:2 (1982), 611–638. DOI: 10.2307/1998901.

J. Harris (1984), "On the Kodaira dimension of the moduli space of curves, II: The even-genus case", *Invent. Math.* **75**:3 (1984), 437–466. DOI: 10.1007/BF01388638.

J. Harris (1986), "On the Severi problem", *Invent. Math.* **84**:3 (1986), 445–461. DOI: 10.1007/BF01388741.

J. Harris (1992), *Algebraic geometry: A first course*, GTM **133**, Springer, New York, 1992. DOI: 10.1007/978-1-4757-2189-8. Corrected reprint 1995.

J. Harris and I. Morrison (1998), *Moduli of curves*, Graduate Texts in Mathematics **187**, Springer, 1998. DOI: 10.1007/b98867.

J. Harris and D. Mumford (1982), "On the Kodaira dimension of the moduli space of curves", *Invent. Math.* **67**:1 (1982), 23–88. DOI: 10.1007/BF01393371.

R. Hartshorne (1977), *Algebraic geometry*, Graduate Texts in Mathematics **52**, Springer, New York, 1977. DOI: 10.1007/978-1-4757-3849-0.

R. Hartshorne (1980), "On the classification of algebraic space curves", pp. 83–112 in *Vector bundles and differential equations* (Nice, 1979), Progr. Math. **7**, Birkhäuser, Boston, 1980. DOI: 10.1007/978-1-4684-9415-0_5.

R. Hartshorne (1982), "Genre de courbes algébriques dans l'espace projectif (d'après L. Gruson et C. Peskine)", pp. 301–313 in *Bourbaki Seminar, Vol. 1981/1982*, Astérisque **92-93**, Soc. Math. France, Paris, 1982. https://eudml.org/doc/109994.

R. Hartshorne and C. Polini (2015), "Divisor class groups of singular surfaces", *Trans. Amer. Math. Soc.* **367**:9 (2015), 6357–6385. DOI: 10.1090/S0002-9947-2015-06228-X.

D. Hilbert (1890), "Über die Theorie der algebraischen Formen", *Mathematische Annalen* **36** (1890), 473–534. DOI: 10.1007/BF01208503.

A. Hirschowitz and S. Ramanan (1998), "New evidence for Green's conjecture on syzygies of canonical curves", *Ann. Sci. École Norm. Sup.* (4) **31**:2 (1998), 145–152. DOI: 10.1016/S0012-9593(98)80013-X.

K. Hulek (1995), "The Horrocks–Mumford bundle", pp. 139–177 in *Vector bundles in algebraic geometry* (Durham, 1993), London Math. Soc. Lecture Note Ser. **208**, Cambridge Univ. Press, Cambridge, 1995. DOI: 10.1017/CBO9780511569319.007.

C. Huneke and I. Swanson (2006), *Integral closure of ideals, rings, and modules*, London Mathematical Society Lecture Note Series **336**, Cambridge University Press, Cambridge, 2006. https://www.math.purdue.edu/~iswanso/book/SwansonHunekeCUP06.pdf.

A. Hurwitz (1891), "Ueber Riemann'sche Flächen mit gegebenen Verzweigungspunkten", *Math. Ann.* **39** (1891), 1–61. DOI: 10.1007/BF01199469.

A. Hurwitz (1901), "Ueber die Anzahl der Riemann'schen Flächen mit gegebenen Verzweigungspunkten", *Math. Ann.* **55**:1 (1901), 53–66. DOI: 10.1007/BF01448116.

A. Iarrobino (1985), "Compressed algebras and components of the punctual Hilbert scheme", pp. 146–165 in *Algebraic geometry* (Sitges, Barcelona, 1983), edited by E. Casas-Alvero, LNM **1124**, Springer, Berlin, 1985. 10.1007/BFb0075000.

N. Jacobson (1989), *Basic algebra*, vol. II, 2nd ed., W. H. Freeman, New York, 1989. Reprinted Dover, 2009.

C. Jordan (1870), *Traité des substitutions et des équations algébriques*, Gauthier-Villars, Paris, 1870. Reprinted by Jacques Gabay, Sceaux, 1989, ISBN 2-87647-021-7.

B. Kadets (2021), "Sectional monodromy groups of projective curves", *J. Lond. Math. Soc.* (2) **103**:1 (2021), 314–335. DOI: 10.1112/jlms.12375.

H. Kaji (1986), "On the tangentially degenerate curves", *J. London Math. Soc.* (2) **33**:3 (1986), 430–440. DOI: 10.1112/jlms/s2-33.3.430.

C. Keem (1990), "On the variety of special linear systems on an algebraic curve", *Math. Ann.* **288**:2 (1990), 309–322. DOI: 10.1007/BF01444535.

C. Keem (1994), "Reducible Hilbert scheme of smooth curves with positive Brill–Noether number", *Proc. Amer. Math. Soc.* **122**:2 (1994), 349–354. DOI: 10.2307/2161023, https://doi.org/10.2307/2161023.

C. Keem, Y.-H. Kim, and A. F. Lopez (2019), "Irreducibility and components rigid in moduli of the Hilbert scheme of smooth curves", *Math. Z.* **292**:3-4 (2019), 1207–1222. DOI: 10.1007/s00209-018-2130-1.

M. Kemeny (2021), "Universal secant bundles and syzygies of canonical curves", *Invent. Math.* **223**:3 (2021), 995–1026. DOI: 10.1007/s00222-020-01001-5.

G. Kempf (1971), "Schubert methods with an application to algebraic curves", Cwi preprint, math. centrum, amsterdam, 1971. https://ir.cwi.nl/pub/7262.

V. Kharlamov and F. Sottile (2003), "Maximally inflected real rational curves", *Mosc. Math. J.* **3**:3 (2003), 947–987, 1199–1200. DOI: 10.17323/1609-4514-2003-3-3-947-987.

F. Kirwan (1992), *Complex algebraic curves*, London Mathematical Society Student Texts **23**, Cambridge University Press, Cambridge, 1992. DOI: 10.1017/CBO9780511623929.

S. L. Kleiman (1976), "*r*-special subschemes and an argument of Severi's", *Advances in Math.* **22**:1 (1976), 1–31. DOI: 10.1016/0001-8708(76)90135-3.

S. L. Kleiman (2005), "The Picard scheme", pp. 235–321 in *Fundamental algebraic geometry*, Math. Surveys Monogr. **123**, Amer. Math. Soc., Providence, RI, 2005.

S. L. Kleiman and D. Laksov (1972), "On the existence of special divisors", *Amer. J. Math.* **94** (1972), 431–436. DOI: 10.2307/2374630.

S. L. Kleiman and D. Laksov (1974), "Another proof of the existence of special divisors", *Acta Math.* **132** (1974), 163–176. DOI: 10.1007/BF02392112.

F. F. Knudsen (1983), "The projectivity of the moduli space of stable curves, III: The line bundles on $M_{g,n}$, and a proof of the projectivity of $\overline{M}_{g,n}$ in characteristic 0", *Math. Scand.* **52**:2 (1983), 200–212. DOI: 10.7146/math.scand.a-12002.

L. Kronecker (1882), "Grundzüge einer arithmetischen Theorie der algebraische Grössen", *J. Reine Angew. Math.* **92** (1882), 1–122. DOI: 10.1515/crll.1882.92.1.

V. S. Kulikov and E. Shustin (2015), "Duality of planar and spacial curves: new insight", *Eur. J. Math.* **1**:3 (2015), 462–482. DOI: 10.1007/s40879-015-0052-6.

E. Kunz (2005), *Introduction to plane algebraic curves*, Birkhäuser, Boston, 2005. DOI: 10.1007/0-8176-4443-1.

E. Larson (2017), "The maximal rank conjecture", arxiv preprint, 2017. http://arxiv.org/abs/1711.04906.

H. K. Larson (2021), "A refined Brill–Noether theory over Hurwitz spaces", *Invent. Math.* **224**:3 (2021), 767–790. DOI: 10.1007/s00222-020-01023-z.

E. Larson and I. Vogt (2023), "Interpolation for Brill–Noether curves", *Forum Math. Pi* **11** (2023), Paper No. e25. DOI: 10.1017/fmp.2023.22.

E. Larson, H. Larson, and I. Vogt (2008), "Global Brill–Noether theory over the Hurwitz space", arxiv preprint, 2008. https://arxiv.org/abs/2008.10765.

R. Lazarsfeld (1986), "Brill–Noether–Petri without degenerations", *J. Differential Geom.* **23**:3 (1986), 299–307. http://projecteuclid.org/euclid.jdg/1214440116.

R. Lazarsfeld and P. Rao (1983), "Linkage of general curves of large degree", pp. 267–289 in *Algebraic geometry: open problems* (Ravello, 1982), Lecture Notes in Math. **997**, Springer, 1983. DOI: 10.1007/BFb0061648.

F. Lê (2020), " "Are the *genre* and the *Geschlecht* one and the same number?" An inquiry into Alfred Clebsch's *Geschlecht*", *Historia Math.* **53** (2020), 71–107. DOI: 10.1016/j.hm.2020.04.002.

J. Lüroth (1875), "Beweis eines Satzes über rationale Curven", *Math. Ann.* **2** (1875), 163–165. https://eudml.org/doc/156693.

F. Macaulay (1900), "Extensions of the Riemann–Roch theorem in plane geometry", *Proceedings of the London Mathematical Society* **32** (1900), 418–430. DOI: 10.1112/plms/s1-32.1.384. The given DOI is for a document that contains multiple articles at the end of the issue in question.

Y. I. Manin (1986), *Cubic forms*, 2nd ed., North-Holland Mathematical Library **4**, North-Holland Publishing Co., Amsterdam, 1986. Algebra, geometry, arithmetic, Translated from the Russian by M. Hazewinkel.

H. H. Martens (1967), "On the varieties of special divisors on a curve", *J. Reine Angew. Math.* **227** (1967), 111–120. DOI: 10.1515/crll.1967.227.111.

T. Matsusaka (1972), "Polarized varieties with a given Hilbert polynomial", *Amer. J. Math.* **94** (1972), 1027–1077. DOI: 10.2307/2373563.

B. Mazur (1986), "Arithmetic on curves", *Bull. Amer. Math. Soc. (N.S.)* **14**:2 (1986), 207–259. DOI: 10.1090/S0273-0979-1986-15430-3.

J. Migliore (1986), "Geometric invariants for liaison of space curves", *J. Algebra* **99**:2 (1986), 548–572. DOI: 10.1016/0021-8693(86)90045-1.

J. Milnor (1968), *Singular points of complex hypersurfaces*, Annals of Mathematics Studies **61**, Princeton University Press, Princeton, N.J., 1968.

R. Miranda (1995), *Algebraic curves and Riemann surfaces*, Graduate Studies in Mathematics **5**, American Mathematical Society, Providence, RI, 1995. DOI: 10.1090/gsm/005.

S. Mukai (1995a), "Curves and symmetric spaces, I", *Amer. J. Math.* **117**:6 (1995), 1627–1644. DOI: 10.2307/2375032.

S. Mukai (1995b), "New developments in Fano manifold theory related to the vector bundle method and moduli problems", *Sūgaku* **47**:2 (1995), 125–144.

S. Mukai (2010), "Curves and symmetric spaces, II", *Ann. of Math. (2)* **172**:3 (2010), 1539–1558. DOI: 10.4007/annals.2010.172.1539.

S. Mullane (2023), "The Hurwitz space Picard rank conjecture for $d > g - 1$", *Adv. Math.* **430** (2023), Paper No. 109211, 27. DOI: 10.1016/j.aim.2023.109211.

D. Mumford (1962), "Further pathologies in algebraic geometry", *Amer. J. Math.* **84** (1962), 642–648. DOI: 10.2307/2372870.

D. Mumford (1971), "Theta characteristics of an algebraic curve", *Ann. Sci. École Norm. Sup.* (4) **4** (1971), 181–192. DOI: 10.24033/asens.1209, `http://www.numdam.org/item?id=ASENS_1971_4_4_2_181_0`.

D. Mumford (1974), "Prym varieties, I", pp. 325–350 in *Contributions to analysis: a collection of papers dedicated to Lipman Bers*, Academic Press, New York, 1974.

D. Mumford (1975), *Curves and their Jacobians*, University of Michigan Press, Ann Arbor, MI, 1975.

D. Mumford (1977), "Stability of projective varieties", *Enseign. Math.* (2) **23**:1-2 (1977), 39–110. `https://www.dam.brown.edu/people/mumford/alg_geom/papers/1977a--StabilityLecturesIHES-Swiss.pdf`.

D. Mumford and K. Suominen (1972), "Introduction to the theory of moduli", pp. 171–222 in *Algebraic geometry* (Oslo, 1970), Wolters-Noordhoff, Groningen, 1972. `https://www.dam.brown.edu/people/mumford/alg_geom/papers/1972d--IntroModuli-NC.pdf`.

H. Nasu (2008), "The Hilbert scheme of space curves of degree d and genus $3d - 18$", *Comm. Algebra* **36**:11 (2008), 4163–4185. DOI: 10.1080/00927870802175089.

N. Nitsure (2005), "Construction of Hilbert and Quot schemes", pp. x+339 in *Grothendieck's FGA explained*, Mathematical Surveys and Monographs **123**, American Mathematical Society, Providence, RI, 2005. DOI: 10.1090/surv/123.

M. Noether (1873), "Ueber einen Satz aus der Theorie der algebraischen Functionen", *Math. Annalen* **6** (1873), 351–359. DOI: 10.1007/BF01442793.

M. Olsson (2016), *Algebraic spaces and stacks*, American Mathematical Society Colloquium Publications **62**, American Mathematical Society, Providence, RI, 2016. DOI: 10.1090/coll/062.

R. Osserman (2011), *Poetry of the Universe: A mathematical exploration*, Anchor, 2011.

O. Perron (1941), "Über das Vahlensche Beispiel zu einem Satz von Kronecker", *Math. Z.* **47** (1941), 318–324. DOI: 10.1007/BF01180964.

C. Peskine and L. Szpiro (1974), "Liaison des variétés algébriques, I", *Invent. Math.* **26** (1974), 271–302. DOI: 10.1007/BF01425554.

N. Pflueger (2018), "On nonprimitive Weierstrass points", *Algebra Number Theory* **12**:8 (2018), 1923–1947. DOI: 10.2140/ant.2018.12.1923.

R. Piene and M. Schlessinger (1985), "On the Hilbert scheme compactification of the space of twisted cubics", *Amer. J. Math.* **107**:4 (1985), 761–774. DOI: 10.2307/2374355.

A. Prabhakar Rao (1978/79), "Liaison among curves in $\mathbf{P}^3$", *Invent. Math.* **50**:3 (1978/79), 205–217. DOI: 10.1007/BF01410078.

J. Rathmann (1987), "The uniform position principle for curves in characteristic p", *Math. Ann.* **276**:4 (1987), 565–579. DOI: 10.1007/BF01456986.

J. Rathmann (2022), "Geometric Koszul complexes, syzygies of K3 surfaces and the Tango bundle", arxiv preprint, 2022. `https://arxiv.org/abs/2205.00266`.

M. Reid (1988), *Undergraduate algebraic geometry*, London Mathematical Society Student Texts **12**, Cambridge University Press, Cambridge, 1988.

B. Riemann (1857), "Theorie der Abel'schen Functionen", *J. Reine Angew. Math.* **54** (1857), 115–155. DOI: 10.1515/crll.1857.54.115.

G. Roch (1865), "Ueber die Anzahl der willkurlichen Constanten in algebraischen Functionen", *J. Reine Angew. Math.* **64** (1865), 372–376. `doi:10.1515/crll.1865.64.372`.

M. Sagraloff (2018), *Special linear series and syzygies of canonical curves of genus 9*, Ph.D. thesis, Universität des Saarlandes, 2018.

G. Salmon (1852), *A treatise on the higher plane curves: intended as a sequel to A treatise on conic sections*, Hodges and Smith, Dublin, 1852. `https://catalog.hathitrust.org/Record/000381547`.

G. Salmon (1879), *A treatise on the higher plane curves: intended as a sequel to A treatise on conic sections*, 3rd ed., Hodges, Foster and Figgis, 1879. `https://archive.org/details/117724690`. Reprinted Chelsea, New York, 1960.

U. Schreiber et al. (2011–), "Six operations", web page, 2011–. `http://ncatlab.org/nlab/show/six+operations`.

F.-O. Schreyer (1986), "Syzygies of canonical curves and special linear series", *Math. Ann.* **275**:1 (1986), 105–137. DOI: 10.1007/BF01458587.

F.-O. Schreyer (1991), "A standard basis approach to syzygies of canonical curves", *J. Reine Angew. Math.* **421** (1991), 83–123. DOI: 10.1515/crll.1991.421.83.

P. Schwartau (1982), *Liaison addition and monomial ideals*, Ph.d. thesis, Brandeis University, 1982.

J.-P. Serre (1955), "Faisceaux algébriques cohérents", *Ann. of Math.* (2) **61** (1955), 197–278. DOI: 10.2307/1969915.

J.-P. Serre (1955/56), "Géométrie algébrique et géométrie analytique", *Ann. Inst. Fourier (Grenoble)* **6** (1955/56), 1–42. `http://aif.cedram.org/item?id=AIF_1955__6__1_0`.

J.-P. Serre (1979), *Local fields*, Springer, New York, 1979. DOI: 10.1007/978-1-4757-5673-9.

D. I. Smyth (2013), "Towards a classification of modular compactifications of $\mathcal{M}_{g,n}$", *Invent. Math.* **192**:2 (2013), 459–503. DOI: 10.1007/s00222-012-0416-1.

M. Trott (1997), "Applying `GroebnerBasis` to three problems in geometry", *Mathematica Educ. Res.* **6** (1997), 15–28.

K. T. Vahlen (1891), "Bemerkung zur vollständigen Darstellung algebraischer Raumcurven", *J. Reine Angew. Math.* **108** (1891), 346–347. DOI: 10.1515/crll.1891.108.346.

R. Vakil (2006), "A geometric Littlewood–Richardson rule", *Ann. of Math.* (2) **164**:2 (2006), 371–421. DOI: 10.4007/annals.2006.164.371.

R. Vakil (2023), "The rising sea: foundations of algebraic geometry", book in progress, to be published by princeton university press, 2023. `http://math.stanford.edu/~vakil/216blog`.

A. M. Vermeulen (1983), *Weierstrass points of weight two on curves of genus three*, Ph.D. thesis, Universiteit van Amsterdam, 1983.

C. Voisin (2002), "Green's generic syzygy conjecture for curves of even genus lying on a $K3$ surface", *J. Eur. Math. Soc.* (*JEMS*) **4**:4 (2002), 363–404. DOI: 10.1007/s100970200042.

C. Voisin (2003), *Hodge theory and complex algebraic geometry*, vol. II, Cambridge Studies in Advanced Mathematics **77**, Cambridge University Press, Cambridge, 2003. DOI: 10.1017/CBO9780511615177.

C. Voisin (2005), "Green's canonical syzygy conjecture for generic curves of odd genus", *Compos. Math.* **141**:5 (2005), 1163–1190. DOI: 10.1112/S0010437X05001387.

C. Voisin (2007), *Hodge theory and complex algebraic geometry*, vol. I, English ed., Cambridge Studies in Advanced Mathematics **76**, Cambridge University Press, Cambridge, 2007.

R. J. Walker (1950), *Algebraic curves*, Princeton Mathematical Series **13**, Princeton University Press, Princeton, NJ, 1950. Reprinted by Dover, 1962, and Springer-Verlag, 1978.

J. Weyman (2003), *Cohomology of vector bundles and syzygies*, Cambridge Tracts in Mathematics **149**, Cambridge University Press, Cambridge, 2003. DOI: 10.1017/CBO9780511546556.

O. Zariski (1931), "On the non-existence of curves of order 8 with 16 cusps", *Amer. J. Math.* **53**:2 (1931), 309–318. DOI: 10.2307/2370785.

Index

Notation is indexed under the initial letter or abbreviation; Greek letters appear in spelled-out order.

In certain entries a boldface page number indicates where the term is defined; however, the use of that convention is not systematic.

SELECTED PUBLISHED TITLES IN THIS SERIES

For a complete list of titles in this series, visit the AMS Bookstore at **www.ams.org/bookstore/gsmseries/**.